THE RACE TO EUROPEAN EMINENCE

WHO ARE THE COMING TELE-SERVICE MULTINATIONALS?

THE RACE TO EUROPEAN EMINENCE

WHO ARE THE COMING TELE-SERVICE MULTINATIONALS?

edited by

ERIK BOHLIN
OVE GRANSTRAND
Department of Industrial Management & Economics
Chalmers University of Technology
Göteborg, Sweden

1994
NORTH-HOLLAND
AMSTERDAM • LONDON • NEW YORK • TOKYO

ELSEVIER SCIENCE B.V.
Sara Burgerhartstraat 25
P.O. Box 211, 1000 AE Amsterdam, The Netherlands

ISBN: 0 444 81961 4

This book is printed on acid-free paper.

Printed in The Netherlands.

Contents

Internationalization in Continental and Southern Europe

Internationalization in the Nordic Region

Preface

Today, there is a common understanding that a fundamental restructuring of global, especially European, telecommunications is taking place. Telecommunications carriers are becoming more and more multinational in scope but chances are that only a handful will survive as full-fledged Multinational Corporations (MNCs) in the 21st century.

Consequently, a race for European eminence among telecommunications carriers is emerging where a truly pan-European market place is gaining momentum. Seizing the new business opportunities increases the pressure for innovation, lower rates, an expanded range of functionalities and new services among traditional telephone carriers as well as the pressure for new forms of regulation. The stakes are high and the potential impacts are great - on regions, on companies, on network and service capabilities as well as on customers and suppliers.

Whereas European carriers in the past could choose the technologies they wanted to deploy, plan the evolution of their networks, and safely select services offered within their national boundaries, today they are forced to respond to user needs and to internationalize - despite regulatory inertia. The race to European eminence is accelerating where strategies pursued and alliance clusters are emerging more clearly, involving not only European players.

The winds of change are stirring up a host of critical questions that need to be addressed by telecommunications companies, researchers, consultants, regulators and policy makers with an interest in the European market. For instance, what will it take to be a pan-European telecommunications carrier in the year 2010? Which regulatory and institutional frameworks will serve a new era of increasing international competition in telecommunications services?

In order to address these fresh and pressing issues, a European Regional Meeting of the International Telecommunications Society (ITS) was convened by Chalmers University of Technology under the auspices of ITS. The conference was held at Stenungsund in the Western archipelago of Sweden, June 20-23, 1993. The title and main theme of the conference was: "The Race to European Eminence - Who will be the Multinational Telecommunications Carriers in the Future Europe?". It turned out that the conference theme was timely. Internationalization of telecom service companies and markets is a current phenomenon, still in its infant stages. A couple of years earlier there would hardly have been sufficient interest or material to even hold a coherent conference.

A number of country papers had been invited and prepared along the theme of inward/outward foreign direct investments (FDIs) in telecommunication services. In addition a number of papers addressed various issues and subthemes interconnected with the competitive race in Europe.

This book is an edited selection of revised papers, originally prepared for the conference. It is particularly aimed for industrialists, strategists, researchers, consultants, regulators and policy makers, but should hopefully also be useful for students and the general public with an interest in an area that will surely affect us all as consumers.

The editorial work has given priority to creating comprehensive coherence and relevance, with updated country information, introductory material on internationalization and MNCs, summaries for easy overview and some theoretical speculations, all part of exercises in preparing for an uncertain future.

In organizing the conference and editing this book, we as editors want to mention and sincerely thank many persons and institutions that have been helpful in numerous ways. We would like to thank all ITS members who served on the conference committee and as theme coordinators, especially Alain de Fontenay, who was also co-chairing the conference. Furthermore, we would like to thank the ITS Officers - Chairman James Alleman, Vice-Chairman Koichiro Hayashi, Treasurer Hans Levenbach, and Managing Editor Noemy Wachtel - and Ms Gloria Patchinston-Houston of the ITS Administrative Office as well as Dr Jürgen Müller, the European Coordinator of ITS. The collected experience of the involved ITS members and their advice was critical for the conference success.

Naturally, we are much indebted to the contributors who took the pains of revising and updating their chapters and did so speedily. We would also like to thank all the corporate and regulatory representatives and their respective staff who ensured relevant and timely plenary sessions during the conference.

We would also like to thank our colleagues at the Department of Industrial Management and Economics, Chalmers University of Technology, who in various capacities have been helpful in carrying through the conference and producing this volume. Mr Sven Lindmark was admirable in carrying out many central administrative tasks during the conference with persistence and enthusiasm. We are indebted to Ms Ulla Anson, Ms Eva Burford and Ms Gullvi Nilsson for their excellent secretarial services, to Ms Annika Svärd-Olofsson for accounting management, and to Ms Annika Jarnehammar, Mr Joakim Philipsson and Ms Annika Rickne for their help at the conference site. We thank Lic. Ola Johansson for help with index compilation and Mr Jon van Leuven for language corrections.

The financial support from Telia AB (formerly Televerket) was critical for the conference and the book publication and is gratefully acknowledged. From Telia, we particularly acknowledge the support and interest of Director General Tony Hagström, Director Bertil Thorngren and Manager Hans Bergendorff.

Göteborg, Sweden, May 1994

Erik Bohlin Ove Granstrand

Acknowledgment

For book and conference endeavour, the generous support of Telia
is gratefully acknowledged.

For providing a conference environment of open and critical exchange, the International Telecommunications Society (ITS) is gratefully acknowledged.

List of Committees

James Alleman, Chairman ITS, University of Colorado, USA

David Allen, East West Center, USA

Loretta Anania, CEC, DG XIII, Brussels

Cristiano Antonelli, University of Torino, Italy

Jacques Arlandis, IDATE, France

Hans Bergendorff, Telia, Sweden

Erik Bohlin, Chalmers University of Technology, Sweden

Alain de Fontenay, TELA Group, USA

Henry Ergas, OECD, Paris

Ove Granstrand, Chalmers University of Technology, Sweden

Koichiro Hayashi, NTT, Japan

Ferenc Kiss, TELA Group, USA

Don Lamberton, Australian National University, Australia

Hans Levenbach, ITS Treasurer, Delphus Inc., USA

Robin Mansell, SPRU, University of Sussex, United Kingdom

Jürgen Müller, DIW, Germany

Gérard Pogorel, INT, France

Almarin Phillips, Econsult Inc., USA

Walter Richter, ITU, Geneva

Abbreviations

ADC	American Digital Cellular
AT&T	American Telephone & Telegraph
B.L.T.	Build-Lease-Transfer
B.O.O.T.	Build-Own-Operate-Transfer
B.O.T.	Build-Operate-Transfer
BERI	Business Environment Risk Index
BOC	Bell-Operating-Company
BOT	Buy Operate Transfer
C&W	Cable & Wireless
CAD	Computer Aided Design
CAP	Competitive Access Provider
CATV	Cable TV
CDMA	Code Division Multiple Access
CIS	Commonwealth of Independent States
CMEA	Council on Mutual Economic Assistance
CO	Central offices
CoCom	Coordinating Committee for Multilateral Export Controls
CPE	Customer Premises Equipment
CPR	Country Potential Rating
DBP	Deutsche Bundespost
DC	Developed Countries
DEL	Direct Exchange Line
DTI	Department of Trade and Industry
EDI	Electronic Data Interchange
EPT	First Pest-City Telephone Company
ERMES	European Radio Message System
ESA	European Space Agency
FDI	Foreign Direct Investment
GDP	Gross Domestic Product
GNS	Global Network Service
GSM	Groupe Speciale Mobile or, Global Systems for Mobile communications
GVPN	Global Virtual Private Network
HDTV	High Definition Television
IRR	Internal Rate of Return
ISDN	Integrated Services Digital Network
ISL	Inter Satellite Link
ISO	International Satellite Organizations
IVAN	International Value Added Network Services
JAM	Joint Account Management
JV	Joint Venture
KTS	Key Telephone Systems
LAN	Local Area Network
LDC	Less Developed Countries
LEC	Local Exchange Carrier
MDN	Managed Data Network
MDNS	Managed Data Network Service
MEO	Medium Earth Orbit
MIGA	Multilateral Investment Guarantee Agreement
MNC	Multinational Corporation
MNE	Multinational Enterprises
MNS	Managed Network Service
MPT	Ministry of Posts and Telecommunications
MSS	Mobile Satellite Services
MTO	Multinational Telecommunications Operators
NCC	New Common Carriers
NMT	Nordic Mobile Telephone System
NPV	Net Present Value
NTT	Nippon Telephone Telegraph
OBP	On Board Processing
ONA	Open Network Architecture
OSS	One-Stop-Shopping
PBAX	Private Branch Automatic Exchange
PBX	Private Branch Exchange
PCN	Personal Communications Network
PCS	Personal Communication Services
PDN	Public Data Network
PDP	Problem Determination Procedures
PIN	Personal Identification Number
PMR	Private Mobile Radio
PSS	Packet Switched System
PSTN	Public Switched Telephone Network
PTO	Public Telecommunications Operators
PTT	Posts, Telegraph and Telephone
RBOC	Regional Bell Operating Companies
RIGA	Reciprocal Investment Guarantee Agreements
RPI	Retail Price Index
SBU	Strategic Business Unit
SCI	Satellite Communications Industry
SNA	Systems Network Architecture
SPCE	Stalinist Centrally Planned Economy
TACS	Total Access System
TDMA	Time Division Multiple Access
TEL	Trans Europe Line
TISA	Telefónica Internacional Sociedad Anónima
TLC	Telecommunication Carrier
TO	Telecommunication Operator
USW	Ultra Short Wave
VAN	Value Added Network
VANS	Value Added Network Services
VAS	Value Added Service
VPN	Virtual Private Network
VSAT	Very Small Aperture Terminal

List of Tables

List of Figures

PART A: OVERVIEW

The Race to European Eminence
E. Bohlin and O. Granstrand (Editors)

Introduction and Overview

Ove Granstrand and Erik Bohlin

Chalmers University of Technology, Department of Industrial Management and Economics, S-412 96 Göteborg, Sweden

1. INTRODUCTION

1.1 Background

Let us then make the assumption—sadly far from current reality—that we are in a paradisiac one-country world—or a one-country Europe for that matter. Would the most efficient industrial organization of its telecommunications service sector coincide with the system of the early 1980s with national monopolies and regionally dominant firms? Hardly.

The ongoing liberalization of telecommunications service markets could thus be regarded as one of the major economic experiments in the 20th-century world economy. Its outcome and impact will be surely felt in the 21th century but difficult to forecast beyond saying that we will be surprised. But uncertainty and turbulence favor those who are well-informed.

1.2 Purpose of this book

The brief background above is not indisputable. Already the title of this book and its underlying metaphor of a race raise questions like:

- Is there actually a race going on?
- How did it start?
- Will the race go on or be stopped and who stands to lose?
- Will there be winners?
- What does the crown of victory look like?
- What are the rules of the race and who sets them?
- What will happen with the spectators and those aspiring to set the roles?
- Are there other races?
- What does the race track look like?
- What is the role of the racing equipment and its supply?

This book aims to give a closer picture of the on-going internationalization experiment in tele services and to probe some questions for its future monitoring. It is focused on Europe for several reasons. First, European industry in general, especially in the smaller countries, has a long history of internationalization. Second, despite Europe's lagging in a number of techno-industrial areas compared to Japan and the US, her telecommunications industry and technology is quite strong with a considerable potential for international competitiveness (see e.g. Freeman et al. 1991, cf. also the case of GSM). Third, the general European integration process is highly interesting and important, as is the European process of liberalization of telecommunications. The difficulties for national and supra-national authorities to align these two processes may very well turn the telecom area into an important test case for both. An emerging pan-European market may also become a battleground for telecommunications companies from around the world, perhaps for one of the largest commercial battles in economic history. To the extent that the European market(s) will be precursory, the winning European telcos stand to gain an important competitive edge also in world markets.

1.3 Chapter outline

It is always tempting to overly emphasize the specific features of a particular industry, especially when viewed from within. Telecommunications is no exception to this; on the contrary. This chapter will first try to put internationalization of telecommunications into its broader context. This is done by overviewing the concepts, history and theories of MNCs in general and then relating them to telecom more specifically. The chapter then gives an overview of the structure and content of the book.

2. OVERVIEW OF CONCEPTS

2.1 The concept of a technology-based service industry

2.1.1 General

The literature on internationalization and MNCs has mostly emerged in connection with internationalization of extraction and manufacturing industries rather than service industries. It can therefore not be taken for granted that received concepts, theories and models of internationalization are immediately applicable to the telecommunications service industry. However, there are close links between products and services, also growing over time. Every product provides a bundle of services for the user and most services require for their provision a bundle of products and capital equipment (buildings, instruments, tools etc.). The distinction between service and manufacturing industries is related to a traditional organization and conceptualization of industrial activities that often becomes fuzzy, perhaps also increasingly so.

We define a technology-based service industry here to mean an industry that essentially provides services, the provision of which requires embodied and/or disembodied technology to a considerable extent. Technology is here taken to typically mean engineering and applied natural and science knowledge. The R&D to sales ratio, i.e. the R&D intensity, is then one indicator of the degree to which an industry is technology-based.

2.1.2 Telecom

For the telecommunications service industry there are two important common denominators with extraction and manufacturing industries, namely the capital and technology intensities. Thus, we may regard the telecommunications service industry as being in particular a *technology-based service industry*.

Table 2.1 utilizes the denominators capital and technology intensity to provide an overview of service industries in general and technology-based ones in particular and an illustration of the concept of technology-based service industry in the context of telecom services. Note that the (physical) capital base embodies technology in its "hard" form. However, capital intensity and technology intensity are not perfectly correlated since there are low-tech/high capital

Table 2.1
Illustrative classification of service industries[1]

Capital intensity[2]	Technology intensity[2]		
	High-tech services	Low-tech services	or "no-tech" services
High capital intensive services	Telecom services	Rescue services	Rental services
	Medical services (surgery etc.)	Entertainment	Hotels and accommodation services
	Surveillance and security services	Retailing/wholesale	Religious services
	Defense services	Gambling	
	Energy services	Disposal services (sewage, garbage etc.)	
	Transportation services	Water supply	
		Financial services etc.	
Low capital intensive services	Universities and higher educational services	Entertainment	Social services[3]
			Insurance services
	Consultancy services (engineering, data, etc.)	Financial services	Restaurants
	Medical services etc.		Consultancy services (management, legal, advertising, accounting etc.)
		Police services	Travel agencies Cleaning, housekeeping
			Gambling Prostitution[4] etc.

Notes:

1) Some services are not in everyday language associated with 'industries' but nevertheless encompass production and transactions of some sort.

2) The principal classification variables used here are capital intensity and technology intensity referring to the relative degree of hardware, and engineering and applied natural and science knowledge employed, respectively. These variables are continuous but for illustrative purposes here divided simply into high and low. Obviously, this distinction can be disputed, but still be useful for illustration. In several border cases, the technology intensity is rather non-existent than low, that is, the corresponding service industry is not technology-based, although possibly based on other professional competencies. Most service categories vary considerably along the variables and thus could be found at several places in the table, which is illustrated in some cases. A third important variable is information intensity, which could have been used as well (cf. the notion of 'information industries'). Technical information is then a special case. Additional variables for the classification of service industries are private/public (the public sector contains many services); legal/illegal; location-bound/unbound etc.

3) Typically in the public sector.

4) This is included as an extreme case to illustrate the span of services, also in a legal dimension.

intensive services as well as high-tech/low capital intensive services (e.g. higher science and technology education and engineering consultancy). Moreover, the technology and capital intensity may change over time. Especially the technification of many services is noteworthy, that is, they become increasingly technology-based. In this connection telecom services increasingly penetrate other services, e.g. medical, education and entertainment services.

The recognition of telecom services as technology-intensive is important, since technology is an important factor in many theories of MNCs and internationalization. At the same time telecom services depend upon a network or infrastructure for their provision which makes telecom services a network industry (see e.g. David 1992), a type of industry that is not particularly recognized in the literature on MNCs and internationalization.

2.2 Concepts related to internationalization

2.2.1 General

The term *internationalization* of a company is normally taken to mean the process of expansion of certain operations to other nations as well as the outcome of that process. Internationalization of an industry and of a market, refers to a) the process by which the companies operating in the industry, and on the market, respectively, are internationalizing, and also to b) the outcome of that process. Thus, internationalization may refer both to processes pertaining to companies, industries and markets, and to a state of those processes at a specific point in time. In the latter case, one can talk about a degree of internationalization. By a *multinational corporation* (MNC), one usually means a corporation which has reached a certain degree of internationalization. There are several ways to assess the degree of internationalization or the degree of multinationality. For example, Aharoni (1971) distinguished between definitions of *multinationality* based on:

a) Structural variables. (For example, the number of nations in which the corporation has located operations of different kinds or the number of nations represented among owners or corporate management.)

b) Performance variables. (For example, absolute or relative measures of sales, assets, value added or employees related to operations in different nations.)

c) Behavioral variables. (For example, national or multinational orientation in top management.)

There are moreover several ways to classify MNCs into different types, depending upon how different nationalities influence the conduct of operations. Robinson (1967) distinguished between *international*, *multinational*, *transnational* and *supranational* corporations. The latter two types cannot be assigned a nationality, and supranational corporations are, in addition, controlled by international legislation and authorities. In a similar vein of thought, Perlmutter (1969) distinguished between *ethnocentric*, *polycentric*, and *geocentric* corporations, the latter type having no clear nationality.

A common projection in the wake of interest in MNCs in the 1960s was that internationalization of companies would eventually lead to the emergence of MNCs without any national character or loyalty, being stateless and "footloose". However, almost all corporations, which are multinational by some standard, still in the 1990s have a single or double nationality in a clear sense and much xenophobia around MNCs derives from this fact. As internationalization of functions such as ownership and management takes place in the sense that there will be more nationalities among owners and managers—that is, the multinationality increases regarding the behavioral variables under c) above—an MNC would gradually lose its nationality. We define *denationalization* here to mean the process by which national features are lost, as well as the outcome of that process. When denationalization takes place the often-used distinctions between home and host countries and domestic and foreign markets subsequently lose their validity. Although denationalization may follow upon internationalization, it is not a necessary consequence, and if it follows, it is apparently with a considerable time lag behind internationalization. Also nationalization in some form may occur, thereby possibly reversing an internationalization process. To some extent this once happened in telecom services (see below and also the final chapter).

Moreover, Porter (1986) defines *multi-domestic industry* (rather than corporation) as an industry in which competition in each country is essentially independent of competition in other countries, while a *global industry* in Porter's terms is rather the opposite.

Ohmae (1985) introduces the notion of a triad power, and more specifically a *triad corporation*, as an MNC that orients its operations in a triad of regions, encompassing the USA, Europe and Japan. In this connection, the concept of *triadization* can be introduced, referring to internationalization oriented towards such a triad.

Globalization of an industry then means the process of growing interdependence of competition in various countries. The notion of globalization of industries and firms has grown popular (see Bartlett et al. 1990), although the concept of a global corporation goes back at least to the 1950s in writings on international business. A *global corporation* is normally held to mean an MNC that takes an internationally integrated and global view of its location decisions in order to take advantage of coordinating on a global scale, with or without global coverage of input and output markets.

Finally, the simple fact must be noted that the geographic and political division of the world into nations is changing as well. For instance, the formation of new nations in the former USSR may increase the degree of internationalization for a company already operating there, and a full unification of Europe into one nation would strictly speaking decrease the degree of internationalization of a pan-European company, everything else being equal.

2.2.2 Telecom

The basic concepts of internationalization of telecommunications service companies (here called telcos for short) do not appear to differ fundamentally from those related to the internationalization of companies in extraction and manufacturing industries. However, the operationalization of the concept of internationalization of telcos is complicated by the fact that the telcos are linked together in a global network as a common physical means for producing services - tele services - which are often not produced or delivered at a single, well-defined point of location (cf. the notion of network industries mentioned above). The mere concept of the location of production and distribution of tele services is elusive. Moreover, the difficulty in many cases of clearly distinguishing between ownership and control of a network makes the use of ownership variables difficult as a basis for operationalization. The interdependence among telcos through the common network creates a complex mix of competitive and cooperative behaviors that further complicate conceptual distinctions. It is quite conceivable that competing consortia of cooperating telcos will be formed and that a consortium of different telcos, with separate nationalities, will cease to have a clear nationality in itself, that is, it will be denationalized. The latter feature would be strengthened through cross-ownership and cross-management arrangements.

As a consequence of conceptual difficulties and the newness of the phenomenon in the telco industry, internationalization has been used in a variety of senses. Some authors use the volume of international calls as indicating a form of internationalization. More correctly, we think, 'internationalization' is to be used when some kind of foreign direct investment (FDI) is involved, i.e. a resource transfer across national boundaries takes place rather than mere international trade in telecom services. This is in line with the traditional concept of an MNC (although the concept has been expanded in the 1980s, see Casson (1987) and Buckley et al. 1992). Foreign direct investments may be undertakings in telcos for establishing sales offices, providing a cellular network or operating cable television abroad or for some other kind of foreign business. A special concept is "foreign domestic business" which is sometimes used to denote the operation of telecom services in a foreign country, without direct links in the form of international traffic with the home country.

3. OVERVIEW OF HISTORY

3.1 General

Internationalization is certainly not a new phenomenon and far from a new concept. International trade and colonialism, for example, have a long history. The English philosopher Jeremy Bentham (1748-1832) is said to have pioneered the concept of 'international' in the late 18th century. The nation state in a modern sense emerged in this century, especially in connection with the American revolution (1776) and the French revolution (1789). The business company in its modern form[1] emerged as an economic institution in the 19th century. The multinational company (MNC) emerged as a phenomenon closely thereafter, although initially atypical and not widely recognized as a business concept. There were several typical MNCs in the 19th century, internationalizing not only sales but also production of products and services, for example International Harvester, and the Nobel-Dynamite Trust Company. Industrial R&D, at first mainly in the German chemical industry, also emerged and to a minor extent internationalized as well. The creator of the Nobel Prize, Alfred Nobel, ran a truly multinational network of R&D organizations on a European basis in his dynamite trust, established in the 1880s.

Many European companies, especially from countries with small domestic markets (e.g. Holland, Sweden, Switzerland), started to internationalize early on and there were several full-fledged MNCs before World War I. Internationalization of various companies, industries and markets slowed down in the interwar years but was further spurred after World War II. World trade grew rapidly to significant proportions in the 1950s. The international economic integration was in general strengthened after World War II, as it had been strengthened after major international crises before in history. However, the MNC was really not recognized on a broader scale as a concept and a phenomenon until the 1960s. Then it was mainly the US

[1] That is, as a limited-liability joint-stock company.

MNCs and their FDIs in Europe that came into focus. At the same time for this and other reasons, fears grew that Europe was slipping behind the US, technologically and economically. The debate around MNCs broadened further and included fears that MNCs were unduly exploiting developing countries; that MNCs were weakening their home country economy; that MNCs would distort international trade; that MNCs would grow into powerful international monopolies; that MNCs would weaken the nation state as an institution and usurp uncurbable political powers etc. Various initiatives were taken, including the initiation of investigations and research on MNCs, in turn leading to various models, theories and concepts of internationalization and MNCs, e.g. the international product life cycle model of R. Vernon at Harvard, and concepts like 'transnational corporation', 'geo-political firm', 'footloose industries' etc. The debate as well as behavior gradually sobered to some extent in the 1970s and internationalization in general was not halted; on the contrary it grew on an average, aided e.g. by further advances in transportation and telecommunication technologies.

The penetration of Western markets in the 1980s by East Asian goods through exports and MNCs, especially globalizing MNCs from Japan, led to further debate and initiatives, but now with some new dimensions, especially political, related to investment and trade friction between US and Japan and between Europe and Japan. In the 1990s, it seems the trend over several centuries towards increasing internationalization and international economic integration will continue despite protectionism and xenophobia, although its pace may be disputed. The MNCs, since the 1980s increasingly global in outlook and also increasingly engaging in strategic alliances with each other, are still growing and developing as a group with no clear signs on an average of economic and organizational weakness or political threats. However, truly denationalized global corporations, that is, MNCs coordinating operations on a global scale with no special home nationality, have yet not emerged in any significance, and the often reiterated scenario of globalization and denationalization in this sense is still just a scenario. Almost all MNCs, by the behavioral standards mentioned earlier, still have a nationality. A few MNCs are binational, i.e. they have acquired two nationalities, as in the case of the Swedish-Swiss 1986 cross-border merger ABB and the Dutch-UK 1907 cross-border merger between Royal Dutch and Shell. Some MNCs downplay their original nationality in their local subsidiaries. However, a true denationalization resulting from internationalization is still far ahead, and will probably remain so for a considerable time.

Thus, viewed over at least a century-long period, the driving forces behind the internationalization of companies, industries and markets appear to be fundamentally strong.

3.2 Telecom

Internationalization and MNCs are an old phenomenon in the telecommunication sector as a whole as well. In the telecommunication equipment industry, internationalization started in the late 19th century, following upon the invention of the telephone by Alexander Graham Bell, patented in 1876. Despite a considerable amount of vertical integration or quasi-integration and nationalistic procurement of equipment in the telecom sector in the 20th century, there are many full-fledged MNCs in the telecom equipment industry in the 1990s. To a considerable extent their internationalization has been spurred by procurement in emerging markets, where exports for various reasons have been less feasible and any indigenous industry has been weak.

Internationalization of telecom service companies also occurred already in the late 19th century and continued to occur in the 20th century as well, albeit only marginally on the whole until the 1980s.[2] The Bell Company operated some networks abroad early on, but was forced to divest these operations in 1925 by agreement with the US government. The Bell Company foreign telecom service operations were transferred to ITT instead. (See Hausman 1993 on the early Bell Company internationalization.) There were also equipment suppliers such as Ericsson that operated some networks. (See Attman et al. 1976 and Sobel 1982.) Cable & Wireless and its antecedents have perhaps the longest history of international operations in telecom. However, the prevailing industrial organization in telecom services until the 1980s consisted of national operating companies, mostly public monopolies, as is well known. (For histories of the establishment of this particular, nationalized industrial organization, see for instance Barty-King 1979, Brock 1981, Cowhey 1990 and Noam 1992.) Despite some antecedents it is fair to say that internationalization in the telecom service sector started in an essential way first in the 1980s, with developments in the US and UK leading the way. A brief chronology is given in Table 3.1. Regulation clearly has played a decisive role, in both impeding and stimulating internationalization. The political drift towards the political "right" in many Western advanced countries, starting in the 1980s, has economically meant an increased reliance upon market mechanisms and a move towards transformation of the public sector, liberalization of markets, privatization of property, and subjecting public companies such as PTTs to competition. Internationalization of telecom service markets and companies has then emerged more or less as a consequence although not solely caused by these changes.[3]

Technological changes have opened up possibilities to circumvent regulation and have thereby been highly influential in conjunction with changes in customer demand, especially among corporate customers and MNCs in other industries. Thus, to some extent internationalization in telecom services has been consequential also to the internationalization of its major customers. At the same time internationalization in the telecom equipment industry has not (so far) been influential in spurring the internationalization of telecom services.

In the mid-1990s the telecom service sector is perhaps in a greater flux than ever before, due to regulatory, market and technological changes. It is difficult to say something about the future of the sector, more than to say that its developments will certainly not be surprise-free. Material for the identification of trends and scenarios will hopefully be found in this book.

---

[2] Internationalization of telecom traffic, which has gradually evolved, is not considered a typical case of internationalization here, as mentioned earlier. It does not typically involve FDIs but is rather a special kind of service export-import.

[3] For instance, the BT case may include more factors than a "consequential internationalization" of telecom services, since the British government intended from the start of the liberalization and privatization process to establish BT as an international 'flagship', in view of the failures of the British equipment industry (cf. 'System X'). The chapter by Dang-N'guyen and Phan develops this point further.

Table 3.1
Chronology of some recent internationalization and liberalization events

Time	Event
1981	BT plc and BT International formed. Privatization C&W
1984	AT&T divestiture carried out. Regulation of RBOCs stimulates them to go abroad
1984	BT privatized
1984	Action lines for European Commission's telecommunications policy presented
1985	BellSouth International formed
1985	NTT International started. NTT Public Corporation was privatized. Market liberalization legislated in Japan
1987	EC Green Paper on European telecommunications
1988	Infonet Services Corp. formed as a jointly owned company by e.g. several European telcos and some Asia-Pacific ones
1989	BT acquires Tymnet
1989	AT&T acquires Istel
1989	Swedish Telecom International started
1990	EC Service Directive
1990	Project Iridium of Motorola announced—global mobile telecommunications through satellites
1990	MCI buys 25% of Infonet
1990	Sprint launches first international VPN service
1991	Duopoly review allowing foreign entry into CATV
1991	Maastricht Treaty, Title XII, making trans-European networks (TEN) a political objective
1991	France Telecom becomes an autonomous operator under public corporate law
1991 (Sept.)	Syncordia started by BT, inviting foreign companies to participate (NTT, DBT, IBM, among others)
1991 (Oct.)	Eucom started by France Telecom and DBT to provide VANS
1991 (Oct.)	Unisource started (alliance of PTT Telecom and Telia, formerly Televerket)
1992 (March)	BT goes alone with Syncordia
1992 (March)	Eunetcom started (alliance of France Telecom and DBP Telekom)
1992 (June)	NTT invests heavily in Thailand's domestic telecommunications
1992 (Sept.)	Project Utel—a joint venture of AT&T, DBP Telekom and PTT Telecom to develop telecommunications in Ukraine
1992 (Dec.)	Edinburgh meeting of the European Council providing financial instruments for TEN
1993 (March)	BT applies for operating license in the USA
1993 (May)	Worldsource started (AT&T and 8 carriers in the Pacific Rim)
1993 (June)	BT-MCI deal announced
1993 (Fall)	A group of some 20 large users formed to collectively source international VPNs (group called European VPN Users Association)
1993 (Dec.)	Project 'Atlas' announced (working name), merging DBP Telekom's and France Telecom's international businesses
1998	Year, targeted in June 1993, for the liberalization of the European Union telecommunications service market

Sources: Analysys (1993), Bradely et al. (1993), CommunicationsWeek International (various issues, 1990-1994), Cordaro (n.d.), Rietbroek (1993).

4. OVERVIEW OF THEORY

4.1 General

Relevant theories here, apart from general economic theories, are primarily theories about MNCs and their internationalization, but also to some extent international trade theory. Starting with the latter, this body of theory has developed rapidly in the 1980s, and has especially taken dynamic factors (e.g. increasing returns or positive feedback) and technological change into account. The doctrine of free trade, with national endowments and comparative advantage in a static framework as an analytical base, has been changed into strategic trade theory with arguments justifying certain types of protectionism under certain conditions. The traditional—and disputed—infant industry argument from development economics has thereby attained more theoretical credibility, and in fact more applicability also to developed countries, trying to catch up after having lost a leadership position in some area. (See further Scherer 1994 and Soete 1994.)

Turning to MNC theory, a number of models of internationalization patterns and theoretical explanations have been developed, with theoretical work starting as late as in the 1960s with the pioneering work of Stephen Hymer (posthumously published 1976). (For an overview of theories of MNCs and FDIs, see Caves 1982, Dunning 1988a, 1988b, Buckley 1989, Cantwell 1989 and Casson 1990, the latter containing most seminal works in the area.) Transaction cost theory and related theories about internationalization and appropriation have come to occupy a prominent role among theories of MNCs. In a Coasian sense, an MNC substitutes for cross-border transactions on the market. Williamson (1975) deemphasizes the role of technology in his transaction cost framework, but more recent contributions emphasize the role of firm-specific assets, including technological and managerial skills, in internalizing cross-border transactions (Teece 1986, Casson 1987). Since Hymer (1976) economic theory on FDI has also emphasized the role of firm-specific technology as an important element in the international growth of companies. Generally speaking, a company's competencies or know-how (technological, managerial, commercial, political etc.) are particularly important assets for an MNC since they do not only exhibit significant economics of scale and scope in their repeated utilization, but also are particularly suitable for appropriation by internalizing cross-border transactions, especially if joined with tangible complementary assets.

The theoretical literature on MNCs since Hymer has generally recognized three broad categories (with slight variations) of factors advantageous to MNCs and jointly favoring the internationalization of companies, factors pertaining to a) ownership or control of company-specific assets that gives the company a competitive or strategic advantage over indigenous companies, b) favorable location-specific conditions and c) high cross-border transaction costs (or high benefits from internalization). Dunning has in particular systematized and developed these factor categories into what he calls an OLI (ownerhsip-location-internalization) paradigm as a basis for an eclectic theory of the MNCs; this paradigm has a central place in the literature. Figure 4.1 gives a general overview of factors commonly considered in a causal analysis of the internationalization of companies, i.e. the emergence and development of MNCs.

Many theories and models of MNCs have been criticized for not paying sufficient attention to dynamic considerations and evolutionary patterns of internationalization. However, there are exceptions. Johansson and Wiedersheim-Paul proposed the concept of *the establishment chain* as a specific sequence of the functional types of operations a given corporation is likely to follow in locating operations in a given nation over time. They conclude: "The establishment chain—no regular export, independent representative (an agent), sales subsidiary, manufacturing—seems to be a correct description of the order of the development of operations of the firms in individual countries" (Johansson and Wiedersheim-Paul 1975, p. 321). The establishment chain model may then be extended in the sense that manufacturing operations are likely to be followed by R&D or at least D operations (see Granstrand et al. 1992). On the other hand, the establishment chain model is not valid in all respects. Progression through the sequence may be rapid, partial or disrupted. This is especially so if internationalization is supply-driven rather than demand-driven, and even more so if separate products are considered by an MNC, coordinating on a global basis.

Vernon's *international product life cycle* model is another model of stages in the internationalization process. This model (in its original version) briefly states that the product is first developed, manufactured, and marketed in the parent country. Then it is exported to the host country. Later manufacturing operations start in the host country for the host country market.

Finally, the product is manufactured in the host country and exported back to the parent country (Vernon 1966). This model has later been modified and generalized (Vernon 1979).

The international expansion process of MNCs has commonly had a gradual component, although it has decreased in relative importance over the years. The gradual expansion into more geographic areas has been influenced by the *psychic distance* to foreign markets, a distance measure weighing together geographic distance and cultural dissimilarities (Johansson and Wiedersheim-Paul 1975).

In general, theories about MNCs are increasingly paying attention to dynamic factors, partly because there is a general trend towards building evolutionary economic theories, including theories of the firm, partly because of increasing attention paid to technological change, which in itself epitomizes a dynamic factor. A good example is the approach of Cantwell (1989), who emphasizes the role of technological accumulation in MNCs.

The organization of MNCs, finally, has in general traditionally conformed to *the hub model*, i.e. a domestic headquarter functions as a hub to which the foreign subsidiaries are linked with lateral connections. However, the hub model is increasingly replaced by a subsidiary network structure (Hedlund 1986).

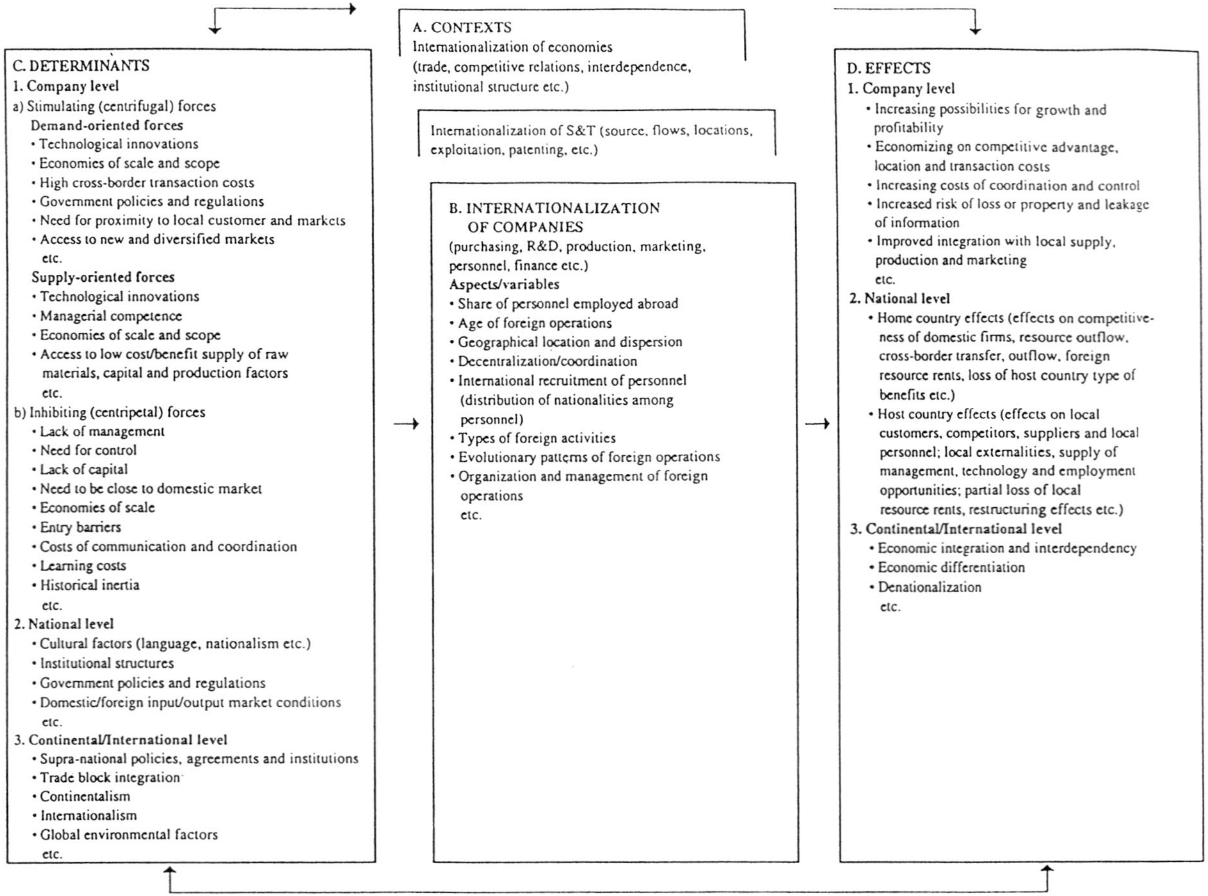

Figure 4.1. Overview of internationalization of companies, its contexts and causality

4.2 Telecom

The literature dealing specifically with service MNCs and the internationalization of service industries in general is growing but still scarce compared to the general MNC literature. Some examples are Dunning (1988a, 1989, 1991), Enderwick (1989) and Buckley et al. (1992).

Internationalization of service industries is also less developed than internationalization of manufacturing and extraction industries.

Many service industries serving industrial or professional customers have initially been influenced in their internationalization process by the internationalization process of their domestic customers. In general, it appears as if FDIs in services follow rather than lead FDIs in manufacturing industries (Dunning 1991). Technological and managerial skills, together with entrepreneurial spirits, have played decisive roles also for the internationalization of service companies (management consultants, banks, insurance companies, travel agents, hotels etc.). Some service companies have also internationalized on their input markets, e.g. software producers who have opted for cheap programming in countries with or without important local customers.

Literature on telecom service MNCs and internationalization of telecom services is even more scarce. Some early examples include Sharkey (1987), Ergas and Pogorel (1987), Noam (1989) and Salso (1990). Internationalization of the telecom service industry has developed far less (in terms of FDIs) than in many other industries, including some service industries. It is quite natural that there is no telecom-specific theory of MNCs and internationalization, then. It is also quite natural to ask to what extent received theory is applicable to the telecom service industry. The validity of traditional theory and models of internationalization in other industries for the internationalization in the telco industry could in fact be questioned on a number of grounds. First, the initial conditions for internationalization of the telco industry, with a network of regional and partially artificial monopolies being quite rapidly subjected to liberalization-induced competition, are very different from the initial conditions for internationalization in other industries. Second, the global systems interdependence among telecom service providers is high. Internationalization strategies and behavior of various actors therefore become interlocked in a complex mix of competition and cooperation. Random events, early mover advantages and increasing returns are for this and other reasons likely to highly influence the courses of internationalization. Third, the strategic importance of telecommunication in national economies is well recognized, which is indicated by large international asymmetries of national regulatory frameworks and telecom policies. Fourth, received theories and models largely derive from non-service industries while the telco industry is a service industry. The capital intensity and the importance of technological innovations in the telco industry would, however, make the telco industry more like manufacturing industries in some respects, as mentioned earlier. Besides, it may be argued, as Dunning (1991) as well as Buckley et al. (1992) do, that received theory for manufacturing MNCs is by and large adaptable to service MNCs. Finally, some traditional models of internationalization, like the (original) international product life cycle model, are themselves losing some of their validity in general.

Nevertheless, the empirical observations so far may not strongly disprove all received models and theories of internationalization and may in some cases in fact provide partial support for them.

We will return to this question in the final chapter of this book.

5. OVERVIEW OF THE BOOK

This book is essentially composed of two sets of chapters. The first is a set of country-oriented chapters, focusing a) on foreign entries and inward FDIs in a country market, and b) on foreign activities and outward FDIs by telecom service companies (typically the PTT) with the country as a home market. The second set of chapters raises various issues and challenges for regulation, policy and management across countries and actors, posed by the current industrial restructuring and internationalization.

Turning to the first set of country chapters, Table 5.1 gives the contents of these as suggested to the respective authors in the preparation of this book. As can be gathered from the table, the suggested contents are leaning towards a descriptive and inclusive focus. At this early stage of the internationalization process, in-depth empirical assessments seemed necessary in order to enhance a critical understanding of the new phenomenon. Quite understandably, it has not been possible to strictly adhere to this list of contents for each country and author. To reduce the potential for bias, the contributors are primarily informed members of the academic community and of industry research institutions with an even spread of author countries. With a broad sample from independent contributors, chances are that the internationalization phenomenon can be put in a proper perspective.

Table 5.1
Suggested contents for country papers on foreign direct investment (FDI)

Aim:	To describe and analyze the situation of FDI in telecom service provision with respect to the European market and players active in Europe (covering a market and actor dimension)
Delineations:	Authors dealing with players originating outside of Europe would focus on activities in the European market. Authors with players of European origin would treat the player's FDI in general with special focus on European activities as well as FDI made by others in the European player's home turf.

Issues:

- Strategy of internationalization
- Drivers and motives of internationalization
- Technologies, services and competencies of primary importance for FDIs
- Regulation impacting FDI
- Measures of FDI and degree of internationalization (for instance, employees abroad, investments abroad, revenue abroad, international presence through offices abroad etc.)
- Organization and structure of international operations
- Alliances and joint ventures
- Scenarios for the future

The choice of countries has been made to obtain as wide European coverage as possible on the different types of market: the internationalized vs non-internationalized markets, the advanced vs less developed markets, and large vs small markets, as well as the northern, southern, continental and eastern markets. Table 5.2 provides some aggregate statistics on the country markets in question compared with a sample of countries not included. With respect to telecommunications, the table makes it clear that there is an even spread from advanced countries to the less developed, as measured by penetration levels. Likewise, the largest telecommunications country markets in Europe (i.e. Germany, the United Kingdom, France and Italy) are included together with smaller markets. Furthermore, the figures show that a span is provided in terms of international dependency and internationalization in general, as measured by total trade per inhabitant. Non-European markets were excluded by coherence and focus considerations. Given the precursory importance of the UK market and actor strategies, the UK situation is treated in four chapters.

Table 5.2
Some statistics on European country markets

	(a) Percentage main lines per inhabitant[1)]	(b) Total telecommunication services turnover (MUSD)[1)]	(c) Total trade flow per inhabitant (exports + imports) (USD)[2)]
Country markets covered:			
Czech Republic[3)]	n.a.	n.a.	n.a.
Denmark	56.6	2356	13057
Finland	53.6	2222	10926
France	49.8	18913	7840
Germany	40.4	25117	9527
Greece	38.9	1293	2706
Hungary	10.9	523	1724
Italy	38.8	16666	6272
Netherlands	46.4	5480	19757
Norway	50.3	2560	4691
Poland	9.3	520	630
Portugal	24.1	1480	4253
Sweden	68.3	5308	13417
United Kingdom	44.2	23364	7380
Remaining European OECD country markets:			
Austria	41.8	2821	12054
Belgium	39.2	2694	n.a.[4)]
Luxemburg	48.2	158	n.a.[4)]
Iceland	51.4	84	15864
Ireland	28.1	1292	13490
Spain	32.4	8363	3797
Switzerland	58.0	4894	20853
Average of European OECD markets	43.3	6691	11133

1) Source: OECD, 1993.
2) Source: UN, 1993. Trade data are from 1990 and population data from most recent cencus.
3) Figures from the CFSR are indicative: penetration level 15.7, telecommunication revenues 542 MUSD, trade flow per capita 1635 USD.
4) Belgium-Luxemburg are counted as a trade union in the UN statistics. For Belgium-Luxemburg, per capita trade 1990 was USD 23371.

Within each country market chapter, several actors have been covered, mainly the incumbent (sometimes former) PTT. Some chapters are more geared towards an analysis of actor strategies, particularly chapters with a bearing on the UK market (i.e. the chapters by Dang-N'guyen/Phan on BT, Williams/Taylor on BT and Kramer/NiShuilleabhain on C&W).

To facilitate an overview of the country market studies, Table 5.3 provides the chapter order, markets and main actors covered. Since the chapters are fairly equal in design and focus, each chapter will not be recapitulated here. Instead, the reader is referred to the separate summary chapter at the end of the book with its tabulations and brief analysis.

Table 5.3
Country chapter coverage

Chapter	Title	Author(s)	Area(s)	Main actor(s)
3	Foreign Entry and Competition for Local Telecommunications Services in the UK after the Duopoly Review	M. Cave Y. Sharma	UK	BT Mercury RBOCs
4	Competition in the British Telephony Market	G. Dang-N'guyen D. Phan	UK	BT Mercury (C&W)
5	Competencies and Diversification: The Strategic Management of BT since 1984	H. Williams J. Taylor	UK	BT
6	Cable & Wireless: Services, Investments and Prospects	R. Kramer A. NiShuilleabhain	Europe Pacific Region North America	C&W
7	Internationalization of the German Telecommunications Service Market and Strategic Behaviour of DBP Telekom	D. Elixmann Th. Schnöring	Germany	DBP Telekom
8	Foreign Direct Investment in Telecommunications In and Out of France	G. Dang-N'guyen D. Pouillot	France	France Telecom
9	The International Economics of Telecommunications Services in Italy	C. Antonelli A. Geuna	Italy	IRI-STET Group
10	Internationalisation of Telecommunication Service Provision and the Greek Privatisation Debate	Y. Caloghirou T. Darmaros	Greece	OTE
11	Internationalization and Foreign Direct Investment in Telecommunications Services in Portugal	D. Lucena C. Seabra	Portugal	Marconi TLP Telecom de Portugal
12	A Note on Internationalization in the Netherlands	O. Johansson	Netherlands	PTT Telecom

Table 5.3 (cont.)
Country chapter coverage

Chapter	Title	Author(s)	Area(s)	Main actor(s)
13	Internationalization of the Swedish Telecom Services Market	O. Granstrand O. Johansson	Sweden	Telia The Kinnevik Group
14	Internationalisation of the Norwegian Telecommunication Market	R. Brekke S. Jensen C. Riis	Norway	Norwegian Telecom
15	Inward and Outward FDI in Danish Telecommunications	A. Henten K.E. Skouby	Denmark	Tele Denmark
16	The Internationalization of Finnish Telecommunications: Motivations, Strategies and Activities	J. Karpakka	Finland	TeleGroup Finland Finland Telecom
17	Transformation of Poland's Telecommunications and Foreign Investment	A.T. Jarmoszko	Poland	TP SA
18	Foreign Investment in Hungarian Telecommunications	F. Kiss	Hungary	HTC
19	Options for the Reduction of Deficiencies in Telecommunications in the Czech Republic	J. Cas	Czech Republic	SPT

Turning to the second set of chapters, Table 5.4 provides an overview of the main issues that surface. By design, the chapters in this section are shorter and raise various types of issues related to internationalization and FDI. This was deemed appropriate given the recent emergence of the FDI activities—allowing for a probing and open-ended search rather than providing ready answers. As such, the contributions have in general comparatively less emphasis on detailed empirical analysis and presentation, stressing instead various arguments, questions and possible implications, sometimes building the point on casual observations. Again, the reader is referred to the summary chapter for an account of each chapter.

Finally, there are two non-editorial chapters that are quite general in scope, meriting separate treatment in the book. First, the opening chapter by Kurisaki offers a broad overview of internationalization options and strategies pursued by telcos, complemented with a general discussion on the ensuing policy implications. Second, the final chapter by Johansson provides an empirical analysis of the internationalization behavior of 25 telcos, in which he finds patterns of strategic evolution in the internationalization process. In terms of regions analyzed, Johansson provides a truly global overview of FDI and internationalization.

Table 5.4
Issue chapter coverage

Chapter	Title	Author(s)	Main issue
20	The Race to New International Mobile Satellite Services	P. Bonanzinga F. Castelli C. Leporelli	Regulatory challenges posed by new international mobile services
21	Direct Foreign Investment in Telecommunications: Comments	E. Noam	Political and regulatory barriers to FDI
22	International Competitive Dynamics: Implications for Domestic Regulators	D. McCarty	The tensions between domestic and international regulation
23	Competing Visions for World Telecommunications: Europe and the Global Evolution of Industry Structure	D. Allen	Policy options to ensure innovative network evolution in an international environment
24	The Internationalization of Telephone Operators: Survival in an Integrating World	D.E. Ioannidis	Strategic options and prospects for internationalization in a nationalistic context
25	Asymmetries, Competition and Internationalization: Telecommunications Operators and the Future Single European Market	P. Barbet L. Benzoni	Assessment of the natural monopoly concept against the background of internationalization and national interests
26	Reorganization of European Telecoms: The Cases of British Telecom, France Telecom, Deutsche Telekom	R. Pospischil	Recent organizational changes and trends in BT, DBT and FT
27	Managing Network Investments: Illustrations from Telcos in Europe, Japan and the United States	E. Bohlin	Organization of network planning in an increasingly internationalized environment
28	Foreign Direct Investments of Telecommunication Operators in Less Developed Countries: Application to the Telefonica Experience	J.J. Duran F. Gallardo	A managerial framework for the process of FDI in LDCs
29	International Business Enterprises and Telecommunications: A User Perspective	E. Roche	The needs of large multinational users

REFERENCES

Aharoni, Y., 1971, On the definition of a multinational corporation, Quarterly Review of Economics and Business, Autumn.

Analysys, 1993, Global service provider. Market strategies. Handbook Series (Analysys Publications, Cambridge).

Attman, A., J. Kuuse and U. Olsson, 1976, The pioneering years struggle for concessions crisis 1876-1932, LM Ericsson 100 Years, Vol. I.

Bartlett, C.A., Y. Doz and G. Hedlund, eds., 1990, Managing the global firm (Routledge, London).

Barty-King, H., 1979, Girdle round the earth (Heinemann, London).

Bradely, S.P., et al., 1993, Globalization, technology and competition: the fusion of computers and telecommunications in the 1990s (Harvard Business School Press, Boston).

Brock, G., 1981, The telecommunications industry. The dynamics of market structure (Harvard University Press, Cambridge).

Buckley, P.J., 1989, The multinational enterprise. Theory and applications, MacMillan, London.

Buckley, P.J., C.L. Pass and K. Prescott, 1992, The internationalization of service firms: a comparison with the manufacturing sector, Scandinavian International Business Review, Vol. 1, No. 1, 39-56.

Cantwell, J., 1989, Technological innovation and multinational corporations (Basil Blackwell, Oxford).

Casson, M.C., 1987, The firm and the market (Basil Blackwell, Oxford).

Casson, M.C., 1990, Enterprise and competitiveness: a systems view of international business (Clarendon Press, Oxford).

Caves, R.E., 1982, Multinational enterprise and economic analysis (Cambridge University Press, Cambridge).

CommunicationsWeek International, various issues 1990-1994.

Cordaro, G., n.d., Towards 1992: the European community's telecommunications policy, mimeo, Commission of the European Communities.

Cowhey, P.F., 1990, The international telecommunications regime: the political roots of regimes for high technology, International Organization, 44, 2, Spring, 169-199.

David, P., 1992, Information network economics: externalities, innovations and evolution, in C. Antonelli, ed., The economics of information networks (North-Holland, Amsterdam).

Dunning, J.H., 1988a, Explaining international production (Unwin Hyman, London).

Dunning, J.H., 1988b, Multinationals, technology and competitiveness (Unwin Hyman, London).

Dunning, J.H., 1989, Multinational enterprises and the growth of services: some conceptual and theoretical issues, Services Industries Journal, Vol. 9, No. 1, 5-39.

Dunning, J.H., 1991, The internationalization of the production of services: some general and specific explanations, GSM Working Paper, 91-11 (Graduate School of Management, Rutgers University, New Jersey).

Enderwick, P., ed., 1989, Multinational service firms (Routledge, London).

Ergas, H. and G. Pogorel, 1987, Multilateral cooperation in international telecommunication: sources and prospects, Research Working Paper, Columbia University, Center for Telecommunications and Information Studies.

Freeman, C., M. Sharp and W. Walker, eds., 1991, Technology and the future of Europe (Pinter Publishers, London).

Granstrand, O., L. Håkanson and S. Sjölander, eds., 1992, Technology management and international business. Internationalization of R&D and technology (John Wiley & Sons, Chichester).

Hausman, J.A., 1993, The Bell operating companies and AT&T venture abroad while British Telecom and others come to the United States, in S.P. Bradely et al.

Hedlund, G., 1986, The hypermodern MNC—a heterarchy?, Human Resource Management, 25, 9-35.

Hymer, S., 1976, The international operations of national firms (MIT Press, Cambridge, MA).

Johansson, J. and F. Wiedersheim-Paul, 1975, The internationalization of the firm—four Swedish cases, The Journal of Management Studies, October, 305-322.

Noam, E.M., 1989, International telecommunications in transition, in R.W. Crandall and K. Flamm, Changing the rules: technological change, international competition and regulation in communications (Brookings Institution).

Noam, E.M., 1992, Telecommunications in Europe (Oxford University Press, Oxford).

OECD, 1993, Communications outlook 1993, Organisation of Economic Co-Operation and Development, Paris.

Ohmae, K., 1985, Triad Power. The coming shape of global competition (The Free Press, New York).

Perlmutter, H., 1969, The tortuous evolution of the multinational corporation, Columbia Journal of World Business, Vol. 4, No. 1, Jan.-Feb.

Porter, M.E., 1986, Competition in global industries: A conceptual framework, in M.E. Porter, ed., Competition in global industries (Harvard Business School Press, Boston).

Rietbroek, J., 1993, EC initiatives on trans-European networks, Information Economics and Policy, 5, 287-294.

Robinson, R., 1967, International management (Holt, Rinehart & Winston, New York).

Salso, E.L., 1990, How PTT's are going multinational, Long Range Planning, Vol. 23, No. 1, 136-146.

Sharkey, W.W., 1987, Oligopoly in international telecommunications, Research Working Paper Series, Columbia University, Center for Telecommunications and Information Studies.

Scherer, F.M., 1994, Competing for comparative advantage through technological innovation, in O. Granstrand, ed., Economics of Technology (Elsevier, Amsterdam).

Sobel, R., 1982, ITT, the management of opportunity (Truman Talley Books/Times Books, New York).

Soete, L., 1994, International competitiveness, trade and technology policies, in O. Granstrand, ed., Economics of Technology (Elsevier, Amsterdam).

Teece, D.J., 1986, Transactions cost economics and the multinational enterprise. An assessment, Journal of Economic Behavior and Organization, Vol. 7, 21-45.

UN, 1993, Statistical Yearbook, United Nations, New York.

Vernon, R., 1966, International investment and international trade in the product cycle, Quarterly Journal of Economics, May, 190-207.

Vernon, R., 1979, The product cycle hypothesis in a new international environment, Oxford Bulletin of Economics and Statistics, Vol. 41, No. 4, 255-267.

Williamsson, O.E., 1975, Markets and hierarchies: analysis and antitrust implications. A study in the economics of internal organization (The Free Press, London).

The Race to European Eminence
E. Bohlin and O. Granstrand (Editors)

Challenge to the Globalisation of Public Telecommunications Operators: Toward a New Framework of Telecommunications Policy

Yoshiko Kurisaki

Information, Computer and Communications Policy Division Directorate for Science, Technology and Industry, OECD 2, rue André-Pascal, 75775 Paris, France

The views expressed in this paper are those of the author and do not necessarily reflect those of the OECD or its Member countries.

This paper is intended to identify newly emerging forms of international expansion of activities of public telecommunications operators (PTOs)[1] and to analyse their national and international policy implications. Particular focus will be placed on the following three forms of globalisation: (1) new service options in the provision of existing international telecommunications services; (2) foreign direct investment (FDI); and (3) off-shore services provided by alliances between PTOs (e.g. so-called "one-stop shopping" and "outsourcing").

1. INTRODUCTION

The increasing globalisation of corporate activities (both manufacturing and service sectors) in the OECD Member countries has spilled over onto the telecommunications sector since the end of the 1980s. This fact is creating fundamental changes in the paradigm of international telecommunications service provision. An increasing number of public telecommunications operators (PTOs) are extending their business activities, either explicitly or implicitly, to countries *outside* of their home countries. Although this is a fairly new phenomenon, speed of expansion of the activities has been dramatic. The absolute value of such investment may be relatively small, when compared with the whole PTO investment which includes that in installation and improvement of network facilities. The strategic importance of PTOs, however, should not be overlooked.

[1] PTO in this paper means an operator and provider of public switched voice telephony services on local, long-distance and international levels. Other service providers, such as sole providers of value-added and/or mobile services, are excluded.

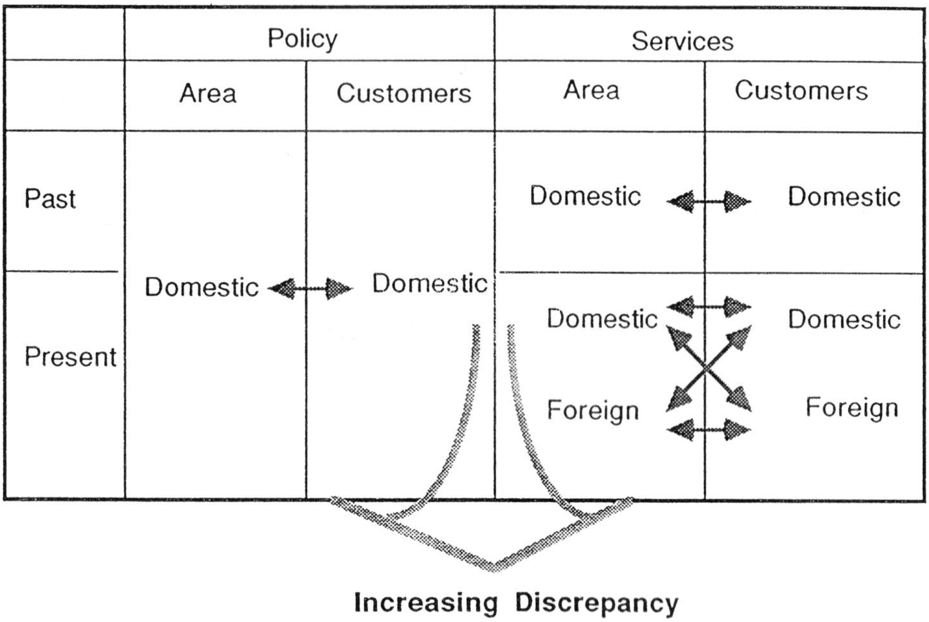

Figure 1.1. Mismatch between recipients of national telecommunications policy and services.

There is a growing level of discrepancy between those to whom national telecommunications policy and services are addressed (Figure 1.1). In the past, both telecommunications policy and services are addressed to national users in the country. While the basic idea of the policy has remained the same to date, the services are extending their reach by including both national and foreign users in national and foreign markets. A number of policy issues arise from this discrepancy. Existence of asymmetry in the level of liberalisation between OECD countries increases economic impacts of PTO globalisation, giving birth to a set of policy issues relevant to all countries. National telecommunications policy now has broader cross-border implications than it had previously.

2. NEW FORMS OF INTERNATIONALISATION

2.1 New service options

PTOs are developing a variety of service options and geographical coverage of their provision. Discount in voice telephony services using simple resale of the international leased circuits is the simplest example of a price option.[2]

[2] For example, Telia (Swedish ex-monopoly PTO) was granted a licence by the UK authority to provide discounted voice telephony services between the UK and Sweden by the international simple resale. It was reported that Telia planned to do the same with the US. (Source: Tariffs Undermined, Communications Week International, 1992). The same application is under consideration between the UK and the US. (This information is as of November 1993.)

PTOs are offering various options in billing and callset-up services that help customers to use the international voice telephony service. Examples of these include international toll free service (such as "International Green Numbers" by PTT Netherlands), and an automated reverse charge service (those "Country Direct Services", such as "OTC Country Direct" by Telstra (Australia), "Japan Direct" by KDD, and "USDirect Service" by AT&T).

Calling card services have opened more variation in calling options for users. An automated credit call service (such as "OTC CallCard" by Telstra, "Calling Card" by AT&T and similar service by MCI and Sprint) was developed to enable users to charge the fee of calls originating in foreign countries to their account in their home countries. Some US PTOs are, furthermore, providing a third-party calling service. This service is made available technologically as a mere extension of the calling card service, with "World Connect Service" by AT&T and "MCI Card" by MCI being examples. In this service, a holder of the calling card in Country A may call from Country B to Country C *via* Country A (Figure 2.1). The card holder first calls Country A, then has the PTO in Country A call Country C. The calling procedure is the same as the automated credit call terminated in Country A, but the call is simply forwarded to a destination outside, not inside, the country.

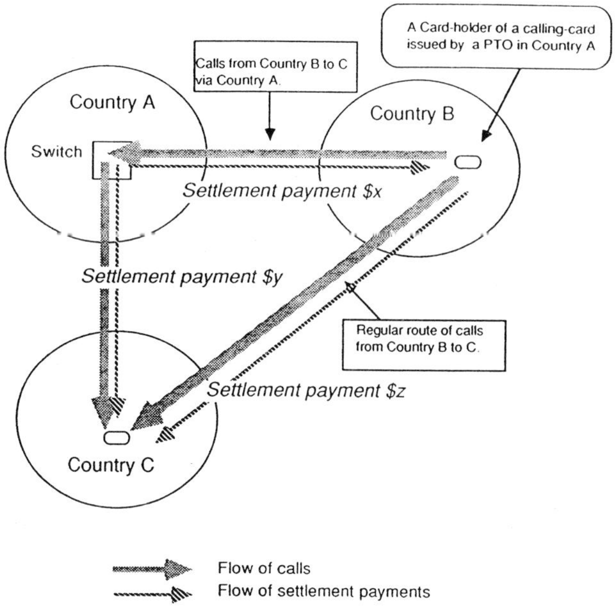

Figure 2.1. Third-party calling service.

In this service, flow of the payment between PTOs in Countries A and B goes contrary to the direction of the call. A PTO in Country A pays the accounting rate to a PTO in B for those calls originated by calling card holders in B and addressed to the third country via A. This is because of the existing International Telecommunications Regulations. The Regulations determine that credit calls originated in a foreign country are treated as outgoing calls from the subscribers of this service's home country: in the example above, Country A. The Regulations also require that an accounting rate must be paid by the country which originates traffic over a route not previously agreed: in the example above, a call from B to C routed via A. Benefits of existing PTOs are safeguarded in this way.

The market size of the third-party calling service is, in fact, estimated to be negligibly small in relation to the whole volume of international traffic between any pair of the two "by-passed" countries (such as countries B and C in Figure 2.1). Customers to whom this service is most appealing would be those who subscribe to the calling card service and who are travelling abroad. It is unrealistic to assume that the volume of international traffic generated by such customers is large enough to result in a significant loss of revenue for the "by-passed" PTOs.

Third-party calling service has, however, some potential elements of competition:

- This service enables a PTO in Country A to participate in telephony services markets between B and C, from which in the past the PTO in A was excluded. The PTO may obtain revenue from collection charges from the calls from B to C via A. If the accounting rate from B to C ($z in Figure 2.1) is higher than the one from A to C ($y in Figure 2.1), C might lose some income opportunities.
- Impacts of third-party calling service on the corporate image of a PTO may be substantial. The idea of the "third-party" calling service by using a credit card call would look "smart" to customers who are used to encountering problems when preparing coins or a number of prepaid telephone cards when making an international call from foreign countries. It is a psychological effect but makes a difference for the "brand image" of a PTO.
- Expansion of the card holders outside of the issuer PTO's home country is, in theory, possible. Alternatively, the PTO may link this service with existing commercial credit cards issued by financial service firms (e.g. VISA, Master, American Express).[3]
- If the calling cards are widely made available to customers regardless of their nationalities, telephone users may wish to hold cards that enable them to use a PTO that offers the lowest price.

All the effects stated above may create some feeling of *competitive pressure* on the "by-passed" countries (Countries B and C in Figure 2.1), despite the fact that the third-party calling service is operated in the realm of credit call service based on bilateral agreements in the framework set by the International Telecommunications Regulations. Discrepancy in interests between PTOs will arise.

Increase in the number of countries that allow simple resale of international leased circuits will give birth to a carrier that does not own facilities but leases international circuits and provides the same service as PTOs.[4] PTOs are no longer the only players in international telecommunications. A situation described below is possible with the international simple resale:

---

[3] This point was discussed in Staple (1992).

[4] Such a carrier is called a "Light Carrier" as opposed to "Heavy Carriers" that own facilities and provide services, in Staple (1992).

Let us imagine that the original condition is exhibited in Figure 2.2.(1). A PTO in Country A has a cable for transoceanic traffic between Countries A and B. The PTO obtains profit by providing toll call transmission service to international calls from the Countries C, D and E to B through the cable. A charges this services to PTOs in C,D and E. The PTO in Country A thus obtains markets for international traffic both from its domestic market and from foreign countries. In a new situation illustrated in Figure 2.2.(2), a telecommunications operator (TO) Z enters the transoceanic telecommunications market. Z leases a large capacity of the cable built by a PTO in Country P, uses an international reverse charge service to attract users in Countries C, D and E, and provides international telecommunications services between these three countries and Country B. In this case, Z is competing with PTOs in P and A, since users in P may also take advantage of the same service. The operator Z may further aggressively try to attract more users in C, D and E by using international toll free services for customers who access its switches located in P.

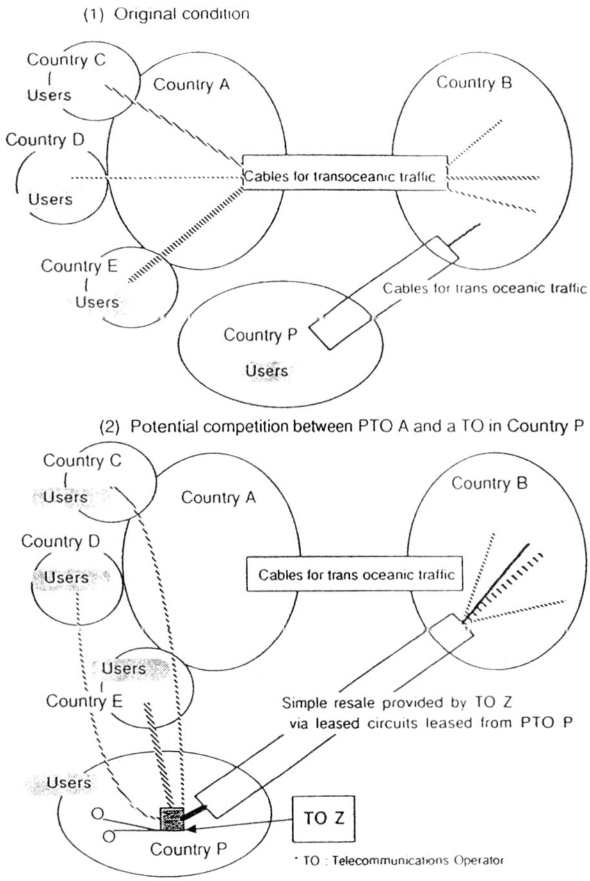

Figure 2.2. Simple resale.

Simple resale will place *downward pressure* on the charges of existing international telephone services. In the example above, Country A may have to reduce the toll call transmission charges to maintain traffic from C, D and E.

2.2 Foreign direct investment

Foreign direct investment (FDI) by PTOs is increasing in terms of both the number and level of investment. FDI takes various forms: purchase of shares of TOs (both PTOs and other service providers), establishment of joint ventures (JVs), obtainment of franchise for the operation of telecommunications service, and mergers and acquisitions (M&A) of providers of enhanced communications services other than TOs. Direct investment has an advantage for PTOs in that they become globally known (i.e. global presence) in a relatively short time and is a means of rapidly entering the telecommunications market. Examples of recent large-scale direct investment include the purchase of shares of the PTO in New Zealand (Telecom Corporation of New Zealand Limited, or TCNZ) in 1990 by a consortium of Ameritech and Bell Atlantic (both from the US) and the purchase of 49 percent of the shares (with voting rights) of Mexican PTO, Telefonos de México (Telmex) by a consortium including Southwestern Bell (a Regional Holding Company in the US) and France Télécom (FT, France) in 1991.

Foreign direct investment gives birth to several policy issues in an interface between existing rules for trade and international capital movement (external factor) and telecommunications regulation that controls the level of openness of the domestic markets (internal factor). This is because of existing asymmetry between country rules for foreign ownership of PTOs and the levels of openness of telecommunications markets.

1. Competition issues

Where do monopoly (or near-monopoly) PTOs obtain financial resources to invest abroad? Is there a possibility of *subsidised competition* when these PTOs compete with non-monopoly PTOs in the telecommunications market in third countries (i.e. those countries that *receive* investment)? What rules do we need to avoid subsidised competition?

2. Trade issues

Is the same level of *accessibility to the market* in the country concerned assured for both domestic and foreign PTOs (i.e. non-discrimination in market access)? Are current regulatory systems opened to foreign-owned PTOs in terms of the availability of information needed to access national markets, such as network information and minimum customer data?

Acceptance of FDI has implications for trade *telecommunications equipment*. If new entrants are limited in their options in equipment, such conditions may work as a de facto entry barrier to the service market. Although it has been a fact that PTOs often had certain manufacturers who provide equipment almost exclusively to them, such practices will have to end. Liberalisation in telecommunications equipment interconnection needs to be undertaken.

3. Beneficiaries of FDI

Who are the *beneficiaries* of FDI by PTOs?

Viewed from the *investing* countries, can PTOs use profit from the home country to invest abroad, instead of better meeting Universal Service Obligations, such as improvement of services, network facilities and reduction of tariffs to domestic users? Viewed from the countries *receiving* investment, what mechanism is needed for the public to take the best advantage of profits from telecommunications markets in its own country? Is there a need for a mechanism to prohibit the profits (some or all) from going abroad?

2.3 Alliances and outsourcing

Co-operation between PTOs is a major means to serve international customers, supplementing geographical coverage of PTOs with each other. The co-operation started in a form of joint account management (JAM) agreements in the late 1980s between major PTOs in Member countries and evolved to form closer co-operation.

Under JAM, two PTOs in different countries take care of the partner PTO's corporate customers that have branches in each other's country. In this system the customers may construct and maintain international corporate network systems only with the support of their home country's PTO.

The development of JAM soon grew to create a more organised form of PTOs' alliances, i.e. so-called "one-stop shopping" (OSS). In OSS, a PTO that becomes a one-stop shop (co-ordinating carrier) takes all the actions needed for corporate network systems on behalf of a customer firm concerned. While JAM is a bilateral agreement merely for the customers of the countries concerned, OSS providers are less concerned about the customers' nationalities.

"Outsourcing" services have become more strategic to PTOs when inefficiency of OSS in network management became evident. A concept of outsourcing includes various levels of involvement of users' networks, from a simple management of data communications networks to the entire handling of information network systems, including planning, construction and operation of networks and information processing (Figure 2.3). A variety of services are included in outsourcing, such as so-called "managed data network service" (MDN), global virtual private network (GVPN), international frame relay service and "bandwidth-on-demand".

Outsourcing may or may not imply ownership of communications nodes and switches. Some PTOs install this equipment owned by themselves in major sites of global corporate networks and thus build a single network, whereas others do not, but provide the service through networks provided by other PTOs. For example, Unisource Business Network, a branch of Unisource (The Netherlands, Sweden and Switzerland, as explained below) has installed nodes in Paris and Brussels, and BT, all over the world for the provision of its MDN, Global Network Service (GNS). For Telstra, Australia, in contrast, MDN is a consultation service that does not involve the construction of its own facilities abroad.

Regardless of the ownership of nodes and switches, many PTOs seem to feel they need alliance with other PTOs in order to ensure geographical coverage required by the targeted users and, in fact, have developed partnerships with each other. Examples of such partnerships include Eunetcom (FT and DBP Telekom, who each own a 50 percent stake, started in 1992), Unisource (PTT Telecom in the Netherlands and Telia, started in 1992 and joined by Swiss PTT in 1993) and World Source (AT&T, KDD and Singapore Telecom joined by several other PTOs in 1993). Alliances between MCI (the US) and Stentor (Canada) to serve corporate network systems of Chrysler Corp. (an MNE in automobile manufacturing based on the US)[5], and BT and MCI[6], are also aimed at global outsourcing services.

[5] Source: Chrysler maps out traffic route, Communications Week International.

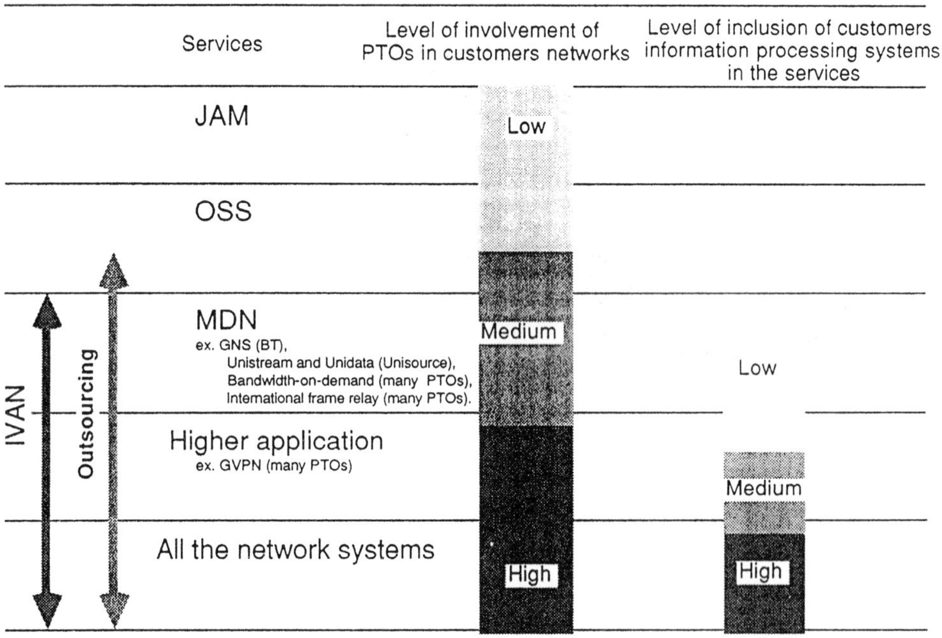

Figure 2.3. Evolution of PTO alliance.

PTOs are increasing their involvement in international value added network services (IVANs) through acting as integrators of various voice and data communications systems (Figure 2.3 above). Large-scale corporate networks usually have multi-purposes and transmit both voice and data. A processing system for a large volume of data, such as a Computer Aided Design (CAD) system, is sometimes interconnected with and transmitted over international private networks. Provision of outsourcing for these customers means that PTOs are eventually enhancing their function as IVAN providers.

There are, however, a number of IVAN providers that originated from a variety of sectors other than the telecommunications sector. PTOs are only some of these competitors. Major examples of the competitors include: specialised firms (e.g. Electronics Data Systems Corp.), computer system vendors (e.g. IBM and DEC), VAN providers (e.g. Philips[7] and GEIS) and subsidiaries of accounting firms (e.g. Andersen Consulting).

Distinction between large users and PTOs is becoming unclear in IVAN. This is a result of on-going liberalisation in the usage of leased circuits. Some large users are partly but increasingly in competition with PTOs in both data and voice telephony services. Emergence of intersectorial competition is recognised by users of large corporate networks.

[6] This proposed alliance includes the creation of a new joint venture company ("NewCo"), the purchase of a 20 percent interest in MCI and the purchase by MCI of BT North America (Source: BT Annual Report on Form 20-F 1993).

[7] Philips Communications & Processing Services International, a Philips VAN unit.

The Société Internationale de Télécommunications Aéronautiques (SITA), an association of the airline sector in the world set up to meet the telecommunications needs of the industry, i.e. IVAN, is an example of this. SITA has been a large user of international leased circuits in relation to PTOs in many countries. The organisation is increasingly in competition with those PTOs who intend to obtain customers in the airline sector. The same relationship between PTOs and large users applies to voice service in those countries where simple resale of leased circuits for the service is allowed. Some large users are starting to provide voice telephony services to users outside of the companies on a commercial basis by using their own corporate networks.[8] Large users may not be mere users for PTOs any more.

Policies that ensure a fair playing field of *inter-sectorial competition* will have to be developed. There is a possibility of abuse of competition if those IVAN providers originally in the private sector compete with PTOs that may cross-subsidize competitive business from revenue in originally or currently monopoly (or near-monopoly) markets, such as voice telephony services.

Relationships between PTOs themselves are also changing in IVAN in the context of outsourcing. A PTO needs to lease circuits from other PTOs to provide MDN over a number of countries. The former is a large customer of leased circuits in the latter's home countries. PTOs are thus increasingly in a two-fold situation with each other, i.e. both competitors and customers.

A type of outsourcing services in which a PTO owns facilities, as described above, may have implications for telecommunication *infrastructure policy*. PTOs providing outsourcing services have their own switches and/or nodes in other countries and even those that are not a member of the alliance. MDN is a penetration of foreign PTOs' infrastructure in the host country. The ownership of infrastructure, however, is not an important issue as viewed by users. The users' priority for the selection of a service provider is the best quality at a low price, not the nationality of the provider. Thus traditional reasoning that justifies government ownership of telecommunications infrastructure, "the government should own telecommunications infrastructure to guarantee the services' quality", may have to be reconsidered, including a possibility of its relaxation.

If a large number of PTOs form an alliance, such movement may have the same effects as a cartel. *Anti-competitive concerns* may arise in international markets.

2.4 Implications for the existing regime

An additional observation arises from all the above analysis: i.e. *The representation system of users' interests* in international telecommunications organisation for technology standards and rule setting of the services, such as the International Telecommunication Union (ITU), will have to be reconsidered. Observations of interests of users of global corporate networks suggest that a perception which supports the ITU concept, that one country's PTO represents interests of this country's users, has a decreasing value.

PTOs, however, still need the existing regime of international telecommunications. The newly emerging international services are based on a number of existing bilateral and multilateral agreements. Examples of the former are rules for landing rights of trans-oceanic cables and transmission rights of cables; and of the latter, those rules established at ITU, including a

[8] For example, many Type II carriers (those TOs that provide services by leasing circuits from Type I carriers) are formulated originally from a telecommunications service branch of large users of corporate network systems.

variety of agreements on international service provision, e.g. accounting agreements on both private and public circuits, credit call services, and other services. These agreements, in fact, still are the basis of the current development of various international services by PTOs. It should not be overlooked, in addition, that ex-monopoly PTOs have obtained benefits from these rules. Mutual familiarity and credibility developed through long-lasting relationships between PTOs are still substantial assets, useful and needed by them for further development of international activities. In short, PTOs are partly leaving the ITU regime towards competition in international markets, while trying to keep existing benefits.

3. TWO DIMENSIONS IN TYPES OF GLOBALISATION

Global activities by PTOs discussed above have two conceptual dimensions: one in which a PTO stretches its service provision to other countries, i.e. an "outgoing" dimension, and another in which PTOs are potentially in competition with each other, an "incoming" dimension. The "outgoing" dimension means PTOs' visible globalisation activities. Investment in the telecommunications sector abroad, provision of service abroad, and corporate alliances are examples of this dimension.

The "incoming" dimension, although not always visible, is created by penetration of effects of other PTOs' globalisation into domestic markets. The "incoming" dimension, in other words, means a situation in which a PTO is eventually placed in competition with another PTO. One PTO's gain in this case is generated from another PTO's loss of its potential gain.

Efforts currently made by many PTOs to invite international corporate network users to locate their hubs in the PTO's home country are an example of the "incoming" dimension. The PTO expects to obtain high profit from the lease of circuits with large capacity and increased international traffic. Economic gain from activities associated with installation of the hubs is also expected, since hubs are often located in the regional centre of corporate activities of user firms. Several PTOs are, in fact, competing with each other by offering their advantages to customers, including level of liberalisation, low price, quality of services and geographical locations.[9,10]

Another example is the offer of competitive rates in transmission charges. In Figure 3.1.(1), the original condition is the same as the one illustrated in Figure 2.2.(1) above. For MNEs, the location of the transoceanic cables will give incentives for them to locate their communications hubs in Country A. What happens if another Country P constructs its own cable to connect with Country B, and offers a transmission rate that is less expensive than the one paid to Country A (Figure 3.1.(2))? PTOs in Countries C, D and E may wish to route their international outgoing traffic via Country P, rather than A. MNEs may locate their communications hubs in Country P, not A. PTOs in Countries A and P are in competition in trying to obtain traffic and MNES from the Countries C, D and E. This example may look similar to the above example of simple resale. The major points to be seen in this transmission charge example are: (1) even if both Countries A and P prohibit simple resale of leased

[9] This type of competition is currently increasing between major PTOs in the Asia-Pacific area and Europe.

[10] During the interviews undertaken for this study, a number of users of international private networks in various sectors of economy (i.e. finance, information service, manufacturing, etc.) pointed out the critical importance of these factors in the selection of the location of communications hubs.

circuits, competition between the two may take place, and (2) PTOs in A and P compete for other PTOs in neighbouring countries (C, D and E), not individual users.

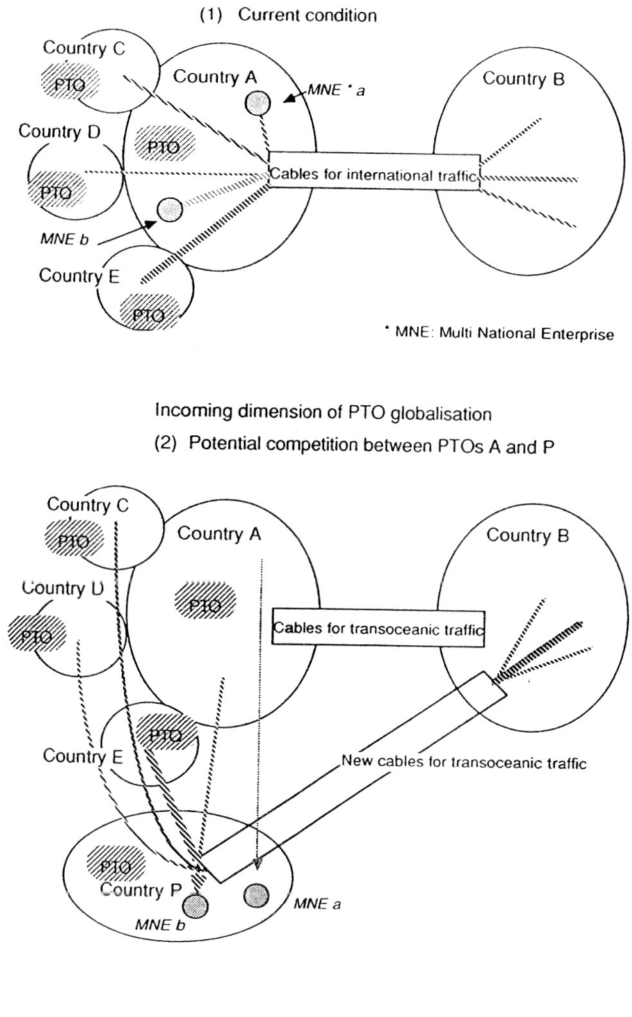

Figure 3.1. Incoming dimension of PTO globalisation.

All the above examples indicate that PTOs are increasingly in competition with other PTOs, *regardless* of their interest in extending their activities abroad. All these phenomena will put pressure on PTOs to secure at least their current customers. In the hub example above, if a PTO wishes to obtain benefits from hubs, telecommunications policy-makers and a PTO will have to further liberalise usage of leased circuits, improve service quality and reduce service price. In other examples, downward pressure will work on collection and transmission charges of international telecommunications. If PTOs and policy-makers do not respond to "incoming" effects of PTO globalisation originated in other countries, costs for the maintenance of current regulation will be paid by loss of both existing and potential incomes.

4. SUMMARY AND CONCLUSION

Increase in PTOs' efforts to extend their business activities outside of their home countries is leading to a paradigm change in international telecommunications service provision. Globalisation of PTO activities is originally a reaction by PTOs to economic globalisation in general and on-going liberalisation of telecommunications policy.

Increasing competition between PTOs based in different countries has substantial implications for national telecommunications policy in their home countries. All PTOs are increasingly in a situation in which they have to enhance attractiveness of the domestic telecommunications market for both national and international customers. Telecommunications tariffs have to become competitive when compared with other PTOs. Regulations on the usage of leased circuits need to be reduced, and waiting time for the preparation of circuits must become much shorter than now. Type approval systems have to become simpler, more rapid in providing approval and liberalised. Interconnection with public switches has to be non-discriminatory between domestic and foreign-owned carriers, and the procedure to obtain permission for interconnection has to be comprehensive, simple and less time-consuming. If PTOs continue to make such efforts, it will eventually lead to the liberalisation of telecommunications in all countries.

A task for policy-makers is to formulate the framework for the development of policies that accommodate globalisation impacts. A policy vacuum in telecommunications is seen in handling questions such as those listed above. Existing rules in telecommunications are only partly able to deal with the issues that have increasing international interface.

International co-operation is needed to resolve the newly emerging policy issues related to PTO globalisation. No one country alone may be able to provide effective solutions. These issues are in the cross-section between originally different policy areas, i.e. telecommunications, trade, competition and international capital movement. Common rules for the telecommunications sector on an international level should be formulated based on cross analysis between telecommunications and three other, originally different policies. Such a process might necessitate some adjustment in existing national telecommunications policy to newly created rules.

REFERENCES

BT Annual Report on Form 20-F 1993.

Chrysler maps out traffic route, CommunicationsWeek International, p. 25, 14 Dec. 1992.

Staple, G.C. 1992, Winning the global telecoms market: The old service paradigm and the next one, in TeleGeography 1992, p. 132, International Institute of Communications, London.

Tariffs undermined, CommunicationsWeek International, p. 1, 14 December 1992, UK

PART B: FOREIGN DIRECT INVESTMENTS IN TELECOMMUNICATIONS

Internationalization in the United Kingdom

The Race to European Eminence
E. Bohlin and O. Granstrand (Editors)

Foreign Entry and Competition for Local Telecommunications Services in the UK after the Duopoly Review

Martin Cave and Yogesh Sharma

Brunel University, Uxbridge, Middlesex UB8 3PH, United Kingdom

1. INTRODUCTION

The abandonment of the duopoly policy in 1991 has opened the floodgates to competition in UK telecommunications markets. Considerable strides had already been made in mobile telecommunications services over the period of the duopoly policy, with two cellular networks in operation and licences issued to three PN operators (two of which have now merged). These are in addition to seven paging networks and two national private mobile radio operators.

But the termination of the duopoly policy has also opened up wire-based competition. Cable television operators are now allowed to offer telecommunication services on their own account, and currently provide 317,000 lines. In addition, about 30 new licences for telecommunication services have been issued by the end of May 1993. While some of these, such as that issued to Energis, a subsidiary of the National Grid, intend to offer long-distance services, other operators will provide local access, using either wireless or wire-based technologies.

It is significant, moreover, that the newly liberalised entry conditions have attracted investment or expressions of interest from overseas companies. Thus the RBOCs have invested heavily in UK cable licences, and overseas companies have applied for licences to operate other wire-based networks in the UK. In few other countries are the possible effects of the internationalisation of telecommunications seen so clearly.

As a consequence of these developments, the UK provides the leading illustration in Europe (and possibly in the northern hemisphere) of competition in telecommunications markets, including local access. The aim of this paper is to explore the possible implications of the development of competition. We first give a brief account of local access technology in Section 2. Section 3 then makes some observations about the cost structures the technologies exhibit. This is followed in Section 4 by an account of the current regulatory environment for the development of competition in local markets. Section 5 focusses upon the provision of telephony by cable television companies, and examines the role of international, especially American, operators in that segment of the market. Section 6 summarises the paper and considers possible competitive interactions among the operators.

It should be recognised at the outset, however, that even though local access markets have been opened up, BT still remains the dominant supplier of wire- based connections. It accounts for over 95% of wire-based lines and 90% of all lines (including mobile). As a result, our paper contains rather more speculation about how competition is developing than an analysis of a mature competitive process.

2. LOCAL DELIVERY TECHNOLOGY

The dominance of copper technology in the local access network, established for over a century, has only recently come under challenge. The interaction between organisational forms and technological choice in the telecommunications industry has not been much investigated, but it is clear that the organisation of the telephone system as a publicly owned monopoly suited the establishment of a wire-based network and created little incentive to investigate alternatives.

Recent technological advances, notably signal compression techniques, have extended the functionality of copper into other service dimensions, including packet data transmission and video. These and other developments are constantly altering the balance of advantage between copper and fibre. BT's current policy seems to be based upon the maintenance of copper in large parts of the access network for residential subscribers and its progressive replacement by fibre for larger business customers. However, these choices are likely to be affected by the regulatory restriction which prohibits BT from delivering entertainment programmes to the home—a restriction which is not due for review until 1998 at the earliest. The lifting of this restriction, much desired by the company, would be likely to alter the balance between the two technologies.

UK cable operators now have freedom to provide telecommunications services on their own account. They have the option of either running a separate telecommunications cable down the same duct as that used for the delivery of entertainment services, or of seeking for electronic integration between the two services. Although the latter is technically possible, the former is currently preferred: this is not surprising as cable companies are sometimes directing their telecommunications services at business customers, while entertainment services are purchased almost exclusively by households.

A further wire-based local access technology now available is the construction of a technologically advanced fibre optic network through urban areas and business parks, known in the US as Competitive Access Providers (CAPs). These networks now run through the overwhelming majority of the largest metropolitan service areas in the US, offering service highly competitive with that provided by the RBOCs. Although no such networks now operate in the UK, roughly the same elements can be found in a licence recently issued to COLT, a company proposing to provide a fibre-based network within the London metropolitan area. The more radical recent departure from traditional local access technology involves the use of wireless communication. Since the two cellular networks started operation in 1985, they have obtained well over a million subscribers. The coverage of their networks extends to 98 percent of the UK population.

The two cellular networks were augmented in the late 1980s by networks constructed by four telepoint licencees offering a portable (but not mobile) service: subscribers could make outgoing calls only if located within a specified range of a base station. This service initially failed, was revived, and failed again.

The more interesting recent development is that of Personal Communication Networks (PCNs), which offer mobility and two-way calling from any location. PCN technology uses digital radio operating at double the frequency of cellular radio, implying improved call quality and digitally encrypted calls. Mercury's PCN system called "One-2-One" was launched in late 1993 to customers within the London metropolitan area. The second operator—Hutchison Microtel—intends to introduce a wider service in early 1994.

The final radio-based technology considered here is non-mobile radio. A notable example of this are the technologies to be employed under the licences granted by the Department of Trade and Industry to Ionica and Millicom. These networks can interconnect with national trunk operators. Calls will be routed to base stations from the trunk network at which point radio takes over. Point-to-point radio systems in the 1.3 to 4.5 GHz band achieve the final connection to end users as signals are conveyed to roof-top antennae units. The link from the roof-top to the hand-set is essentially the same as for existing fixed services.

The development of these alternatives is a powerful illustration of how the relaxation of regulatory restrictions can, in a comparatively brief period, elicit new technical alternatives. But it is important to recognise, even before a discussion of the costs of alternative local access technologies, that the alternatives offer different types of service. These will clearly affect the willingness of different groups of customers to substitute one for another. Particularly important are the following characteristics:

(a) *national availability*: is the technology available everywhere, or only in selected parts of the country?

(b) *mobility*: is it possible to use the technology whilst moving around?

(c) *portability*: is it possible to carry the instrument from place to place? Put another way, is the technology location-specific or person-specific?

(d) *flexibility*: can the technology provide other services than simple telephony?

(e) *frequency requirements*: is radio spectrum required to operate the service?

(f) *two-way calling*: can the technology provide both incoming and outgoing calls?

Table 2.1 shows the extent to which the various technologies discussed above exhibit these characteristics.

Table 2.1
Characteristics of local access technologies

	PSTN	Cable Operator	Cellular	Telepoint	PCN	Fixed Radio
National availability	Yes	No	Yes	Yes	Yes	Yes
Mobility	No	No	Yes	Limited	Yes	No
Portability	No	No	Yes	Yes	Yes	No
Other services	Yes	Yes	Limited	Limited	Limited	Limited
Frequency required	No	No	Yes	Yes	Yes	Yes
Two-way calling	Yes	Yes	Yes	No	Yes	Yes

3. THE COSTS OF ALTERNATIVE LOCAL DELIVERY TECHNOLOGIES

There are broadly three methods of investigating the costs of telecommunications technologies. The first involves the estimation of cost functions from a cross-section or time series of observations. This approach achieved considerable popularity at the time of the break-up of the Bell system, when rival econometricians produced estimates proving or disproving the existence of economies of scale and scope. Much subsequent work has focussed upon properties of the functions of the local exchange carrier. For example, Shin and Ying (1992) undertake tests on a pooled cross-sectional sample of 58 LECs from 1976 to 1983. (They conclude that the cost function is definitely not sub-additive, suggesting that the benefits of breaking up the monopoly outputs of existing LECs substantially outweigh the potential inefficiency.) Unfortunately, the appropriate data for the UK telecommunications systems are available only to BT which (not surprisingly) has not published the results of a similar analysis.

The second approach to investigating telecommunications costs starts from the operators' management accounting data. Thus in jurisdictions where cost allocations are submitted to the regulator and placed in the public domain, it is possible to identify costs which can be attributed to particular services. For this process to work, the investigator must have access either to the raw data or have a detailed knowledge of cost allocation procedures, for example, through the publication of a cost allocation manual which the operator must abide by in presenting data. The simple publication of fully allocated cost data is not enough, as they are subject to the vagaries of the overhead allocation system. Unfortunately, in the United Kingdom the only data available on the cost attributed to BT's various services are contained in its Financial Returns By Service (FRBS) statements for 1988/89 to 1990/91, published by OFTEL in January 1992. But it is impossible for investigators to strip out the overhead component from the reported results and thus derive an indication of causally attributed costs.

The third method of investigating costs is to construct an engineering-economic model of alternative telecommunications systems. An exercise of this kind was undertaken by consultants to OFTEL and the Department of Trade and Industry as part of the duopoly review (see Bell and Cave, 1992). The aim of the exercise was to investigate the cost-benefit trade-off of liberalised entry, but the studies also cast significant light upon the cost structure of the alternative technologies.

For our purposes, the principal conclusions of these and similar American studies (such as Reed, 1992) are as follows:

(a) The costs of a wire-based telecommunications network built from scratch are heavily skewed towards access related costs. In particular, trench and duct costs account for between 25% and 40% of total annualised costs per line, depending upon location. As a first approximation, the cost function comprises a large fixed element common to all homes passed, a component reflecting the connection of a particular subscribers to the local loop and a relatively small additional usage cost, much of it associated with switching.

(b) In the case of a cable system, whether telecommunications and entertainment services are electronically integrated or whether separate wires are employed, a high proportion of systems costs are common to the two services. As a consequence, the incremental cost of the supply of telecommunications services by an existing cable operator is substantially less than the stand-alone cost.

(c) PCN and other radio-based technologies exhibit a different structure, with a smaller proportion of access-related costs and a higher proportion of usage-related costs. The evidence of the consultancy work undertaken for OFTEL and the DTI suggested that PCN exhibits substantial cost advantages over fixed link alternatives irrespective of population density. However, some caution is appropriate in interpreting these figures because at the time PCN was not yet fully developed. PCN (but not the non-mobile radio systems such as that proposed by Ionica) also imposes additional costs on the subscriber in terms of the acquisition of a more sophisticated hand-set.

(d) The costs of radio-based systems depend in important respects upon the trade-off chosen between the number of base stations, the weight and cost of the hand-set and the quantity and nature of the spectrum employed. If spectrum availability and the size of the hand-set are not constraints, only a small number of base stations are required. As spectrum becomes scarcer, more base stations are required but the receiving instrument may also be smaller.

(e) A more recent US study of the costs of PCN suggests the existence of important economies of scope between PCN and telephone, cable television and cellular services.

In summarising this discussion we draw attention to three points in particular. The costs of a new wire-based network are heavily skewed towards access: usage costs are correspondingly low and diminishing. However, many of the access-related costs of an existing network are sunk or non-salvageable. Second, radio-based systems exhibit a different cost structure, even though there is continuing debate about the level of costs which they impose. Third, spectrum costs are potentially important. In the United States, procedures for charging realistic amounts for PCN spectrum are well advanced. This stage has not been reached in the UK, although the introduction of spectrum charges could readily alter the economics of competition between alternative delivery mechanisms.

4. THE REGULATORY ENVIRONMENT

We noted in the Introduction the new arrangements for structural regulation of the UK telecommunications industry following the end of the duopoly policy. The number of mobile operators of each type is determined through a regulatory decision-taking process. There are currently two cellular network operators and two PCN licencees, with no current plans to readvertise the third licence. Cable television operators may apply for a licence for telecommunications services through a straightforward procedure. And any applicant may apply for a fixed link telecommunications operating licence. Although such applications are considered "on their merits" severe criteria are not applied.

Regulation of conduct is determined above all by BT's licence and the associated price control regime. The licence imposes service obligations and puts bounds on price discrimination. In particular, any quantity discounts have to satisfy a complex set of conditions, even though in the longer term incremental cost pricing will be permitted.

As far as the overall level of BT's prices is concerned, the key element of price control is a price cap linked to the Retail Price Index. The severity of this price cap has increased from RPI-3 (1984–89), RPI-4½ (1989–91), RPI-6¼ (1991–93) to RPI-7½ (1993–97). The coverage of the basket has also increased over the relevant period, notably by the inclusion of international calls in 1991. From 1993 BT may not count for the purposes of satisfying its price cap quantity discounts in addition to those available before August 1993.

The RPI-X price cap for the tariff basket as a whole is augmented by side constraints on the prices of particular services. Since 1984, BT has been restricted from increasing its quarterly rentals by more than 2% above the RPI. As the overall cap has tightened, this restriction has permitted increasing and substantial rebalancing between access and usage charges. Nonetheless, the company asserts that developments in technology are making a progressively higher proportion of costs access- rather than usage-related, and that in consequence the losses the company is experiencing on its access business are increasing.

The company demonstrates this proposition by the Financial Results by Service data discussed above. A summary of the results for 1990–91 is shown in Table 4.1.

Table 4.1
BT's financial results by service (FRBS) 1990/91 (£m)

	Access Calls	Local Calls	National Calls	International Calls	Access & Local Calls
Turnover	2157	2057	2837	1721	4214
Return	-650	1069	1843	802	419
Capital Employed	5926	2463	2480	979	8389
Return on Capital Employed (%)	-10.9	43.4	74.3	81.9	5.0

These calculations are subject to the standard difficulties associated with the fully allocated costs—notably the element of arbitrariness in allocating overhead costs among different services. Nonetheless, as they stand they demonstrate a substantial negative return on access (of -8% for business access and -11.8% for residential access) and a return in 1991 on a combination of access and local calls of 5.0%. Assuming a cost of capital of 16%, this represents a loss of £923 million for BT's local business as a whole.

In response to BT's representations about the access deficit shown by the FRBS, OFTEL agreed in 1991 that it should be shared with other operators interconnecting with BT, through the payment by the latter of access deficit contributions (ADCs). These are calculated proportionately to the profitability of the service sold by the interconnecting operator. Thus a competing operator supplying a minute's international call would pay to BT a larger access deficit contribution than would be paid by the same operator supplying a customer with a minute's local call. However the Director-General of Telecommunications has the power to waive in whole or in part the access deficit contribution payable by a "small" operator. An operator supplying less than 10% of the relevant market may have the ADC waived, provided that BT's share of the relevant market is not less than 85%. The definition of the relevant markets is made by the DGT. Waivers cease in 1997, except for PCN operators, which may retain them. In doing so it granted Mercury substantial waiver of access deficit contributions (OFTEL, 1993b).

The final major element of the regulatory environment relevant to the development of local competition concerns the principle upon which interconnection charges are established. Under BT's licence condition, these are established on a fully allocated cost basis including the attribution of appropriate overheads and the use of an appropriate cost of capital. The DGT recently redetermined the interconnection charges paid by Mercury to BT.

This complex set of arrangements governing interconnection is subject to much current debate. In June 1992, the DGT issued a statement expressing the view that accounting separation between different BT businesses was necessary if competition were to develop and if there were to be public confidence that BT was not abusing its dominant position. OFTEL asserted that accounting separation would achieve the objectives of transparency and non-discrimination: transparency through the exposure of costs relating to each of BT's businesses, and non-discrimination by ensuring that services offered by one BT business to another are offered on similar terms to non-BT businesses. In June 1993, OFTEL published a statement indicating that it proposed to achieve these aims by requiring BT to publish separate retail and wholesale accounts for its activities (OFTEL, 1993a). A statement on the implementation of accounting separation was expected shortly in early 1994.

With BT's current dominance of local telecommunications markets, the form of regulation of its conduct is critical to the development of competition.

Given its dominance of the local market, the regulatory pricing of BT's services creates an umbrella beneath which competing suppliers can or must shelter. To persuade customers to desert BT in their favour, competitors would normally have to offer a price advantage sufficient to match the cost of switching to an alternative supplier, which—given the absence of number portability—may at present be substantial. Given this fundamental background condition, OFTEL is seeking to strike a balance between BT and its competitors in two ways: by imposing accounting separation upon the company and by requiring the sharing of BT's access deficit, although the latter is qualified by the use of waivers as an explicit entry-promoting device. These arrangements provide a complicated and shifting background against which competitors for local telecommunication services have to position themselves in the market. The next section contains a discussion of how these incentives have affected cable television companies in the provision of telephony.

5. FOREIGN ENTRY AND THE IMPACT OF CABLE

The abandonment of the duopoly policy in 1991 also had the effect of opening UK telecommunications markets to foreign entrants. Most foreign entry has taken the form of investment in the cable television industry, where the US Bell operating companies now play a dominant role. However, overseas companies also participate significantly in fixed link and mobile licences. We discuss each of these in turn.

As part of the 1991 Duopoly Review the cable industry was given freedom to provide voice telephony services either independently on their own account or in conjunction with BT or Mercury through interconnection. Telephone licences were consequently issued to cable companies under the terms of the 1990 Telecommunications Act, designating each company as a "Public Telephone Operator" (PTO). The freedom to offer telephony added valuable cashflow revenue possibilities for cable companies in addition to returns from entertainment.

Cable operators now pass over 3.1 million UK homes using 10,000 miles of laid cable. There are currently 610,000 subscribers including over 300,000 telephony subscribers. The Cable Television Association (CTA) estimates that 6 million homes will be connected to a cable service of some sort by the end of the decade, out of 12 million homes passed by cable. The current penetration rate is around 20% of homes passed. Independent sources suggest that cable companies need 30% penetration levels to break even and 40% to adequately reward investors (CSFB, 1993). The US-owned Diamond Cable estimates a 35–45% market capture of residential telephone subscribers (and 25% of the business market) by its industry from BT by 1998.

In the 1990s, a series of rationalisations swept the cable industry. This reflected the fact that the only way to achieve scale benefits was for operators to serve contiguous franchise territories. Multi-territorial franchises allowed the cable operator's costs, in particular those of video headend electronics, to be spread over the widest possible base of users. This contrasts with the 1980s where the trend of policy seemed to be towards issuing cable licences covering medium-sized territories of up to 150,000 homes. By contrast, the experience in the United States suggests, according to some estimates, that at least 500,000 homes are required for any cable franchise aspiring to earn profits.

The Cable Television Association (CTA) in the UK estimates that approximately £2bn has already been invested in the UK cable network infrastructure. With annual infrastructure investments currently amounting to £600-700m, it is believed that commitments to direct funding into the UK cable industry by 1997 amount to at least £5bn, of which at least 90% is foreign direct investment.

Table 5.1 indicates the extent to which American Bell Holding companies dominate ownership of cable franchised. The first column shows total homes in all the franchises with which an operator is involved; the second adjusts this to take account of joint ownership, and yields each company's "equity homes". Ownership concentration of cable homes has increased gradually in the light of the corporate reorganisation and rationalisation sweeping the industry in the last two years.

Table 5.1
Cable ownership

Cable Company	Total Homes	Equity Homes	Homes Passed	Total Conected	Built %	Penetration %
NYNEX Cable (US)	2738	2738	168	30	6	18
Tele-West (US)	2955	2216	541	140	18	26
SW Bell (US)	1156	1156	215	54	19	25
General Cable (France)	1368	841	203	55	15	27
Insight (US)	953	828	73	5	8	7
Videotron (Canada)	1135	722	313	70	28	22
CUC Broadcasting (Canada)	612	216	49	29	22	20
Jones Cable (US)	1058	539	54	10	5	18
Encom (Canada)	611	519	54	10	9	18
Diamond (US)	501	501	14	3	3	22
Comcast (US)	1079	448	211	64	20	30
English Cable (US)	351	351	30	8	9	27
Post Newsweek (US)	364	326	31	14	8	46
Maclean Hunter (Canada)	328	310	0	0	0	0
Bruncor/US Cable (Canada/US)	260	260	260	0	0	0
Pacific Telesis (US)	226	226	0	0	0	0
IVS (US)	214	211	12	2	5	21
Devanha (US)	360	204	250	41	70	16
Starstream	202	202	20	3	10	15
BT (UK)	120	120	57	11	47	19
Singapore Telecom (Singapore)	184	76	33	7	18	21

Source: The Cable Television Association

On average, cable telephony tariffs continue to undercut BT by around 10 to 15%. Table 5.2 shows that the service is gaining favour with subscribers who seem to be more responsive to entertainment connection in franchises offering both services as opposed to entertainment alone although such comparisons are affected by the maturity of the different systems. According to the UK's largest cable operator, NYNEX Cable, adding telephony to the equation results in enhancements to revenue by up to 50% for an additional capital outlay of only 20%. Given that average gross revenues from telephony (£22) exceed those from entertainment (£20), the cable operators have an incentive to install their own telephone switches and long-distance lines in order to eliminate local interconnect charges they currently pay to BT and Mercury for carrying their telephone calls. The geographical contiguity of franchises will determine the scale benefits of operators constructing their own telephone networks. Even in the short term, there are prospects for more favourable interconnection rates for the cable operators once new long-distance services are provided entrants into that segment.

Table 5.2
Telephony impact on cable television

	No. of Franchises	Homes Connected	Homes Passed	Television Penetration (%)
Top 44 Franchises, of which	44	427,571	2,008,326	21
Franchises offering telephony	30	265,300	1,158,058	23
Franchises not yet offering telephony	14	162,271	850,268	19

Source: The Independent Television Commission

Their presence in the UK is testimony to international differences in regulation which permit foreign cable ownership in the UK. International companies have been able to siphon their domestic profits into aggressive cable network infrastructure programmes. Out of over 130 cable franchises operating in 62 conurbations, only 13% are owned or controlled by UK companies. The majority of franchises are controlled by US companies (50%), followed by Canadian (25%), French (4%) and a small remainder of jointly-controlled operations (8%).

There is, however, a new competitive threat for UK cable operators. Although BT is prohibited under the terms of the duopoly review from offering broadcast entertainment services until 1998 at the earliest, it has recently become established that video on demand is not a broadcast (point to multipoint) service but a switched (point to point) service. Accordingly, BT is permitted to offer it under its licence conditions. Recent technological advances now appear to make video on demand possible, although doubts remain about its cost. BT has announced plans for trialling the system and for making further substantial investment. The Cable Television Association has expressed anxiety that the announcement of these plans may cause funding for cable developments to dry up as backers will be concerned that they may not achieve the penetration rates necessary for profitability. So far, however, there is no firm evidence that the flow of capital into cable developments has been affected.

Thus cable television is the aspect of the UK telecommunications and broadcasting industry which has particularly attracted foreign investment. The reasons for this are various. In the first place, the United Kingdom offers more opportunities for foreign direct investment than almost any other country, including the United States. Secondly, telecommunications operators in the United States have been prevented by regulation from providing entertainment services in the same area; yet there is a widespread expectation that these restrictions will shortly be removed. The United Kingdom thus offers a valuable testing ground for investigating the technological and commercial aspects of offering telecommunications and entertainment services in combination.

However, foreign participation is not limited to cable. A number of overseas operators have made substantial investments in other telecommunications activites for which fixed link, mobile or international simple resale licences are required. These are listed in Table 5.3.

Table 5.3
Principal foreign participation in UK telecommunications licences (excluding cable)

Overseas Company	UK Licencee	Type of Licence
Bell Canada (Canada)	Mercury (minority shareholder)	Fixed link
Fidelity Management & Research Corporation (US)	Colt (majority shareholder)	Fixed link (competitive access provider)
MFS (US)	MFS	Fixed link (competitive access provider)
US West (US)	Mercury One-2-One (50% participation)	PCN
Hutchinson Telecom (Hong Kong)	Hutchinson Microtel (majority shareholder)	PCN
Millicom (US)	Millicom	Fixed link
Worldcom (US)	Worldcom	International simple resale
ACC (US)	ACC	International simple resale

Tables 5.1 and 5.3 demonstrate that over 25 foreign firms now participate in UK telecommunications markets. Nor is the flow of investment in a single direction. Cable and Wireless has a long history of operating predominantly overseas and BT has made a notable entry into US markets, with its 20% purchase of MCI. The motivations for these investments are various, but their effects are to increase substantially the range of choice available to UK customers.

6. WHAT HAPPENS NEXT?

The previous sections have discussed the potential for competition between different operators and technologies in telecommunications markets, focussing upon the local loop.

The scope for such competition is illustrated particularly in the United Kingdom, as government policy since 1991 has permitted virtually any operator to offer service, using a wide range of technologies. Already some subscribers can choose between BT, Mercury, a cable operator, and a PCN operator. In the next few years that choice is likely to be widened and extended to more customers.

Moreover, the United Kingdom has permitted high levels of international involvement in the industry. So far, this has focussed particularly on the cable sector, which is now predominantly owned by overseas, especially American, investors. One of the motivations for such investment is a desire to experiment with new technologies in a liberalised environment. However, lack of symmetry between the UK and the USA in their access to one another's markets has become an increasingly vexed issue.

What we are observing is the unfolding of an unusually interesting competitive struggle in local telecommunications markets. Under the duopoly policy, Mercury began to make inroads into certain of BT's markets, using a pricing strategy designed to appeal to large users. With BT restricted until 1991 from offering quantity discounts, Mercury was able to undercut BT's usage tariffs. At the same time, relatively high access charges prevented low users from switching to Mercury. This was a predictable and straightforward outcome of the regulatory restrictions on BT's pricing. Cable companies however are attracting both residential and business markets, undercutting BT's access and usage charges.

Within the wireless technologies, we already observe a pre-emptive response of cellular operators to the emergence of PCN. Both cellular network operators have recently introduced a new tariff system designed to attract low user business. Customers receive a reduction in their connection charge and monthly rental, but pay approximately twice the standard rate for call minutes. Mercury's tariff structure for its One-2-One PCN service generates call charges lower than those for cellular operators; some local calls are free.

The skirmishes described above essentially involve operators using technologies with broadly the same cost structures. What is more revolutionary is the inter-technology competition between wireless and wire-based systems. The issue here is unusually complex. The normal expectation is that customers would segment according to the cost structures of the operators, which themselves would be reflected in their tariffs. Thus a wire-based technology, with substantial access costs and low usage costs, would be especially attractive to high users, while wireless technologies, with lower access but higher usage charges, would attract low users. Even though it is unexpectedly difficult to demonstrate using conventional economic models of competitive interaction that this outcome would necessarily eventuate, the implied link between costs and tariffs is the likely outcome.

A key element in the system is the degree of regulatory constraint imposed on BT. As noted above, quantity discounts introduced after July 1993 will not "count" for the RPI-7.5% price constraint, but the company will have increasing freedom to bring marginal prices down to incremental cost. Moreover, as many of BT's local network costs are sunk, the company is likely to prefer price cutting to loss of market share. It can combine this with cost cutting, in a fast-expanding market characterised by economies of scale. Thus while we anticipate that BT will lose market share, the company has many competitive advantages in local telecommunications. In June 1992, OFTEL published two estimates of BT's loss of market share by 1997, the end of the next price control period. These are shown in Table 6.1.

Table 6.1
OFTEL's projection of BT's loss of market share

	Low Estimate (%)	High Estimate (%)
Access Lines	5	14
Local Calls	5	14
National (Long-Distance) Calls	15	27
International Calls	25	51
Implied Rate of Return on Capital (%)	18.5	16.5

The past year has seen substantial developments in competition in local service, and the higher estimate now looks more plausible. But much depends upon OFTEL exercising its regulatory discretion, on the speed at which number portability levels the playing field between BT and its competitors, and (more conventionally) the cost efficiency and marketing skills of the operators. The next few years will thus show for the first time whether competition in local telecommunications is sustainable.

REFERENCES

Bell, A. and M. Cave, 1992, Lessons from the UK duopoly review, in Klaver F. and P. Slaa eds., Telecommunication: New signposts to old roads, IOS Press.

Cable Television Associations, 19 November, 1993, The case for cable.

Credit Suisse First Boston, 9 August, 1993, UK Cable update.

OFTEL, 1993a, Interconnection and accounting separation: A consultative document.

OFTEL, 1993b, Determination of terms and conditions for the purpose of an agreement on the interconnection of the British Telecommunications telephone system and the Mercury Communication Ltd system under condition 13 of the licence granted to British Telecommunications under section 7 of the telecommunications act 1984: Interconnection charges and explanatory document.

Reed, D.P., 1992, Putting it all together: The cost structure of personal communication services, OPP, Working Paper No. 22, FCC.

Shin, R.T. and J.S. Ying, 1992, Unnatural monopolies: Local telephone, Rand Journal of Economics, pp. 171–183.

The Race to European Eminence
E. Bohlin and O. Granstrand (Editors)

Competition in the British Telephony Market

Godefroy Dang-N'guyen and Denis Phan

ENST Bretagne, Département Economie-Gestion, BP 832, 29285 Brest Cedex, France

BT's extremely visible strategic actions in past years reveal an extraordinarily strong capacity to renew itself, to raise new challenges and to learn quickly from initial errors. It is still difficult to assess globally the performance of BT so far, in terms of penetration of foreign markets, of leadership as well as of globalization. The prototype of the dominant international company in the telecommunications sector is, and still remains, AT&T. However BT is now clearly a strong challenger, particularly since a strong alliance has been shaped with AT&T's major competitor in the US, MCI. In this paper, we would like to analyze the way BT has gradually unfolded its comprehensive strategy since privatization, along three major business segments: long distance markets, international competition and the local loop. This will constitute the first part of the paper.

In the second part, we draw attention to the consequences of BT's strategy in Europe. In particular, we show that the evolution of the British company already has shaken up the European market as a whole. Several options may be possible at the continental level, which we will examine: the option of establishing trans-European common carriers, the great divide between information transport and servicing, the possibility of imperfect competition (in economic terms, namely differentiated products), as well as the breaking-up of the operators into long-distance and trans-European carriers on the one hand, and local distributors competing with cable networks, on the other hand.

In a nutshell, our analysis emphasizes the limited scope for competition achieved so far in the long-distance network, and tries to explain why. But our main attention is also focussed on the role played by the local loop in the evolution of the structure of the telecommunications industry. Our claim is that it has been perceived as a bottleneck, hence a source of potential profits, but for historical reasons these profits have never been materialized. The availability of new technologies, as well as the inability of the dominant carrier to react quickly to new threats, lead to a completely different picture. Now the bottleneck is becoming a key asset for competitive advantage in the business. Therefore the Duopoly Review in the UK has generated the conditions for the evolution of telecommunications. A shift in technological paradigm, largely unforeseen by the regulatory authorities as well as the key competitors, BT and Mercury, will be achieved, with many options still being opened to telecommunication operators.

1. BRITISH TELECOM AND THE COMPETITION FROM MERCURY

Many papers have been written on the "British Laboratory"[1]. However, most of them focus on the effectiveness of the regulation process. Thus a normative point of view is presented in those papers, whereby it is the regulatory process which is assessed. Our concern is different. We are much more interested in understanding the competitive process itself, and we do not argue whether it would have been improved or worsened by a different regulation.

1.1 Mercury's penetration of the telecommunications market

In 1981, Mercury p.l.c. was created as a subsidiary of Cable & Wireless, Barclay's Merchant Bank and British Petroleum. In 1984, Mercury was fully licensed as a telecommunications operator (PTO), for 25 years. In August, Cable & Wireless took 100% ownership of Mercury, because the two other partners felt that the reward from a £200 million investment would be too delayed.

Services were opened in 1984 (international calls and leased lines) and 1986 (national telecommunication services), once the interconnection change between BT's and Mercury's networks had been settled by Oftel. In May 1987, a full service was offered to residential customers. In May 1992, Mercury had 210,000 customers of which 80,000 were business customers. The financial results of Mercury may be summarized in the following table:

Table 1.1
Mercury's results (millions £)

Year	1985	1986	1987	1988	1989	1990	1991	1992
Turnover	?	?	?	66	181	492	702	915
Profits	?	-14.6	-18.9	-11	18	66	116	155
Cumulative investment	10	90	140	295	510	795	1018	1400
Year's investment	10	80	50	155	215	285	223	382

Evidently, since 1988, Mercury has not been able to finance its need for investment out of the profits earned from its activity. Therefore, the funding comes either from direct finance by the parent company Cable & Wireless, or from debts raised by C&W. It is thus necessary to look at C&W financial results. The sources of finance are given in the following table:

Table 1.2
Cable & Wireless sources of funds (£ millions)

	1987	1988	1989	1990	1991
Internal funds	368	423	587	636	683
Loans	50	0	396	0	3

Source: Own calculations from Northern Business Information data

[1] See in particular: Vickers & Yarrow (1988), Gist (1990) and Cave & Trotter (1990).

This is not the place to investigate why Cable & Wireless did not expose itself to debenture. The point is that this financial strategy led to self-financing, out of their own profits, most of Mercury's investment plan during the period 1987–1992. The availability of Hong Kong Telephone monopoly may have been enough to raise the funds internally. In any case, the penetration rate of Mercury was constrained by its inability to invest more than what it had earned from the previous year.

Another striking element of Table 1.1 is the long timespan taken by Mercury to install its network. Five years after having begun a commercial course, Mercury is not able to earn 8% of BT's revenues. Hence, introducing competition in a market where BT previously enjoyed a monopoly is not an easy task. Obviously, the way in which competition is introduced influences the outcome of this process. Up to the Duopoly Review, Mercury's customers did not have fully equal access to BT's and Mercury's services. They had either to plug in a blue push button or to buy a special socket (the "smart socket" which costs about $100), able to address their PIN (Personal Identification Number) to the local switch, which would route the communication on the Mercury network. Mercury has complained that the unavailability of equal access has prevented a higher penetration of their service within the market.

Other explanations may be provided to justify the poor performance of competition in the British market, up to the Duopoly Review:

1. As mentioned before, it takes time to set up a network, simply because Mercury had to raise a large amount of funds, train the personnel, dig into the street to install the ducts, etc. This is probably the main explanation.
2. Mercury has purposefully targeted big customers with large telephone bills, in order to raise a large amount of cash and not to overexpose Cable & Wireless to debenture. But big customers are not so numerous; hence the market penetration of Mercury is low.
3. The government did not want to threaten the success of BT's privatization, and therefore has made it hard for competition to bite into BT's market share.
4. Customers were not really prepared to switch to Mercury, because of their lack of knowledge of this new network, and because of the fact that the price difference was not so high.
5. BT has reacted more quickly than foreseen to the threat of entry of Mercury. They have modernized their network, rebalanced their tariffs, improved their productivity and their relationship to their customers, etc.

All these reasons may, together, explain the lack of success of Mercury in getting a large share of BT's market. In a companion paper (Dang-N'guyen & Phan, 1993) we have investigated the strategic motives of BT, together with the financial constraint on Mercury's capability to enlarge its network. In this paper, we would like to focus our attention on each of the segments in which competition may have played a role: long distance, local and international networks.

1.2 BT and competition in the international market

When it was privatized, BT was supposed to become the flagship of the British telecommunications industry's conquest of export markets. Hence, the government had assigned to BT the mission to develop a genuine multinational strategy. At a time when multinationalization was not a real option for telecommunications operators, it was difficult for BT to understand where the stakes really were. Therefore, BT's first attempts at multinationalization have been rather shortsighted.

In 1985, BT announced the purchase of Mitel, a Canadian manufacturer of PBX, which was running into financial difficulties. The aim was twofold (Hills, 1986, p. 132): to penetrate the US market where Mitel had a significant position in the small PBX and key systems segments,[2] and to build, in the home market, a vertically integrated company able to raise the cost of entry of British manufacturers in the market of direct sales of small PBX's.[3] Despite the opposition of the British Monopoly and Mergers Commission, the Minister of Trade and Industry, Leon Brittan, approved the merger in January 1986. However, this venture has proved itself to be a major failure. BT's subsidiary never returned to profitability, and BT has tried hard to sell it.

There are many reasons to explain why the acquisition of Mitel was not a success. First, economies of scale have become relevant in manufacturing and distributing PBX, simply because of the rising sunk costs of devising the software imbedded in digital switches. Hence, many regional manufacturers like Rolm, Mitel, Jeumont-Schneider, have disappeared. Second, the vertical integration strategy developed by BT was perhaps an outdated one. It was particularly suited to the former "electromechanical paradigm" whereby the carrier was very active in designing the system it would purchase. With digitization, ICs were welded on bakelite cards and easily replaceable. Therefore the design was no longer an issue and it was no longer necessary for the equipment manufacturer to interact strongly with the network operator. Finally, BT, as much as Mitel, had no experience in marketing equipment abroad. Foreign markets in Europe, were particularly hard to penetrate, while the strong position of Mitel in the US was threatened by AT&T comeback after divestiture, and by Northern Telecom's innovative products.

The second major stroke in the field of internationalization was the acquisition of McDonnell Douglas' subsidiary, Tymnet, the second packet switched network in the US, and a major supplier of international data services, as well as manufacturer of data transmission equipment. This acquisition had not only advantages. In particular, BT has had to integrate its own UK's PSS (Packet Switched System) network, based on Telenet equipment, with Tymnet's. Second, it had to build up a platform on which to provide a wide range of generic value-added services (Dialcom's Telecom Gold, BT's Gold 400, and Tymnet's On Tyme were three electronic messaging systems available to BT which had to be integrated into a single and unique offer). But integrating heterogeneous equipment, implemented worldwide, has given BT a peculiar know-how, which might be particularly valuable for its "global" customers, who generally are used to owning very different equipment scattered all around the world. This explains why BT is so emphatic on globalization, network and facilities management, while its presence in application-specific value-added services (EDI in particular, but also Information Services) and tailorization to the specific business-related needs of the customers are very scant.

---

[2] Ghemawat (1986). This author estimates that Mitel's market share of the US was 35% in the less than one hundred lines segment in 1982, which itself represented 30% of the total PBX market. But in 1985 it lost 9.4 million C$ out of a turnover of 370.8 million C$.

[3] Vertical integration would have made it harder for GEC, Plessey and others to get a foothold on this market. Previously BT had the monopoly of distribution of small PBX's, which were bought to GEC and Plessey.

Out of this deal grew the "Concert" offer: this is a global Network Service software, announced in 1990. Very quickly, talks have been resumed with IBM, in order for the latter to be both a client and a partner in this venture. BT tried to broaden this alliance through the Pathfinder project, renamed later Syncordia. Preliminary talks were opened with Deutsche Telekom, NTT and IBM. The aim of the project was to provide a complete outsourcing service to large, multinational companies like IBM. But the talks failed to lead to a joint venture, and BT went it alone.[4] BT also developed an International Private Network service, Featurenet, equivalent to MCI's VPN or AT&T's SDN, and an international private circuit management service called Primex.

BT has targeted a position at the forefront offering advanced telecommunications services on a global basis and has been helped by the experience gained in conjunction with the purchase of Tymnet. Then it has understood the necessity, as a strategic goal, to reinforce its software capability. At the moment, it has focussed more on generic, network-related software services, like Concert or Dialcom's Gold, than on specific business, applications-oriented packages.

In implementing this strategy, BT has failed to create strategic alliances. Either the partnerships signed by BT are very asymmetric, or they are not completed. BT has been able to licence Syncordia to Singapore Telecom and in Thailand. But BT could not reach an agreement with Deutsche Telekom and NTT for the same venture, because the latter operators were not content with being a mere subcontractor. The difficulty of setting up balanced alliances with potential competitors may be understandable. The company feels itself in advance of other European carriers in their internationalization strategy: it has a private statute, while most European operators are still publicly owned. And although BT's products are not so well developed as, say, MCI's or AT&T's and their customer base thinner, BT is catching up quickly. This strategy is eased by the flourishing financial position of the British company, which enables it to raise cash and to take gambles in the uncertain, volatile and even changing business of advanced data communications.

The third internationalization move of BT was in mobile communications. In 1990, BT took 20% of McCaw Cellular, the second mobile telephony operator in the US. BT had to sell off this subsidiary to AT&T two years later, because it did not generate the expected profits. This failure might be surprising, at first sight. Mobile communications is a market segment enjoying high growth rate and high profitability. However, this picture varies from country to country. While the European market is blossoming and GSM provides a new opportunity for stimulating demand on a European basis, the more mature market in the US is being curbed since the US recession. Also, local regulations are tighter in the US, standards are incompatible with the European ones and there is no scope for synergies between US and European products and services. Finally, BT was limited to a 20% acquisition and could not oversee in detail the management of its subsidiary. So the competitive advantage of having acquired McCaw did not appear.

[4] Syncordia currently has 4 customers including IBM, the European Air Transport Computer Reservation System Amadeus, BP Chemical and a bank. The kinds of services provided by Syncordia are integrated network management, multilingual customer service, customized billing and financial reporting, plus availability of leased lines and infrastructure connecting Syncordia's locations all around the world Lynch (1992).

This quick overview of BT's internationalization strategy shows that a clear pattern does not emerge until the last 2 years. Up to 1991, BT, backed by its strong financial position, has jockeyed for a *portfolio or conglomerate strategy*, with several new businesses opened internationally in order to maximize the expected return, but without many thoughts about the strategic value of those acquisitions. This strategy has proved to be disastrous, with Mitel and McCaw being a burden on BT's finances.

The British government's will to see BT as the flagship of the expansion of the British telecommunications industry abroad may also explain its poor performance. Urged by the government to become a global player, BT has attempted to move as quickly as possible in any direction, without carefully planning its expansion. Finally, one can also say that international expansion is a rather new business for any telecom operator around the world, and the rules of the game have to be learned. The way BT has done this is particularly instructive: it has relied on a vertical and horizontal integration approach, and used extensively the external acquisition procedure, and eschewed the possibility of strategic alliances. Its approach has been very similar to that of AT&T, although they have penetrated different market segments. The outcome of this strategy has been poor indeed.

One can deduce from BT's experience that a vertical and horizontal integration strategy is not the best way now to compete internationally. Since deregulation and the advent of digital communications, a vertical approach generates as many handicaps as it brings potential benefits. With the introduction of competition in the provision of basic infrastructures for data transmission, it is no longer profitable to own those that represent sunk costs. Even when monopoly is maintained, the pressure by the end users to curb the costs of provision of these infrastructures, relayed by the natural distrust of regulatory authorities against the dominant carriers, makes it hard for the latter to earn extra profits. Because the market structures are also more fluid, and because competition in equipment manufacturing is becoming global, setting up a vertically integrated structure, without previous knowledge of manufacturing issues, may be unprofitable. This explains why, in the first phase of internationalization, BT has globally failed.[5]

Recently, however, BT has learned from the market, and they now have a more focused multinationalization strategy. They are building up specific assets (like Concert or Syncordia), technologically founded on software conception and design. Their competitive advantage might be superior, but the risks are higher also. The hallmark of this strategy is surely the announced deal with MCI. But it is too early to say anything relevant about this deal, except that the two companies are rather complementary.

1.3 Competition in the local loop

As said above, BT has not suffered from strong competition from Mercury in the local loop, up to the Duopoly Review. As a matter of fact, we argue that the local loop has acted as both a *bottleneck* and a *strategic asset* for BT.

The local loop has often been described as a bottleneck by the telecommunications experts. The argument is that, until recently, it was not profitable to install optical fibres to the residential customers; hence the bandwidth available to the end user was limited, and a broad range of services could not be delivered to him or her. Even more, because of the historical

[5] As much as AT&T, by the way, whose alliance with Philips and Olivetti failed, as well as its attempt to penetrate the French market through the acquisition of the ailing CGCT (purchased by Ericsson).

development of tariffs, the profits created by the usage of the local loop were very meager. Hence, entry in that market segment did not seem profitable for a cash-hungry operator like Mercury.

On the other hand, the access to the local loop seems invaluable, since it establishes contact with the customer through subscription, billing, etc. With the diversification of services allowed by the digitization of the network, and the probable convergence between cable TV and telecom, it might be of primary importance to set up a strong commercial interface with the customer. So, Mercury would have been very keen to get a foothold in the local loop, once the long-distance network had been installed. And BT would have been anxious to increase its tariffs for local calls, in order to compensate for the tariff decreases in the long-distance and international traffic, where competition by Mercury was putting pressure, making entry in the local loop even more profitable. But the regulator, Oftel in the UK, would prevent BT from rebalancing its tariffs too quickly in order not to put a penalty on the residential customer, who would have suffered at most from a large increase in local tariffs.

So the paradox of the local loop is the following. On the one hand, the regulator(s), DTI and Oftel, did not believe in the effectiveness of competition therein. Hence, they have prevented BT from rebalancing tariffs quickly, and consequently from making more attractive the entry into this market segment. On the other hand, after the Duopoly Review the sustainability of the monopoly in the local loop is no longer their credo. But still they prevent BT from penetrating the cable TV market, in order to make sure that their new credo, the full liberalization of any telecom segments, will become a self-fulfilling prophecy, i.e. that competition is sustainable everywhere.

Mercury, on the other hand, was urged by Oftel to expand its network as quickly as possible, particularly to reach residential customers. The access charge for the interconnection with BT's network was particularly cheap for Mercury, if it could bypass BT's network. It has been estimated that Mercury gets a 60% rebate on normal tariffs when it directly accesses BT's networks at the level of BT's transit exchange. This rebate is 40% only when the access is through a local switch. But Mercury resisted the pressures to gain residential customers, and decided to stick to large users, the City and the long-distance and international networks, while neglecting the households.

BT would have liked to limit competition as much as possible, while it was hesitant about which strategy to pursue, nationally and internationally. It did not pay too much attention to its residential customers, and had no pressure from either Mercury or the regulator to do this, at least up to the Duopoly Review. The customers complained bitterly about the quality of service by BT and subsequently Oftel has undertaken a systematic analysis of BT's quality of service statistics. It appears that they really have worsened considerably in the years 1986–88. A traditional index is the "fault cleared within two days" index which has had the following evolution:

Table 1.3
Fault repair within two days by BT (%)

1985	1986	1987	1988	1989	1990	1991	1992
?	87.1	83.2	83.3	92.5	94.4	96.4	99

Source: Oftel

For the same period, the equivalent index for Switzerland was about 90% (compared to 83% for BT), and about 85–90% for Bundespost. But since 1988, BT has probably caught up with other European operators. This may be explained by several factors: strikes by BT's personnel, the financial collapse of October 1987, etc.; but there are three more fundamental reasons. First, it had to guarantee a fair return on capital for the shareholders, since the success of BT's privatization was grounded on a reasonable evolution of the share prices.[6] BT has reinforced its balance sheet main ratios by underinvesting, compared to the publicly owned carriers, as shown by the following table. This table also shows that BT had an investment policy comparable to those of the Bell Holding Companies:

Table 1.4
Investment strategy of telecommunications operators as percent of their turnover

	1985	1986	1987	1988	1989	1990
DB Telekom	47	45	45	45	46	46
Televerket	42	42	41	35	31	33
BT	24	23	23	27	25	21
NTT	36	31	30	32	30	30
Bell Atlantic	24	25	23	24	23	21
BellSouth	25	25	25	24	23	22
Ameritech	22	22	20	19	19	20

Source: Gerpott & Pospischil (1993), "Internationale Effizienzvergleiche der DBP Telekom", Zeitschrift für betriebswirtschaftliche Forschung, Vol. 45, pp. 366-389.

Despite the political risks, the average share prices evolved satisfactorily:

Table 1.5
Evolution of BT's shares prices

	1984	1985	1986	1987	1991
HIGH	223	265	278	334	422
LOW	166	220	180	204	271

Second, BT had to take time to understand how the markets were evolving, particularly in the new segments (mobile, data transmission, LAN interconnection, Virtual Private Network, ISDN, etc.). Finally, BT was aiming at productivity gains and had to fire some manpower. But once again, it was politically impossible to do this immediately after privatization. As a result, BT did not protect its key asset (the access to the customer) which is now put under competition after the Duopoly Review.

Thus, in this three-player game (Mercury, BT, Oftel), every player seems to have played against his own interests, mostly because of a lack of understanding of what was really at

[6] The shares, which were floated at 130 p, seem to have been highly volatile, due to political uncertainties surrounding BT's future. In the several elections (each won by the Tories) which have taken place since 1982, the Labour Party threatened to renationalize BT if it came to power.

stake: BT neglected its residential customers to the benefit of the big end users and BT's shareholders, Mercury did not want to take advantage of BT's neglect, and the regulator did not believe that this was possible. As a result, competition in the local loop, which was deemed very difficult by each player, was at first considered infeasible, but became perfectly viable after the Duopoly Review.

Several conclusions emerged from this review, summarized by Bell & Cave (Bell & Cave, 1992): (1) Any new application to offer telecommunications services and fixed link will be treated on its own merits, hence everybody will be entitled to apply for a licence. (2) Cable Operators are allowed to provide voice telephony without determination from Oftel, also without using BT's and Mercury's network. They are allowed to interconnect their networks. (3) Mobile operators may operate fixed links, while fixed operators may not provide mobile or telepoint services under their main licence. (4) BT and Mercury are restricted to carrying entertainment signals and providing entertainment services for at least ten years.

Following this loose regulation, a number of candidates have applied for a general licence particularly in the local loop, and cable operators have begun to get subscribers. The major operators with a telephone franchise are now: General Cable, Cable London, East London Telecommunications, Birmingham Cable, Videotron, United Artists Communications, Nynex Cablecable, Diamond Cable, Southwestern Bell, Cambridge Cable. And the cable industry is experiencing a big reshuffle. First, Telewest has been set up as subsidiary of US West, the US Regional Holding Company, with TCI, the first cable operator in the US and in the world. Telewest directly holds several franchises: Avon, Edinburgh, South London (3 franchises) and Tyneside. But Telewest has a minority interest in Birmingham Cable (48%) and Cable Corporation (17%). And the entry of some Bell Holding Companies into the British market is very impressive. Besides US West, Southwestern Bell, Pacific Telesis and Nynex have obtained cable franchises in the UK and may now directly compete with BT and Mercury. As of 1 July 1993, there were potentially 2,327 million households connectable to a cable TV network. But only 473,415 were connected (penetration rate: 20.3%). Offering telephony may be a fantastic boon for the cable operators, who, up to the Duopoly Review, had bleak perspectives in terms of business. On 1 September 1993, there were already 226,871 telephone lines on cable, out of which 200,859 were households and 26,012 businesses.

1.4 Competition in the long-distance network

This market segment has been at the heart of the competitive process between BT and Mercury. Still, entry of the latter has been quite small. In some segments, however, it has been more successful. For example, Mercury had 24% of international switched telephone service between the UK and France in March 1991, 21% of service with Japan, 19% with Switzerland and 15% with the US (Dang-N'guyen & Phan, 1993). Now, if we look at the ratio of Mercury's to BT's turnover, it is still modest:

Table 1.6
Mercury/BT turnover ratio

1988	1989	1990	1991	1992
1.6%	1.6%	4%	5.3%	6.8%

Source: Own computations

Although we have to allow for the international operations of BT, which are not directly related to telecommunications services in the UK, the ratio is still low, 7 years after the opening of Mercury's commercial services. Mercury argues that in order to become credible, it must reach a critical mass of 15% of BT's business. Had the Duopoly Review not allowed for the entry of new operators, Mercury would have reached this target around 1997–98, if we extrapolate the figures set up in the table above. This happened despite Mercury's price advantage over BT on the long-distance network:

Table 1.7
Tariffs of trunk calls > 56 km (low-cost routes)[7]

PRIME/PEAK	3/11/86	7/5/89	6/4/90	1/7/90	22/4/91	2/9/91	31/3/93
BT	11.73	11.73	11.73	11.36	10.52	10.52	10.52
Mercury 2100	10.18	10.69	11.10		10.20	9.60	9.60
Mercury 2300	11.16	11.16	11.16	11.16	10.60	10.60	10.60
STANDARD							
BT	8.80	8.80	8.80	8.52	7.88	7.88	7.88
Mercury 2100	7.74	8.13	8.40		8.10	7.11	7.11
Mercury 2300	8.48	8.48	8.48	8.48	8.18	8.18	8.18
ECO/CHEAP							
BT	4.40	4.40	4.40	5.13	4.76	5.00	5.00
Mercury 2100	4.00	4.20	4.37		4.80	4.80	4.80
Mercury 2300	3.75	3.75	3.75	3.75	3.62	3.62	3.62

So there is a puzzle: since Mercury's tariffs in the long-distance market are cheaper than those of BT, how does it happen that the new rival cannot bite more deeply into BT's market share? Our explanation has been formalized in a companion paper.[8] Following a model set up by Judd & Petersen,[9] we argue that this limited penetration is due to the self-financing policy of Cable and Wireless (C&W), the owner of Mercury. More precisely, C&W have financed the growth of Mercury's network with their own resources, mainly through the receipts gained from their franchise, Hong Kong Telephone. By denying themselves the possibility to look for external borrowing, C&W and its subsidiary Mercury could not invest so much and quickly take a significant market share.[10]

With this behavior of Mercury, it is possible to show that BT can manipulate the trunk call prices (the so-called "limit pricing strategy") in order to maintain Mercury's profits at a low level and restrain its investment strategy. The amount of this manipulation will depend on which kind of regulation BT expects to find at the end of the Duopoly period. If the duopoly

[7] Priced in pence per minute excluding VAT and rounded to the nearest 0.1 pence.

[8] Dang-N'guyen & Phan, op.cit.

[9] Judd & Petersen (1987). Our model draws also heavily on the analysis of Fudenberg & Tirole (1983).

[10] One still may wonder why C&W has not been so aggressive in terms of investment and borrowing policy. Our explanation is the following: C&W had no interest in triggering BT's reaction, which would harm their profits and make their business more risky. By choosing to self-finance the building up of their network, Mercury was making the possible reaction of BT (price war) more detrimental to BT than to Mercury, because BT's receipts would have been much more affected by a price decrease than Mercury's.

had been maintained (or more precisely, if BT had expected that) it would have a strong motive to renew its network as quickly as possible, and therefore to push to the limit its productivity gain, in order to get as great a market share as possible before the second period after the Duopoly Review, so as to reach that period with the highest market power. Only an aggressive pricing strategy during the first period before the Duopoly Review would have made this possible.

If, on the other hand, BT expected to have more competition in the second period, either through the entry of new competitors, or through measures taken by the regulator to force a Bertrand (price) competition between the two, then BT would have been able to get no more than a Stackelberg leadership, a goal more easily attainable. Hence, in the period during the growth of Mercury, BT could keep some profits for distribution to shareholders and maintain high tariffs, or have its productivity gains be delayed and enjoy the tranquil life of a dominant position. In that case, the limit pricing may have been less stringent.

Thus, the competition raised by Mercury during the period 1984–1992 may not have been very harmful to BT, which may have been able to manipulate Mercury's market share. Consequently, the incentives for the evolution of BT have to be found elsewhere. The regulatory process itself may show the reasons for this evolution to a more pro-competitive attitude. In a sense, Oftel and DTI have given BT the time necessary to adjust its management style and corporate culture to the new context. It is as if Oftel had built up a strong rhetoric about the necessity to submit BT as soon as possible to the competitive forces, while underneath everything was done to preserve BT's (and Mercury's) financial security.

One hint about this soft regulatory pressure is the RPI-X formula for price-capping BT's tariffs. Suggested by P. Littlechild to share the benefits of productivity gains by BT among shareholders and customers, it underwent some misinterpretations by the British government. While he suggested price-capping only BT's local tariffs, to avoid a rapid rebalancing and predatory pricing in the long-distance segment, the price cap was applied to a basket of services. Even more important, the evaluation of the "X" was rather arbitrary.[11] In the absence of any references to past situations, X has been fixed at 3%.

This simply reflected the average productivity growth in British industry as a whole. No account was taken of the productivity gains stemming from technical progress in the information technology sector. Even worse, no X-efficiency gains were accounted for by the benefit to the customer. Any such gain would have been captured by either shareholders or manpower (through salary increases or excess manpower) or investment.[12]

Inefficiency originated by a monopoly or dominant position may be divided into two distinct categories: (1) an allocative efficiency, due to a pricing formula in excess of marginal cost pricing or even of second-best Ramsey prices; (2) an X-inefficiency due to a lack of incentive to manage the company optimally.

The allocative efficiency was partly corrected by the RPI-X formula, which gave the incentives to BT to correct this misallocation, because it gained from this correction. The X-inefficiency is much more difficult to handle. In the British context, the regulator has assumed that its correction may come from both private ownership, as shareholders trying to get the most from BT's management, and competition.

The motive for such a benevolent attitude might be the political will of the British Government to complete the privatization of BT with success. The consequence of this has

[11] Vickers & Yarrow, op. cit.

[12] But this did not happen since BT underinvested during the period.

been that, unlike British Airways which has cut costs soon after being privatized, BT could live on a past trend up to 1990. This period of relative peace and lack of competitive pressure has been exploited by the management to convert the Civil Servant culture of the Old Post Office into a more market-oriented culture. It has also been used by BT to make some strategic mistakes internationally, and to learn how to behave in the global markets.

The pressure on BT slightly increased in 1989 when X was moved to 4.5 %; but the real shock happened in 1991 and 1992, during the Duopoly Review, when X reached 6.25 % and now 7.5 %. This has led BT to launch its "Sovereign" programme with a drastic reduction of manpower. This seems to suggest that there has been a progressive understanding by Oftel of the size of the potential productivity gains achievable by BT. Since the arrival of Mercury would boost productivity gains in the long-distance market anyway, the point is whether the same gains could have been expected in the local loop, despite the quasi-absence of Mercury in that market segment.

To summarize, competition in the long-distance market has not been the real spur to an increased efficiency of BT. Instead, the regulation mechanism has played the key role. Initially, there had been almost no pressure on the British group, but things changed completely in 1991–92, when the duopoly was abolished, the price cap moved to RPI-7.5 %, and cable operators entered the market.

2. DEREGULATION IN THE UK: IMPACT ON THE EUROPEAN MARKET

For European regulators, in particular those of the European Commission in Brussels, the British market is often perceived as a "laboratory", where experiments are undertaken that, if successful, may be generalized to the rest of Europe. For example, after the duopoly was set up in 1984, the Commission, in their "Green Paper", recognized that competition in voice telephony may be limited to undertakings with "special rights", so that entry in the telecommunications market may be restricted. Once the Duopoly Review had shown that a further deregulation might be possible (even with licenses), the European Commission set out to establish the principle of abolition of the monopoly of both infrastructures and voice services. What then will be the impact of the liberalization of the British market? We can sketch several scenarios, which rely on our understanding of the way the telecommunications industry might be shaped in the future.

2.1 A contagion scenario

First of all, it is easy to imagine a contagion scenario taking account of the hindsight given by the British experience. What the latter tells us is that entry into the telephone service market may be accelerated with the availability of existing infrastructure. This would suppress the self-financing problem encountered by Mercury. The penetration rate of cable TV in several European countries has reached such a level that upgrading these networks with voice telephony services may be much easier than in the British case. Hence, receipts may be diverted more quickly from the dominant operator. This is particularly true for Belgium, the Netherlands, and, to a lesser extent, Germany.

For such an entry at the local level to be profitable, however, it must be coordinated with the availability of a long-distance network, offering an interconnexion to the local networks, different from the dominant carrier's. Otherwise, cable operators would pay excessive access charges and get little profit from the operation of their local network.[13] A second prerequisite, for the generalization of competition in the local loop, is probably the privatization of the dominant carriers. At the moment, the two biggest operators in Europe, DBP Telekom and France Télécom, are still public companies. SIP, PTT Netherlands, and Telefonica are joint stock companies, but the State still has a dominant role as shareholder. Since a full privatization may take some time in both Germany and France, one cannot expect that liberalization of the voice telephony market may happen before the year 2000, even if formally it could take place in 1998. So the contagion scenario may not be credible in Europe, unless some strong investors, like the Bell Operating Companies, with financial means and technical expertise, take the lead.

2.2 Separation of servicing from infrastructure operation

If full competition between several alternative telecommunications operators is unfeasible in the short term for regulatory and institutional reasons, it might be easier, and more in accordance with the evolution of the business in telecommunications, to separate the provision of infrastructure (essentially transmission infrastructure, and the management of traffic inside the ducts) from the handling and management of services. Digital multiplexers and cross-connect equipment allow for a limited routing and switching function. Packages of network management software are available now, which perform many of the tasks that the operator of a centralized, hierarchical public network would have performed in the past for any of his subscribers.

This idea was expressed several years ago by N. Curien and M. Gensollen.[14] Now it is in the process of being achieved, through the regulatory evolution as much as through the requirement of demand. For instance, the concept of ONA (Open Network Architecture) and its consequence, the collocation of the equipment of the competitors in the carrier's premises, makes it inevitable to split the carrier's activities into two independent businesses: transmission on the one hand, and information processing (including switching) on the other hand. In the same vein, BT has recently announced that the accounts of BT Networks and BT Retail may be distinct. Thus, it is not so big a regulatory leap forward to divest BT Networks from BT Retail.

If the partition of infrastructure and servicing occurred, the carriers would probably be much more active in designing and selling applications services to the customer. This, however, would require an investment in specific assets, such as the knowledge of the clients' businesses and needs. For example, the carriers may be targeting the banks as a major user of telecommunications infrastructure and services. Then the needs of this sector have to be

---

[13] In the absence of data, it is difficult to know how many direct or incremental costs would be generated by the addition of telephony service to cable TV networks. Bell & Cave (1992) have drawn, from a simulation model devised for Oftel, the conclusion that telephony services on cable TV may be more expensive than BT's, but this inefficiency would be far outweighed by the pressures exerted by competition on BT's costs and productivity.

[14] See Curien & Gensollen (1987). However, they made a distinction between network-related services and applications services, the latter being offered on a competitive basis. This distinction is now less useful, since most of the network-related services may be provided through additional and optional pieces of software, which might suppress the motivation for including them in the transmission infrastructure supply.

understood, and the different business segments of a bank have to be mastered by the telecommunications company, in order to design applications just tailored to this client's needs. The acquisition of this knowledge may be easier through the purchase of software companies. However, telecommunications operators may compete increasingly with computer manufacturers, such as IBM or DEC, who are also shifting their activity from hardware to software provision.

2.3 Dividing transport and distribution

If the local loop becomes more and more subject to competitive pressures, it might become economically meaningful, and politically very attractive in Europe, to break up the telecommunications networks, not between transmission and switching as in the previous scenario, but into local and long-distance segments. The local loop, together with cable TV networks, would be regulated by local authorities, and competition at the local level may be organized or controlled by these authorities. Competition in the long-distance and trans-European networks may be organized and regulated by national and European authorities. At the moment, the cable TV networks (most of them using an old-fashioned coaxial cable technology) are potentially (and actually in the UK) threatening the receipts of telephone carriers, through diversification into voice telephony. But given the technological evolution, it is foreseeable that in the near future, the converse will be true: telephone companies, installing optical fibres in the local loop, may provide TV channels on their fibres, and therefore threaten cable TV companies. This scenario has been announced for many years in the US, where the Bell Holding Companies are very keen to enter the cable TV market, and now will be able to do so. And since television is one of the most regulated businesses, one can expect that political pressures will bring local political monitoring of this competition.

On the other hand, competition in the long-distance and trans-European market may also become possible, because the benefits of competition, at the moment, outweigh the duplication of infrastructure and welfare loss that this competition entails. With the privatization programmes all across Europe, the trend to concentration and mergers, it is possible that only a handful of trans-European carriers (limited to long-distance markets) will survive.

3. CONCLUSION

In this paper, we have looked at the way competition has been implemented in the UK, particularly before the Duopoly Review. Our analysis is that BT has been given leeway to diversify into other activities and that competition has been very limited, even in the most lucrative segment, the long-distance market. This "soft regulation" has led to BT's initial mistakes abroad, and to a "small" entry by Mercury, manipulated by BT's tariff policy. The result has been that the government has not been satisfied with the level of competition achieved before the Duopoly Review. Faced with protests from residential customers neglected by BT, the British government, while reviewing the Duopoly of BT and Mercury, has decided to open competition to cable operators, opening the door to foreign competitors, the RBOCs in particular. As a result BT has to cut manpower and to find elsewhere (abroad) the receipts it is losing from the reduction of its national customer base. The policy of manipulating the competitor's entry is no longer possible.

The British model of introducing competition may be exported in the European Union. However, the situation is much more complex there, and the history of telecommunications in each country may prevent a replication of the British "experiment". Several scenarios can be sketched, which delineate a different process of competition, but which all draw upon what happened in the UK during the period 1984–1991.

REFERENCES

Baron, D.P. and P. Myerson, 1982, Regulating a monopolist with unknown cost, Econometrica, Vol. 50, pp. 911-930.

Bell, A. and M. Cave, 1992, Lessons from the UK Duopoly Review, in F. Klaver and P. Slaa eds., Telecommunication: New signposts to old roads (IOS Press, Amsterdam).

Cave, M. and S. Trotter, 1990, The regulation of competition and entry in markets for UK telecommunications services in D. Elixman and K.H. Neumann eds., Communications policy in Europe (Springer Verlag, Berlin).

Curien, N. and M. Gensollen, Mars 1987, De la théorie des structures industrielles à l'économie des réseaux de télécommunications, Revue Economique, Vol. 38 no. 2, pp. 521-578.

Dang-N'guyen, G. and D. Phan, 4-5 June, 1993, Analyse de la concurrence entre operateurs de télécommunications: L'exemple du Royaume-Uni, Xèmes journées de microéconomie appliquée, Sfax.

Fudenberg, D. and J. Tirole, 1983, Capital as a commitment: Strategic investment to deter mobility, Journal of Economic Theory, Vol. 23, pp. 150-182.

Gerpott, T. and R. Pospischil, 1993, Internationale Effizienzvergleiche der DBP Telekom, Zeitschrift für betriebswirtschaftliche Forschung, Vol. 45, pp. 366-389.

Ghemawat, P., 1986, Intecom, Harvard Business School Case Study 386-053.

Gist, P., 1990, The role of Oftel, Telecommunications Policy, February, pp. 26-51.

Hills, J., 1986, Deregulating telecoms (Pinter Publ., London).

Judd, K.L. and B.C. Petersen, 1987, Dynamic limit pricing and internal finance, Journal of Economic Theory, Vol. 39, pp. 368-399.

Laffont, J.J. and J. Tirole, 1986, Using cost observation to regulate firms, Journal of Political Economy, Vol. 94, pp. 614-641.

Lynch, K., 1992, Syncordia going solo, CommunicationsWeek International, 6 July, p. 4.

Vickers, J. and G. Yarrow, 1988, Privatization (MIT Press, Cambridge).

The Race to European Eminence
E. Bohlin and O. Granstrand (Editors)

Competencies and Diversification: The Strategic Management of BT since 1984

Howard Williams[a] and John Taylor[b]

[a]University of Strathclyde, Glasgow, Scotland
[b]Glasgow, Caledonian University, United Kingdom

1. INTRODUCTION

This paper, through its analysis of the strategy of BT to become one of the 'Top Telcos', seeks to explore the potential restructuring of the telecommunications service sector. The argument offered here is that the telecommunications service sector will become bifurcated, with, on the one hand, a small cluster of genuinely global businesses and, on the other hand, a very much larger cluster of companies trading in increasingly competitive large markets, for example, centred upon the local loop or other geographically defined markets. Whilst this view of the future of the telecommunications industry is gaining ground there remains considerable doubt as to the processes through which change will occur. The issues confronting the major PTOs are those of how to diversify into new markets (which by definition will tend to be competitive and non-regulated) whilst escaping from the regulatory pressures in those markets in which they remain dominant carriers. It is in these processes of diversification and change that the greatest uncertainties exist, and the PTOs seemingly confront their most complex problems. Thus, for example, the evidence from BT is that whilst the strategic intent of being a global telco has been unchallenged within the company for almost a decade, the strategies and tactics implemented to achieve this vision have been shaped and reshaped, often at substantial financial loss to BT.

The theoretical underpinnings of this review of the diversification strategies of BT are based upon an understanding of firms as a set of competencies and that these competencies not only provide the skills and the resources to sustain a diversification but also constitute the analytical capabilities of the firm in assessing its strategic direction and diversification options (see, for example, the work of Teece, Rumelt, Penrose, Kay, Taylor and Williams). At the centre of this theoretical framework lies a set of issues about the nature of the competencies of a firm, and how such competencies which have been shaped within one set of technological, regulatory and market conditions are able to respond and adapt to a turbulent environment. Thus, in the context of the telecommunications industry, are the competencies required in a rapidly changing and turbulent environment the same as those required to operate and manage a PTO in a monopoly market? If not, in what ways can these competencies be reshaped and redesigned? Therefore one of the issues central to this paper is an understanding of the nature and changing pattern of competencies within the PTOs, particularly BT.

2. BACKGROUND

The pursuit of a global telecommunications business has been identified as a strategic vision to be cherished by many of the PTOs, particularly those in the OECD countries. However, it is important to differentiate these new aspirations for a global presence from the well established international characteristics of the telecommunications industry. From its origins the telecommunications industry has been an international industry which created and allowed the growth of international institutions. These institutions have supported, fostered, and ultimately regulated the international development of the industry. Most notable of these institutions is the International Telecommunications Union; one of the earliest international institutions of the modern world, and despite the challenges it now confronts, it still performs a crucial role in shaping the development of the industry at an international level. However, the emergence of *global* aspirations of individual telephone companies is a relatively new phenomenon, and in part, can be traced back to the unleashing of some PTOs in the first waves of liberalisation, deregulation (and re-regulation), divestiture, and privatisation that swept across OECD countries in the early 1980s.

In parallel to these changes in the telecommunications service sector patterns of general economic activity have appeared to be changing.[1] The 1980s can be seen as a period in which the notions of a global economy and global organisations began to take shape and become clearly differentiated from those with an international perspective. Changes in the nature of demand, the technological basis of production, research and development costs, shortened product life cycles and the broad regulatory structures (including the role of the State in the economy) created new directions and patterns of economic activity. In particular, the confluence of these forces brought questions of both globalisation and the responsiveness to individual customer needs into sharp focus (see, for example, Prahalad and Doz). These forces of change have manifested themselves in the restructuring and re-organisation of several industries, for example, and pertinent to this paper, the telecommunications equipment industry. The relatively steady-state world of the 1970s and earlier, centred upon national markets for telecommunications equipment and major national companies, has been supplanted by a more turbulent environment and a restructuring of the industry on a global basis (Cawson et al. 1990)

These changes in the general pattern of economic activity coupled with the redefinition of the regulatory and institutional structures surrounding the telecommunications service providers have placed a new focus on the need within the telecommunications sector to develop and implement global strategies for service delivery. In effect, the opening up of hitherto protected domestic markets to new entrants coupled with new regulatory and pricing structures, as well as changes in the nature of demand, have provided the impetus for change as well as the opportunity. An indication and estimate of changes in the competitive structure of the major national telecommunications markets in the world is given below (Table 2.1). In the UK an indication of the changing market conditions is rapid growth in the number of licence applications made since the removal of the duopoly policy. In 1993 there were over 22 formal applications to the Department of Trade and Industry/OFTEL for PTO licences. The willingness to grant new licences and encourage a multi-vendor telecommunication market in

[1] There is a considerable literature on the changing pattern of economic activity in advanced capitalistic societies. Some of this debate is captured in texts such as Piore and Sable (1984), Aglietta (1979) and Best (1990).

the UK has been given added impetus by statements from the Director General of OFTEL, Don Cruikshank. The Director General has argued that the minimum test of a competitive market is a choice between three suppliers and his decisions on interconnection charges and access deficit payments have been designed to encourage new entrants. In brief 'technical' decisions on interconnection charges and the sharing of an access deficit burden have maintained considerable regulatory and competitive pressure on BT.

Table 2.1
Telecoms competition: progress compared

	1993		1997*	
	% of revenues exposed to competition	Share held by competitors	% of revenues exposed to competition	Share held by competitors
UK (BT)	100	10%	100	22%
Germany	6-8	negl.	30	3%
France	6-8	negl.	30	4%
Belgium	6-8	negl.	30	1%
Sweden	80	5%	100	20%
Japan (KDD)	100	17%	100	25%
Japan (NTT)	50	7%	50	15%
US (AT&T long distance)	100	38%	100	45%

*Best estimate
Source: Daiwa

The detail of the telecommunications market in the UK provides an important context within which to place the diversification strategies of BT. The opening up of the UK market through the rescission of the duopoly policy, and the reassertion of a distinct cable TV market, have created new challenges to BT. Importantly, new entrants have begun to appear in the telecommunications market, for example, the electricity supply industry which is exploiting existing resources, such as towers and wayleaves, or companies using new technologies, for example, the intention of Inonica to build a radio local loop in major cities and towns across the UK. Moreover, existing operators, particularly Mercury and the cellular operators, have sought to exploit the advantages of their recent investment in new infrastructures by implementing new competitive strategies, for example Mercury's One-2-One which offers, in exchange for a monthly subscription of £12.50, unlimited off-peak calls in London and the surrounding area. This new service was launched in September 1993 and reported to have over 35,000 users within three months. Further, new tariff levels and pricing packages are being developed by BT and the new entrants in order to maintain a competitive position. Thus, BT has removed distance-based pricing from the weekend tariff and has eliminated the morning (09.00–13.00 hrs) premium on all calls. Similarly, Mercury has sought to offer discounts on the tariffs set by BT, and in January 1994 implemented discounts of 25% on BT's basic tariffs to directly connected Mercury customers (The Independent 24 January, 1994).

Further, the decision to create a distinct cable TV market from which BT is effectively prohibited has, *de facto*, allowed the Bell Regional Holding Companies amongst others to secure a position in the UK telecommunications market. The arrival of these companies in the UK and the terms under which they can interconnect their own networks or interconnect with BT has created a third and powerful telecommunications network which is challenging BT in the local loop.

3. DIVERSIFICATION STRATEGIES WITHIN BT

Since its privatisation in 1984 BT has continually sought to escape from the manacles of a public bureaucracy and to savour the new freedoms of a private company. As such BT has sought to develop and implement business strategies designed to maximise its returns, not just on shareholders' capital but also in terms of the new commercial opportunities. Essentially BT has sought to develop strategies designed to respond to new competitive threats in the UK, and to exploit new overseas opportunities. Within the UK, BT has sought through investment in network modernisation and radical processes of organisational change to reduce its cost base, increase the quality of service and introduce new services.

In terms of network modernisation BT has invested considerable sums to upgrade its network. In the early 1980s BT was operating a relatively unsophisticated network, especially in terms of switching technologies. Its quality of service was largely defined by the limits of an analogue network and as such was increasingly out of tune with the needs of its major customers. The organisation and administration of BT business activities were also perceived to be relatively outdated. For example, the billing systems and the administrative data bases surrounding the network in terms of customer records, line plant records, fault repair and diagnosis echoed the hierarchical and administrative arrangements that had sustained the company as a classic PTT, located within and integral to the civil service. As such these information systems reflected the separate cultures of engineering, administration and traffic that existed within BT as well as the distinctive geography of the company (Taylor and Williams, 1992).

Thus not only has BT invested in new technologies in the network, it has also redesigned and reshaped its organisational structures and decision-making processes. In some ways this process of organisational change and development has become an occupational hazard within BT as numerous re-patternings of the organisation have occurred since 1984. The most recent of these reorganisations, Project Sovereign, has sought to refocus the business in a radical manner, replacing the old hierarchy with an organisation focused around the needs of two customer groups: businesses (reflected in the organisational unit Business Communications) and residential users (reflected in the organisational unit Personal Communications). In the process of restructuring the organisation BT has also sought to remove numerous layers of management, in particular those administrative layers of middle management that typically reinforce an introspective behaviour rather than placing priority on the customer (Taylor and Williams, 1989; Pospischil, 1993).

In parallel BT has sought to re-engineer and upgrade its information systems. The purpose here has been twofold, firstly to enhance the service provided to existing customers, and secondly to create a flexible platform from which future business needs could be supported with the appropriate, timely delivery and analysis of information.[2] Parallel developments can

---

2 Within BT the strength of this relationship between organisational design and corporate information systems was illustrated by an anecdotal story told by the Managing Director of BT, Mike Bett in August 1993. When commenting on the allocation of individual customer accounts to the different divisions, Business Communications and Personal Communications, Mike Bett argued that the single most powerful and reliable discriminator

be seen in other telephone companies as they have sought to re-design corporate information systems (see, for example, Hannock, 1993).

Table 3.1
Network modernisation: exchange type
(% of connections served by type of telephone exchange)

% connections	1985	1987	1989	1991	1993
Digital	0.1	1.5	23.2	46.9	64.0
Electronic	32.9	40.5	40.7	37.1	31.9
Crossbar		16.6	14.8		
Strowger		31.3	21.3		

Table 3.2
Fibre optic cable
(x1000 km in the network)

1985	1987	1989	1991	1993
54	207	610	1441	2337

The restructuring of BT which has been facilitated by network modernisation and the development of new information systems for administrative and network management purposes has allowed the company to "downsize". The headcount within BT fell in 1993 from 210,500 to 164,600 (as of 31/12/93), which makes the current figure some 30% less than the numbers employed in March 1990—the date when the company announced its restructuring programme. The net reduction in staff will continue with an estimated 12,000 further job losses in 1994.

The cost of this restructuring programme has been in the order of some £2.751 bn, of which £1.6 bn was set against the 1993 accounts. However, whilst these one-off costs are substantial the staff costs (excluding the charges for redundancy) fell by 8.1% in 1993. Set against the new price cap that has been estimated to reduce BT's profit by about £100m per annum, these savings in annual operating costs are very significant. One estimate has put the annual savings in overhead costs at £1.1 bn, therefore allowing BT to continue to generate cash in an increasingly regulated and competitive environment (The Guardian 10/6/92).

In summary, within the UK BT has reshaped its organisation, rebuilt its network and reduced its cost base to such an extent that the current regulatory and competitive pressure will have relatively little impact on the financial performance of the company over the next few years. Moreover, should competition develop rapidly and the market shares held by BT diminish, then the access deficit arrangements set up in 1992 as part of the last review of the price cap will ensure a flow of funds into BT. In a rather ironic twist of events the main regulatory challenge in the UK is becoming how to define the balance between the profitability of BT through the control of its prices and, at the same time, allowing sufficient opportunities for

provided by the corporate information systems was the difference in line rentals between business and residential users rather than actual expenditure on telecommunications services.

new entrants to arbitrage between their costs and BT prices. In effect, and despite its protestations, BT has, within the UK, largely managed to draw the sting of the Regulator.

4. INTERNATIONAL INVESTMENTS

Since its privatisation BT has sought to exploit opportunities in overseas markets. This strategy has been fuelled by a belief that ultimately only a few global telecommunications services providers will exist, and that there will be a desire to escape from the regulatory constraint of the UK market and, in part, to retaliate in the home markets of some of the new entrants in the UK.

It is possible to categorise BT's international investments under four main headings, namely

- Equipment manufacture
- PTO services
- Telematic services
- Network Management Services

Apart from investments in equipment manufacturing, these categories offer no useful periodisation of the diversification strategy. The attempt by BT to exploit new opportunities in different markets has continued on all fronts (except equipment) since privatisation. But the detailed analysis of investment shows considerable volatility as strategic investments in one period are hastily sold in a subsequent period, casting aside the original claims for their long-term contribution to the development of BT as one of the top Telcos with remarkable ease. It is in the analysis of these investment patterns that questions arise concerning the competencies that are central to a successful diversification strategy for BT.

4.1 Equipment manufacture

Since privatisation BT has reorganised its own manufacturing resources under its subsidiary Fulcrum, and bought and sold a 51% stake in Mitel, a Canadian exchange manufacturer. Fulcrum was 100% owned by BT but in April 1991 a 75% stake was sold to an associated undertaking. The investment in Mitel is a more complex story which began with the purchase of a 51% stake in the company in 1986 as an integral part of the declared strategy of being a "top class information technology business operating around the world" (George Jefferson, BT Chairman at the time of the purchase of Mitel). At this time Mitel, which had enjoyed enormous success in the early 1980s, was trading in difficult circumstances. In the 1986 BT annual report George Jefferson recognised this position as follows:

> "Mitel continues to be run by its own board though there are now three British Telecom staff who are members of that board. It is taking firm measures to establish a sound basis for the future and I expect to see a real improvement in its trading position in the current year."

However, Mitel continued to underperform, often trading at an actual loss despite some 11% of revenues being diverted to research and new product development, and as such its strategic contribution to BT was diluted. Mitel had failed to help BT secure a strong position in the PABX market outside the UK and build upon the market and technological synergies that underpinned the rationale for the acquisition. Ultimately the strategic value of Mitel was downgraded and the assets were offered up for sale. In August 1992 BT was reported to have sold its 51% stake in Mitel to Schroder Ventures, an international venture capital and management buy-out group. It had taken BT two years to dispose of the investment and BT had to absorb a £120 m exceptional charge on its profit and loss account. The initial payment to BT by Schroders amounted to US$30 m with a subsequent payment of US$13 m, compounding at 10% p.a., up to a maximum of US$22 m. In selling the company BT is quoted as saying "that manufacturing telecommunications equipment is not part of the company's key mission of providing worldwide network-based services".

4.2 PTO services

Since privatisation, BT has made numerous investments in the area of providing PTO type services, whether in regulated or non-regulated markets. In the UK BT has made, and subsequently sold, a substantial number of investments in cable franchises (see Table 4.1). The opportunities opened up by the cable franchises seemed to permit BT access to new markets in the UK as well as facilitate the development of new services for the domestic market. Its range of investments reflected an optimism in this market as well as the hope that the regulatory constraints would be eased, thereby allowing the company to provide services over its existing network. In 1991 the company adopted the view that its investments in cable franchises in the UK were a distraction from its main strategic developments and BT thus began to dispose of its assets. This view by BT was undoubtedly reinforced after the results of the duopoly review which effectively precluded BT from expanding into the cable markets for at least 7 years after the granting of a specific licence. Thus in the 1990 accounts the following statement is made:

> "A strategic restructuring of the group was announced towards the end of the financial year, aimed at serving customers more effectively in British Telecom's chosen markets. It is estimated that the costs of implementing these changes, including those of manpower reduction, and associated provisions needed for refocusing the group's operations, including the exit from certain cable television interests, will be £390m. This has been provided as an exceptional charge. It is expected that the major part of these costs will be incurred in the year to 31 March 1991."

Table 4.1
Review of BT investments in cable franchises
(% holdings)

	1984	1985	1986	1987	1988	1989	1990[1]
Aberdeen Cable Services	17	20	20	30	38	71	
Essex Telecable Ltd	20	nr	nr	nr	nr	nr	
Merseyside Cablevision	4	nr	nr	nr	nr	nr	
Westminster Cable Company	21	20	20 + loan	37	45	45	
Coventry Cable	-	25	74	74	74	75	
Swindon Cable	-	-	100	100	100	100	
MTV	-	-	-	-	25	nr	
Premiere Partnership	-	-	-	-	30	nr	

[1] During 1990 these holdings were sold.

Within the UK, BT also holds a 51% stake in Cellnet, one of the two major cellular carriers and Telepoint licences. This latter market appears to have been stillborn, although recent attempts to revive the market were made by Hutchinsons Telecom with their Rabbit service. The initial success of the re-launch, particularly within the perimeter of the M25 in SE England, was not matched elsewhere or sustained over time and the service was closed down in November 1993 with losses of £122 m. In the end Rabbit had only 9,000 users (Financial Times 6 November, 1993).

Outside the UK BT has acquired the PTO rights in a number of small territories. In 1993 BT held PTO licences in the Isle of Man, Belize, and Gibraltar.

BT has also sought to move into the US PTO market through two major investments, McCaw Communications and MCI. The McCaw story, as with Mitel, has gone full circle and BT has now relinquished its stake in the company. As a strategic venture McCaw seems to have turned sour but BT has at least managed to show an accounting gain for its investment.

In June 1989 BT bought a 20% interest in McCaw Cellular for £840 m cash. Subsequently an additional investment of £67 m was made to maintain the group's 20% interest. In March 1990 McCaw acquired the controlling interests in the LIN Broadcasting corporation and sold (in Feb 1990) its interests in the cellular telephone operations in Southeastern USA. At the time of the acquisition BT claimed that "its (McCaw) principal asset lies in the value of its licences to operate cellular services".

Iain Vallance signalled the significance of McCaw to BT when he declared that the acquisition was the foundation stone of BT's assault on the US market. The company believed that its diversification into Cellnet in the UK could provide both the technological and market synergies necessary to sustain the development of McCaw. Given that the UK market for cellular telephony was advanced relative to that in the US, and that Cellnet had performed more than adequately, this assumption as to the relevant competencies and synergies was regarded as justified. Moreover, the entrepreneurial approach within McCaw was seen as a valued asset to help BT secure its position in the US market.

However, in buying into McCaw BT was accepting a substantial regulatory constraint in that existing US legislation bars foreign firms from owning more than 25% of any American firm that uses the radio spectrum. Ultimately, this regulatory barrier has inhibited BT in its development of its US cellular operations, and coupled with new and potential alliances between other carriers, BT sought to downgrade its investment in McCaw and create yet another alliance on which to attempt to build its global business.

In downgrading its interest in McCaw BT signalled its willingness to sell its stake in the company and in so doing "BT has abandoned one of its mainstays of global diversification". A deal has been arranged between McCaw, BT and AT&T. AT&T will buy 47 m new common shares in McCaw while BT will sell 35.8 m class A&B stock to AT&T. The deal will strengthen both AT&T and McCaw in the cellular and personal communications markets in the US. In this arrangement BT will see an accounting gain of some £200–250 m.

The second attempt BT has made to capture a significant share of the US PTO market is in its recently announced alliance with MCI. On June 2, 1993 MCI and BT announced the "telecommunications deal of the century" (Bert Roberts, MCI's Chairman). In the proposed alliance BT and MCI are investing a combined $1 bn to launch a new joint-venture company of which BT will hold a 75% stake. BT will also purchase a 20% stake in MCI for a reported $4.3 bn of which $830m will be paid immediately for a 4.6% stakeholding. As part of the alliance BT will also transfer to MCI most of its North American assets.

4.3 Telematic services

The migration of the telecommunications services providers and the computing service companies into new markets for information services has long been seen by analysts as inevitable—for example as depicted by the often cited (in the 1980s) clash of the Titans, AT&T and IBM. Yet the evidence of this blurring of the market boundaries and consequent merging of firms to create new organisations is lacking. Whilst the development of new telematic service markets is unquestionable there appears to be no pattern emerging as to the trajectory of firms supplying this market. The analysis here would suggest that the market for new telematic services requires a distinctive set of competencies which do not easily sit within, or derive from, the existing organisational structures and culture of the PTOs or the major computer and software vendors.

In the case of BT the development of a telematic services sector has long been one of its strategic intentions. As with other aspects of its diversification strategy BT has, since privatisation, actively sought to provide value added services to its major customers. In the UK its major attempt to enter this market was in 1985/86 when BT proposed to join forces with IBM under project JOVE. This proposal caused considerable consternation as the prospect of a new and dominant supplier in a potentially highly profitable and newly liberalised market would appear to frustrate the competitive development of the market. In effect the proposal for JOVE was seen as an attempt to create a monopoly position in a liberalised market rather than foster competition. As a consequence the proposal for JOVE was not allowed to proceed by the UK regulatory authorities.

There was, however, another debate surrounding the development of JOVE and that was focused on the competencies (or lack of competencies) within BT. The concern was that BT, although newly privatised, was still inhabited by and constructed around the ethos of a PTT and would not be able to manage effectively its relationship with its proposed partner, IBM. In effect, the concern was that JOVE would see the migration from BT to IBM of high value added services, new market opportunities and BT's most prized customers.

The second major initiative in this area has been the recent attempt to negotiate a strategic stake in EDS. During late 1992 it became clear that General Motors wished to divest itself of non-core activities in order to secure its position in its core businesses. BT attempted to capitalise on this opportunity by seeking to buy a controlling stake in EDS, the $4 bn computer services company.

BT clearly saw customer, market and technological synergies as the foundations of the acquisition. The argument being advanced was that BT would, through a controlling stake in EDS, move into a dominant position in the provision of telematic services to some of the world's major corporations. The acquisition of EDS would have given BT the necessary base from which to develop its global businesses. However, the proposed acquisition failed to materialise due to resistance by EDS management to the BT takeover, BT's reluctance to settle for less than eventual control and tax problems created by the EDS share structure. Negotiations were terminated in April 1993.

From the analytical perspective of this paper the failure of BT to secure a major stake in the telematics services market raises important questions about the nature and portability of its competencies. At one level of analysis obvious market, customer and technological synergies exist between BT's existing businesses and the opportunities in the telematics service sector. Yet the realisation of these synergies and their sustained development have proved to be elusive.

4.4 Network management services

The processes of network modernisation, multiple supplier environments, and the growth of large and complex intra- and inter-corporate networks have created both the technical capability and the market demand for network management services. In effect the modernisation of the network has resulted in the network control functions moving away from the physical entities of the network elements (e.g. switching units and transmission lines) towards the software infrastructure. As such network control has migrated from individual elements to the network as a whole.

This shift in the locus of the control of the network is a relatively recent phenomenon which BT has sought to exploit not only in terms of operating its own network but also in the burgeoning market for outsourcing, particularly amongst major international companies. In order to develop this market BT has invested in a number of projects which to a greater extent have been clearly focused on the competencies within BT; the only exception here is the acquisition of Tymnet in 1989 from McDonnell Douglas. Apart from Tymnet the other two projects are the creation of Syncordia and the related investment in Global Network Services (GNS).

GNS is essentially a global PSS to which direct access is now possible in 62 countries. Investment in the project has focused on its technological characteristics as well as its geographical spread. Time relay services were only available in 12 countries in 1992 at 56/64 kilobits/sec. However, recent developments have seen time relay protocols at T1/E1 speeds (1.544/2.048 megabits/sec) becoming available throughout GNS.

However, the flagship of this network management strategy is Syncordia. This company was established by BT in Atlanta in 1991. It has three specific areas of business: the management of links between networks; management of private network services; and network operations and management services. It represents a £200m plus investment by BT and has attracted customers such as IBM, BP Exploration and BP Chemicals, Amadeus and Firmenrich.

The original intention of Syncordia was that it would be a joint venture with Deutsche Telekom, France Telecom and NTT. However, none of these companies have joined the project and each has sought to exploit the same market through different joint ventures. In what signals the emergence of an intensely competitive market, and one in which the key players are the major PTOs seeking a global presence, two new alliances have been established. AT&T has announced its Worldsource project, a collaborative venture with KDD, Telstra, the Canadian company Unitel and Singapore Telecom. Similarly Deutsche Telekom and France Telecom have created Eunetcom, a consortium which MCI was about to join before signing its agreement with BT.

The joint venture with MCI has re-orientated BT's strategy in this market segment in that the new venture being launched by BT and MCI is being formed around Syncordia, GNS and MCInet services group.

5. CONCLUSIONS

The diversification strategies of BT, although all designed to achieve the well established and unchallenged corporate goal of being a major global Telco, have been highly volatile and beset with uncertainties and failures. The company has sought to move into new business ventures on a number of different fronts and under various types of agreement, yet despite the seemingly obvious synergies in terms of technology, customers and market characteristics, its track record is one that has yet to be proved. BT is not alone in this dilemma and the volatility of these relationships echoes those that characterised the restructuring of the telecommunications equipment industry in the 1980s.

The theoretical focus of this paper has been on understanding the evolution of corporate strategy through an analysis of the competencies of individual firms and how these competencies provide a framework for the identification and subsequent development of new ventures. What is evident from the case of BT is that the merging and synthesis of the competencies of the firms in the joint venture or acquisition is a complex process that is not necessarily influenced by the existence of synergies at the level of technologies, customers or markets.

REFERENCES

Aglietta, M., 1979, A theory of capitalistic regulation: the US experience (New Left Books, London).

Best, M., 1990, The new competition (Polity Press, Bristol).

Cawson, A., K. Morgan, D. Webber, P. Holmes and A. Stevens, 1990, Hostile brothers: Competition and closure in the European electronics industry (Clarendon Press, Oxford).

Charles, D., P. Monk and E. Sciberras, 1989, Technology and competition in the international telecommunications industry (Pinter Publ., London).

Hannock, J.L., 1993, Billing development in a US regional Bell operating company, British telecommunications engineering 11(4), January, pp. 252–261.

Kay, J., 1993, Foundations of corporate success (Oxford University Press, Oxford).

Penrose, E., 1959, The theory of the firm (Blackwell, Oxford).

Piore, M.J. and C.F. Sable, 1984, The second industrial divide (Basic Books, New York).

Pospischil, R., 1993, Reorganisation of European telecommunications: the Cases of British Telecom, France Telecom, and Deutsche Telekom, Telecommunications Policy 17(8), pp. 603–622.

Prahalad, C.K. and Y.L. Doz, The multi-national mission (Free Press, New York).

Rumelt, R.P., 1991, How much does industry matter, Strategic Management Review 12(3), pp. 167–186.

Taylor, J. and H. Williams, 1989, Organisational evolution, policy development and changing capabilities in British Telecommunications Plc and its antecedents, 1977–1988, Report, Network and Resource Management Centre, University of Strathclyde.

Taylor, J. and H. Williams, 1992, Organisational aspects of KBS development: a Comparison of IS development in PacBell and BT, Report, Network and Resource Management Centre, University of Strathclyde.

Teece, D.J., 1992, Competition, co-operation, and innovation: Organisational arrangements from regimes of rapid technological progress, Journal of Economic Behaviour and Organisation 18, pp. 1–25.

The Race to European Eminence
E. Bohlin and O. Granstrand (Editors)

Cable & Wireless: Services, Investments and Prospects

Richard Kramer[a] and Aine NiShuilleabhain[b]

[a]Institute for Information Studies, Stafferton Way, Maidenhead, SL6 1AY UK*
[b]Columbia Institute for Tele-Information, 809 Uris Hall, NY, NY 10027 USA*

Cable & Wireless can reflect on the past few years with considerable pride. The company has maintained its position in Hong Kong, still its main source of profits, and grown its other businesses, notably Mercury in the UK. The performance of these individual companies has been well documented elsewhere. C&W presently enjoys the widest geographic spread of projects and revenues of any major carrier. Is this a product purely of historical accident, opportunism or business design?

Whatever the case, C&W remains an elusive company because it is at once a collection of local monopolies—a well-understood model—and at the same time, certainly Europe's best example of a challenge to that monopoly model. It understands the view from both sides: for example, it is ironic that C&W simultaneously argues the virtues of monopoly and competition in different markets.

Even now the range of priorities and projects seems unmanageable—a previous focus on European expansion was dropped to concentrate on Asia, further expansion of the UK business and on emerging multimedia and mobile markets, especially in the developing world. Yet the group seems to want a return to Europe, to add to a list of demands and commitments far too long to satisfy.

One of the secrets of this juggling may be that at its roots, C&W is a company almost perpetually in transition, currently moving from local monopolies and international voice towards capturing global services business, Asian expansion, mobile services and a long-term growth focus on developing markets. With excess transmission capacity in underseas cables, focus is on services to generate new streams of revenue. Moreover, C&W claims it is not after involvement in alliances, yet has formed partnerships in its major international ventures—Optus, IDC, Tele2, Mannesmann Mobilfunk, etc.—gaining access often because it is not BT or AT&T.

The following is a general review of the range and scope of investments and business across the Cable & Wireless group, with a few remarks on the unusual, often contradictory directions this carrier has taken, and the logic behind those decisions.

---

* Affiliations for identification only. This paper was compiled at relatively short notice, as an addendum to the ITS conference proceedings. It reflects only the personal views of the authors. Errors of fact or judgement remain with them.

1. INVESTMENTS AND MANAGEMENT

Numerous others have written about the history of Cable & Wireless; we shall not recount it here.[1] At present, C&W's operations cover 50 countries, divided into four regions—UK/Europe, Asia/Pacific, the Western Hemisphere, and a smaller division, Middle East/Indian Ocean/Africa. Its market presence is still largely built on colonial and Commonwealth ties.

Under a new management team from 1992, C&W developed a "Federation" concept to describe the relationship among its diverse businesses. Chief Executive James Ross claimed that C&W was already the best example of a global alliance,[2] and would seek deals with multimedia and software providers, not other telephone companies.

C&W remains one of the world's most consistently profitable telcos. Operating margins rose slightly in 1993 to 23.8%, led by profits of 32% for HKT. Earnings in 1993 were aided by cash from BCE (£118 m in capital gains) and HKT warrants (£60 m). Its assets are mostly in UK (50%) due to HKT's rapid depreciation. C&W has 40,000 staff, 40% in HK and around 25% each in UK and the Caribbean. Net cash flow was positive (£83 m) in the first half of 1993, a major step forward for the company.

C&W's financial health relies on two (related) factors: international telephony and Hong Kong. These factors are reflected in the distribution of revenues seen in Figure 1.1. The company still gets close to half of its revenues and over half of profits from international service. In addition, with 25,000 km of cable, and another 17,000 km planned by 1996, international services—already a competitive market with declining tariffs—remain a focus. C&W has an equally heavy financial dependence on Hong Kong, which accounted for 43% of revenues and 75% of profits in 1993. These numbers were down slightly on the past two years, and political uncertainty has receded somewhat with China's economic growth.

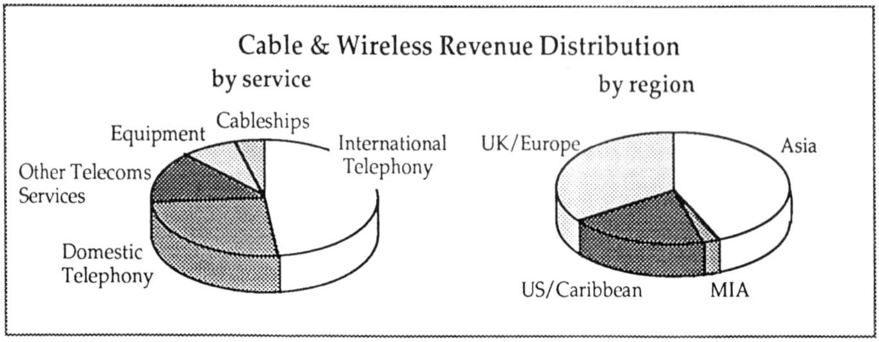

Figure 1.1. Cable & Wireless revenue distribution.

[1] See for example Barty-King (1979), and more recently, Bradley (1993),

[2] Ross has a strong claim in many respects—other "global alliances" such as BT-MCI and Unisource intend to maximise returns from international traffic in the same way as C&W already does, by controlling both ends of calls, making international settlements merely internal transfer payments.

Overall performance has put to rest lingering doubts in the financial community about its capital structure and management. Its stock price has risen consistently since March 1992, going from close to 500pc to 830pc before making a two-for-one split in October 1993. Its vulnerability to exchange rate fluctuations—every 1¢ drop vs. the £ costs C&W an estimated £4-5 mn—has also eased as Mercury's sterling earnings grew to offset non-sterling revenues.

C&W manages an unusually broad spread of investments. The company previously sought equity positions in existing carriers, but its ambitions were constrained by limited capital and an unwillingness to borrow to finance growth. Historically (since de-nationalisation and corporatisation), C&W has been a debt-averse company. Before the first tranche of shares was sold in 1981 C&W needed government approval on investments over £10 mn. In late 1993 –early 1994 its debt-equity ratio was in the 10-13% range, and the company avoided borrowing heavily to reduce the (perhaps unjustified) threat of takeover. Moreover, the locus of management power in the company has often swung between London and the regions (i.e. the local monopolies), which by themselves would tend to make fewer strategic investments.

Half of C&W's capital expenditures of £2.5 bn from 1990–1993 were for the development of Mercury's network, which will require another £1.75 bn through 1995 (including up to £500 m to establish One-2-One), with Hong Kong taking another £750 mn of around £3 bn total investments. Between these demands and others—European expansion, Optus, the Caribbean—C&W may be over-committed. Estimates predict a need for £1.5 bn yearly across the group in 1995, though the growth curve for many of the businesses remains steep and its sale of 20% of Mercury to Bell Canada relieved some pressure.

C&W has spent nearly £400 mn on start-ups since 1990, and will spend another £200 mn by 1995. However, to put these numbers in perspective, start-up costs represent less than 10% of annual profits. C&W has only minimal exposure from these small projects and recently shied away from costlier opportunities like equity stakes in Greece and Hungary. The company values stakes in existing carriers at about £5 mn per percentage point of equity. Its future commitments include Optus, taking £134 mn thus far and up to £250 mn if the consortium holds to the five-year A$2 bn investment programme which won them a license and government approval. Other ventures like Tele2 (£20 mn), Russia (£50 mn over 5 years), Mannesmann, IDC (£40 mn through 1993), etc. are less costly, but C&W must continue investing in the US and Asian markets to keep pace with competitors.

2. REGIONAL BUSINESSES

2.1 Hong Kong and the Asia Pacific region

The jewel in C&W's colonial crown remains its 58% stake in Hong Kong Telecom (HKT), which, as mentioned above, provided 43% of C&W revenues and a staggering 75% of 1993 profits. Although HKT faces competition in mobile services, and loses its local monopoly in 1995, its long-term outlook is strong, based on its international services monopoly until 2006 and the importance of Hong Kong as a hub for traffic to China. HKT offers local and international services, data and mobile services (CSL), data processing and computing (Computasia), holds 33% of AsiaSat and has stakes in 11 underseas cables (a total investment of £500 mn).

Hong Kong enjoys high service levels, a fully digital network completed in 1993, with 48 phones per 100 persons, the world's highest per capita paging market and 30% yearly growth in mobile (HKT has 81,000 subscribers), with four competing services in operation.

Competition is coming to the local loop, but this is not a major threat: local service comprises only 14% of HKT revenues and with unlimited local calls for £6/month, there is little room for new entrants. Other key indicators were strong through 1993—line growth up 8%, international traffic up 26% (providing 60% of revenues), China traffic (now 44% of international volume, or one-quarter of HKT's revenues) up 36%. Also, HKT has customers eager for new Starline value-added services—duplex ringing, call waiting and forwarding, home fax, call etc.—which penetrated 40% of homes in four years. Finally, a reorganisation will add 2,000 sales and marketing people by 1995.

HKT's 1993 financial results were impressive—13% revenue growth, despite a large pension fund contribution, an 8% cut in international tariffs, and a 29% increase in depreciation. HKT has a market capitalisation of US$ 20 bn, a remarkable 11% of the Hang Seng Index. Its balance sheet has little debt, and returns 50% on equity. Margins were 33% in 1993. In 1993, HKT profits increased 22%, on revenue growth of 13%, aided in part by staff reductions of 6% (though these cost £9.1 m in severance payments). Mid-year 1993 results continued the growth: 40% increases in revenues and profits. Capital expenditure by HKT is over $500 mn, half for switching, with transmission taking another 25%. Pending approval to provide CATV service after 1996, HKT has earmarked another £600 mn in investments, beginning with a video-on-demand trial in mid-1994.

Hong Kong's future is clearly hitched to the fate of the Chinese market, which remains highly volatile, though political uncertainty about Hong Kong Telecom has been reduced considerably by the sale of a minority stake (20%) of HKT to the Chinese Investment Trust, CITIC. C&W also arranged for the CITIC to take a 33% stake in AsiaSat, the pan-Asian satellite service, though it is widely acknowledged that C&W manages the venture. C&W further proved its political skills by naming a well-respected Hong Kong executive, Linus Cheung, to replace the deceased Mike Gale as Chairman of HKT.

Perhaps more importantly, C&W is the only foreign firm to own equity in, and operate, telecommunications networks in the PRC. Since 1985, C&W has entered ventures with the Chinese MPT and provincial authorities, mostly in the Guangdong province, neighbouring Hong Kong. Activities include development of digital trunk networks between cities, underseas cable to Hong Kong, establishing mobile services (cellular and paging) and the sponsorship of a research and training centre in Beijing. C&W plans to increase the number of exchange lines in Guangdong (pop. 62 mn) from 2 mn to 5 mn by 1995, and recently instituted a plan to transfer depreciated but useable analogue telephone switching equipment from Hong Kong to China. C&W also controls a majority share of the leased circuits to the country. China accounts for 41% of international traffic routed through Hong Kong, and traffic volume between the two countries has grown consistently around 40% since 1990.

In Japan C&W holds stakes in both Type I (facilities) and Type II (services) carriers. The main venture is IDC (International Digital Communications), where C&W has a 17.2% share in an international carrier with 18% market share in 1993, won by offering 20% discounts on KDD tariffs. C&W also holds 7-8% stakes in new digital cellular providers in Tokyo, Tokai, and Kansai, as well as shares of a VANS provider, Fair-Way Networks, and a mobile data service, Omni-TRACS.

In Australia C&W has 24.7% stake in Optus, which has a duopoly in domestic and international voice, satellite, fixed link and mobile services until 1997 (though there is a third GSM provider). Optus has equal access with favourable interconnect terms, and aims to have 20-25% of the market by 1997. Optus has already gained over 250,000 analogue cellular subscribers reselling Telstra's service with price discounts. Other ventures include Korea (VANS in partnership with DACOM), Taiwan, the Philippines and Malaysia.[3]

With Hong Kong's position as a hub and the strong position in emerging competitive alternatives across the region, Asia must be considered in many respects C&W's "home" market. Again, however, with the exception of control over both ends of international voice connections, the synergies between these disparate businesses are less than apparent.

2.2 Mercury and Europe

The case for fabled synergies is even weaker in Europe, where Mercury's growth has not been matched by success on the Continent. Mercury's history is well-known and discussed elsewhere in this volume. The facts relevant for this analysis are that Mercury's all-digital trunk network represents an investment of £2 bn for C&W, with turnover topping £1 bn in 1993, and strong half-year results to October 1993 of £701 mn in turnover and £99 mn in profits. Mercury faces increasing competitive pressures, but has established itself more firmly in the UK market, where future growth hinges on marketing of mobile, domestic voice and international traffic (an especially profitable business where C&W owns the terminating end). Its market share in international traffic was 24% and domestic long distance was around 15%, with the latter facing considerable pressure from BT price cuts. By revenues, Mercury market share was 10.5%, and the company has turned the corner financially, generating a positive cash flow and strong results.

Recent developments with Mercury include the sale of a 20% stake to Bell Canada Enterprises (BCE) for £480 m, relieving cash flow problems at C&W and providing expertise and links into the UK CATV market (C&W also invested £30 m in BCE's UK CATV interests). The introduction of equal access "Residential 132" service removed the need for special Mercury "blue button" phones, and the launch of One-2-One, a digital cellular (PCN) joint venture with US West, generated publicity by offering free off-peak local calls. Assessing the wisdom of these ventures requires both time and a debate of far greater detail.

By 1994 Mercury had 1.4 mn subscribers, but only 200,000 were direct connections, with yearly traffic increases of 40-50%. Mercury's indirect access services for business and residential users accounted for 82% of new lines, and the company actually saw a drop in the number of direct connections in 1992/3. Residential 132, One-2-One and connections via CATV clearly improve Mercury's access to the residential market, though it still faces complaints from subscribers over billing, service quality and lack of contact with account managers. Management problems aside, direct access—i.e. avoiding BT interconnect costs—is a main determinant of profitability, and is behind the move into cellular (though there are other benefits from experience in wireless in developing world markets). The moves into CATV via the BCE deal and a stake in Leeds Cable are equally cost-inspired; as CATV providers begin switching more of their own traffic, they are demanding cheaper, more flexible interconnect arrangements from carriers like Mercury.

Mercury reorganised in 1993 to focus on three areas; major customers; small business and consumers; and Mercury Enterprises, non-network services such as mobiles, paging, security etc. The company also stated it would focus on emerging "telemedia" markets, meaning greater links with UK CATV providers, adding a full range of information and entertainment

[3] All of these ventures are lengthy tales in and of themselves, deserving far more extensive treatment.

services. The first step towards that path will be the integration of long distance services with One-2-One in 1994.

Other cost pressures come from BT's tariff rebalancing as well as new competitors such as Energis and Ionica whose target markets—Mercury's price-sensitive business subscribers—are clearer than their business plans. Mercury remains heavily dependent on international traffic and large corporate users with increasing options, maintaining an average 4% discount over BT on indirect access calls and 9% for direct access.

In contrast with Mercury's relative success[4] C&W has largely failed to establish itself on the European Continent, closing offices in 1992 and 1993, and appointing new management for Europe in 1994. Revenues from pan-European services for 1993 were estimated at £40 mn, with estimated losses of £10 mn. C&W has been left out of emerging European consortia and still needs a greater market presence to support its service portfolio. C&W has offices in Germany and Italy, but remains a low-key presence in core markets such as France and Spain.

C&W's most significant European venture is a 39.9% stake in Tele2, Sweden's Mercury, for which it paid £5 mn in 1990. Although in the start-up phase, Tele2 has already established a revenue base and is building infrastructure in Stockholm, Goteborg and Malmo and laying cable along Swedish railways track at a cost of £20 mn over 5 years. Interconnect costs 6.5p per end, too high for domestic long distance but economic for international service, a £285 mn market in Sweden.[5] C&W also holds a 5% stake in Mannesmann Mobilfunk, the second German mobile operator whose D2 GSM network had over 300,000 subscribers by 1994 and will invest £1.7 bn by 2000 to provide complete coverage. Other ventures, all on a small scale, include CPE distribution in France and Ireland, rumoured discussions about a 25% stake in Telecom Eireann, fax and messaging service in Italy and 49% of a paging/mobile operator in Hungary.

2.3 North America and the Caribbean

In the United States, C&W consolidated its interests into C&W Inc., encompassing C&W Communications (serving business customers only) and C&W Americas—geared towards multinationals. C&W have 2,000 staff in the US and revenues of $506 mn in 1993 from domestic and international long distance, messaging, and switched private line services. Most are aimed at small enterprises which want control over calling costs and customised billing. CWC averaged 22% traffic growth since 1991, but is still dwarfed by AT&T, Sprint and MCI, and due to foreign ownership, treated under regulations as a "dominant carrier" like AT&T. C&W continues to seek a simple resale license to the UK, Sweden, Australia and Canada. Acquisitions in the US include the GTE Telemessenger voice-messaging service in 1991 and Wang's Network Services packet-switched network in 1992.

In much of the Caribbean, Cable & Wireless remains the monopoly service provider, often with 20-25 year licenses. By 1994 its focus had broadened to include Latin American countries bordering the Caribbean, CATV services and mobiles. C&W is evolving Caribbean networks through three phases: digitalisation of national networks, links via the Digital Eastern Caribbean Microwave System (a £12 mn investment), and connection of the region to C&W's cables and global services. Main holdings include 79% of Telecoms of Jamaica,

---

[4] Here we recognise that there are very much two schools of thought as regards Mercury: one which criticises its efforts as substandard compared with its market advantages, and the other which lauds not only its own successes but the positive (arguably larger) impact it has had upon BT.

[5] Discussed at length elsewhere in this volume.

which accounts for 40% of C&W's Caribbean revenues, 49% of carriers in Trinidad and Tobago, majority shareholding of the two Barbados operators; and a long list spanning operations in Anguilla, Antigua, Bermuda, the British V.I., the Cayman Islands, Dominica, Grenada, Montserrat, St. Kitt & Nevis, St. Lucia, St. Vincent, and the Turks & Caicos Islands.

2.4 Global services, local ventures

C&W built a division—C&W Business Networks (CWBN)—around service provision to multinational customers, but has not promoted it heavily as its competitors have done. CWBN products include the standard range of managed data services, VPNs, fax and messaging services and so on, with marketing agreements for VPNs with Sprint and joint ventures with DEC for outsourcing. Supporting these services, and one of C&W's main assets, is its international cables and satellite infrastructure; the company claims to own 24% of the world market. This position of strength has its problems: C&W is under considerable pressure to sell more services and generate traffic on its network of underseas cables, to maximise returns from investments in large amounts of excess capacity.[6]

C&W was no stranger to the hype surrounding global services and alliances (CWBN is a rather modest product of that participation compared with its peers) but has more steadily and significantly entered a wider range of local ventures than other carriers. In addition to those mentioned above in the main trading blocs, C&W targeted non-OECD nations for expansion, specifically in mobile services and digital overlay networks linked via international gateways to UK or Hong Kong hubs. It announced a joint programme with the ITU to train developing world telecoms managers and built a training college in Cornwall.

C&W has been one of the most active carriers in the Russian Republics, investing $10 mn in JVs with Sakhalin Telecom and Nakhodka Telecom for overlay networks in free economic zones, $4 m in Sovam Teleport and forming a partnership with Intertelcom called Metropolitan Communications, to build a network around Moscow and the Northern oilfields. C&W also committed $7 mn to cellular service in Belarus (40% of BELCEL), entered a JV with Telecom Finland in Latvia (which as announced will require some $160 mn), and 40% of Baltic Communications Ltd. (BCL) in St. Petersburg, which offers digital direct satellite service, using Eutelsat to hub international traffic via Mercury. Further ventures include mobile and paging services in Bulgaria, South Africa, and Pakistan, international services ventures in Bangladesh, Yemen, Sierra Leone, Bahrain and Saudi Arabia, and so on.

---

[6] This is not the place to debate the wisdom of such investments, but simply to note the two camps: one which believes compression technologies make the current amount of bandwidth more than adequate and another which sees image communications requiring ever more capacity. The success of the former—compression technology—would be disastrous for C&W.

3. CONCLUSIONS

The list of projects seems endless, and raises some questions about the value of C&W's strategy: does the range of ventures ultimately drain resources and distract senior management from the core business? Or do they support an image of a partner with experience in markets where its competitors are not trusted? They certainly explain why C&W is especially vulnerable to takeover: the individual valuation of its various stakes in ventures around the world is likely much greater than the Group valuation (strong as that is). This threat may escalate as investment in telecommunications carriers becomes more commonplace and better understood in the wake of ongoing, well-publicised privatisations.[7] However, there is another side to this story: C&W is well-positioned to capitalise on developing world markets wherever they might take off. The growth curve on many of its ventures is long, but will also be steep. The company is more than likely to find itself in the midst of an economic boom as in China or other Asian countries (and its new focus on Latin America completes this picture of growth market coverage).

It is this curious mix of historical accident and opportunistic, small-scale investment which makes Cable & Wireless such an unusual company. Yet when one looks across a seemingly grab-bag range of international ventures, regional businesses and market segments they address, there are some apparent logics to their investments. One can discern a pattern in the models, which fit largely into three general types:

- the national colonial monopolies—and in this we of course include Hong Kong Telecom, for whom competition remains at the margins—still accounting for over half of revenues and over three-quarters of profits across the Group;[8]
- the new competitive entrant, epitomised by Mercury, but following a similar model in Japan, Australia, Sweden and to a lesser extent, the US;
- a combination of digital overlay networks, mobile services and hubbing international traffic in developing world markets, from Hungary eastward, and from the Caribbean southwards.[9]

Granted, this list sounds rather all-inclusive, but there are many other choices C&W avoids: the company does not—indeed cannot afford to—seek large equity positions in foreign operators: it does not have the resource to buy into Hungary, Greece or the other middle-level privatisations.[10] It has also not aligned itself with another carrier, as with Eunetcom or Unisource. Nor has it taken the approach of other successful international players by specialising in one area, as with Pacific Telesis in mobile communications. Cable & Wireless retains a unique—and often poorly understood by analysts—investment profile in

[7] The recent success of flotations as with Singapore Telecom and selling stakes to strategic partners, as in Hungary, provide attractive models.

[8] C&W's strength would be far greater still had it retained other Commonwealth ventures in places like Australia and Canada. Ironically, C&W is reentering many of these same markets as a competing second carrier.

[9] One wonders whether there is a C&W model across the organisation for evaluating opportunities in the various ventures, or whether they are subject to local discretion.

[10] By middle-level privatisations we distinguish between environments where there is network infrastructure but where the telephone density is still half of European standards (i.e. around 20 lines per 100 people).

the industry. Perhaps this lack of specialisation explains the consensus view that C&W is unable to leverage these investments across the group, beyond the financial strength that such a broad spread of projects brings them.

By leveraging investment in underseas cable and sharing traffic among its various national outposts, C&W is foremost an international carrier. Domestic telephony, followed by mobile services, often appear as afterthoughts. On balance, C&W has profited most from controlling both ends of international voice calls, and second, from providing small, concentrated networks—local service of high quality for generally low tariffs—to feed traffic onto the international network. Certainly the company relies on the strength of its collection of local monopolies, underscoring the point that all business is at some point local, and C&W's customer base remains firmly rooted in its historical monopoly territories. Yet amidst all the hype regarding global carriers, alliances and so on, it stands relatively quietly as by far the most diverse collection of businesses in the communications sector, a model which few of its peers seem well-equipped to emulate.

REFERENCES

Barty-King, H., 1979, Girdle round the earth (Heinemann, London).

Bradley, K., 1993, Phone Wars (Century Business, London).

Internationalization in Continental and Southern Europe

The Race to European Eminence
E. Bohlin and O. Granstrand (Editors)

Internationalization of the German Telecommunications Service Market and Strategic Behaviour of DBP Telekom

Dieter Elixmann and Thomas Schnöring

Department of Market Structures and Technology, Wissenschaftliches Institut für Kommunikationsdienste (WIK) GmbH, Postfach 2000, D-53588 Bad Honnef, Germany

Despite the traditional openness of the German economy, foreign direct investment in the sector providing telecommunications services is a rather new characteristic. Likewise, the incumbent in the German market, DBP Telekom, is only gradually entering the international scene. On the one hand, this paper scrutinizes the activities of foreign telecommunications companies in Germany. The opportunities for entry into German telecommunications markets are virtually brought about by recent regulatory decisions. On the other hand, the focus is on cross-border activities of DBP Telekom and an assessment of its international strategy, which can be fully understood only in the context of current business and legal constraints.

1. INTRODUCTION

The German economy traditionally is open. International trade in goods and services accounts for a substantial portion of our gross domestic product and Foreign Direct Investment (FDI) activities are a typical feature of the German market as well as of German companies. According to recent Bundesbank statistics assets abroad held by German firms amount to roughly 250 bill. DM whereas foreign firms hold assets in Germany of about 200 bill. DM.[1]

In the sector providing telecommunications services, however, cross-border activity remains of minor importance. For many decades and still in the early 1980s the characteristic pattern of this sector has been entrenched monopolies with non-contestable home turfs and a steady stream of returns often used also for financing of the State. Despite this national orientation, there have always been international services and international cooperation (e.g. with respect to the worldwide telephony and telex network) between those national players.[2]

However, only recently "internationalization" has made a quantum leap forward in this sector. There are several reasons for this development, like technical progress, regulatory decisions, changes of demand patterns, increasing competition leading to an intensifying search for profitable investments, etc. Thus, for some years there has been much "disarray" in the

---

[1] See Deutsche Bundesbank (1993). This statistic contains direct investments plus cross-border intra-company credits. The German gross domestic product in 1992 equals roughly 3,000 bill. DM.

[2] They are working together e.g. in international satellite organizations, in transcontinental cable activities, in standardization committees, in the ITU etc. We will not focus on this kind of international cooperation in this paper.

telecommunications sector with respect to internationalization compared to the cosy situation before. There are many new forms of transborder activities (acquisitions, mergers, co-operations, joint ventures, etc.) between telecommunications carriers and one can also see the emergence of international links between carriers and firms from other sectors.[3]

In the present paper we are concentrating on FDI, i.e. one special step in the process of internationalization. In this respect we focus on what foreign companies are doing in German telecommunications markets. Vice versa, DBP Telekom is about to develop international activities. Moreover, there are activities of Telekom within Germany—tariff reform, cost reduction, product innovation, organizational reforms—which in some sense can be viewed as reactions to the increasing (threat of) internationalization of Telekom's home turf. Yet, it would be too early today to assess whether the latter activities can be viewed as "strategic reactions" to market entry in Germany. Thus, we are limiting ourselves in this paper to the international strategy of Telekom.

This paper does not tackle the internationalization in the telecommunications equipment market; see e.g. Schnöring (1992). So, if we are loosely speaking about telecommunications markets in the following it always means the market for telecommunications services.

The paper is organized as follows. In Section 2 different aspects characterizing the internationalization of the German market are presented. We focus primarily on the regulatory framework for foreign participation and the activities of foreign companies in various market segments. To understand Telekom's current international activities it is necessary to focus on legal and business constraints, which are presented in Section 3. Section 4 concentrates on the elements of the international strategy of DBP Telekom. Section 5 contains a summary and the conclusions which are drawn.

2. INTERNATIONALIZATION OF THE GERMAN MARKET

In general, the scope of FDI in a country depends on many variables such as its factor endowment (resources, skills, capital stock etc.), the market size, its innovative capacities and its barriers to entry. The relevance of these variables will differ from sector to sector. In telecommunications networks and services it is quite unlikely that differences in factor endowment will be of major importance for FDI. The size of a national market, the innovative power and the competitiveness of the incumbent firm(s), however, will be of greater importance as well as, due to the special situation in the telecommunications sector, the regulatory regime of a country. The regulatory regime defines monopoly areas and constitutes rules for market entry in those segments where entry is possible and, thus, is a vital factor for the intensity of competition in a market.

A special form of FDI in telecommunications is the acquisition of shares of a former state-owned telecommunications operator (TO) by foreign investors. This was an issue in the privatization process of TOs in many countries during the last ten years and there has been a

---

[3] Since the beginning of this decade, increasing research effort has been spent on analyzing the process of internationalization in the telecommunications services sector. Among several others, the papers of Granstrand and Johansson as well as Kramer and Ni Shúilleabháin published in Christoffersen and Henten (1993) represent a good introduction to the many aspects of this topic.

range of different approaches.[4] This paper will not deal with this aspect of FDI in any length because it is dealing with FDI in the German telecommunications market and the current discussion about a privatization of DBP Telekom has not reached this point yet. From our point of view, it is quite clear that shares of a privatized DBP Telekom will be publicly issued and quoted on the stock market, and that there will be a profound interest of the German government in a broad participation of private German investors.[5] Yet, it is quite likely that there will be an international public offering of Telekom shares and that the joint-stock company Telekom, therefore, will be open for foreign shareholders.[6]

Germany has one of the largest telecommunications service markets in the world, estimated by various indicators such as telephone lines, total turnover and others. It is a mature market with a high potential for further growth. Dataquest estimates a revenue per capita of 324 US$ for the German telecommunications service market for 1991 compared to 335 US$ for France, 390 US$ for the UK and 603 US$ for the United States. These figures give an indication of the growth potential of the German market. The German mobile communications market has been underdeveloped and is still lagging behind the development in many other industrialized countries. The German value-added service market has been regarded as underdeveloped for some years (Stoetzer, 1992); however, the German reunification has added a substantial potential for growth.

The German economy is internationally oriented with a great amount of international communication. Thus, it is not astonishing that in terms of outgoing minutes of telecommunications traffic DBP Telekom is the second largest carrier for international telecommunications.[7] International telecommunications is a highly profitable market segment that becomes more and more competitive. Incoming and outgoing international telecommunications traffic to and from Germany certainly is an interesting target for international competition. Moreover, one should keep in mind that the recent EC decision on liberalization of voice telephony in Europe envisaged for the year 1998 will bring about new opportunities for competition for both domestic and international traffic.

DBP Telekom as incumbent in the German market for telecommunications services today is the third largest TO in the world, measured by turnover. Yet, according to various performance indicators it can be shown that DBP Telekom has a substantial potential for improving efficiency, see e.g. Gerpott and Pospischil (1993). Its tariff policy mainly rests on the characteristics of the old monopoly days. There is a lack of customer orientation. Due to its legal status as a separate entity within the State Administration, the staff of DBP Telekom still consists partly of civil servants (roughly 50%). Moreover, there is a special public legal relationship between DBP Telekom as employee and the people employed, which exhibits charac-

---

4 In some cases the government has sold minority stakes of its former telephone companies; in some cases, however, a 100% stake was sold in one single step to foreign investors.

5 Thus, it can be excluded that the German government will opt for a strategic minority stake of one single investor from Germany or from abroad.

6 Investors from abroad primarily will come from the institutional sector, e.g. banks or pension funds. Due to the openness of the German capital market it can be expected that there will be no restrictions with respect to the maximum percentage of shares of any single shareholder or the percentage of shares owned by foreigners. However, several German companies have a limitation of voting rights. Yet, it would be too early to assess whether the latter instrument will be used in the privatization of DBP Telekom.

7 According to Staple (1993) the ranking of carriers as to outgoing minutes of telecommunications traffic in the year 1992 was as follows (in millions): AT&T 6,984, Telekom 4,087, France Télécom 2,449, BT 2,188, MCI 2,083.

teristics not known under the private legal terms of the competitively organized part of the economy. These factors are grounded in DBP Telekom's history as monopoly service provider and public administration.

Summing up, market factors as well as the weak points of the incumbent make it clear that the German market is an attractive location for FDI in the telecommunications service business.

Figure 2.1 gives an overview of the size of the German market for telecommunications services and various market segments from the year 1991.[8] The monopoly market segments still account for more than 85% of the total market, leaving less than 15% for those market segments open to competition (and FDI). It is expected that the competitive market segments will grow much faster during the coming years and that the share of the monopoly market segments will decrease despite any change of the regulatory regime. Nowadays DBP Telekom's market share can be estimated at about 90% of the whole telecommunications service market in Germany.

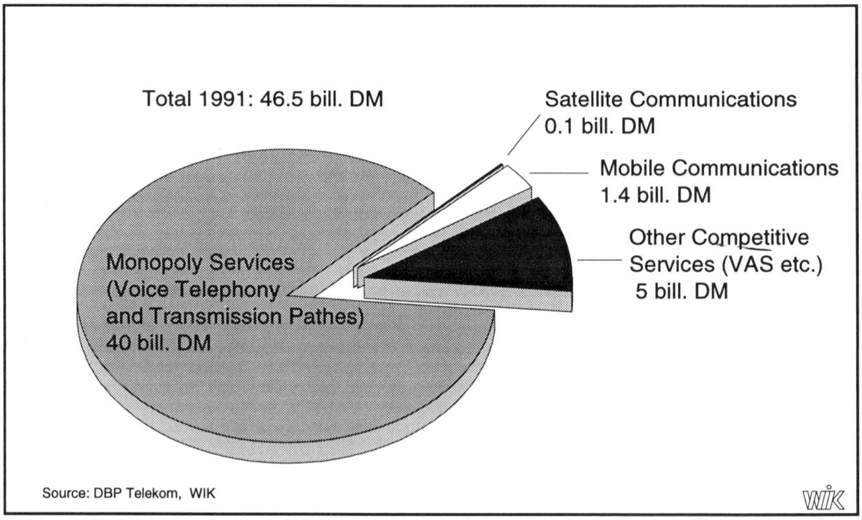

Figure 2.1. Telecommunications services market in Germany, 1991.

In the following sections we will firstly analyze the openness of the German telecommunications market for FDI from the perspective of the regulatory regime. After that, we will describe and evaluate market entry through FDI in the various segments of the German telecommunications market.

---

[8] This figure focuses on the market for services, i.e. rentals for terminal equipment are neglected.

2.1 Regulatory framework for foreign market entry

In a broad perspective the German antitrust law shapes the framework for entrepreneurial undertakings.[9] The scope of participation of a foreign as well as any private German company in the German telecommunications network and services market, however, is primarily defined by the regulatory rules of the German telecommunications market. The "Telecommunications Installations Act" (Fernmeldeanlagengesetz) and the "Postal Constitution Act" (Postverfassungsgesetz), both effective since 1989, define the legal framework for entry into the telecommunications market. Since then German telecommunications policy is based on a new principle: Competition shall become the rule, monopoly areas shall be the exception and the implementation of monopoly areas has to be duly justified.[10]

Currently the rules constitute two monopoly areas: a network infrastructure monopoly—which is a monopoly of providing transmission paths[11]—and a monopoly for voice telephony service. All other telecommunications services are open to (regulated) competition. The resale of leased lines is allowed. The two monopolies are granted to DBP Telekom.[12] The network monopoly, however, has two major exceptions: one is the area of mobile communications and the other is the area of satellite communications. These two areas are open to licensed market entry.

The (renamed) Ministry of Posts and Telecommunications (Bundesministerium für Post und Telekommunikation or BMPT), the regulatory body of the German telecommunications market, is executing the rules laid down in the above-mentioned Acts. Thus, on the one hand it regulates Telekom with respect to prices, fulfillment of infrastructural obligations, quality of service etc. On the other hand it is also responsible for fair competition in the market, ensuring e.g. that the incumbent does not abuse its market power against competitors.[13]

Under the current rules the BMPT has a substantial degree of freedom in formulating its policy about market entry in those market segments that are in principle monopoly areas. The BMPT adopted an approach resting mainly on a different number of licenses for (monopoly) services opened up to competition and regulatory requirements which have to be met by market participants.[14] Following this philosophy, a couple of market segments, apart from those

---

9 The German antitrust law mainly focuses on unfair practices ("Gesetz gegen den unlauteren Wettbewerb") and on restrictions of trade ("Gesetz gegen Wettbewerbsbeschränkungen"). Referring to these issues EC rules have to be taken into consideration, too.

10 There are at least two other important changes brought about by the DBP reform of 1989 in Germany. The first one is the separation of the regulatory and operational functions formerly performed entirely by the Ministry. The second one is the "divestiture" of the public administration "Deutsche Bundespost" into three operational entities for postal services, postal banking services, and telecommunications services.

11 With respect to cable TV it deserves to be stated that Telekom has the exclusive right to deploy the cables; however, it may not be engaged in the provision of entertainment services. These are offered by independent private companies whilst the content of the programmes is regulated by the German states (Länder).

12 In a formal sense, the monopoly areas are assigned to the State (Bund) which has the right to grant them to third parties, i.e. in this case Telekom.

13 The BMPT has defined rules that shall safeguard fair competition between Telekom and private competitors. These rules concerning competition partly are contained in the ministerial ordinances for DBP Telekom and partly they are subject to the licenses. Some focus e.g. on tariff principles or tariffs and on internal transfer prices for monopoly services within Telekom. Others oblige DBP Telekom to supply certain categories of leased lines.

14 In principle, the requirements which have to be met by the incumbent Telekom and new entrants can differ. Moreover, it is not a priori clear that Telekom is allowed to participate in every area opened up to competition; see the example of PCN outlined in more detail in section 2.2.1.

which are identified as competitive per se by the law, have been opened up to entry by private competitors, which means in particular those that are open for FDI.

Thus, market entry has already been or is possible in the segment of mobile as well as satellite communications. Just recently, by a decision in line with the EC legislation the BMPT enabled entry in the market for "corporate network" services whereby these networks can also carry voice telephony.[15] Eventually, all other (non-monopoly) market segments (e.g. value-added services) are open for market entry without any licensing procedures since the reform of 1989. Service providers only have to register with the BMPT; this means in particular that there is no discrimination against foreign service providers in any sense.

BMPT's license policy can be described as liberal and pro-competitive compared to most other European countries. In particular, it does not discriminate against foreign ownership in any sense. For example, in the context of the licensing procedure for a second GSM network provider the ministry has declared that participation of international investors would be an irrelevant criterion in the decision process. So far, there have been no restrictions for foreign ownership in any other licensing procedure of the BMPT and it is unlikely that this attitude will change in the future. In this respect the German telecommunications market is more liberal than many others. The USA, Japan and other countries have restrictions for foreign ownership in the telecommunications sector that limit the possibilities for FDI.

2.2 Activities of foreign telecommunications service companies in Germany

Since the reform of 1989, foreign telecommunications carriers and others have become active in Germany both in those markets which have been identified as competitive per se and in those which have been opened up for entry by regulatory decisions.

2.2.1 Mobile communications

The backward German mobile telecommunications market[16] has been the most interesting market segment for FDI so far. Many foreign telecommunications companies have tried to enter the market. Since 1989, the license policy of the BMPT has already opened windows for market entry for GSM, PCN and public mobile radio networks. Foreign companies, especially TOs, have actively tried to get the licenses mostly as part of larger consortia led by German firms. A call for bids concerning the second provider for mobile data networks was published in October 1993.[17] It was also announced since long ago that bids for two licences in the paging market focusing on the European radio messaging system (ERMES) will be invited

---

[15] Integrated companies (in the sense of § 18 of the German law for joint-stock companies (Aktiengesetz)) can get a general authorization for operating facilities enabling closed-user-group telephony. A single authorization will be needed for the operation of telecommunications facilities which provide services for an a priori limited group of users; however, the main purpose of this group must not be the provision of switched telephony for third parties.

[16] Until 1992 there were two analogue mobile networks in Germany, the so-called B-network and the C-network. Today, only the C-network is operating and has around 800,000 subscribers; however, it is working at the capacity limit. German penetration rates are considerably lower than in many other industrialized countries. An empirical appraisal of the determinants of penetration rates in mobile communications can be found in Stoetzer (1993b).

[17] The final decision can be expected in May 1994. Currently mobile data services are solely provided by Telekom through its service "Modacom".

soon. Recently, however, some severe technical difficulties have been detected[18] which might cause a substantial postponement or even a stop in the introduction of ERMES services in Germany.[19]

Ten consortia applied for the license for the second GSM network (the "D2 license") in 1989.[20] Each consortium had foreign partners and many of the big international TOs were engaged—BT, C&W, Pacific Telesis, Bell South, Ameritech, US West and others. The Mannesmann Mobilfunk consortium won the license. Pacific Telesis (26%) and Cable&Wireless (5%) have an equity share in the "Mannesmann Mobilfunk GmbH". Pacific Telesis brought its technological expertise into the consortium and was an important partner during the planning and construction phase of the D2 network. The total investment of the D2 provider amounts to around 4.2 bill. DM. Both GSM networks started commercial business in the summer of 1992.

The two D-operators are establishing, operating and selling the network whereas the distributional and selling activities with respect to the end user are also performed by independent private service providers. Competition between these service providers will be very intensive and it can be expected that the number of providers will decrease in the future either by market exit or by mergers. In October 1993 there were 14 service providers. Seven of them either are partnerships of German firms with partners from abroad or are wholly owned by foreign companies; one is a subsidiary of a German company the majority of which is in foreign hands.[21]

There are a couple of market forecasts of the future subscribership for cellular networks. As can be expected these are not consistent as a point estimate; rather, they deliver a range of the potential future levels. Thus, there is an estimation of 4 to 6 mill. subscribers for cellular networks in the year 2000; another one estimates about 12 mill. subscribers for the period 2003/2005 in Germany. The development of the GSM subscriber base has been very dynamic since the summer of 1992. The actual development overtook the expected one by far. D2, the network of Mannesman Mobilfunk, did particularly well until the spring of 1993. In March, it

---

18 It was discovered that the proposed German system would cause interference with existing cable services; for more details see Mobile Communications, March 25, 1993.

19 ERMES services are to introduce competition into the segment of paging services currently provided solely by Telekom. Telekom's turnover in this segment amounts to around 135 mill. DM in 1992. Telekom currently offers two paging services: the "Cityruf" with more than 250,000 customers and a turnover of around 70 mill. DM and the "Eurosignal" with about 200,000 subscribers and a turnover of around 65 mill. DM (as of Sept. 1993). Cityruf is a regionally oriented mobile service focusing mainly on the services sector, the construction sector, the energy sector, and the health sector of the economy. Eurosignal is more internationally oriented, enabling roaming to some countries adjacent to Germany. The subscribership of Cityruf is increasing whereas that of Eurosignal is decreasing.

20 The D1 provider is Telekom.

21 Hutchison Telekom has a 68% majority stake in the service provider named "Hutchison". Cofira, the French mobile communications company, has a 60 % stake in "Dekratel Mobilfunk". Nynex has a stake of 15% in "Debitel Kommunikationstechnik" led by Daimler-Benz subsidiary Debis. Martin Dawes Communications Europe (51% of which is owned by France Télécom, the rest of 49% being held by Martin Dawes Communications Ltd. in London) owns 100% of Martin Dawes Telecommunications GmbH. Millicom (active in the cellular business in the U.S.) has a stake of 8% in "Unicom Mobilfunk Service" led by Rheinelektra (coal mining, electricity) and Alcatel-SEL. Securicor (the UK-based cash carrying group) has a 5% stake in "Proficom Kommunikationsdienste". Motorola owns 100% of the service provider of the same name. The German car manufacturer Ford recently has established a subsidiary "Ford Mobitel 2000". We are very grateful to our colleagues at WIK Matthias Stoetzer and Markus Weinkopf, for providing us with these data stemming from another WIK research project.

had about 200,000 subscribers and a market share of about 61%. Telekom's D1 network was only second with around 135,000 subscribers. This was a shock for the incumbent. It speeded the creation of DeTeMobil, a private subsidiary responsible for all mobile services. Telekom wanted an at least 50 percent market share and increased its marketing activities. In October 1993, 14 months after starting the business, the GSM subscriber base has leapt up to more than 700,000. D2 still has a lead with 53% market share, but D1 has narrowed the gap with 340,000 subscribers yielding a 47% market share.

The development of the GSM network in Germany is a very clear sign for the benefits of competition. Network roll-out was a lot faster than with the former monopolistic analogue C-network. National coverage is in reach. After roughly one year of operation, the subscriber base is reaching a figure the C-network achieved at the end of its life cycle. Moreover, the prices of the equipment and the service in GSM are coming down fast.

The development of the GSM market and in particular of Mannesmann Mobilfunk will be of great importance for the future development of the entire German telecommunications service market for a number of reasons. It is a fast-growing and big market segment. It shows the benefits of competition and foreign direct investment to the customers, to the politicians, and to the general public. It proves that market entry is possible and economically viable in competition with DBP Telekom and it indicates to DBP Telekom, the postal union as well as the regulator that the incumbent firm has to improve its efficiency and market orientation. For foreign investors it shows that it is possible to get a foothold in a vital part of the German telecommunications market. This may have a signaling function for other market segments and could be a stimulus to other foreign TOs.

In the summer of 1992 the BMPT announced the licensing of a third cellular network operator (the "E1 license") in competition with Telekom and Mannesmann.[22] Under regulatory terms, the PCN network is to use the DCS 1800 standard. In contrast to the D2 license only two consortia—E-Star led by BMW (car manufacturing) and E-Plus led by Thyssen (steel, machinery) and Veba (energy)—applied for the license. Both consortia had foreign partners. The US carriers GTE and US West were partners in the E-Star consortium; Bell South and Vodafone are members of the E-Plus consortium. The final decision was announced in the beginning of 1993 and was in favour of E-Plus.

Thyssen and Veba together have a majority in the E-Plus consortium; each company is holding a stake of 28%. Bell South has a stake of 21% and Vodafone has 16%.[23] It is expected that installation of the E-Plus network will lead to investments of about 4 bill. DM up to 1997 in the almost nationwide E1 network.[24] E-Plus hopes to break even by 1997 and envisages a market share of about 35% (of the total cellular market in Germany) by the turn of the century. To be economically successful it may be expected that E-Plus cannot only follow a strategy of winning customers from the two D-networks; rather, it has to find new customers especially in the segment of private users. Given the dynamic head-start of D1

22 Telekom and Mannesmann are not allowed to offer services in this segment until at least the end of 1994 or so, when a regulatory assessment of the competitive situation in the market for cellular services in Germany is scheduled. So far, the E1 network is competing directly with the two providers in the GSM segment.

23 The rest of the E-Plus consortium consists of the French savings bank "Caisse des Depots et Consignation" (2%) and a remainder of 5% allocated to 6 mid-size companies from East Germany.

24 In total the planned investments of E-Plus amount to 4.6-4.8 bill. DM for the next 15 years. Economic experts in Germany are estimating the additional induced investments at about 3 bill. DM. It can be expected that half of the total amount of roughly 8 bill. DM will be invested during the next 5 years; see Funkschau April 2, 1993, pp. 8-9.

and D2 it will be difficult for E-Plus to live up to its expectations. E-Plus is starting its network roll-out in East German agglomeration areas, which indicates that E-Plus tries to benefit from the on-going shortage of terrestrial subscriber lines in East Germany.

Apart from licenses for cellular mobile networks, BMPT has also issued licenses for private mobile radio trunking (PMR licenses) in competition with the PMR service of DBP Telekom named Chekker. In the period 1991–92, 28 PMR licenses have been issued for 14 agglomeration areas in Germany. The public bidding process has been open for foreign companies. Usually consortia applied for these licenses and some winners had foreign partners like Motorola, Alcatel SEL and Digital Mobile Communication Ltd. From the German perspective the PMR market seems to be one of the few areas where equipment manufacturers take the chances of liberalization to diversify into the network operators' market and thereby take the risk of direct competition with their major customers.

Roughly 1 mill. users of PMR in Germany are estimated for the year 2000 and an aggregated market size of 370 mill. DM for 1995 (60% for services and 40% for equipment).[25] The actual development is rather slow; Telecom's Chekker service had roughly 35,000 users in 1992. Private competitors do not publish their subscriber base but it can be estimated at around 20,000.

The results of the licensing for PMR show that there is some interest of foreign investors in the German PMR market and that they have gained entry in some regional markets. But the interest has been rather limited compared to the cellular mobile market. From the perspective of a foreign TO the relatively small size of each regional PMR market may not be worth the effort of getting a license.

One can expect a licensing process for mobile data networks in the near future. For the time being, Telekom will be the only provider in this segment: its service "Modacom" (mobile data communications) was to be introduced as a national service on June 1, 1993; regional field-trials were already underway. According to Telekom sources the operation of around 250,000 mobile data terminals is estimated for the year 2000 with a turnover of around 150 mill. DM.

[25] Telekom's revenue in the fields of public accessed trunked mobile radio amounted to around 10 mill. DM in 1992. An average annual growth of 35% is expected for the coming years (see Handelsblatt, April 23, 1993).

2.2.2 Satellite communications

The postal reform of 1989 defined satellite communication as a second exception to the network monopoly and allowed open market entry under a licensing policy approach.[26] BMPT developed a license policy with no restriction for foreign ownership of licensees. In 1991, under the impression of the enormous shortage of telecommunications services in East Germany directly after the German reunification, the BMPT started to issue licenses for satellite voice telephony services from and with East Germany, expiring in 1997.[27] Since May 1992, private competitors of DBP Telekom are allowed to lease space segment capacity for the German market from foreign signatories of the international satellite organizations (EUTELSAT, INTELSAT and INMARSAT) and not only from DBP Telekom. The license policy approach adopted by the BMPT is publicly regarded as liberal compared to European standards. Therefore, satellite communications are the second network-based market segment in Germany that has been widely open to international competition and FDI since 1990.

The market development, however, was not as dynamic as was expected. The demand for licenses increased rather slowly. Up to October 1993 BMPT had issued 29 licenses. Table 2.1 gives an overview of the licensees:

Table 2.1
Satellite communications licenses (Status: October 1993)

ANT Nachrichtentechnik GmbH	Neef Elektrotechnik GmbH & CoKG
Infoware Gesellschaft für Informationstechnik mbH	MBB Messerschmitt-Bölkow-Blom Deutsche PhoneSat GmbH
British Aerospace Communications Ltd.	RTL plus Deutschland Fernsehen GmbH & Co. Betriebs-KG
Preussen Elektra AG	NDR Norddeutscher Rundfunk
Mecom Medien-Communikations-Ges. mbHu. Co.KG	RIAS Berlin
Hessischer Rundfunk Anstalt des öffentlichen Rechts	Voice of America Alpha Lyracom Space Communications Inc.
D-SAT	British Telecom Deutschland GmbH
SEL Standard Elektronik Lorenz AG	Washington International Teleport
Mobile Satelliten Communications Incorporated	Maxwell Satellite Communications Ltd.
Locstar S.A.	GE American Communications Inc.
OHB-System GmbH	Spaceline Communications Services GmbH
Teleport Europe GmbH	France Télécom
VESATEL B. V.	Euro DATA
ZDF	

Source: BMPT

[26] Details concerning the regulatory framework, the supply side, and the structure of the German satellite communications market are described in Szafran (1993). Much of the following information is based on this paper.

[27] In principle this is an exception to the voice telephony monopoly.

Figure 2.2 shows that foreign companies have shown a strong interest in the liberalized German satellite communications market; 38% of the licensees are held by foreign companies and 21% have a mixed German and international ownership. From the group of TOs BT, France Télécom, and PTT Netherlands together with Swedish Telecom (VESATEL) are holding licenses.

Application for a license in the satellite business indicates that the licensees have a sound interest in entering the German market. Yet, a license granted to a company is still just an option for market entry and, thus, is not identical to real economic activity. Unfortunately, there is a lack of reliable information on to what extent licensees have concrete business plans or have already started offering services in Germany. The total amount invested in hardware by mid-1992 is estimated at 75 mill. DM.[28] In all likelihood there is only a small number of firms which currently account for the main portion of this sum.

The size of the German market of satellite communications is estimated at around 100 mill. DM in 1992.[29] Broadcasting activities, either as distribution of TV programs or in the form of satellite news gathering, account for more than two thirds of this market volume, leaving the rest for point-to-point connections and VSAT networks, respectively.

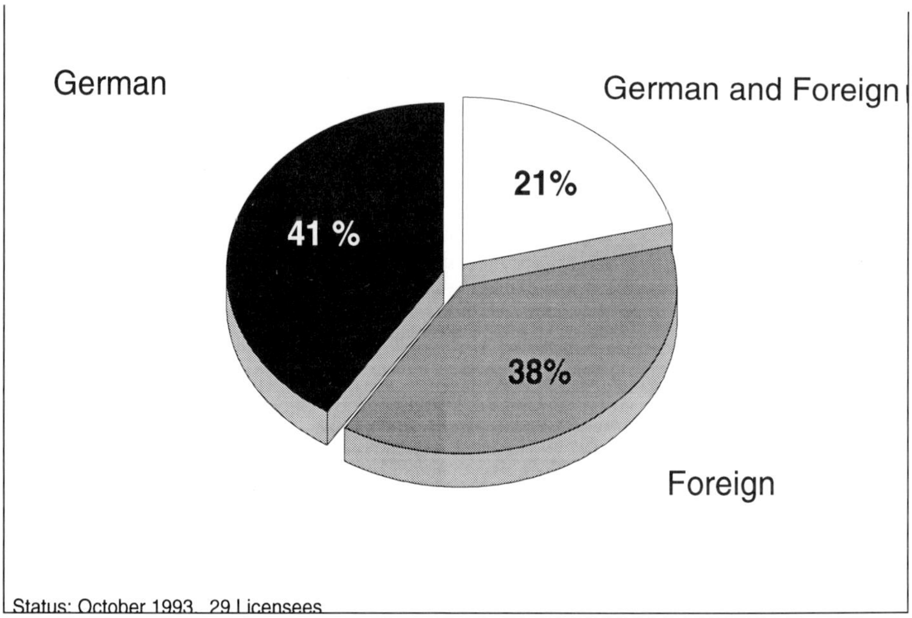

Figure 2.2. Ownership structure of satellite licensees.

[28] There is an estimation by BMPT that the total cumulative amount which will be invested in private satellite networks in the period 1991-1998 is equal to roughly half a billion DM.

[29] Activities of Telekom are not taken into account in this figure.

DBP Telekom provides services in each segment of the market for satellite services. Telekom has had a competitive advantage because it could launch services years before private competitors were allowed to do so. Today, Telekom still has a dominant position in many segments; however, the market for satellite services in Germany exhibits strong dynamics allowing private competitors to catch up.[30]

2.2.3 Value-added services

In the market for value-added services, entry is totally liberalized. Service providers need no license; they just have to register at the BMPT. By regulatory terms, potential service providers have to register and classify the services offered into rather broad categories.[31] The register shows more than 220 service providers, among them many international companies. Unfortunately, the information contained in the register is not detailed enough to derive the size and the structure of the German VAS market as well as "hard data" concerning the participation of foreign service providers.

Table 2.2 gives an overview of such major foreign VAS providers offering services in Germany. Companies are classified as foreign when a substantial part (normally 50%) of the equity capital is owned by foreign shareholders.

One can see that specialized international VAS providers such as EDS, GEIS and REUTERS are active as well as foreign TOs like AT&T, BT, France Télécom, C&W, PTT Netherlands, Swedish Telecom and others. A third group of service providers comes from the equipment manufacturing sector—IBM, Alcatel, DEC, PKI and others.

Two of the recent developments in the German market deserve some special attention: firstly, France Télécom's strategy towards the German market, and secondly recent changes within the organization of Unisource.

On the one hand France Télécom is engaged in strategic alliances with DBP Telekom (EUCOM, Eunetcom).[32] On the other hand France Télécom is increasingly involved in activities competing with DBP Telekom in the German market. Transpac, a subsidiary of France Télécom, has recently taken the majority in Info AG which is one of the middle-sized German-based VAS providers competing with Telekom and which is focusing its business on outsourcing for national and international clients. Info AG has plans to extend its network substantially. This situation is viewed as an impediment for the further development of the Eunetcom joint venture for international services and has to be sorted out in some way.

---

30 At the end of March Germany, France, the Netherlands, and Great Britain reached an accord to permit mutual recognition of satellite licenses. As a first step a "one-stop-shopping" solution has been implemented for VSAT networks and satellite news gathering becoming effective on May 1, 1993; see BMPT Pressemitteilung April 29, 1993. At least with respect to regulatory aspects, market entry from abroad will be facilitated by this procedure.

31 These categories are: transmission services, data processing services, information services, managed network services, telemetry services, transaction and reservation services, and protocol conversion services.

32 Eucom and Eunetcom will be focused on in more detail in section 4.2.2. See also the assessment of the strategic alliance with France Télécom in section 4.2.4.

Table 2.2
Major VAS providers in Germany with foreign partners

Company	Year of entry	Major owners	Empl. in Germany	Turnover in 1992	Nodes in 1993	Entry areas
			Telecommunications Operators			
AT&T	1989	AT&T (USA)	150			Int'l Data Services, Calling cards, Telephone sets
BT Germany		BT (UK)	130	14 Mill. ECU	9	Int'l Data Services
Cable & Wireless		Cable & Wireless (UK)			7	Int'l Data Services
Sprint International		Sprint (USA)		12 Mill. ECU	21	Int'l Data Services
MCI						Int'l Data Services
Unisource		33.3% PTT Netherlands, 33.3% Swedish Telecom, 33.3% Swiss PTT	70	2 Mill. ECU		Int'l Data Services
Info AG	1989	95.5% Transpac (France), 4.5% others	160	12 Mill. ECU	40	
EUCOM	1990	50% France Télécom and 50% DBP Telekom				Value-Added Services
Infonet Germany		80% DBP Telekom 20% Infonet	25		20	Int'l Data Services

Table 2.2 cont.
Major VAS providers in Germany with foreign partners

Company	Year of entry	Major owners	Empl. in Germany	Turnover in 1992	Nodes in 1993	Entry areas
		Specialized VAS Providers				
EDS Germany		General Motors (USA)	1400			
GEIS Germany		100% General Electric (USA)		21 Mill. ECU	15	
Swift	1977	About 2,000 banks		17 Mill. ECU		
Reuters						
GSI		Various	400	46 Mill. ECU	3	
		Telecommunications Equipment Manufacturers				
DEC		DEC (USA)				
IBM		IBM (USA)		40 Mill. ECU	63	
Alcatel Network Services		Alcatel (France)				
Northern Telecom		52.4% BCE (Canada), others				
PKI		Philips (Netherlands), others				

Unisource, the joint partnership of PTT Netherlands, Swedish Telecom and Telecom Switzerland focusing primarily on international communication needs of European business users,[33] has located its international headquarters in Frankfurt. This is an interesting move because during the last decade it has frequently been argued that London would be the "natural" location for the European headquarters of communications companies. The decision of Unisource in favour of Frankfurt indicates the importance of German customers and the German market.

During the past years, Managed Network Services (MNS) have become a major focus of the industry. TOs, specialized VAS providers and equipment manufacturers have started to offer sets of services grouped around transmission services under the heading MNS. The services offered respond in particular to the rapidly growing demand for international communications. The market vision is global service provision for the customer. Headings like one-stop-shopping and one-stop-billing characterize the needs of international business customers. Therefore, it is quite natural that MNS providers are internationalizing their business in the form of cooperations, joint ventures and FDI (see e.g. Analysys 1993). This general development can also be observed in the German MNS market (Stoetzer, 1993a).

Estimates of the market for MNS differ substantially due to differences in definitions.[34] A recent WIK study estimates an annual total market size of 1.5 bill. DM for MNS in Germany.[35] The German MNS market is international in two aspects—on the services side and on the shareholders side. Services are increasingly offered on an international scale and many service providers either are part of an international telecommunications company or have foreign partners. A WIK survey of private MNS providers in Germany indicates that more than 50% of the firms are already cross-border oriented in the sense that they see their major markets outside of Germany, and roughly 50% of the MNS providers were engaged in cooperation with foreign partners (Stoetzer, 1993a).

The German MNS market is simultaneously in a phase of expansion and trial and error. New technologies (ATM) and new services (Frame Relay) are entering a growing market. New competitors, alliances and cooperations are rapidly changing the market structure. Only in the second half of 1992, two major competitors—INAS and Unisource—have started business in Germany and others like Info AG have strengthened their international cooperation with Transpac and Sprint. At the same time another competitor, C&W, has decided to withdraw its MNS center from Frankfurt that was opened just a year before.

2.2.4 Organization of the activity of TOs in Germany

Several North American carriers and those from the Far East as well as European carriers have by now established distinct entities bundling their international activities. The entry of carriers from abroad into the German market as an element of their internationalization efforts often has been part of a broader European strategy. Thus, some have diversified their worldwide

---

[33] In October 1993 Unisource signed a cooperation with KDD (Financial Times, October 16, 1993). Moreover, it is said that Unisource can be characterized as cooperating with Sprint; however, we are not aware whether it is already a deal. Thus, Unisource is a regional provider aiming to get a global reach.

[34] CIT Research estimates German MNS market volume as 31.5 mill. DM. OVUM estimates a market volume of 1,239 mill. DM. This illustrates the tremendous differences.

[35] According to Stoetzer (1993a) there is a market volume of 300 mill. DM for private MNS providers whereas DBP Telekom accounts for the rest of 1.2 bill. DM. The total turnover of Telekom in the fields of value-added services amounts to around 7.5 bill. DM in 1991 (see Handels Zeitung, March 4, 1993).

activities by establishing special European or even German subsidiaries. Even in those cases where the foreign strategy of a carrier is guided and organized by the domestic headquarters, more and more foreign carriers are about to establish at least offices and branches in Germany. Table 2.3 gives an overview of foreign TOs with offices, branches or subsidiaries in Germany.

AT&T has several subsidiaries focusing on the German market. As AT&T is vertically integrated with respect to services and equipment manufacturing, respectively, it is not surprising that there are activities in Germany with respect to both lines of business.[36] GTE is not present in Germany; however, there is a branch in Belgium. Teleglobe Canada recently has opened a so-called "liaison office" serving as headquarters for Europe and Africa.

Among the RBOCs, today only Pacific Telesis International and Bell South are represented in Germany by subsidiaries. The branches of Nynex and US West have been closed recently. Ameritech and Southwestern Bell were never present in Germany; we are not aware of the organization of their European activities. Bell Atlantic's European activity is led by the European headquarters in Rome.

NTT's German subsidiary serves as a headquarters for Europe.

To the best of our knowledge Telefonica, Stet/Sip, PTT Netherlands, and Televerket have no special German office or branch. However, the Swedish-Dutch-Swiss company Unisource has its headquarters in Frankfurt.

2.3 Evaluation and outlook

The German telecommunications services market is in a transition period exhibiting elements of an old monopolistic regime and liberalized market segments. The latter mainly comprise mobile services, satellite communications services and VAS. Foreign companies have already penetrated these markets substantially. The cellular mobile market has attracted the most interest and seems to become the major driving force in the internationalization process of the whole German telecommunications service market, as is the case in many other European countries, too (see Schnöring 1992). German satellite communications have also attracted foreign investors but the development is slower than expected at the beginning of the process of liberalization. Many foreign companies are positioning themselves on the different market segments of the German VAS market.

---

[36] AT&T covers the German market through both regional offices of its business units and subsidiaries. For example the "Global Business Communications Systems" unit is active in Germany as well as the subsidiary AT&T Network Systems Deutschland GmbH and also NCR. Primarily focusing on services, there is the subsidiary AT&T Deutschland GmbH. NCR activities in Germany and those of AT&T Deutschland GmbH are probably going to be merged in the future (see Frankfurter Allgemeine Zeitung, November 4, 1993).

Table 2.3
Main foreign TOs with offices and subsidiaries, respectively, in Germany

Company	Year of entry	Major owners	Offices in Germany	Empl. in Germany	Entry focus
AT&T Deutschland GmbH	1988	AT&T (USA)	5	400	supply of telco services, cooperation with domestic customers
Bell South Holding GmbH	1989	Bell South Corp. (USA)[1]	1		administration of assets
Bell South Mobilfunk GmbH	1992	Bell South Corp. (USA)[1]	1	30	market research, consulting (E-Plus)
BT Deutschland GmbH	1987	BT (UK)	6	100	supply of global services
Cable & Wireless	1991	Cable & Wireless (UK)	1	25	supply of international telco services
France Télécom Deutschland GmbH	1987	France Télécom (France)	2	20–30	supply of international services
KDD Deutschland GmbH	1977	KDD Corp. (Japan)	2	7	market research, supply of telco services, consulting, cooperation with domestic customers

Table 2.3 cont.
Main foreign TOs with offices and subsidiaries, respectively, in Germany

Company	Year of entry	Major owners	Offices in Germany	Empl. in Germany	Entry focus
MCI International	1983[2]	MCI Int'l (USA)	2	10	supply of telco services, cooperation with domestic customers
NTT Deutschland GmbH	1991	90% NTT Corp., 5% NTT-DATA Comm. Systems Corp., 5% NTT-Int'l Corp. (Japan)	1	5	furthering int'l cooperation, promoting global network activities, fostering int'l procurement activities
PacTel GmbH	1989	PacTel Int'l (USA)	1	5	support of cellular mobile radio activities
Sprint Datenservice GmbH	1990	Sprint Int'l (USA)	1	10	supply of telco services
Teleglobe Canada	1993	Teleglobe Can. Inc. (Canada)	1	5	"carrier for carriers"

1) Both subsidiaries of Bell South in Germany are also reporting in part to Bell South International and Bell South Europe (Brussels), respectively.
2) Prior to 1983, Western Union (Int'l), taken over by MCI, has had an office in Germany.

Today, Telekom still is the dominant player in nearly all market segments. An exception is the second GSM network. Competition will, however, undoubtedly be intensified. Further market entry can be expected: the trend towards outsourcing[37] will fuel this development as will further steps to liberalize the voice telephony market and the network monopoly.[38] Moreover, the corporate network decision of the BMPT of December 1992 that allows private voice telephony service between locations of a corporation spread over the country may give a boost to private communications networks and outsourcing in the German market. Thus, it seems fairly plausible to expect the share of the revenues in monopoly areas to decrease, and viceversa the share of revenues in competitive areas to increase, which in all likelihood will also lead to an increasing weight of foreign players in Germany.

Almost every big international player has at least established a bureau in Germany. Some TOs have already started to offer services. Most of them are following a strategy of joint ventures with German or foreign partners; some are doing it on their own and some are using different ways at the same time. Market entry of foreign TOs is still in a state of infancy and trial and error. Specialized international service providers have penetrated the market earlier and some of them are well established in the German market.

Up to now, foreign investors have taken the chances offered by the liberalization in Germany for market entry and FDI. Further steps of market liberalization will accelerate the process of market restructuring. It seems to be clear that the role of foreign companies will become more important in the future. But today, it is hard to forecast which of the different market segments will play the leading role. This will very much depend on the regulatory policy (in Germany as well as in the EC) as well as on Telekom's ability to operate efficiently and its strategic reactions to the threat on its home turf.

3. LIMITING FACTORS OF INTERNATIONAL ECONOMIC ACTIVITIES OF DBP TELEKOM

3.1 Current business constraints

DBP Telekom is the third largest TO worldwide with revenues of about 54 bill. DM in 1992. Obviously, the turnover of a TO with monopoly rights depends very much on the size of a country. Telekom is looking for rapid growth in national and international markets and wants to raise revenues to about 80 bill. DM by the year 2000. In achieving this goal Telekom is facing a number of great challenges—some are common for almost every former monopolistic PTT and some are special to the situation of DBP Telekom.

---

37 Outsourcing can have very different dimensions, ranging from a turn-key solution to specific tasks defined by a firm to be outsourced. In between, one can think of outsourcing in the form of conceptualization, realization, operation, and administration performed by a TO, or the take-over of a subscriber-owned network where operation is performed on behalf of the customer.

38 EC's present plan sets a deadline of 1998 for all telephone monopolies to end. Moreover the EC recommends the measure of easing the access of closed-user group networks (e.g. railways) as well as of cable-TV networks by service providers. It remains to be seen how the EC Council of Ministers will decide on liberalization of national and cross-border telephony.

3.1.1 Regulatory environment and performance of Telekom

Rapid changes in technologies, in customers' demand patterns, and in the competitive environment are common to many other TOs. The separation of the regulatory function and operating function, and the divestiture of the Deutsche Bundespost into three separate companies in 1990, have put a burden on the German telecommunications sector in the short term because the transformation absorbed a substantial amount of scarce management resources.[39] Yet, these steps were a necessary condition for DBP Telekom to act as a company in a globalizing market.

In the meantime, it became obvious that the steps taken in 1989 were too short-sighted. DBP Telekom needs equity capital that the government cannot provide. It needs more freedom from public sector laws and rules governing its personal management, as well as needing certainty regarding the legal framework of its international strategy. Therefore, a debate about a second major reform of the telecommunications sector leading to a privatization of DBP Telekom came up just two years after the last reform. The constitutional framework of DBP Telekom remains quite unstable and the company still has to adapt to the structures created by the first reform.[40] The relatively slow and stumbling reform of German telecommunications policy is a problem for the current, and in particular for the future, business opportunities of DBP Telekom.

Studies analyzing the strength and weakness of the company at the end of the 1980s showed that Telekom has many strong points but weak points as well. The company is lagging behind in efficiency compared to other competitors. Capital and labour productivity is lower measured by different indicators and the usage of the telephone network is a lot lower than for example in the USA.[41] Additionally, the company lacks market and customer orientation.[42] Telekom has started to tackle these problems. Sales and services were reorganized, and new financial controls are being developed to enable the company to be run on a more commercial basis. DBP Telekom is introducing a new organizational structure since early 1993 in a move to be more customer-oriented.[43] Moreover, all mobile communications activities have been brought into a separate subsidiary wholly owned by Telekom ("DeTeMobil GmbH").

So far, the Telekom management has been modest in announcing major job cuts for the coming years. The company is trying to increase its relatively low labour productivity per-

---

39 Moreover, the tax regime as well as the obligation to cross-subsidize the postal and the postal banking services are unfavourable for Telekom. For most of their revenues the three entities of the Deutsche Bundespost are currently freed from paying the value-added tax and their assets and profits underlie no local business taxes or income taxes, respectively. Instead they have to pay an annual levy to the treasury which is more or less revenue-based and, thus, has to be paid irrespective of losses. Telekom, which is the only entity with profits, has to pay the deficits of the sister companies. Only recently, Telekom was forced to pay a special levy to the Treasury to solve German budget problems.

40 This certainly is a competitive disadvantage compared to some competitors like BT that had the chance to start the necessary transformation process from a monopolistic public administration to a market-oriented private company much earlier.

41 See e.g. Tenzer (1992) or Gerpott and Pospischil (1993).

42 One of DBP Telekom's weak points is e.g. relatively low customer satisfaction in the large business customer segment which accounts for 40% of its turnover (Gallist, 1993).

43 The three main divisions focusing on the customer will be the private customer division, the business customer division and the division for system customers. These divisions are supported by the network division. This reorganization takes place in various management layers of the company affecting both the headquarters and the regional entities.

formance by a strong growth of turnover (Gallist, 1993). Time will show if this is a realistic solution. In the longer term, the company may find that major job reductions are unavoidable when growth objectives cannot be reached and profitability becomes a more important goal. Recent announcements of 30,000 job cuts up to the year 2000 point in that direction.

Traditionally, Germany has a strong telecommunications equipment industry led by Siemens and Alcatel-SEL. In 1980, Germany had been the largest exporter of telecommunications equipment with a world market share of 16%. During the 1980s the national industry lost some of its competitive edge while the competitive position of Siemens and Alcatel-SEL as companies with worldwide activities remained strong. The world market share of German exports dropped to 9% in 1990 (Schmoch and Schnöring, 1993). Today, Telekom is increasing its pressure on the equipment suppliers to bring down prices to reduce capital costs. Thus, German manufacturers are facing a difficult situation given the fact that in the German market R&D expenditures for new equipment are not paid separately as is the case in France and other countries, and given that the industry also has to compete with vertically integrated companies like AT&T or Northern Telecom (Grupp and Schnöring, 1992). Therefore, the liberalization of telecommunications service markets in Germany forces equipment manufacturers to adjust. The German Reunification created a boom for the German equipment industry. Turnover and profits increased substantially in the short run and the reunification will enlarge their home market in the long run (Schnöring and Szafran, 1992). Both factors should help the industry to overcome its handicaps.

3.1.2 German reunification

In principle, the challenges mentioned are common to many other TOs on their way from being a public entity to a joint-stock company with private shareholders. The challenges may differ a bit between companies because they had different starting points that put them into a better or worse position. However, there is one unique challenge for DBP Telekom which cannot be compared to others and that is the German Reunification. It has brought about a lot of risks for the development of Telekom in international markets. However, it offers opportunities as well.

The rapid development of the East German telecommunications infrastructure[44] and the merger of the two German TOs[45] absorbs tremendous amounts of scarce resources that could have been used otherwise, for instance, for a faster internationalization of Telekom. Telekom is planning to invest 60 bill. DM in East Germany up to 1997. The annual operating profits of Telekom are totally consumed for the levy and subsidization of sister-companies. Consequently, there are virtually no allotments to own resources leading to a reduced equity ratio. It can be expected that Telekom has to borrow 75% of the funds for the overhauling and modernization in East Germany. The result is that there is an imbalance in the capital structure of the company compared to other TOs—one major argument in the German discussion about privatization of Telekom.

The catch-up race in East Germany also absorbs scarce management capacity and it provokes a tendency to stress an inbound and national perspective within the company that is traditional for public TOs but is an impediment for the necessary internationalization of Telekom.

---

[44] This is a vital pre-condition for the economic recovery in the Eastern part of Germany.

[45] At the time of the merger DBP Telekom had about 200,000 employees and the Deutsche Post Telekom of the former GDR had about 40,000 employees.

The German reunification was also an impediment for the implementation of the separation of the political and regulatory functions from the business functions in day-to-day practice which was intended by the Postal Constitution Act of 1989. The structures created by the new law started to work on the first of January 1990, roughly two months after the breakdown of the Berlin Wall and 10 months before the political and legal reunification. Therefore, politics had a major impact on Telekom's ability to perform its business in ways not intended by the law, and this has slowed down the transformation process of Telekom into a market-oriented company.

On the other hand, the German reunification creates chances for the company. The home market is increasing. Building a network more or less from scratch offers opportunities to experiment with new technologies (fiber in the loop, radio-based fixed links etc.) as well as with new methods (turn-key projects performed by equipment suppliers on behalf of Telekom etc.) and structures. This will enlarge the future competitiveness of the company. A successful Telekom 2000 program for the "Neue Bundesländer" will also be a good reference for business in Eastern Europe as soon and as far as these markets develop.

3.2 Current legal constraints

Legally DBP Telekom is a special asset of the (federal) State ("Sondervermögen des Bundes"), i.e. it is still a public administration. The scope of its activities is limited by the German Constitution and the Postal Constitution Act. DBP Telekom cannot expand its activities just for commercial reasons; rather, diversifications must stay in line with its legal status as special asset. This fact creates some uncertainty regarding the admissible scope of Telekom's international activities. The Ministry of Posts and Telecommunications has to approve the foundation of foreign subsidiaries and the participation in foreign companies.

The examination of these activities is scrutinized and judged with reference to the legal status of Telekom. According to German experts in constitutional rules Telekom is allowed to make FDI with respect to the provision of international and national telecommunications services as long as there is a link with the purpose of the company in Germany defined by its legal status; see Badura (1991) and Scholz (1993). As long as the FDI is linked to the provision of international telecommunications services from or with Germany there is no dispute. This may be different when it comes to the operation of a telecommunications network in a foreign country. However, Telekom's undisputed participation in telecommunications infrastructure projects in Ukraine, Kazakhstan and other East European countries indicates that this kind of FDI is also possible and that there is room for interpretation of the rules.[46]

Anyway, in Germany the restrictions on Telekom regarding FDI and the uncertainty about the interpretation of these restrictions are generally seen as an impediment to the internationalization of the company and as a competitive disadvantage. This, in turn, is used as a major argument for a privatization of Telekom.

From an economic point of view these restrictions hardly make sense at a time when telecommunications markets become both competitive and international. Companies should have a level playing field in this respect and DBP Telekom may be seen as more restricted regarding FDI than BT, AT&T, the RBOCs or C&W for example. Yet, there are TOs like NTT whose situation is worse. With the increasing internationalization of the market, the creation of a level playing field for FDI in telecommunications services, with respect to both the rules

---

46 At least under those special circumstances prevalent in Eastern Europe.

for market entry and the rules for FDI of TOs with special rights, will be one topic on the agenda of international telecommunications regulation for the coming years.

4. INTERNATIONAL STRATEGY OF DBP TELEKOM

4.1 Role of international activities within corporate strategy

Despite the current business and legal constraints outlined above, Telekom has already formulated basic entrepreneurial objectives and guidelines in 1992 which are to play an essential role in the future development towards a customer- and market-oriented company.

Telekom's mission takes on firstly the current status as a (quasi) public utility in the fields of communication and information whose activity is governed by the market as well as infrastructural tasks. Secondly, it is stated that the global orientation of telecommunications markets calls for an active participation of Telekom in the international arena. Thirdly, the market shall decide on the range of services. The focus is to be on the customer to improve the benefits of telecommunications and to open up new applications. Fourthly, competition is viewed as a challenge and Telekom wants to take the chances brought about by competition.[47]

Thus, it is fair to state that Telekom has realized that the age of the entrenched monopolies with an almost exclusively national focus is ending also in continental Europe, and that it must try to strengthen itself for competition. However, it is also obvious that due to the overhauling of the East German infrastructure there is a "national bias" within Telekom's short-term entrepreneurial perspective, and international aspects are necessarily ranked second on the agenda in the current day-to-day business.

From a specific viewpoint Telekom has long been internationally oriented. In fact, the provision of international telecommunication services has been a major source of revenue for DBP Telekom for a long time. Telekom is the second largest carrier for international telecommunications measured by outgoing minutes of telecommunications traffic; see (Staple, 1991). According to estimates of Dataquest, revenue from international services in the German market was 4.5 bill. US$ (17% of total telecommunications service revenue) in 1991 compared to 3.4 bill. US$ for France (17%) and for the United Kingdom as well (17%). In Europe only three relatively small countries have higher shares of their revenues in international service: Netherlands 31%, Belgium and Switzerland 22%.

Table 4.1 shows the top 20 international telecommunications service providers ranked by their international business from a different source. The results are quite similar.

International business has been highly profitable in the past and still is, despite tariff cuts and increasing competition from new entrants as well as among the members of the old TOs cartel. This fast-growing and highly profitable market segment is a major focus of various forms of international competition (tariff cuts, call-back service, call-direct services) and market entry. Therefore, the figures not only indicate the importance of the market for international services for DBP Telekom's revenue and its profitability but also indicate the vulnerability of the company to increasing international competition. Telekom has reacted to the increasing competition. The company has reduced its tariffs and increased its efforts with new services and joint international ventures in many ways.

---

[47] There are another four principles in Telekom's mission saying that (5) Telekom acts on the basis of economic principles and (6) needs profits to secure viability of the company. Moreover, (7) each employee contributes to the success of the company and (8) Telekom takes on social responsibility.

Table 4.1
Top 20 international communications service providers

Rank	Company	International Communications Service Revenue		Total Revenue (mill. US$)
		mill. US$	% of total revenue	1991
1	AT&T (United States)	5,500*	14%	38,805[1]
2	DBP Telekom (Germany)	4,500*	16%	28,442
3	BT (United Kingdom)	3,146	14%	23,275
4	France Télécom (France)	3,058	15%	20,382
5	Cable & Wireless (United Kingdom)[2]	2,095	46%	4,588
6	KDD (Japan)	1,787	100%	1,787
7	PTT Telecom (Netherlands)	1,358	26%	5,193
8	ASST (Italy)	1,300*	29%	4,426
9	MCI (United States)	1,300*	15%	8,433
10	OTC (Australia)	1,281	100%	1,281
11	Telefonica (Spain)	1,198	12%	9,701
12	Telmex (Mexico)	1,163	22%	5,391
13	Swiss PTT (Switzerland)[3]	1,093	20%	5,400
14	Stentor (Canada)[4]	1,000*	0.8%	12,981
15	Saudi Telecom (Saudi Arabia)	912	36%	2,500*
16	Belgacom (Belgium)	890	32%	2,787
17	Televerket (Sweden)	864*	15%	5,692
18	Sprint (United States)	700*	0.8%	8,780
19	OPT (Austria)	650*	26%	2,534
20	Italcable (Italy)	546	100%	546

Notes: All figures converted to U.S. dollars at 1991 average exchange rates. Figures are for fiscal year 1991. BT, C&W and KDD figures for fiscal year ended 31 March 1992.

* = Estimates.
1 = Excludes product revenues.
2 = Includes holdings in foreign operators.
3 = Excludes non-telecoms business.
4 = International service to U.S. and Mexico only; formerly Telecom Canada.

Sources: Communications Week International, on the basis of The Yankee Group Europe, Watford, England; see Communications Week International 23 Nov.1992.

While Telekom has a long tradition in the provision of international telecommunications services from and to Germany, its activities in foreign markets via offices, subsidiaries, joint ventures or other forms of FDI have started only recently and on a relatively low scale compared to BT or other international carriers.

According to recent statements of the management[48], DBP Telekom's current international strategy follows two major objectives:

48 See e.g. TelekomVision 2/92; DBP Telekom Strategy Unfolds, Interview with Rolf-Dieter Leister, chairman of the supervisory board, Communications Week International 22 June 1992; H. Ricke, Chairman of the Board (1993); Gerd Tenzer, Member of the Board, in Dataquest Perspective (1993).

- to meet the increasing demand for "full service" of international customers and
- to become a key partner for the development of telecommunications infrastructure and services in Eastern Europe and the former Soviet Union.

The first point is often described as follow-your-customer strategy and can be seen as a reaction towards customer demand and increasing competition in the highly profitable market segment of international communications of large businesses. So far, DBP Telekom wants to meet the challenge mainly on the basis of joint ventures, alliances and contractual arrangements with other carriers and not so much by the establishment of a global international physical network.

DBP Telekom's emphasis on Eastern Europe and the Commonwealth of Independent States (CIS) is virtually based on the geographical location of Germany and DBP Telekom's experience from its investment programme in Eastern Germany.

4.2 Scope and focus of international activities

The Deutsche Bundespost (DBP) has always been an active partner in established international institutions like ITU, CEPT, ETSI etc. and has supported the development of telecommunications in Third World countries.

In 1977, DBP enlarged its international activities by creating "Detecon",[49] a subsidiary mainly active in consultancy for telecommunications projects in developing countries. Recently, Detecon has broadened its business focus towards the German and West European market.

DBP started to increase its international presence beyond the notion of development aid and participation in international institutions in the mid-1980s. At first, DBP specialists attached to German Embassies in Washington and Tokyo were to analyze the telecommunications policy development in the USA and Japan, and communicate German telecommunications policy. These activities were mainly policy- and not market-oriented. Market-oriented international considerations and activities started to increase parallel to the divestiture of DBP at the end of the past decade.

4.2.1 International offices

In 1990, DBP Telekom established five international offices located in London, Paris, Brussels, New York and Tokyo, followed by another office in Moscow 1992. So far, these offices are relatively small units with just a few employees. They shall mainly represent DBP Telekom and analyze the development of markets, competitors and telecommunications policy in those countries. They shall serve DBP Telekom customers abroad and shall make contact with foreign business customers having major communication links to Germany. So far, these offices do not actively market DBP Telekom's services abroad. The emphasis of the different activities varies between the offices according to local conditions. The presidents of the international offices report to the international division of the DBP Telekom headquarters.

It remains fairly open in which direction the international offices will develop once DBP Telekom's international strategy and activities have unfolded somewhat further. One possible

---

[49] Detecon consists of the following shareholders: Telekom 30%, Deutsche Bank 25%, Dresdner Bank 25 %, Bau- und Handelsbank 20%. However, Telekom has an option for 51%. With respect to the day-to-day business the role of the banks is not so much focused on the operating and entrepreneurial functions; rather, they primarily cover functions as to the financing of Detecon and its projects.

outcome would be regional headquarters for all activities in certain foreign markets. Another one would be if business activities for national and foreign markets are organized along certain business categories (customer groups and services) at home, with the international offices acting as bridgeheads in foreign markets as they do nowadays.

4.2.2 Product focus

DBP Telekom's management has declared that the emerging global customer with his special multinational communications needs is the major focus of DBP Telekom's new international activities. Full service—one-stop shopping, one-end billing, managed links, the provision of tailor-made private networks, of network management service and facility management—for international key accounts will be a prime objective. One-stop shopping and one-end billing arrangements with other carriers, the development of an efficient transnational network infrastructure within Europe, joint ventures and alliances for the provision of new services are elements of DBP Telekom's new international activities.

In 1988 DBP Telekom's international activities gained momentum. Since then, DBP Telekom is a partner of Infonet Service Corporation, an international managed network service (MNS) and value-added service (VAS) provider. DBP Telekom owns 16.2% of Infonet, along with a number of other TOs, including major shareholder MCI. Infonet is one of the established international VANS providers (Table 4.2) with subsidiaries in many countries. In 1992, DBP Telekom took an 80% share in Infonet Network Services Deutschland GmbH which will market Infonet's global communications services in Germany. Infonet's future was and still is subject to speculative discussions. The debate centres on its relationships with some of its shareholders and their global ambitions apart from Infonet (Unisource, Eunetcom etc.). The recent cooperation of BT and MCI gave rise to further speculation about Infonet.

Also in 1988, DBP Telekom and France Télécom formed Eucom, a 50/50 joint venture for the provision of application-specific VAS in Europe. Eucom aims at VAS markets beyond the traditional scope of the two TOs by establishing joint ventures for specific VAS together with other partners. Eurotop (Eucom 60.2%, GSI-TT, IBM France) and Eurolog (Eucom 74.9%, Digital Equipment Corporation and Technology Management Group) are examples of subsidiaries of Eucom active in VAS for the travel business and transportation sector. Other VAS products are in the process of development. Eucom itself is acting as a holding company for different application-specific VAS provider companies (Telekom Vision 1/92).

During late 1990 and early 1991, DBP Telekom was being associated with BT and NTT over the creation of a consortium, to be known as Pathfinder, to offer global outsourcing services to large multinationals. The Pathfinder initiative was stalled and BT, the major driving force, renamed its initiative Syncordia, to create a global service provider devoted to the outsourcing of global corporate networks. Talks with DBP Telekom and NTT did not succeed.[50] In March 1992, France Télécom and DBP Telekom announced plans to launch Eunetcom, a global managed network service company. The Eunetcom initiative is seen as a reaction to the stalled negotiations about DBP Telekom's participation in BT's Syncordia. Eunetcom shall offer international VAS and outsourcing of managed networks to a handful of mega-customers.

---

[50] The reasons for the failure of BT's efforts to come to grips with DBP Telekom and NTT regarding Syncordia was presumably the emphasis of BT on holding a majority stake. This was in contrast to efforts of DBP Telekom to give a share in Syncordia also to France Télécom, with which DBP Telekom has close ties.

Table 4.2
Top five international value-added network service providers (1991)

Rank	Company	VANS Revenue (mill. US$)	Employees	Parent Company
1	Sprint	550	2,000	Sprint (United States)
2	GEIS	540	2,500	General Electric (United States)
3	BT GNS	525	3,000	BT (United Kingdom)
4	IBM Information Network	320	NA	IBM (United States)
5	Infonet	145	500	Carrier Consortium (U.S.-based)

Source: Communications Week International, November 23, 1992

So far, Eunetcom consists of a financial holding company in Amsterdam (Eunetcom BV), a management holding company (Eunetcom SA) in Paris and a technical support centre (Eunetcom Betriebs GmbH)[51] in Frankfurt, but the company still has to sign up its first customer. Managers have been chosen and operations began in September 1993 with about 20 employees. In October 1993 Eunetcom announced that it is negotiating to run part of IBM's European network. So IBM could become its first customer just as it was the first client of BT's Syncordia last year.[52] According to industry analysts, Eunetcom's service portfolio and strategy remain vague and the initiative is still looking for a North American or Japanese partner to bolster its presence outside of Europe.[53] Eunetcom is still very much in an embryotic status and has some way to go before it will become a major driving force in the global outsourcing market.

The provision of international services especially for large global customers has been the main product focus of DBP Telekom's new international activities since 1988/89. However, DBP Telekom has also started international activities of a different nature. The company has been part of two consortia applying for mobile communications network licenses —Phonepoint for a Telepoint license and Unitel for a PCN license—in the United Kingdom (UK) in 1988 and 1989. Later, DBP Telekom withdrew from both consortia for commercial reasons.

Both participations can be viewed as an attempt to get technological and commercial expertise in new market segments which had not been opened in its home market. Likewise, the participation of Detecon, the DBP Telekom subsidiary, in Eurobell Ltd. which is engaged in broadband activities aiming to provide all sorts of telecommunications services can be interpreted in a similar way—getting technological and commercial expertise in new market segments. Apart from the activities mentioned so far, DBP Telekom has directly and indirectly

[51] Eunetcom Betriebs GmbH is a 100% subsidiary of Eunetcom SA.

[52] See Wall Street Journal, October 8, 1993.

[53] See e.g. Analysys (1993), Communications Week International February 1 and August 16, 1993.

applied for mobile communications network licenses in a number of cases. Some were successful, some were not.

Table 4.3 presents an overview of the FDI activities of DBP Telekom and Detecon in the market segments of VAS, mobile services, terrestrial services and satellite services.

4.2.3 Regional focus

The targeted product focus, provision of international services especially for large customers, determines the regional focus of the international activities of DBP Telekom. This is the market of the highly industrialized regions—Western Europe, USA, Japan—with a special focus on Western Europe according to Telekom's home turf. Telekom's international offices are mostly located in these regions. Eucom and Infonet are oriented towards these markets but so far Telekom has been reluctant to invest substantial sums in operators or service provision companies active in national markets of foreign industrialized countries.

Eastern Europe and the former Soviet Union are an exception in Telekom's international activities in a number of ways. DBP Telekom is trying to play a specific role in the modernization and upgrading of the telecommunications infrastructure in these countries. This move to the East is strongly supported by the German government on the political level. DBP Telekom's guiding principle in East-West relations is to support economic development in these countries. Thus, the company takes a long-term perspective for its investment in Eastern Europe and not a short-term profit maximizing approach. This approach differs substantially from the approach of many other Western carriers which are much more reluctant to invest in infrastructure projects with long pay-back periods.

DBP Telekom is currently active in Ukraine, Belarus, Russia, Kazakhstan and Hungary. DBP Telekom is looking for participation in joint ventures and consortia with national and other international partners to involve local operators and local capital and to share the risks (Table 4.3).

Overall, there are around 30 projects in which Telekom is involved as of September 1993. The financial means in each project are said to amount to several 10 mill. DM, see (Tenzer, 1993), so that we would estimate the order of magnitude of the total investment of Telekom in Eastern Europe to roughly 1 bill. DM.[54] Telekom's commitment to these markets is substantial compared to other Western carriers and indicates that DBP Telekom wants to become the central hub for international communications with Eastern Europe.

4.2.4 Strategic alliances

Cooperations, joint ventures and the acquisition of equity shares in international companies are necessary elements of TO's strategies in internationalizing telecommunications markets. It is hard to follow all initiatives being announced. Sometimes, one has the impression that the announcements themselves seem to be more important than the actual activities being announced. Market structures are very much in a transition period with a lot of trial and error. This is the case for every large TO and for DBP Telekom, too.

So far, DBP Telekom's new international activities are mainly based on cooperations, joint ventures and the acquisition of minority shares in service companies. They are not based on the acquisition of majority shares in or the establishment of network operators outside its

[54] Other sources speak of several hundreds of millions of Deutschmarks (Dataquest Perspective, 1993) in Central and Eastern Europe.

home market. DBP Telekom is cooperating with many TOs but from the very beginning the cooperation with France Télécom has played a special role. Eucom and Eunetcom are the most prominent examples of this alliance. Both DBP Telekom and France Télécom refer to this partnership as a key element of their international strategies (Communications Week International 1 February, 1993), but the actual strategic positions of the two carriers in the international arena are quite different from each other.

France Télécom has been much more active in FDI in European and overseas markets. It has taken stakes in the Mexican TO Telmex and participated in a takeover of Argentina's northern networks.[55] It has stakes in the German service provider Info AG and Olinet, the Italian subsidiary of Infonet. The acquisition of the London Regional Transport network as well as the acquisition of Maxat providing satellite links gave France Télécom a strategic presence in the UK. At the end of October 1993, US West International and France Télécom announced that they were joining forces in the bid for a minority stake of state-owned Hungarian Telephone Co. (MATAV), thus competing with the DBP Telekom—Cable & Wireless—Ameritech consortium. In addition to these activities, Transpac, France Télécom's data networking subsidiary, has been quite active in foreign markets even within the home turf of DBP Telekom. The acquisition of a majority stake in Martin Dawes gave France Télécom a position in the German market of mobile service provision.

Therefore, the partners of the strategic alliance—DBP Telekom and France Télécom—have quite different positions in the international arena. With the more aggressive approach of Transpac in the German market and the liberalization of corporate networks in Germany the alliance has entered a difficult phase right now, because it increases the likelihood of direct competition between two partners in an important market segment.

The recent joint venture of BT and MCI further increased the pressure on DBP Telekom and France Télécom to take action. One option being discussed is an equity swap between the two once they are privatized.

From a pure business point of view as well as due to the effects on the competitive situation in a once liberalized market, a number of questions regarding the pros and cons of such a close partnership of two of the largest European TOs need to be raised. Where are the synergies of the two companies on the procurement side, on the production side, in the field of R&D, of marketing new services or in the field of market access in third countries? To what extent will the cooperation increase the mutual competitive strength of both companies in the coming global market place, reducing the likelihood that the two compete head-on with each other? What does a close cooperation of two of the largest European carriers mean for the development of true competition in a liberalized European market after 1998? These are important issues which have to be addressed by the management of the companies and telecommunications policy makers in the respective countries.

---

[55] Partners of France Télécom in the Argentine have been, among others, STET and J.P. Morgan.

Table 4.3
Foreign direct investment of DBP Telekom[1]

Company	Year of participation	Total investment	Share	Product focus	Regional focus	Partners
			VAS			
Infonet	1988	100 US$	16.2 %	MNS, VAS	global	MCI, France Télécom, PTT Netherlands et al.
Eucom	1988	total equity: 150 Mill.DM	50%	application-specific VAS	Europe	France Télécom
Eunetcom	1992	40,000 Dutch Guilders	50%	outsourcing for global MNC	global	France Télécom
Teleport Iwanowo	1993		(DFKG)[2]	International Teleport	Russia	Intertelekom
Financial Network Association (FNA)	1991	250,000 Belgian Francs	12.5%[3]	Data networks and services for the financial services sector	world-wide	MCI, France Télécom, Mercury, Stentor et al.

Table 4.3 cont.
Foreign direct investment of DBP Telekom[1])

Company	Year of participation	Total investment	Share	Product focus	Regional focus	Partners
			Mobile			
Phonepoint	1989/91		10%	Telepoint Consortium	UK	BT, France Télécom, Nynex, STC
Unitel	1989/90		15%	PCN Consortium	UK	STC, US West, Thorn
--	1992			Analog Cellular Mobile	Ukraine	PTT Netherlands, TeleDanmark, National TOs
--	1993		(Detecon)[4]	GSM Consortium	Turkey	Siemens, Simkor, Alcatel, Teletas
--			1993	GSM Tender Consortium (not successful)	Hungary	BT, France Télécom
Mobile Teleystems (MTS)	1993	300 Mill.DM (total), 60 Mill. DM for Siemens/Telekom	47% Siemens/ Telekom, 53% MGTS	GSM Network	Russia (Moscow)	MTGS (Moscow municipal TO), Siemens
			Terrestrial			
TEL	1991			Trans-European Line	Central Europe	National TOs
Dekatel	1992	15 Mill. DM (Telekom)		Overlay Network	Kazakhstan	Kazak PTT

Table 4.3 cont.
Foreign direct investment of DBP Telekom[1]

Company	Year of participation	Total investment	Share	Product focus	Regional focus	Partners
			Terrestrial			
Utel	1992			Overlay Network	Ukraine	AT&T, PTT Netherlands
--	1992			X25	Russia (Moscow)	
Eurobell	1993		20% (Detecon)[4]	Cable TV and Broadband Communication	UK	Helsinki Telephone and others
--	1993			Tender Consortium for Hungarian Telecom (MATAV)	Hungary	C&W, Ameritech, and others to be announced
InfoTel	1993	total equity: 10.5 Mill. DM	25%	Data Communication Network	Russia (20 cities)	three Russian partners (e.g. MGTS)
"50 times 50"	1993	1 Bill. US$ (total)		Overlay Network (MOU)	Russia (50 cities)	Intertelekom, others pending
			SAT			
Romantis	1992		51%	Satellite Network	Russia	Sviaz, DASA, Bosch Telecom

1) Realized and planned
2) Deutsche Fernkabelgesellschaft, subsidiary of DBP Telekom
3) Not identical with voting rights
4) Partnership of Detecon, subsidiary of DBP Telekom

4.2.5 Evaluation

Given the size of the company, its technological expertise, and the size of its home market, DBP Telekom has the potential to stand as one of the top network operators in the global telecommunications market of tomorrow. But the company still has some way to go to achieve its own goal to be one of the leading players in the future liberalized European and worldwide market. So far, many industry experts question the commitment of DBP Telekom to become a global service provider.

The company is under strong pressure in its home market and this pressure will increase further. DBP Telekom has to cut costs, to increase its efficiency, to become a lot more market-oriented, and it has to invest large amounts of resources into the upgrading of the East German infrastructure at the same time. The political environment is unclear with a debate about a privatization of DBP Telekom and a further liberalization of market entry in voice communication and network operating. All these factors make it difficult to predict the future international strategy of the company and its success. Obviously, both will depend very much on how these "inner German" factors are sorted out.

If Telekom makes good progress in resolving its adjustment to a market-oriented, competitive service company, it has, especially due to its large international customer base and its home market, a good chance to become a strong and vital competitor in the future global market. If German politics finally realize in account the changing rules of the international telecommunications market and create a telecommunications policy framework that fits into these changes, the chances of DBP Telekom are even better.

5. SUMMARY AND CONCLUSIONS

Germany has the third largest telecommunications service market in the world with a huge amount of in- and outgoing international traffic. Thus, on the one hand the mere level of economic activity in Germany would be an incentive for foreign companies to become engaged in Germany. Currently, the German telecommunications services market is in a transition period. Elements of an old monopolistic regime still exist but are eroding more and more. Market prospects are therefore increasingly attracting foreign investors.

The rules of the reform of the German telecommunications sector of 1989 constitute two monopoly areas: a network infrastructure monopoly of providing transmission paths, and a monopoly for voice telephony service. All other telecommunications services are open to (regulated) competition. The network monopoly has two major exceptions, mobile communications and satellite communications, so these areas are also open to market entry.

There are no special restrictions for FDI in the German telecommunications market. Foreign and domestic private investors are treated equally and the scope for market entry is defined by regulatory rules as well as the licensing policy of the BMPT, which is fairly liberal by continental European standards.

Foreign companies have already penetrated German telecommunications markets substantially. The cellular mobile market has attracted the most interest and seems to become the major driving force in the internationalization process of the whole German telecommunications service market. The development of the German satellite communications market is slower than expected. Yet, there are many foreign TOs and other foreign companies engaged.

Many foreign companies are also positioning themselves on the different market segments of the German VAS market.

Telekom is still the dominant player in nearly all market segments. An exception is the second GSM network. Competition will, however, undoubtedly be intensified. Further steps of market liberalization will accelerate the process of market restructuring. It seems to be clear that the role of foreign companies will become more important in the future. With the decision to liberalize market entry for voice communications in corporate networks, an important new market has been opened. This decision may boost the development of private national and international competitors as the provision of outsourcing services will become more viable in the German market. Anticipating the full liberalization of the voice telephony in 1998, foreign competitors may increase their activities in the German market to achieve a good starting position.

Today, it is hard to forecast which of the different segments will play the leading role in the internationalization of the German telecommunications market. This will very much depend on the regulatory policy (in Germany as well as in the EC) as well as Telekom's ability to operate efficiently and its strategic reactions to the threat on its home turf.

So far, DBP Telekom's responses to the threat of competition have been rather slow to materialize. The reform of the German telecommunications sector in 1989 and the German reunification in 1990 have put DBP Telekom under enormous pressure: the company had to develop a market-oriented business strategy. Huge investment programmes for East Germany had to be launched while a great deal of rationalization is required at the same time. DBP Telekom started to introduce a new organizational structure in early 1993 which is to improve flexibility and market orientation. A further rebalancing of tariffs is necessary while DBP Telekom faces increasing competition on its home market. The debated privatization of DBP Telekom will absorb further resources.

DBP Telekom is looking for rapid growth and aims to raise turnover from 54 bill. DM 1992 up to 70 bill. DM in 1998 (5% p.a.). Considering the threat of extended market entry in Germany—for national as well as international communications—DBP Telekom must increase its efforts to attract international traffic and must increase its business outside of Germany. Therefore, it seems likely that DBP Telekom will try to enter telecommunications markets outside of Germany in the Triad more aggressively than in the past.

It is DBP Telekom's stated objective to be one of the leading players in the European and world-wide telecommunications service market. Telekom wants to be a global player; however, the commitment of the company to this objective may be questioned. Internationalization of DBP Telekom does remain handicapped by constitutional uncertainties. On the other hand, it is fair to state that the company has been rather cautious with international activities.

Overall, the trends in internationalization and globalization are realized by the firms being active in the German telecommunications market and by some politicians. However, a broad part of the public debate about the future of telecommunications policy in Germany is still characterized by a domestic perspective, showing that the emerging international nature of market forces has not been fully recognized.

REFERENCES

Analysys, 1993, Global service provider market strategies, Cambridge.

Badura, P., 1991, Zulässigkeit und Grenzen der Gründung von Tochtergesellschaften und des Erwerbes von Beteiligungen durch die Deutsche Bundespost, insbesondere durch die Deutsche Bundespost Telekom: Archiv für das Post- und Fernmeldewesen, 43 Jg. Nr. 4, p. 389.

Christoffersen, M. and A. Henten eds., 1993, Telecommunication—limits to deregulation? Proceedings of the 7th European Communications Policy Research Conference, Soro Storkro, Denmark, October 21–23 (IOS Press, Amsterdam).

Dataquest Perspective 1993, Tough times for telekom, Interview with Gerd Tenzer, Member of the Board, DBP Telekom, Telecommunications Europe, March 3, pp. 1–5.

Deutsche Bundesbank, 1993, Die Entwicklung der Kapitalverflechtung der Unternehmen mit dem Ausland von Ende 1989 bis Ende 1991, Monatsbericht April, pp. 33–49.

Gallist, D., 1993, Gefährdungspotential minimieren, TelekomVision 2/93, pp. 22-23.

Gerpott, T.J., and R. Pospischil, 1993, Internationale Effizienzvergleiche der DBP Telekom, Zeitschrift für betriebswirtschaftliche Forschung, H.4., pp. 366–389.

Grupp, H. and T. Schnöring, 1992, Research and development in telecommunications: National systems under pressure, Telecommunications Policy, January/February 1992, pp. 46–66.

Ricke, H., 1993, Mehr Kommunikation untereinander, mehr Erfolg miteinander, TelekomVision 2/93, pp. 20–21.

Schmoch, U. and T. Schnöring, 1993, Wie steht es um die Wettbewerbsposition der Telekommunikationsindustrie in Europa? Patente und Weltmarktanteile, WIK Diskussionsbeiträge, Nr. 104, Bad Honnef.

Schnöring, T., 1992, Entwicklungstrends auf den europäischen Telekommunikationsmärkten, Auf halbem Wege von monopolistischen nationalen zu wettbewerblichen internationalen Telekommunikationsmärkten, WIK Diskussionsbeiträge, Nr. 102, Bad Honnef.

Schnöring, T. and U. Szafran, 1992, Umsatzboom bei den inländischen Herstellern von Telekommunikationsgeräten durch die deutsche Einheit, WIK Newsletter, Nr. 9, pp. 8–10.

Scholz, R., 1993, Tochtergesellschaften und Unternehmensbeteiligungen der Deutschen Bundespost Telekom im Ausland, Archiv für Post und Telekommunikation, 2/93, pp. 103-173.

Staple, G.C. ed., 1993, TeleGeography 1993—Global Telecommunications traffic statistics and comments, Telegeography, Inc., Washington, D.C., in cooperation with the International Institute of Communications (IIC), London.

Stoetzer, M., 1992, Value-added services—problems of definition and data, Telecommunications Policy, July 1992, pp. 388–400.

Stoetzer, M., May, 1993a, Netzmanagementdienste und Corporate Networks in Deutschland: Marktstruktur und Marktentwicklung, WIK Diskussionsbeiträge, Nr. 109, Bad Honnef.

Stoetzer, M., 1993b, Customer behaviour in mobile communications: A cross-country analysis. Paper presented at the 20th Annual EARIE-Conference, September 4-7, Tel Aviv, Israel.

Szafran, U., 1993, Der Markt für Satellitendienste, WIK Diskussionsbeiträge Nr. 107, April, Bad Honnef.

Telekom Vision 1/92, Die Mehrwert GmbH, pp. 30–32.

Telekom Vision 2/92, Kampf um die globalen Märkte, pp. 22–25.

Tenzer, G., 1992, Neuer Schwung für die Netze, Telekom Vision 3/92, pp. 20–23.

Tenzer, G., 1993, Telekom-Partner für Osteuropa, Telekom Vision 8/93, p. 18.

The Race to European Eminence
E. Bohlin and O. Granstrand (Editors)

Foreign Direct Investment in Telecommunications In and Out of France

Godefroy Dang-N'guyen[a] and D. Pouillot[b]

[a]Département d'Economie-Management, ENST Bretagne, BP 832, 29285 Brest Cedex, France
[b]Département Analyses Industrielles, IDATE, La Garrigue-Avenue de Castelnau, 34830 Clapiers, France

1. INTRODUCTION

Until recently, France Telecom was operating exclusively on the national scene, its international presence being limited to a few countries, historically close to France, such as the former French colonies. FCR (France Câbles et Radio) and Sofrecom, two subsidiaries of the French network operator, have long developed activities in African countries in particular. Outside these traditional international activities, specific plans were devised for the promotion of new systems or services, when it was deemed profitable to export the French know-how. In particular, Minitel (the French videotex system) has been highly promoted in foreign countries, with limited success because the marketing concept of Minitel (to give away the terminal equipment) has never been replicated elsewhere, due to cost motives. But, besides those limited export strategies, no overall strategy had been conceived, to expand the French operator's activity abroad. Conversely, the absence of a true liberalization policy from French authorities, and the dominant position of France Telecom in its home country, have prevented foreign investors from doing business in France. Recent changes have occurred, however, due to the implementation of the European Directives, in particular the so-called "Service Directive". This makes it mandatory for the French, as for any other European government, to authorize the competitive provision of data services of any kind. Also, the liberalization of some market segments, particularly in mobile communications, has given some opportunities to foreign investors to get minority interests into some new ventures. But the foreign presence is still very limited. Therefore, the bulk of our paper will be devoted to France Telecom's foreign direct investment, although from case to case, we will discuss the limited FDI in France.

France Telecom's strategy changed at the end of the 1980s. The threat of deregulation, the expansion of BT in France,[1] and the requirements of large customers seemed to call for a more dynamic attitude to multinationalization. When appointed as the chief for DAII (Direction of Industrial & International Affairs), Mr. Damlamian declared that "France Telecom's position in the world (was) very limited and (that the company had to) plan larger operations in order to increase its international turnover to 10% by the year 2000". A very ambitious programme indeed, with regard to the 0.6% ot FT's overall turnover, the proportion

[1] When BT took over McDonnell Information System, the British company came to grips with a French subsidiary of about 250 persons and this made the presence of BT in France much more visible.

of FT's international operations then. In total, 2.8 billion FF were invested in foreign operations in 1991, and this represents a quantum leap when compared with the amounts of the previous years.

Today, France Telecom international operations account for 3–4% of its total turnover; this is still low compared with the ambitious goal set up by Mr. Damlamian, but it is increasing. In any case, FT's ambitions are large and diversified, with the management saying "we cannot see any specific orientation, except the global need for gaining a stronger international presence".

Actually, the position and dynamics of France Telecom's international strategy can be analysed through different factors. First, the French trade performance was largely geared towards developing countries (Africa and the Middle East in particular) for high-tech products. In these sectors, France is generally considered as an intermediate country, importing from advanced industrial countries and exporting to less developed ones. International telecommunication contracts have, for a long time, been part of this picture. The export of telecommunications equipment is often a part of the multipurpose agreements between France and other developing countries (Morocco, Saudi Arabia, etc.). France has sold arms and infrastructure equipment (power plants, telecom networks, trains and civil engineering) in exchange for raw materials (oil, etc.). It is probably very recently that this picture has changed, in particular after the takeover of ITT's European subsidiaries by Alcatel. From then on, the French telecommunications industry had a true multinational character, focusing on Europe.

Second, the French telecommunications network has been completely redesigned over the past 15 years. While efforts and investments were totally focused on the national digitization programme and other parallel developments in specialized networks and services (data networks, ISDN, videotex, etc.) up to the mid-1980s, the French network has become one of the most advanced in the world, at least at the moment. France Telecom was recognized to have acquired know-how in the process, which could be exported (see operations in Latin America).

The layout of the paper is the following: in the first part, we describe what have been the major attempts of France Telecom to get a foothold in the international arena. We symmetrically present the effort of foreign firms to penetrate the French market. In the second part, we examine whether technology has been the key factor of success of France Telecom's strategy. In the third part we look at the marketing of France Telecom. We scrutinize its way of gaining experience in foreign markets. In the conclusion, we try to understand the internal and external factors of this company, which may explain or place a limit on its ability to be a "global player".

2. THE FRENCH SITUATION

2.1 Traditional relationships with developing countries

Traditionally, France has built strong relationships with overseas territories, in particular African countries. Therefore, it is used to participating in their development by negotiating large support contracts. Generally, these contracts include financial aid (credit lines) and cover different sectors, from industry to services: network infrastructures (roads, railways, electricity, telecommunications) have often been involved. In most cases, countries have a fixed financial budget constraint and they have to choose between the different opportunities so as to elaborate a priority programme. Telecommunications have often been taken into account. In this context, manufacturers (Alcatel; TRT, the French Philips subsidiary; Matra; JS Télécom, now owned by the German group Bosch) and operators are working closely together, although the bulk of the deal goes to the equipment manufacturer.

To give a few examples: France Telecom's subsidiary, France Câbles et Radio (FCR), has often run, in cooperation with the local authorities, the international networks of the former African colonies. After independence, at the beginning of the 1960s, it formed joint ventures together with FCR, called for example SIT (Société Internationale de Télécommunications) as in Cameroun. Gradually, the local authorities took over the ownership of these joint ventures, but still experts from FCR are hired to help run the international network. This traditional cooperation programme of France Telecom began after the former French colonies got their independence in the 1950s and still continues. But it is increasingly overshadowed by the actions undertaken since 1989 in other parts of the world.

2.2 Large operations in public networks

In targeting international opportunities, France Telecom has tried to build up a presence in opening areas such as Eastern European countries and Latin America. Foreign direct investment in Eastern Europe has been part of a broader movement of Western business to get a foothold in the so-called "countries in transition towards the market economy". A particular focus has been put on Poland and Rumania, for historical reasons, because these two countries were traditionally (i.e. before the communist regimes of the postwar period) influenced by French culture and diplomacy.

Eastern Europe

By the end of the 1980s East European countries had begun their move towards the market economy. Telecommunications were considered, along with other basic infrastructure networks, as a high priority; thus, several countries immediately called for foreign participation to help install new networks. In particular infrastructure for mobile telephony was seen as a quick way to give access to modern communications services. France Telecom was certainly not in the best position to compete with other international carriers: its own experience was and still is limited in that segment, particularly when compared to Nordic countries. Nevertheless, it could hardly miss the opportunities and, therefore, decided to gear its efforts towards Poland and Rumania. In Poland, France Telecom has been granted a 25-year mobile licence, to operate in cooperation with Ameritech and local PTT, PSTA. Centertel, the new local mobile network, entered operation during 1992, primarily developed in Warsaw, with an initial capacity of 2,000 users.

France Telecom was also interested in a mobile licence in Hungary, where it had planned to join BT and DBP Telekom, and has been unsuccessfully bidding in the privatization of the national public operator MATAV, which has been conceded to a consortium led by Deutsche Telekom.

Latin America

Since the end of the 1980s Latin America has also experienced a move towards democracy. Concurrently, the new regimes have launched privatization programmes, either to boost their own financial situation, or due to the need of technical support. Chile, Argentina, Mexico, Puerto Rico and Venezuela are all now engaged in this process. France Telecom was particularly interested in their telecommunications modernization programmes due to its own experience: the French operator has proved to be able to set up and redesign a network efficiently and rapidly.

France Telecom has succeeded in Argentina and Mexico. In both cases, the French operator has been given the task of operating and modernizing the network whilst the partners (STET in Argentina or Southwestern Bell in Mexico) are in charge of marketing and financial issues. In 1990, the level of equipment in both countries was the same as in France in 1970; the digitization rate was also very low.

Argentina was its first attempt in these parts. Together with the Italian STET, France Telecom was finally chosen to operate the telecommunications network in the northern part of the country. STET and France Telecom own 32.5% each of the Nortel consortium, which itself holds a majority interest in Telecom Argentina (60%). The remaining shares have been sold to private local investors (30%) and to employees (10%). Telecom Argentina has subsequently become a prosperous company. More important than network capacity development has been the improvement of the quality of service, which the company tackled before anything else, with immediate positive results. Today, however, the investment programme is now well in place; whilst in 1991 the amount of capital expenditures was only 15% of the turnover, this figure grew to 50% in 1992. Digitization is progressing rapidly (11.3% when privatized to 32% in September 1992) and new advanced services have been introduced (call forwarding etc.).

Mexico was the second target for France Telecom and, thus, it became associated with the American Southwestern Bell in a scheme which is somewhat different from that of Argentina. In this case, the owners are not directly involved with the local operator; they manage a holding which controls Telmex. Southwestern Bell and FCR/France Telecom own 24.5% each of the holding (20.4% of shares but more than 50% of the voting rights). The Telmex investment programme amounts to $12 billion over a period of 5 years (the renewal of 470,000 lines; development of fiber optics and the mobile network).

In more global terms, Latin America seems to be a promising area for the French Telecom operator. After being eliminated from Venezuela, where another American/European consortium (AT&T/GTE/Telefonica) won, France Telecom is bidding for the Uruguayan telecom operator, together with STET.

Western Europe

France Telecom is also interested in the privatization plans being devised in Western Europe. It has, in particular, bidden for a 33% stake in the Greek telecom operator, OTE.[2] As a member of the "Panafon" consortium, it has already been awarded a licence to operate a mobile telephony network in Greece (its partners are Vodafone from the UK, holding 45% of the shares, Databank from Germany and Intracom, a local manufacturer, with 10% each). The group paid $156 million to acquire the licence (the second licence to be awarded in Greece, after that of the Italian STET) in August 1992. They are supposed to build a complete GSM network in Athens within 18 months and to serve 85% of the Greek population within 6 years. The global investment amounts to $850 million for both the Panafon and STET consortia.

France Telecom is also active in the United Kingdom, mainly through two subsidiaries: MaxSat, a VSAT operator, created by FT at the beginning of 1991, and Maxwell Communications. One year later, FCR bought Maxwell's shares so that it now fully owns the

[2] Since then, the new socialist government of Mr. Papandreou has postponed the privatization programme.

subsidiary, which had sales of around FF 20 million in 91/92. Transpac Network Services (see "diversified operations") was also established in 1991 and has planned to invest £20 million (around 180 million French francs) in network development. About 12 contracts had been signed at the end of the first year of operations with companies such as Marks & Spencer, Ladbroke Racing and London Buses. Furthermore, France Telecom holds a 20% interest in RAM Mobile Data, one of the British licensees for mobile data network. Finally, France Telecom has taken a participation in Martin Dawes, one of the airtime resellers in the mobile market.

Pacific Basin

In Asia, where France is already present through its manufacturers' equipment, France Telecom also intends to participate in new privatization operations. Given the poor economic conditions in this region, local governments are calling for BOT ("Buy Operate Transfer") operations, which offer them the opportunity to take control of operations in the medium term with a limited investment (see below).

Other operations have been launched by France Telecom in these areas during recent years. In 1991, it joined a consortium led by Hutchison Telecom (Hong Kong) to bid for Aussat, the second Australian operator, which has just been privatized. They did not succeed, the Optus consortium (Bell South/Cable & Wireless) finally winning.

2.3 Diversified operations

Besides operations within the area of public telephone networks (fixed and mobile), France Telecom has also attempted to enter diversified areas, in particular data transmission and information services.

In data transmission, Transpac has been given a specific mission. The French data network has been running since 1978, making it one of the oldest X25 networks in the world. Its success has, in particular, been due to the development of national videotex which uses Transpac as a transmission relay. The number of users has rapidly grown to the extent that they now number around 110,000. Organized as a subsidiary, Transpac was called on to export its know-how. Three methods have been used to approach markets abroad:

- Joint venture: Interpac was set up through the grouping of international X25 operators. This joint venture was necessary to interconnect several national X25 networks, to meet the customers' needs.
- Creation of foreign subsidiaries: Transpac Network services was established in the United Kingdom to develop operations locally (and to respond to the strong presence of BT/Tymnet in France).
- Acquisitions: Transpac bought Info AG, a German information services company specialized in X25 facilities, a potential competitor to Deutsche Telekom, and this has puzzled the Germans. Transpac has also acquired 51% of Olinet (Olivetti), which runs the Italian part of the Infonet network (now Interpac Italy).

The French operator is also involved in the field of information services. For a long time now, it has been developing competence in telematics and value-added services (Télésystèmes, equity interests in GSI Erli, GSI Tecsi, etc.). Its involvement was strengthened in 1992 through its entry into the capital of the Sema Group, the second largest European software company. Mr. Roulet, the chairman of FT, declared at the beginning of the year that his group was looking for diversified operations in information services for several reasons: firstly, to secure a strategic sector which needs capital; secondly, to be more involved in software development: "(FT) is the sole large operator that has not until now designed management software itself"; and thirdly, it can emphasize the internationalization process.

The acquired interest in the Sema Group offered, in particular, an opportunity to strengthen FT's position in the international arena, most notably the UK (44% of Sema Group's turnover). France Telecom has invested $750 million in the controlling holding of the Sema Group (50.1% Paribas/49.9% FT). Besides the financial agreement, FT and Sema Group have hinted at two further ways of cooperating: the development of software for telecommunications and facilities management (2 joint ventures).

Globally, France Telecom's participation in foreign operations is now large and diversified:

- by the fields covered: basic network infrastructures and services, mobile telephony, data networks, value-added services, information services;
- by the nature of operations: acquisitions, joint ventures, equity participations;
- by geographical coverage: Western and Eastern Europe, Latin America, Asia.

Table 2.1
Synoptic table of France Telecom's foreign operations

Operator	Country	Type of operation
Telecom Argentina	Argentina	Participation in the northern public operator, alongside STET and a local group
Telmex	Mexico	Participation in the national public operator, alongside SW Bell and the local group, Carso
Centertel	Poland	Participation in the first mobile network, alongside Ameritech and the local PTT, PSTA
Panafon	Greece	Participation in the second mobile network, alongside Vodafone (Leader), Databak and Intracom
Transpac Network Services Interpac Italia Info AG	UK Italy Germany	X25 operation subsidiaries in UK, Italy (51% of the former Olinet) and Germany
RAM Mobile Data	UK	20% interest in UK mobile data network
MaxSat	UK	VSAT operator
Sema Group	France/UK	18% interest in the 2nd largest European information services company
In progress or projected:		
OTE	Greece	Bid for a 35% interest in the Greek public operator
Uruguayan Telecom	Uruguay	Joint bid with STET for the privatization of Uruguayan telecom operator
MATAV	Hungary	Bid for Hungarian public operator, MATAV
Failed:		
Kalori Communications	Australia	Bid for the privatization of Aussat, second Australian public operator

Source: IDATE (France)

We can also notice that the ventures set up over the 2–3 past years are totally different but are, at the same time, somewhat complementary to those previously initiated, through the type of activity (from engineering to network operations or service provision) and through the countries concerned (from traditional cooperative countries to new areas, including industrial countries).

We must add that the France Telecom Group is not the only telecommunications operator to invest abroad. Compagnie Générale des Eaux, a diversified holding company,[3] has set up a subsidiary, General Cable in the UK, to run, alone or with partners, the franchises obtained for the operation of a cable and telephony network. General Cable has franchises in the district of Sheffield and in Birmingham. All in all, General Cable is the 7th largest cable operator, in the UK.

Générale des Eaux is the main competitor to France Telecom in France. It controls Cofira, the holding which owns SFR (Société Française de Radiocommunications). The latter has obtained a licence to operate an analog and a GSM network for mobile telephony in competition with France Telecom. Cofira is a holding where Compagnie Générale des Eaux has 41% of the shares, Crédit Lyonnais 19.9%, TDF (a subsidiary of France Telecom!) 7.5% and the rest is divided between Bell South, Racal Telecom, Magneti Marelli (Fiat group), Tractebel (a major Belgian holding, owning cable TV companies) and other French investors.

Générale des Eaux is also operating cable networks in different parts of the country. The total amount of connectable households is quite high, although most of them are technically run by France Telecom, while Générale (similarly to Lyonnaise des Eaux-Dumez, the other cable operator) is only the commercial company. Therefore, the experience gained by Générale des Eaux in running their cable and telephony network in the UK may be particularly useful if they are allowed to do the same in France. In other words, Foreign Direct Investment by Générale des Eaux is a step towards direct investment in France, by a French competitor to France Telecom, which could be allied with foreign companies.

2.4 Inward FDI in French telecommunications

The French telecommunications market has traditionally been very closed. Some opening occurred in the 1980s, especially on the equipment side; the liberalization of the terminals market made it possible for foreign manufacturers to enter it (Siemens, Northern Telecom, etc.). Some opportunities also appeared in public equipment (AT&T). Nevertheless, the opening up of the services market was extremely limited; only a few foreign operators are now active on the French market, mainly through minority interests in independent networks or through diversified activities. The major operations cover:

– Some RBOCs:

Bell South, in cooperation with France Telecom, holds interests in one of the two national mobile data networks; they are also minority shareholders in ComDev (cable network operator).

Pacific Telesis, in cooperation with ETDE (a subsidiary of Groupe Bouygues), is active in trunk networks.

---

[3] Générale des Eaux has franchises to run water networks in France (hence the name) but is also a real estate promoter and a civil engineering firm, and is now diversifying in communications, through participation in Canal Plus and operation of cable TV franchises and mobile telephony (SFR), as well as environment protection (wastage recycling).

Finally, US West is a minority shareholder in Lyonnaise Communications (cable network operator).

– AT&T:

Besides its historical relationships regarding the international connection, AT&T has recently succeeded in bids over two medium-sized companies. The first one is a software company, Dataid, which AT&T with its dedicated subsidiary, Istel. The second one, is Barphone, a manufacturer of private switches.

– BT:

Since its combination with Tymnet, BT has installed in France a commercial base of 250 people responsible for the engineering, installation and management of corporate networks. In addition, the British operator has been granted a VSAT licence.

Taking also into account the presence of large international value-added network operators (GEIS, EDS and more recently Unisource) as well as the sales offices of foreign operators (Deutsche Bundespost Telekom), it is obvious that the number of players is large but their sales remain low. With a 1992 turnover of over FF 220 million, BT/Tymnet is certainly far more advanced than all others.

Table 2.2
The major worldwide operators on the French market

Trunks	Networks	Corporate Networks	Cable Networks	Data Processing	VANs Sales Office
AT&T				■	■
BOCs	■		■		
BT		■			■
DBP Telekom					■
NTT					■

All together, Table 2.2 provides an overview of inward FDI.

3. TECHNICAL ISSUES FOR FRANCE TELECOM'S FDI

France Telecom has earned its fame in mastering technical problems since the 1970s. First, when the first fully digital exchange in the world was conceived, and operated in 1970, and the know-how subsequently transferred to Alcatel; but more surely when the crash programme to equip France with a modern large telephone network was achieved in a remarkably short timespan, from 1974 to 1980. In 1974, there were 6.239 million telephone main lines in France. In 1980 this number had reached 15.89 million, that is, more than 2.5 times.

From that period, FT has maintained its reputation of excellence by developing new and innovative services: Minitel, Transpac (X25 packet switched data network), Numeris (ISDN). It is noteworthy that all these network-based services are relying heavily on the use of, or interconnection with, the public telephone services. For example, Minitel, the French videotex service, relies on both the public switched network and Transpac to interconnect information services providers and customers.

When, from 1989 onwards, France Telecom had to invest abroad, technical issues came to the fore, by establishing the competitive advantage of the company in comparison with other telecommunications operators. Hence, we can distinguish, from FT's point of view, two broad categories of FDI: business areas where a technical expertise had been developed inside France Telecom, and other domains where the technical expertise was lacking but FT considered that they had to invest, given the demand of the market and business opportunities.

3.1 Business areas in which FT has a technical expertise

We can divide these into two broad categories: public network services and data communications services.

Public network services

Let's come back to France Telecom's FDI in Mexico and Argentina. The Telmex venture is particularly significant. When the Mexican company got privatized in December 1990, the quality of service was particularly poor: the installation delay was averaging 3 years, the lag for repairs could reach several months and the density was less than 6.1 lines per 100 inhabitants, (Wall St. J., 1992). Two years later, things have improved dramatically due to a tough cost reduction attitude adopted by the new management. However, due the absence of direct involvement of France Telecom in the management of Telmex, it is difficult to understand from outside how the transfer of know-how from France Telecom and Southwestern Bell to Telmex management has performed. France Telecom experts are probably acting as advisers.

The Argentina venture (Telecom Argentina) is more demanding, since FT, together with STET, has the direct responsibility to install and operate the network in both the north of the country and Buenos Aires. At the moment, the team is performing quite well. Employment has been reduced from 19,000 in 1990 to 17,000 at the end of 1991. Investments have been boosted (603 million dollars against 131 million in 1991). The network has been digitized at a level of 32% in September 1992, against 12% in September 1991 and 11.5% in November 1990. New services have been launched, such as call warning, call forwarding, detailed billing. Similar services have been provided on the French telephone network for several years, and are very standard inasmuch as they are pieces of software installed in the switches.

It is clear from these two examples that FDI in public networks may require different levels of commitment by the investor. But whatever the type of involvement (advisory position as in Telmex, or full control as in Telecom Argentina) the end result seems to be very positive, for both the local carrier and France Telecom. The transfer of the technical expertise seems to be working smoothly, despite a very different economic and social context. Part of the explanation may rely in the nature of know-how being transferred. In fact, the availability of digital technology has enabled the operators to write down many of their administrative and management tasks (for example billing, or tracking bad payers) into a piece of software, now imbedded in the switches of the operation centers. This clearly makes more easy the transfer of know-how, once this software has been debugged and adapted to the local context.

Given the success of the operations in both Mexico and Argentina, France Telecom would like to extend its participation to other countries. Together with STET, it has applied for a licence to take over the Uruguayan Telecommunications Operator. France Telecom would also like to take participation in Southeast Asian countries,[4] but there, the privatization process does not follow the traditional path. Public calls for tenders are set up under the name of so-called "Buy Operate Transfers": a number of lines are adjudicated to a foreign partner, with the obligation for the latter to sell back this portion of the network to the local operator after a limited number of years of maintenance and exploitation.[5] This formula may offer FT an opportunity to team up with Alcatel for bidding for such Buy Operate Transfers. We have not been aware of any successful bids by the two partners so far.

In Africa, France Telecom has a traditional mission of cooperation with the ex-French colonies. FT would thus be very eager to take participation in privatized carriers of African countries. It has already taken one in a carrier of Equatorial Guinea.

France Telecom has shown its capability to operate a public telephone network in the international context. The technical issues involved may have not been so difficult to master and, in the meantime, France Telecom has earned a reputation of installing and operating a public network. So the bottom line seems positive.

Public data networks

FT has recently exported the installation and operation of public data networks. Transpac, the subsidiary of FT, is supposed to run the largest X25 network in the world.[6] Recently, Transpac has attempted to export this know-how abroad, in particular to the UK, Italy, Germany and Sweden. Some participations have been acquired especially in British companies. The liberalization of public data networks throughout Europe creates scope for increased competition in this domain.

Transpac clearly follows a multinational (albeit regional) strategy. It may evolve after the big agreement announced between Deutsche Telekom and France Telekom, to provide end-to-end services to big customers. Part of the agreement may lead to the merger between FT's Transpac and Deutsche Telekom's Datex-P.

Table 3.1.
Foreign activity of Transpac

Participations of Transpac	Activities	Turnover
Transpac Network Services (UK)	X25 services, facilities management	30 million francs
Info AG (Germany)	back-up for computer centers, X25 services, facilities management, EDI	240 million francs
Transpac Scandinavia	Outsourcing, X25 services	28 million francs
Comco Transpac (Switzerland)	X25 services	
Interpac (France)	X25, EDI, VPN	65 million francs
Interpac (Luxemburg)		4 million francs
Interpac (Italy)		14 million francs

Source: France Telecom.

4 India, Indonesia, Pakistan, Thaïland, Vietnam.

5 This method is very similar to the leasing process; it requires that a manufacturer and an operator may join their efforts in order to win the tender.

6 Part of this large size comes from the traffic flow generated by Minitel, the French videotex service, which is carried over by Transpac's network.

Transpac clearly follows a multinational (albeit regional) strategy. It may evolve after the big agreement announced between Deutsche Telekom and France Telekom, to provide end-to-end services to big customers. Part of the agreement may lead to the merger between FT's Transpac and Deutsche Telekom's Datex-P.

Public data networks are undergoing big transformations, due to the continuous evolution of the technologies and the increased needs of customers for large bandwidth. The whole issue turns around which standards to implement, which technology to adopt. It is clear that the bit rates delivered through the public X25 networks are insufficient to cater for the LAN to LAN interconnexion, for example. However, there is not yet a consensus among telecom operators, equipment manufacturers and users alike, on whether TCP/IP, extended X25 or Frame Relay might be better suited for the future needs. Basically, Frame Relay is a stripped-down version of X25, which allows a management of higher bandwidth. However, proponents of X25 say that the latter may deliver bit streams up to 2 Mbits/s with a better quality of service than Frame Relay. On the other hand, TCP/IP is a well established communication protocol, which provides managed services beyond the basic transport, protocol transparency, and network management (see Pisky, 1993).

France Telecom, in that context, is like other European TOs, lagging behind their US competitors, and the alliance with Deutsche Telekom may be seen as an effort to catch up. However, FT may be, among the European carriers, one of the most advanced in this area. A 256 kbits/s X25 service has been opened in France since April 1993, and Frame Relay interfaces to their X25 networks might be available. This may explain why FT has taken the bold step to invest in Europe through its subsidiary Transpac, and now with Datex-P.

At the border between multidomestic and global markets, there is the videotex service. Developed successfully nowhere but in France, the Minitel concept may not be exportable elsewhere. At the moment, there are several gateways between the Minitel network and other videotex services, but the only significant attempt to export Minitel has been the venture with US West in Colorado and Nebraska which has yet to prove commercially successful.

Conclusion

France Telecom is claiming that in the business area where it holds a first-hand expertise (telephone services, public data services, videotex) it is aiming at a position of main carrier, with or without majority control, (see Hirsch 1991). According to FT, the FDI in public network services represent approximately one third of its investments abroad. This figure may seem relatively small, given the size of its public network, and the many opportunities which have been opened with the privatization of many foreign telecommunications operators. But this business is rather new for FT, and despite the first successes, the company may still have to learn how to proceed for a successful bid.

Public telephone services and public data services require different know-how, technically and commercially. While public telephone services are in the realm of multidomestic activity where the relationship with local authorities is of foremost importance, public data services pertain to the kind of global service that a telecommunications company would like to offer to its big customers.

In multidomestic markets, France Telecom together with its partners has obtained agreements that the monopoly given to operation of the public network in Argentina is extended to 7 years (from 1991 onwards) and may last ten years. In Mexico, Telmex has a monopoly on local distribution networks for 35 years, while the long-distance network may be open to competition from 1997 onwards. In both cases, the preservation of the monopoly has been obtained through a lobbying of the bidders with the local authorities. In the same fashion, price regulation, hence the financial results of the operator, may be the outcome of a bargaining process between it and the regulator. So a good relationship with the local authorities may be at the heart of the competitiveness of FDI in public telephone services.

In public data networks, the know-how required is very different. Customers may be less sensitive to pricing, but much more to the quality of the service. Companies like Transpac have to struggle for quality of service, global (end-to-end) solutions for their customers, alert marketing. The speed of evolution of the customers' needs in those market segments may overwhelm the expertise accumulated by Transpac (even with Datex-P) over the last 15 years. Hence, in public data communications, an alertness to customers' needs, the evolution of de facto standards like TCP/IP, may be much more important than the accumulated experience in running X25 networks.

Basically, in the face of the mounting internationalization of carriers such as AT&T, the Bell Holding Companies and BT, France Telecom has favored an alliance with Deutsche Telekom. This alliance promises to lever the commercial strength of the second and the fourth telecom operators in the world, to create a nucleus of expertise inside Europe, able to respond to large customers' needs.[7] Also, the alliance strategy is often a prerequisite for business, in an industry characterized by the necessity of communication, i.e. interconnection, interoperability and the like. But it has still to prove its efficiency outside Germany and France.

3.2 Business areas where FT has no previous competitive advantage

In some market segments, FT is not at the forefront of technology and market development, for different reasons: lack of understanding of the market (ISDN), technical difficulties (mobile telephony), lack of demand in France (Frame Relay). Nonetheless, it has to invest in foreign countries, even in those segments, because the foreign opportunities seem appealing. Two market areas are relevant in that respect, mobile telephony and value-added services. In fact, each of these segments represents one third of FT's FDI, a level comparable with its investments in public network services.

Mobile telephony

The development of analogue cellular services has been particularly cumbersome in France. The national system, Radiocom 2000, was developed by Matra for trunk services, and subsequently adapted to full radiotelephony usage. Initially, the system had no roaming and no handover facilities, which have been added afterwards. This has resulted in an increase in the cost of the service. Also, the congestion problems in Paris, due to a lack of availability of frequencies, has prevented the market from taking off. The introduction of a competitor, SFR (Société Française de Radiodiffusion) in 1989, has not altered significantly the state of development of the service, although it has prompted FT to improve marketing strategy.[8] Radiocom 2000 has over 350,000 subscribers, while SFR has about 100,000. This is a hint of the

---

[7] The outcomes of this alliance are Eucom, the holding company taking participations in value-added service firms, and Eunetcom, the joint venture designed to provide facilities management to global companies, in competition with BT's Syncordia.

[8] SFR is a subsidiary of Cofira, a holding company, the main shareholders of which are: Compagnie Générale des Eaux (41%), Crédit Lyonnais, TDF, Bell South and Racal Vodafone, plus some minority shareholders (Coditel, a Belgian firm; Magneti Marelli, subsidiary of the Fiat group; Compagnie de Navigation Mixte).

difficulty that a foreign investor in France may encounter. The good brand image of FT has to be tackled either through very attractive tariffs and/or through a dynamic investment policy.

Things ought to have changed after the introduction of digital GSM systems in 1992. Up to now, however, the lack of good equipment and terminals provided by Alcatel to both rivals has not enabled the GSM to take off in France as much as in Germany. A broad consultation undertaken by the regulatory authority, Direction Générale à la Réglementation, for the introduction of new licences in Personal Communication services (PCN), may further increase the scope for FDI in France.

Despite the poor technical achievement of Radiocom 2000, FT has been very keen to penetrate some foreign markets. The argument put forward is that this market is likely to soar in the near future. In the year 2000, FT should represent the second market segment in telecommunications, with 20% of the business, against 70% for fixed telephony according to some (optimistic) forecasts, (see Hirsch, 1991) Hence, an operator who does not master this technology may be confined to a marginal position. So FT has competed to obtain licences to exploit cellular systems in several countries. It has been successful in Greece (together with Vodafone), Poland (together with Ameritech and the Polish Telecom Operator), India (in the city of Bombay), and, for paging, in Hungary and Czechoslovakia. Another licencing agreement has been signed by France Telecom's subsidiary EGT (Entreprise Générale des Télécommunications) with Martin Dawes, a British airtime reseller, in order to provide digital mobile services (Itineris) in neighbouring countries.

But once again, FT is not the only investor abroad, even in mobile telephony. The archrival Générale des Eaux has taken control of Talkland, a major airtime reseller in the UK. It is the biggest service provider which is not affiliated to either Cellnet or Vodafone companies, and distributes the service of the two competitors. This investment by Générale des Eaux may be perceived as a learning experiment; the company is gaining in the UK the knowledge of the marketing strategies which have made the British mobile telephony market so successful in the recent year. Afterwards, Générale des Eaux may then apply those techniques to its French subsidiary, SFR, in order to better compete with FT.

Unlike the ventures in public telephony, those in cellular telephony put France Telecom in an inferior position vis à vis its partners: in Greece for example, France Telecom is teaming with Vodafone, which has much larger experience in marketing mobile services. Vodafone's network in the UK is by far the largest in Europe, and the British company is particularly active in tailoring their service to the customers' needs, with a very appealing tariff strategy. Similarly, Ameritech may be in a better position to show its capability for cellular services, due to longer experience in Chicago. So we can expect France Telecom to be learning a lot in establishing these ventures in terms of operational experience, but not without being able to gain a financial profit comparable to that of the partner.

Value-added services

The market for value-added services is a kind of "Loch Ness Monster" that everybody claims to have seen at least once, but nobody agrees on how to describe its form, size and content, (see Stoetzer). France Telecom has undertaken to capture a sizeable part of this market in Europe, through its subsidiaries, Transpac, Télésystèmes and Eucom, the latter being a joint venture with Deutsche Telekom.

Télésystèmes has taken participations in ATEL (Spain), Datamont (Ferruzzi Group, Italy), IOP (Germany) and Intersoft (Tunisia). Télésystèmes is a computer service firm, which may overlap with France Telecom participation in the French company CAP-Sesa, the European number one in computer services, in which France Telecom has a minority interest, together with the German group Daimler through its subsidiary Debis.

The participations of Transpac in the UK, Germany, Sweden, Switzerland and Italy involve, besides network-based X25 services, other value-added services. In the UK, Transpac Network Services offers Facilities Management. In Germany, Info AG, the German subsidiary of FT, delivers electronic mail on a countrywide basis, back-up systems for big computer systems (for the banking and insurance sectors), as well as EDI (Electronic Data Interchange). In Sweden, outsourcing of private networks is provided for Swedish firms.

However, the most impressive actions undertaken by France Telecom refer to its joint holding company, with Deutsche Telecom, called Eucom. The latter has taken a 30% share in one major value-added service provider, GSI Transport Tourisme, which also has a subsidiary in Germany, called Danet. The motive behind such acquisition is that FT, like Deutsche Telekom, has not the expertise to offer directly specific value-added services to large customers, and that an external acquisition of know-how may be the safest way to get this expertise. In the same vein, FT has taken 30% of the French computer services company Sema Group, which makes 50% of its turnover in the UK.

Value-added services and mobile telephony are two areas where FT is not a technological leader. But FT had to invest in order to remain both competitive and "credible". To some extent, the strategy of FT in these two segments mirrors BT's: the latter has also invested abroad in mobile telephony (by acquiring 20% of McCaw Cellular, later sold to AT&T) and in value-added services (by acquiring Tymnet). However, while BT targets both the US and European markets, FT had their investments more scattered across Europe for either mobile telephony (Greece, Poland) or value-added services (UK, Germany, Spain, Sweden, Italy). Also, the amounts of money are not comparable: BT paid 110 million dollars (more than 600 million francs) for gaining 20% of McCaw Cellular and 355 million dollars cash (1.6 billion francs) for Tymnet. The two markets are very different, indeed. Value-added services are very much a fragmented market, where each segment possesses its own peculiarities: economies of scale and scope, asset specificity and brand image. Mobile telephony, on the other hand, is very much a homogeneous market, where sunk costs are very high, and therefore the commitment of the suppliers is also very important. It is not surprising that consortia are built up in order to share the burden of the sunk costs, to mobilize technical and market capacities so as to reduce the period of exposure to risks of a financial commitment in anticipation of receipts.

In reaction to these different structural conditions, France Telecom has adapted its strategy. In value-added services, FT is very keen to buy market shares abroad, in order to get a fair return on its know-how accumulated in-house. But in some market segments, it is much more important to have a customer base than to show technological capability. And because FT has not bought or taken a participation in market leaders, the risk is then not to get a fair return on its investment, because the client base is too narrow. For this reason FT has also committed itself to an investment in larger companies such as SEMA and GSI Transport/Tourisme (together with Deutsche Telekom). However, it is not sure whether this will be enough to compensate for the commercial deficit of a company more used to national customers and an administration-oriented rather than customer-oriented culture.

4. IS MARKETING THE KEY FOR SUCCESS OF FRANCE TELECOM'S FDI?

4.1 A lack of know-how in international markets

France Telecom is supposed to be lacking in know-how concerning international commercial issues. But this is the general situation of any TO which has been prepared to invest in foreign countries. Even AT&T and British Telecom, who, among the major TOs in the world, have a strategic vision of their multinationalization strategy, are lacking in ability to penetrate foreign markets, and trying to learn as quickly as possible.

Both BT and AT&T have attempted to cater for this absence of know-how through strategic alliances or acquisitions of local partners. BT has bought McDonnell's Tymshare to get a foothold in the US data networks, and McCaw Cellular for their mobile business, and recently MCI, to get access to the US long-distance telephone market. AT&T had signed agreements with Olivetti and Philips in 1983, to get an entry point into the European market. The success of both BT and AT&T has been limited, however. AT&T had to end its partnership with Olivetti in 1988 and Philips in 1991, while BT had to divest from McCaw in 1992. So the handicap at France Telecom is not so high compared to its competitors, but on the other hand, its efforts to fill up this gap will probably be harder than foreseen.

The difficulty may be compounded by several features, specific to the French context. First, France Telecom may be missing the good references which make a customer sure about the credibility of the supplier. This is particularly important for advanced services and big customers, where reputations may be a clue for choosing a particular supplier. It is so important in that respect, that France Telecom, only together with AT&T and BT, has won an order from General Electric for outsourcing their global private telecommunications network. But this example also shows that competition may take the form of cooperation, when necessary.

Also, France is not traditionally a country reputed for the ability to export or to invest abroad. The French are used to having a pattern of exports less specialized than countries like Germany or Japan, and to an export base limited to countries with lower growth potential, typically African countries, (see Turpin, 1989). In order to cope with this competitive disadvantage, the French firms have recently been very active in trying to invest in countries where the quality of demand, and the access to distribution channels, might be of foremost importance. Therefore, French companies have acquired firms in the US, the UK and Germany. But still, the level of FDI of French firms in the US is much smaller than that of the UK or Germany, although they are trying to fill this gap. In 1988, French firms bought US firms for an equivalent of 7.2 billion dollars, against 2.3 billion for German firms, (see Crespy, 1990, p. 63). And the trend has continued in 1989 and 1990, but has been halted by the general recession which has affected the world economy since the Gulf War. In any case, this traditional handicap of the French firms may also be apparent for France Telecom.

It is too early to see whether this recent attempt of French firms at multinationalization will be successful. However, this cultural deficit is also present in telecommunications services. The difficulty of multinationalization in this sector stems from the fact that TOs have to move quickly from a national market towards globalization, side-stepping as quickly as possible the transitional period of multinationalization.

4.2 Some recently acquired new trumps

France Telecom has been traditionally handicapped in the foreign markets where business takes precedence over diplomacy. This handicap is common to many French firms, as has been said above, but things may change quickly.

First of all, FT has been able to take advantage of the support of the French government when necessary. This, sometimes, may appear to be decisive, when a purely commercial approach is deemed insufficient. For example, when the bid for Telecom Argentina was made, the FT/STET proposal came second to Ameritech's. But the failure of Ameritech's bank, Manufacturers Hanover Trust, to be committed to its financial proposal, forced this offer to collapse. And, FT/STET was chosen as second best, because the French and Italian Governments could give the guarantee to France Telecom and STET to insure their financial commitment.

A second trump may be a possible closer cooperation with Alcatel. The French manufacturing group, which is now a world leader, has expressed, through its CEO, Mr. Suard, an interest in taking a participation in a privatized France Telecom. For years, Alcatel's business has been dependent on France Telecom's orders and R&D support. But after the acquisition of ITT's European subsidiaries in 1986, the French manufacturer had shown the will to "go global", and to rely less heavily on FT. The latter, on the other hand, has introduced a competitor to Alcatel, Matra Ericsson, to supply FT with central offices. Finally, the evolution of technology, in particular the emergence of the concept of Intelligent Networks, made it possible for Alcatel to address directly large end users, to offer them a private solution with Alcatel's equipment, to some extent bypassing France Telecom's public network. Hence, Alcatel and FT were becoming more and more competitors, and less and less vertically related.

However, Alcatel is changing its mood, because it probably recognizes that the manufacturing sector is much more competitive, hence less profitable, than the service sector. Thus, getting a foothold in France Telecom would probably guarantee a certain level of profits. If the links of FT and Alcatel become closer, then a formidable competitor in the global market may appear, comparable to AT&T but with a less risky business overall, because less subject to competitive pressures.

Another resource for change in FT's commercial position is its reputation in running foreign networks and investing abroad. The first moves of FT have been perceived, up to now, as rather successful but limited in scope. France Telecom could take stock of this recent reputation to increase its participations abroad.

To conclude, France Telecom still has commercial handicaps, but could target markets where these handicaps are less felt. Simultaneously France Telecom could try to set up joint ventures and find partners with complementary assets (like Vodafone in Greece) to overcome these handicaps.

4.3 Conclusions

France is reputed to be a dirigist country, with a natural propensity to protectionism, national champions and a weakness in international trade. At a time when international competitiveness is of foremost importance in many industries, these long-lasting handicaps may be present in telecommunications services too. France is also known for mastering the technologies where the intervention of the State is decisive: aircraft and space industries, weapons, nuclear power plants and, until recently, telecommunications. With the retreat of the State from the telecommunications networks operation, one may wonder whether French firms in this sector, and in particular France Telecom, will be able to cope with the increasing globalization of the telecommunications markets.

The recent efforts of France Telecom to go abroad bring the first answers to this question. First, the ventures of the French carrier are rather limited in scope, but comparable to those of its major competitors. At the corporate level, there is a clear understanding of the stakes raised by the internationalization of telecommunications markets. Therefore, FT has been able to seize its chance when it had to.

However, it seems that the decision-making process inside France Telecom, which leads the company to internationalize its activity, reflects the way it is managed. This may lend an original touch compared to the competitors. Basically, France Telecom is run by an élite of engineers, who have graduated from the same schools, the so-called "Grandes Écoles". These clever people are quickly learning the global game of international competition. When faced with a new opportunity to invest abroad, they will analyse carefully the pros and cons of the venture, evaluate the financial, but more often the technical, risks of the investment. This analysis may take a long time, but once the decision has been taken to go ahead, they pursue their aim very consistently and may be able to win in the end, despite initial difficulties. In both the Mexican and the Argentinian bids, the proposal of France Telecom did not come out as the first best. Nonetheless, they won thanks to their tenacity and their ability to bear high financial risks. This attitude typically reflects the management style of those "enlightened" engineers who are aware of the strategic difference between this multinationalization attitude and that of, say, BT's management, which is more concerned with the evaluation of risk and financial returns.

BT is responsible for short-run profits to satisfy its shareholders. Therefore, it tries to be very alert, to select more carefully its customers in order to serve in priority those who are ready to pay the most. The globalization of BT then results, because these customers want a global offer: the Syncordia project simply reflects their will to be a carrier in the US, and if possible in Japan or Southeast Asia as well as in Europe. France Telecom, instead, is very much interested in multinationalization *per se*. It thinks that it has accumulated specific knowledge of its basic activity, the operation of a telecommunications system. Its management is ready to sell this expertise abroad, because as engineers, they consider that this experience is valuable.

Furthermore, what makes these people more reactive than, say, the Germans, is their ability to combine economic and technical issues. What they have learnt during the so-called "catching-up" period is that a good commercial offer always has to complement a good technical infrastructure. In fact, it is their ability to serve the residential customer which has been praised during the years 1975-1982. The Germans, instead, have long lived on the vision of the public monopoly, technically perfect and acting as a bureaucracy of the all-mighty State. Now the Germans have changed their mood after the Post Reform of 1990, but plagued with the problems of reunification, they are not able to devote financial resources to adventurous activities in the Third World countries.

Another interesting feature of FT's multinationalization strategy is its limited ability to invest in Europe. At the moment, the regulation of telecommunications activity does not allow for cross-country investment; only in the UK do the European TOs have a chance to get market opportunities. However, they have not seized these opportunities, which have been left to American Bell Operating Companies and AT&T. The reason for this lies probably in latent fear of retaliation. If France Telecom were to invest in, say, Mercury, BT would press its government to get a symmetric position in France. Therefore, the tacit gentlemen's agreement between European TOs, which may be the heir of the late CEPT, is beneficial to outsiders like the Bell Holding companies, and detrimental to the Europeans.

Also, the context of France Telecom's investment strategy may be changed if the French TO gets privatized. Many options are possible in that perspective, and the choice is still open: closer relationship with Alcatel, or exchange of shares with Deutsche Telekom if the latter also gets privatized, or exchange of shares with smaller TOs like PTT Netherlands or Televerket, which would bring some net cash to FT, given the difference of size between FT and the partner. Suppose for example that FT exchanged 5% of shares with PTT Netherlands. Then, given the latter's smaller size, they would have to compensate with some financial contribution, the difference of value between the two participations.

In any case, it is difficult at the moment to detect a well-defined globalization strategy. France Telecom is, at best, adopting a multidomestic strategy, seizing opportunities when they happen. It remains to be seen whether FT will be exporting its know-how in industrial countries, as quickly as it had been able to export its know-how of public network operations to developing countries.

REFERENCES

Crespy, G. ed., 1990, Marché unique, marchés multiples (Economica, Paris).

Hirsch M., 1991a, La course autour du monde, Revue Française des Télécommunications, no. 77, May, pp. 37-38.

Hirsch M., 1991b, Projets stratégiques et réalités tactiques, Revue Française des Télécommunications, no. 77, May 1991, p. 40.

Pisky, D., 1993, So close yet so far, Communications Week International, 15 February, pp. 14-15.

Stoetzer, New telecommunications services in Europe: Which factors influence their adoption?, Communications et Strategie no. 8, quatrième trimestre, pp. 13-32.

Turpin E., 1989, Le commerce extérieur français: une spécialisation industrielle fragile, Economie et Statistique, février.

Wall Street Journal, 19 February, 1992.

The Race to European Eminence
E. Bohlin and O. Granstrand (Editors)

The International Economics of Telecommunications Services in Italy

Cristiano Antonelli[a] and Aldo Geuna[b]

[a]Laboratorio di economia dell'innovazione, Dipartimento di economia, Universita' di Torino, Via Po' 53, 10124 Torino, Italy
[b]Aldo Geuna, MERIT, University of Limburg, P.O. Box 616, 6200 MD Maastricht, The Netherlands

A comprehensive analysis of the Italian case would require more than a few pages. The Italian telecommunications system is undergoing a complex process of change. The analysis of the reorganization of Italian telecommunications represents a fascinating case where a variety of different issues are intertwined. In order to understand basic constraints that feature the behaviour and consequently the possible choices a very selective approach is necessary. It seems that attention should be focused on four aspects: (i) the historic background that helps in understanding the complexity and variety of interests within Italian telecommunications; (ii) the vertical integration relationship within STET between SIP and ITALTEL; (iii) the effects of new information technologies on the demand for telecommunications services and their implications for the Italian case; (iv) the extremely fast process of internationalization that the telecommunications services industry has recently shown.

1. INTRODUCTION

The revolutionary technological change introduced in the telecommunications services sector since the mid-1970s, together with the processes of liberalization and privatization, are drastically altering the functioning of a market characterized by a strong multilateral oligopoly. Therefore, the scope for international growth of domestic telecommunications services firms is emerging as the critical issue of international telecommunications. The most appropriate way to understand such dynamics seems to be first a careful reconstruction of the process that led to the new emerging situation, and secondly the analysis of the international strategies followed by the leading telecommunications carriers in the last ten years.

In so doing, a special effort must be made to apply the tools of the economics of multinational firms and international trade to the new emerging economics of international telecommunications services.

The main goal of the paper is thus to apply the international economics of telecommunications services to the Italian case, giving special attention to outward and inward direct investments.

In the first part (sections 2.4) we shall analyze the historical background and the main idiosyncratic features of the Italian telecommunications sector. In the second part (sections 5.6), we shall give a brief presentation of the evolving institutional framework and of the

Italian telecommunications services market; then we shall discuss size, orientation and strategy of outward direct investments of the Italian carrier and of inward foreign direct investments in Italy.

2. THE HISTORICAL BACKGROUND

The Italian telecommunications services were inaugurated in 1881. In the period 1881–1925 the system grew rapidly with the entry of many new operators. In 1925 there were 73 telephone companies operating on a regional basis, authorized by the Ministry for Posts and Telecommunications. Trunk communication, both domestic and international, was provided by ASST (Azienda di Stato per i Servizi Telefonici). The 73 companies operated under a "Convention" with the Ministry, which regulated the service, defined the access charges for the trunk service with respect to local lines, and authorized all changes in tariffs.

In 1925 the original "Conventions" expired and the system was reorganized into five regional companies: STIPEL, Societa' Telefonica Interregionale Piemontese e Lombarda; TELVE, Societa' Telefonica delle Venezie; TIMO, Telefoni Italia Medio Orientale; TETI, Societa' Telefonica Tirrena; SET, Societa' Esercizi Telefonici. TETI and SET were owned by "Le Centrale", a large holding company diversified into a variety of financial services that had strong relations with Ericsson and ITT. STIPEL, TELVE and SET were owned by SIP, a large electric company based in Torino. SIP went bankrupt in 1933 with the collapse of the Italian financial system.[1] Together with many other corporations characterized by high levels of capital intensity and financial illiquidity, SIP was rescued by IRI which retained direct control of all the telephone activities, while SIP specialized in the electrical core business. IRI kept a strong minority interest in SIP which still remained by all means an independent company whose shares were traded on the Stock Exchange.

In 1925, the new convention with the five regional companies established that each was responsible for the local traffic within the region and for large parts of the trunk service within its regions. ITALCABLE was given the monopoly for all extra-European communications. ASST was responsible for all inter-regional traffic, for important parts of the trunk service within the regions, and for the European international service.

In 1933, STET was created as a subholding within the IRI group to manage the three telephone companies STIPEL, TELVE and TIMO. In 1950 STET acquired control of SIT-SIEMENS, the former German SIEMENS manufacturing branch for telecommunications switching and transmission in Italy. Eventually, also ITALCABLE, originally created to operate the transatlantic cables to Latin America, entered the STET group, yet retaining the character of an independent company with shares currently traded on the Stock Exchange. The Italian system in 1933 was, after all, very close to the present one in the USA.

In 1957 the thirty-year-old convention was renewed. As a consequence La Centrale was obliged to sell TETI and SET to STET. The new convention in fact established that all telephone companies should be controlled (not necessarily fully owned) by the State.

[1] The three largest Italian banks, Banca Commerciale Italiana, Credito Italiano and Banco di Roma, went bankrupt because of the high levels of long-term financial credits to the main Italian manufacturing groups. In order to avoid the generalized collapse the Italian Government created the public holding IRI, where both the banks and the manufacturing companies indirectly controlled were concentrated. Since then IRI became the strong arm of the Italian industrial policy.

As in 1963 the electrical industry was nationalized, SIP was obliged to sell all its electrical infrastructure to ENEL. TETI, STIPEL, TIMO, SET and TELVE, all fully owned by IRI, merged into the rich and fully liquid SIP which consequently became a majority-owned company within the STET group.

In 1964 a new convention was signed under which SIP received the responsibility for the design of the national telecommunications system. ASST, however, did retain full control of the operations of the trunk system.

This structure did not basically change until 1992, when ASST was "transferred" to IRI-STET[2] and FINSIEL was sold by IRI to STET.[3]

3. THE VERTICAL AND HORIZONTAL INTEGRATION

Guglielmo Reiss Romoli, the chairman of the Group for more than thirty years, shaped STET, closely following AT&T. STET remains today the telecommunications company with the highest levels of vertical and horizontal integration worldwide. In fact, besides the four telecommunications services companies (SIP, ITALCABLE, SIRM for marine radiocommunications and TELESPAZIO for satellite telecommunications) STET owns:

- *ITALTEL* (the old SIT-SIEMENS) specializing in manufacturing switching and transmission equipment as well as terminals of all kinds;
- *SIRTI*, specializing in the installation of cables and switching centers;
- *ILTE*, a printing company specializing in telephone directories;
- *SEAT*, originally the unit for collecting information about directories, merged into STET a few years ago, but still very active, as a division within the group in collecting advertisement;
- *FINSIEL*, the largest software company in Italy (i.e. an Italian CAP-GEMINI, but specializing in State-owned corporations and local as well as central government agencies).

Research and development activities are mainly carried out by CSELT which is incorporated as an independent company within the group.

The cohesion of the group was clear at the time Reiss Romoli built the group in the thirties and later in the fifties. The dynamics of demand and especially of technology in the recent years, however, put a lot of strain on the original vision. The change from electromechanical into electronics technology in manufacturing switching and transmission equipment has been a source of major problems as well as great opportunities.

---

[2] The word "transfer" here refers to the lack of a clear legal definition of the terms of the contract. In fact, ASST, now renamed IRITEL, remains a State company whose management has been assigned to STET.

[3] FINSIEL is the leading Italian company for software. FINSIEL specializes in software for large corporations, mainly public: State ministries, defence, large state-owned retirement funds and corporations.

The manufacturing of telecommunications equipment had been traditionally split among the five companies operating in the country:

- *ITALTEL*, the STET company;
- *TELETTRA*, of the FIAT group;
- *FACE-STANDARD*, of ITT;
- *FATME*, of the Ericsson group;
- *GTE-Telecomunicazioni* of the GTE group.

TELETTRA specialized in transmission, ITALTEL mainly in switching and terminals as well as GTE, FATME and FACE-STANDARD. The procurement of SIP and ASST, technically advised by SIP, was evenly shared among the five companies. Italy (not Gallia) was divided in four parts (not in partes tres): the East was equipped with Ericsson machinery, the South with ITT, the Center with GTE and finally the West and large parts of the Center with ITALTEL. Traditionally the switching equipment was shared between ITALTEL with 50% of the total, and the three multinationals with 15% each. Transmission equipment was split between Telettra (80%) and ITALTEL.

Telecommunications equipment, in the electromechanical age, had high levels of durability and consequently high requirements in terms of compatibility among different vintages of capital equipment. Consequently the original technical choices made as early as in the thirties (somebody even claims before) had long-lasting effects on eventual technical choices. Hence the choice of Ericsson machinery as well as GTE or FACE-STANDARD created a "lock-in" effect. The large employment figures of the three subsidiaries, wisely concentrated in the south and near Rome, were a second strong argument in favour of the long-lasting division of labour.

STET never quite accepted the division of labour argument, and over time a growing share of the huge SIP procurement shifted from ITT and FATME towards ITALTEL. A gentlemen's agreement between ITALTEL and Telettra was part of the attempt to reduce over time the share of the three foreign subsidiaries. At one point a joint venture had been created with GTE as a step in a strategy of further reduction of the influence of ITT and Ericsson. The shake-up in world telecommunications equipment in the eighties with the collapse of ITT and GTE, purchased respectively by Alcatel and Siemens, reopened the debate further enriched through the acquisition of Telettra by Alcatel.

Technological change also created the opportunity for a discontinuity in the technical lock-in, with respect to the different "makes" of equipment. The temptation for STET to rediscuss dramatically the "division of labour" became all the stronger. The integration of ASST within STET is likely to make the contrast even sharper.

A second part of the story, however, needs to be told. The transition to digital switches makes it possible to create finally a technically homogeneous network and to achieve a full integration between SIP and ASST. This is basically the essence of a story which lasted almost twenty years. The first attempts to manufacture a digital switch within ITALTEL date back to the early seventies. Eventually ITALTEL managed to manufacture UT 100—which is jokingly spelled, in Italian, Ultimo Tentativo 100 (Last Attempt)—with a delay of more than ten years.

In the same years the demand for telecommunications services and new lines exploded in Italy with the consequence that almost all the South has been reequipped with ITALTEL electromechanical equipment. The pressure on SIP to provide the new lines became in fact unbearable, as did the need to support ITALTEL financially and commercially while it was finally trying to deliver a UT. This is a story about the limits and the dangers of vertical integration that can be further enriched by the next steps.

At the end of the eighties ITALTEL finally could deliver a UT 100 system. However, in order to be economically profitable ITALTEL needs to sell more than 1 million lines per year, more than SIP, and ASST now want to purchase also buckeyes of the large stock of technically but not economically obsolete electromechanical switches installed in the South until the late eighties. The joint venture with AT&T, with the entry of AT&T into the equity of ITALTEL (20% of the shares) and the entry of ITALTEL into the equity of AT&T-NSI (the original joint venture between AT&T and Phillips) with 20% of the shares, is the outcome of the attempt to commercialize UT internationally and to obtain the AT&T technology in order to integrate the UT technology in a full network system. The asymmetric character of the joint venture is well highlighted by the direct entry of AT&T into the equity of ITALTEL, contrasted by the entry of ITALTEL into a minor affiliate of AT&T in Europe. A similar role would be played by the soft credits to Russia for three-four million lines in Saint Petersburg and the rescue of Nortel in Argentina.

4. DATA COMMUNICATIONS

Data communications are emerging as the lion's share of telecommunications services demand. New information technologies make data telecommunications essential to large and small corporations, to networking and to effectively operating large bureaucracies worldwide as well as on the national level.

The demand by data-networking companies has consequently been growing very fast all through the eighties, at a time when SIP was crippled by the vertical integration ties with ITALTEL and its problems with UTs. The creation of a national data communications infrastructure was delayed, still awaiting the famous UT and ISDN.

In order to face the rising concern of large corporations SIP, also to raise extra-funds, was generous in providing leased lines to everybody. The business of leasing lines quickly became a major source of income for SIP. It seems that almost 4% of total sales today are accounted for by the revenue from leasing. Leased lines became the infrastructure of a variety of private networks that are operated with a comprehensive understanding of the notion of "third-party". (Any FIAT subcontractor, with no legal ties to any FIAT company, can today enter the FIAT network from anywhere and go everywhere worldwide.)

The economics of private networks built upon leased lines are the basic economics of fixed costs and consequently the economics of marginal costs with a negative slope. This amounts to saying that, once the large private networks have been created, there is a huge incentive to widen the scope of application and eventually to create new telecommunications service companies: the marginal costs of additional calls in fact are necessarily much lower than with any commutated public service.

In the Italian case large corporations had already established their own networks as soon as the requirements of break-even quantities for voice communication were satisfied: that is, the private networks built upon leased lines were created as soon as the total costs of leased lines became lower than the costs for commutated services. The need for technological sophistication of the service in terms of data communication was perceived later.

Large corporations had the opportunity to rapidly improve the quality of the private networks already in place in order to reap the advantages of data communications. At the end of the eighties all the large manufacturing groups as well as banks, retailing and insurance companies had efficient private networks built on leased lines. Eventually the lack of a public data communication network forced many medium-sized firms to follow the large corporations and create their own data networks using leased lines.

The final arrival of ITAPAC and the RETE-FONIA-DATI[4] in the late eighties, that is, the supply of data communications services on public specialized networks, found a demand largely satisfied by the "hidden" supply of the private networks.

The recent acquisition of FINSIEL by STET can significantly change the situation. FINSIEL is a leader in the software for large multiplant corporations and masters the technology of data networking based on leased lines. As of January 1, 1993, the national regulation for data communication expires; consequently SIP might legally and substantially favour FINSIEL as a large customer of leased lines both directly and indirectly (when the actual customer is a customer of FINSIEL) both in terms of prices and access conditions (for instance CENTREX might be authorized for FINSIEL usage). Consequently FINSIEL, that is STET, might gain in the marketplace the sales of data communication it lost to private networks.

5. THE NEW INSTITUTIONAL FRAMEWORK

In Italy, until 1992, the Ministry for Posts and Telecommunications was responsible for the management and control of telecommunications services and network. The Ministry was directly involved in the management of E-mail and telegraph services. Moreover, through ASST, it was responsible for the interregional traffic and for the traffic and connections with European and North African countries. The production of other services had been exclusively licensed to companies of the state holding STET. The control over the activities was exercised through two regulatory bodies, the National Institute for Posts and Telecommunications and the General Inspectorate for Posts and Telecommunications, both subordinated to the Ministry for Posts and Telecommunications.

The relevant anomalies of the Italian system were:

1. the coexistence within the same entity (the Ministry for Posts and Telecommunications) of management and control functions;
2. the overlapping of the different activities carried out by SIP, ITALCABLE and ASST; this chaotic "division of labour" has caused great inefficiency.

[4] ITAPAC is a package switching network; FONIA-DATI is a circuit switching network. Approximately, for low traffic and short distances it is more convenient to use FONIA-DATI; when the distances are growing it becomes more economical to use ITAPAC. Finally, for high traffic and long distances it is always cheaper to use leased lines.

With the law number 58 of January 1992 the Italian government has defined a programme for the restructuring of the telecommunications sector. The Ministry is now responsible only for the control and regulation of the production of the various services. The services that were previously managed by the Ministry, directly or through ASST, are now carried out by the IRITEL. This institutional framework (see Figure 5.1), first step of the restructuring process, is finally able to overcome one fundamental anomaly of the Italian system. Now, in fact, the control and regulation are completely separated from the management.

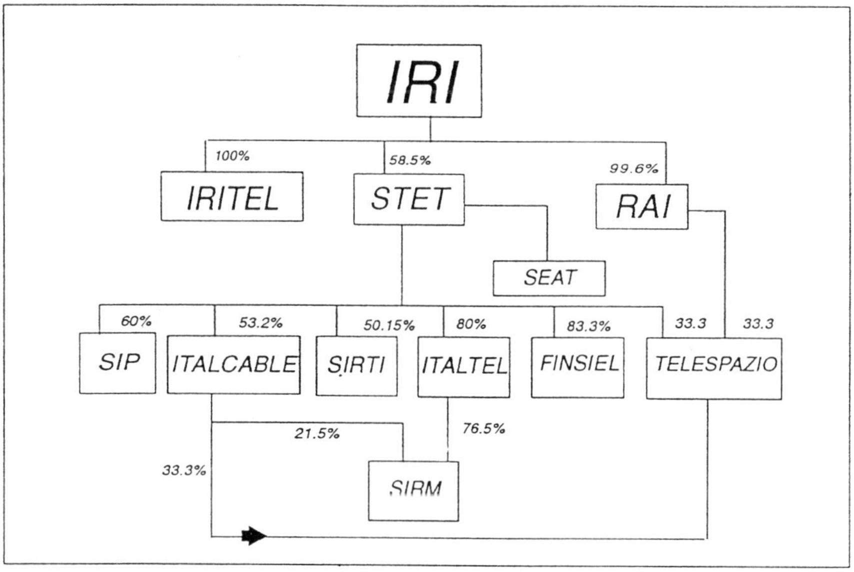

Shares at 31 March 1993

Figure 5.1. The institutional framework.

With the decision enforced by CIPE in April 1993 and the following "plan for the restructuring of the telecommunications sector" carried out by IRI in June 1993, the second anomaly will also be solved. The project stipulates the creation of only one big carrier, *Telecom Italia,* that takes its origin from the merging of all the licensees[5] controlled by IRI. Telecom Italia will be structured in different businesses since, at a further time, there will be the spin-off of the businesses that are managed in a competition regime[6]. In Figure 5.2 we have shown the possible future institutional framework.

Moreover, it is worthwhile mentioning the CIPE's recommendation that stresses the importance of the separation between the manufacturing activities and the management of the services. We think that this points to the future creation of an autonomous company charged with the management of all the manufacturing activities.

[5] Iritel, SIP, ITALCABLE, TELESPAZIO and SIRM.

[6] See for example the case of mobile telecommunications.

For more than a quarter of a century the structure of the Italian telecommunications sector had not changed; then in a few years a restructuring process has given a completely new shape to this sector. The main driving forces are: technological change, the opening of telecommunications services to a competition regime, and the globalization of the market[7]. In the next section we try to analyze the interaction between the competitive situation and the process of internationalization.

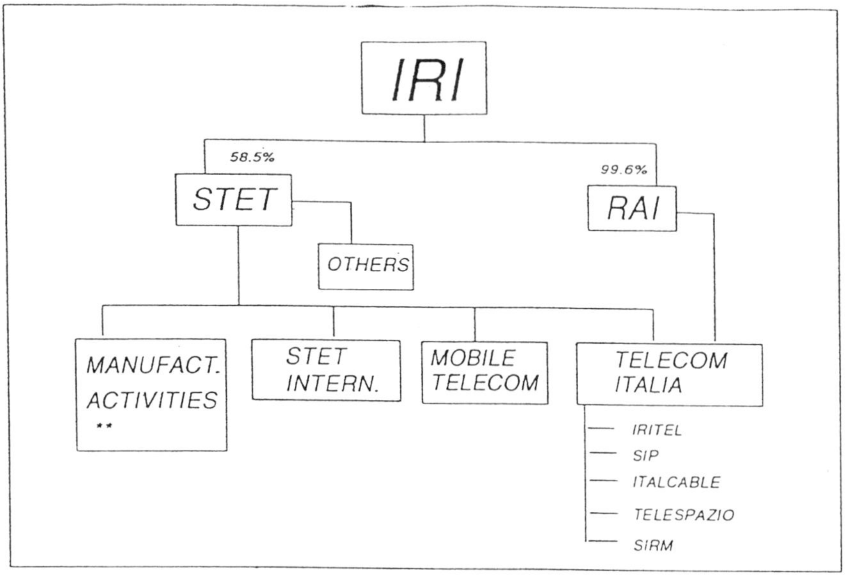

** It still does not exist but its creation is quite probable.

Figure 5.2. The future institutional framework.

6. THE PROCESS OF INTERNATIONALIZATION: INWARD AND OUTWARD DIRECT INVESTMENTS

6.1 The Italian market for telecommunications services

The Italian market for telecommunications services has reached a considerable dimension, following a more than satisfactory trend of growth (see Table 6.1). In the years from 1970 to 1988 the state-owned carriers have undergone a dramatic shortening of their profits[8] and a consequent restriction on their investment programmes. There were three main consequences. First, a serious slowing down in the expansion of capacity; secondly, a technological gap with respect to other European countries; and finally, a system of cross-subsidies which heavily

[7] It is important to stress again the interrelatedness of these factors. In fact, we can state that they originate in each other.

[8] Due to the regime of subsidiarity under which the service was offered.

penalized firms and business users on behalf of families. Since 1988 the national carriers have been involved in an important effort aimed at reducing the gap through a deep process of modernization and upgrading of the infrastructure. The investment in the network has almost doubled in four years, passing from 5,068 billion lire in 1988 to 10,939 in 1991.

The success of this growth of investments is evidenced by the improvement in most quality indicators. The increase in subscribers, net of a large number of contract terminations (+30.8% on 1990), came to 721,000 bringing total subscribers to 23,071,000 at the end of 1991.[9] This produces a telephone density of 39.9 subscribers per 100 inhabitants (41.5 in the first half of 1992) which is close to the European average. The degree of digitalization of the network (40.8% in 1991) is now comparable with that in the United Kingdom and slightly smaller than in France.

Of particular interest is the mobile telephone service, which has experienced a booming growth, with Italy now representing the second largest market in Europe after the United Kingdom. SIP is still the only licensee for the service, even though a second licence should be granted before the end of 1993. In the competition for the second licence are four consortia, all characterized by the presence of foreign operators.[10]

Finally, Italy is lagging well behind in data transmission and in VAN services. If we make, for example, a comparison with France, the leader in Europe in this field, the results are very discouraging. In Italy at the end of 1990 there were 285,601 PARD whereas in France there were 647,480 PARD.[11]

In Table 6.1 we have shown the data concerning the performances in the telecommunications services industry of the STET group in the period 1990–1992. As almost 99% of the Italian market is controlled by the state-owned group STET (Boschi & Mutinelli, 1993)[12] we can consider these data a good approximation for the Italian market.

Table 6.1
Telecommunications services. Main indicators

	1990	1991	1992
Sales	17,598	20,289	22,716
Capital expenditure	9,901	10,939	9,703
Workforce numbers	91,643	93,792	94,007
Telephone subscribers	22,350,000	23,071,000	23,430,000[c]
Telephone density	38.7	39.9	41.5[c]
Portable telephone	266,000	568,000	691,000*[c]

Source: STET Reports 1991, 1992.
Value in billion Lire.
[c] First of half 1992.
* In June 1993 there were one million subscribers.

[9] At the end of the first half of 1992 the total subscribers were 23,430,000.
[10] Omnitel: Olivetti (51%), Bell Atlantic (16.6%), Cellular Communications (14.7%), Swedish Telecom (9.7%), Shearson-Lehman (8%). Unitel: Fiat (40%), Fininvest (40%), Racal (20%). Pronto Italia: Pronto Italia Holding (60%), Pacific Telesis (40%). Mti: Telefin (n.a.), Millikom-Kinnevik (n.a.). Source: Boschi & Mutinelli (1993).
[11] For an interesting analysis of the reasons for this lagging behind, see Boschi & Mutinelli (1993).
[12] In the 99% is also included the share carried out by IRITEL.

6.2 The internationalization process

From a dynamic evolutionary point of view, exports on international markets are not only the main but also the first form of international growth. According to the evolutionary theory of international growth of the firm, in fact, the second step in exploiting international competitive advantage for the firm is the multinational growth that takes place by means of direct entry into other national markets and foreign direct investments (Dunning, 1981).

Multinational growth of telecommunications carriers since the mid-eighties has increased at an astonishing pace: the total of main acquisitions of foreign telecommunications companies, as calculated by the U.N. Programme on Transnational Corporations (UNCTC, 1993), has increased twenty-fold from 399 million U.S. dollars in 1985 to 16,539 million U.S. dollars in 1990. Multinational growth of telecommunications services firms has received a strong push since the late eighties from four factors (Antonelli, 1993b):

- privatization of former state-owned companies;
- liberalization, with consequent growing vertical disintegration of the telecommunications affiliates and product specialization;
- growing international tariff asymmetries with new opportunities for specialized entry;
- globalization of multinational firms and their growing reliance on outsourcing procedures.

International carriers have responded to the new opportunities with different strategies of international growth. There are three main groups. First, the telecommunications services firms from France, Italy and Spain that have acquired important stakes in the national telecommunications services industry as a whole in such countries as Argentina, Chile, Venezuela, Mexico, Poland, Hungary and Ukraine through the process of internationalization *via multidomestic strategies.*[13] The primary factor of international competitive advantage is the technological capability that makes it possible to update technically obsolete networks, formerly managed by State agencies unable to meet the new requirements in terms of technical competence and heavy investments necessary to adopt the new digital technologies. Such a model of international growth, based upon foreign direct investments in either the general-purpose telecommunication network or specialized segments such as the mobile telephone, is likely to lead to the transplantation into the telecommunications services industry of the multidomestic firm (Antonelli, 1993).

Secondly, the telecommunications carriers, mainly from the United Kingdom and USA, that have developed a process of internationalization *via services* (Bianchi, 1991). These telecommunications services firms have been the first to experience the competitive regime which follows the deregulation. They can benefit from the experience acquired by operating on competitive markets, from the lower levels of international telecommunications tariffs in the home country, and from specific advantages in terms of technological and commercial know-how. Therefore, they have focused mostly on the supply of applied, high technology and integrated services, to big customers like multinational corporations. Finally, it seems that some of the process of internationalization can be interpreted by the *cooperative networking model* based on a variety of non-equity agreements such as joint ventures and commercial

[13] A very similar kind of approach is the one *via network*. For its definition see A. Bianchi (1991).

alliances, where each partner swaps the access to its own domestic network and to its own installed basis of international companies.

6.3 Outward direct investments

Focusing attention on the Italian outward direct investments, we can highlight how the STET group is following a process of internationalization via multidomestic strategies. Starting from 1990, STET has proved to be one of the most active carriers on the international scene, taking over national telephone networks or acquiring licences (see Table 6.2). This process of internationalization "via network" relies on great expertise in managing national networks that are going to privatize, but still under a regulated regime. The primary factor of international competitive advantage, the technological capability, together with the strength in the integrated management result from the structure of the group—historical involvement in the production of all the base services and vertical integration with producers of telecommunications equipment.

Particularly important is the case of Argentina and the other South American countries. In fact, the extremely high and successful participation in the supply of telecommunications services can also be seen as a consequence of the historical linkages that STET, through ITALCABLE, has already developed with these countries.

An interesting case is also that of the mobile telephone. SIP developed this service long after the other European carriers. This situation, instead of being a disadvantage, turned out to be a strong advantage. In fact, SIP has been able to learn from the others' experiences and then to adopt innovative procedures. The specific know-how that it has developed in this way has enabled SIP to win the competition for the second licence for the mobile telephone in Greece.

Finally, it is worthwhile mentioning that a vertical disintegration is taking place with the withdrawal of the STET group from the equipment manufacturing industry. Selling its controlled operating companies, STET and its owner, the Italian government, have accepted EEC regulations about the liberalization of the telecommunication equipment industry. In so doing STET is also free to better focus its efforts and strategies on the growing market of telecommunications services.[14] The first step in this direction has been the creation in 1992 of the company STET International (51% STET, 25% SIP, 15% ITALCABLE, 9% TELESPAZIO) in charge of the management and coordination of all the foreign activities of the group.

As Table 6.2 illustrates, before 1990 the foreign activities of STET were restricted to participation in the three international consortia for satellite communications and to ITALCABLE USA, a commercial representative in the US. Then, from 1990 the process of internationalization has experienced a dramatic growth.

In November 1990, the holding Nortel Inversora (STET 32.5%, France Télécom 32.5%, Perez Companc 25%, J.P. Morgan 10%) acquired 60% of *Telecom Argentina*, the telecommunications services carrier for the north part of Argentina—in this area is included Buenos Aires. Telecom Argentina owns the 50% of *Telintar*, licensee for international telecommunications services, telex and telegraph; 50% of *Startel*, licensee for value-added services; and, from 1992, 50% of *Movistar*, licensee for the second mobile telecommunication licence in the area of Buenos Aires. Moreover, in 1990, the STET group through ITALCABLE bought

[14] The disintegration of the manufacturing capacity will enable STET to increase the financial capability necessary to enter foreign markets.

24.26% of *Impsat*, licensee for the satellite communications—VSAT technology—and for the services of the private digital network in optical fibres.

Since 1991 the STET group has also been in Brazil. ITALCABLE and TELESPAZIO each own 12.5% of *Victori Comunicacoes*, a company for development of telecommunications services with particular attention to VAS. TELESPAZIO also owns 49% of *Telespaco* which carries out activities of telemonitoring and telemeasuring. Moreover, from 1992 ITALCABLE possesses 99.99% of *ITALCABLE do Brazil*, the commercial representative of ITALCABLE for the area.

In 1991, ITALCABLE bought 35% of *VTR*, a company that manages telex and data transmission services in Chile.

Coming to the European countries, there are the two already mentioned participations in satellite communications consortia; one in France, *Consortium Eutelsat,* and one in UK, *Consortium Inmarsat.* Furthermore, in 1992 ITALCABLE started in London the company *TMI Telemedia UK* which offers value-added services.

Table 6.2a
International investments of the holding Stet (End 1992)
Telecommunication services

Company	Country	Year	Company	Share	Entry areas
Nortel Inversora*	Argentina	1990	Stet	32.5%	Telecom. Holding
Telecom Argentina**	Argentina	1990	Nortel	60%	Telecom. Services
Telintar	Argentina	1990	Telecom Argentina	50%	International Telecom. Telex
Startel	Argentina	1990	Telecom Argentina	50%	VAS
Movistar	Argentina	1992	Telecom Argentina	50%	Mobile communications
Impsat	Argentina	1990	Italcable	24.26%	Satellite communications
Victori Comm.	Brazil	1991	Italcable Telespazio	12.5% 12.5%	Telecom. Services VAS
Telespaco	Brazil	1991	Telespazio	49%	Telemonitoring
Italcable Do Brazil	Brazil	1992	Italcable	99.99%	Commercialization
VTR	Chile	1991	Italcable	35%	Data services Telex
Gulf Telespazio	Bahrain	1992	Telespazio	33%	Telemonitoring

Sources: STET Reports.
* Profit 80 billion Lire in 1992.
** Sales US $1.25 billion, profit US $150 million in 1992.

Table 6.2b
International investments of the holding Stet (End 1992)
Telecommunication services

Company	Country	Year	Company	Share	Entry areas
Consortium Eutelsat	France	<1991	Telespazio	8.11%	Satellite communications
Consortium Inmarsat	United Kingdom	<1991	Telespazio	2.74%	Satellite communications
TMI Telemedia	United Kingdom	1992	Italcable	100%	VAS
STET Hellas Telecom.	Greece	1992	Stet International	74.8%	Mobile communications
Astelit	Russia	1991	Italcable Telespazio	25% 25%	Satellite communications business user
Intelcom	San Marino	1992	Stet International	70%	Telecom. services
Italcable USA	US	1979	Italcable	100%	Commercialization
IPSP	US	1991	Stet	8.33%	Satellite communications
Consortium Intelsat	US	<1991	Telespazio	2.47%	Satellite communications
TMI US	US	1990	Italcable USA	100%	VAS

Source: STET Reports.

Much more important is the licence obtained in Greece for the second mobile telephone. *STET Hellas Telecommunications*, owned 74.8% by Stet International, manages the mobile telecommunications service GSM in the whole country from 1992 onward.

In 1992, STET International also bought 70% of *Intelcom*, the company that handles the telecommunications services in San Marino.

In 1991, the STET group entered the market of the Eastern countries. ITALCABLE and TELESPAZIO each own 25% of *Astelit*, a joint venture with the Russian Astra for the management of satellite communications for business users in Moscow and Saint Petersburg.

In 1992, the STET group made the first investment in an Arabian country. TELESPAZIO owns 33% of *GULF TELESPAZIO*, a firm that realizes telemonitoring in the Gulf countries.

Finally, in the USA, the STET group has owned the already mentioned ITALCABLE USA since 1979 and has a participation of 2.47% in the *Consortium Intelsat*. Moreover, in 1990 ITALCABLE started the company *TMI US* that offers value-added services and in 1991 the STET group bought a participation of 8.33% in *IPSP*, a company for the launch and management of private satellites.

6.4 Inward direct investments

Turning our attention to inward FDI in Italy we find a clear time subdivision. Before 1990, there were only two main foreign actors, *GEIS* owned by GE and *Intesa* owned by IBM and FIAT. Only after 1990 did the main European and US carriers start to enter the Italian market. As with the other European markets, the Italian became interesting only when the government decided on liberalization and privatization of the telecommunications services. Moreover, the high tariffs for international calls from Italy enabled carriers like AT&T and BT to offer a very competitive low-tariff service to the business users with international relations. For both periods we can state that the underdevelopment of data transmission and VAS, which characterized the Italian market, led to a weak supply of business services from the public carriers. Therefore, we can describe all the different entry strategies in the Italian market with the general pattern of internationalization via services.

Table 6.3 gives an indication of the FDIs that have been made in Italy as of 1992. The second wave of FDI is led by the US carrier AT&T. In 1990, *AT&T Italia* started the activity. It offers both voice and data services from and to the US. Noteworthy is the calling card service that, by-passing the national carrier, exploits the tariff asymmetries. In 1990 *BT* opened the first office in Rome and in 1991 a second one in Milano. The aim is to obtain a portion of the data traffic of multinational users working in Italy or in border countries. BT provides the access to Global Network services that can be delivered on package switching networks. In 1990 also the second British carrier, *Cable & Wireless,* started an office in Milano and one in Rome. C&W is following the same strategy as BT. Through its own network Global Digital Highway it is furnishing data services and international fax service. In 1992 *France Telecom* entered the Italian market by acquiring from Olivetti the company Olinet, now renamed *Interpac*. Interpac is the Italian terminal of the telematic network Infonet. In this way France Telecom can give access to its networks—Transpac and Numeris—and to the Minitel service for Italian users. In 1992 the Dutch *PT Telecom* opened an office in Milano with the aim of giving data support to Dutch multinationals working in Italy. This office is also the representative of the Swedish carrier Televerket.

Finally, it is worth mentioning the potential entry of other foreign operators through the management of the second licence for the mobile telephone—with GSM technology. As we have already pointed out in section 6.1, four consortia are participating in the competition and in each of them there are foreign investors.

Table 6.3. Principal international actors in the Italian market

Telco	Year of entry	Major Owners	Offices	Employees	Turnover 1991 bill. Lire	Entry strategy	Entry areas	Nodes
GEIS	1982	60% GE (US) 40% STET*	n.a.	105	36.9 1990	VAS on leased lines	VAS	n.a.
INTESA	1987	50% IBM 50% FIAT	n.a.	250	35.0	Customized services	VAS	n.a.
AT&T Italia	1990**	100% AT&T (US)	2	n.a.	n.a.	Exploitation of differential of price	Calling cards, Communication services	n.a.
BT	1990	100% BT (UK)	2	30	n.a.	International customers	Data services	4
CABLE & WIRELESS	1990	100% C&W (UK)	2	n.a.	n.a.	International Customers	Data Services, International fax	n.a.
INTERPAC	1992	100% France Telecom (F)	n.a.	20	2.0	Acquisition	Data services	n.a.
PT Telecom NL***	1992	100% PT Telecom (NL)	1	n.a.	n.a.	Dutch related customers	Data services	n.a.

Sources: STET Reports and Boschi & Mutinelli 1993.
* Up to 1988 100% GE, then in 1989 STET bought 40% of the shares.
** In 1986 there was already an office of AT&T in Rome.
*** The office in Milano is also the representative of the Swedish carrier Televerket.

7. SUMMARY AND CONCLUSIONS

A new economics of international telecommunications services is emerging out of the revolutionary technological and institutional changes that have drastically reshaped the domestic telecommunications services industry, with important opportunities for export of international telecommunications services and for multinational growth of firms that supply them.

The fall of the monopoly in many domestic telecommunications markets, connected with the processes of liberalization and privatization, and the new opportunities provided by low-cost satellites and optical fibres—*e.g.* data transmission highways—in delivering international telecommunications services, have fragmented the established structure. Therefore, the growing asymmetries in both technological capabilities and telecommunication service prices, stemming from this transformation, are shaping the strategies of the different national carriers.

Our analysis has shown that in the Italian case the public carrier is following a process of internationalization via multidomestic strategies. while FDIs in Italy are mainly, if not only, expressions of the so-called global strategy. Each strategy has its own elements of weakness and strength (Antonelli, 1993b):

- Multidomestic strategies require high levels of FDIs but are very efficient tools of valorization of technological and commercial knowledge acquired in managing the transition to digital technology in the home networks. Second and most important, they make it possible to enter domestic markets, with a strong share of final demand from households, which is likely to be highly revenue-elastic and weakly price-elastic with good prospects for future profitability.
- Global strategies take advantage of the strong complementarity between export and multinational affiliates and of the dynamics of network externality. They require low levels of resources to fund FDIs, but high levels of technological competence. Their limits consist first of all in the high levels of price elasticity of their demand, mainly a derived demand, and second in the key role played by international tariff asymmetries. Consequently they are exposed to changes in present trends of the political economy of international telecommunications.

Even if the public carrier is trying to improve the quantity and quality of data transmission services and VAS, it seems that the most probable scenario for the near future will be characterized by a substantial division of labour with three distinct markets: the basic network for voice telecommunications, the cellular services and the data communication services. On the one hand, the public carrier will take care of the final demand from households and, after the dismissal of the manufacturing activities from the STET group, it will be able to develop an international growth in the same area of demand. On the other, private companies, mainly foreign, will fulfil the business demand for data communications and mobile services.

This lock-in of the public carrier in the demand from households is due not only to the incapacity to offer business services for a long time but also to the strategy of leasing lines to companies, enabling them to build up their networks and thereby lowering the future demand for public services.

REFERENCES

Allen, D., 1993, European networks in the global evolution of industry structure, Paper presented at the International Telecommunications Society European Conference, Stenungsbaden Yacht Club, 1993.

Antonelli, C., 1984, Multinational firms, international trade and international telecommunications, Information Economics and Policy, Vol. 1, No. 4, 1984.

Antonelli, C., 1991, The diffusion of advanced telecommunications in developing countries, OECD Development Centre, Paris, 1991.

Antonelli, C. ed., 1992, The economics of information networks (North Holland Elsevier, Amsterdam).

Antonelli, C., 1993a, Externalities and complementarities in telecommunications dynamics, International Journal of Industrial Organization, Vol. 11, No. 3.

Antonelli, C., 1993b, Technological change and multinational growth in the new international telecommunications services economics, C.I.T.I. Columbia University, New York.

Antonelli, C., 1994, The economics of localized technological change and industrial dynamics (Kluwer Academic Publishers, Dordrecht and Boston).

Bianchi, A., 1991, Reti e servizi di telecomunicazione: transizione verso la concorrenza ed internazionalizzazione degli operatori, Nomisma, Acquisizioni Fusioni Concorrenza N. 2, December.

Boschi, D. and M. Mutinelli, 1993, I servizi di telecomunicazione, in Cominotti R. & Mariotti, S. (eds.), Italia Multinazionale nei Servizi. Radiografia degli Investimenti Diretti Esteri in Italia, ETAS Libri.

Casson, M.C., 1987, The firm and the market (Basil Blackwell, Oxford).

Caves, R.E., 1982, Multinational enterprises and economic analysis (Cambridge University Press, Cambridge).

Dang-N'guyen G. and D. Pouillot, 1993, Foreign direct investment in telecommunications in and out of France, Paper presented at the International Telecommunications Society European Conference, Stenungsbaden Yacht Club.

Daniels, P.W., 1982, Service Industries: Growth and location (Cambridge University Press, Cambridge).

David, P.A., 1987, Some new standards for the economics of standardization in the information age, in P. Dasgupta and P. Stoneman, eds., Economic policy and technological performance (Cambridge University Press, Cambridge).

Dunning, J.H., 1981, International production and the multinational enterprise (Allen and Unwin, London).

Dunning, J.H., 1987, Multinational enterprises and the growth of services: Some conceptual and theoretical issues. University discussion papers in international investment and business studies, December, No. 14.

Elixman, D. and T. Schnöring, 1993, Internationalization of the German telecommunications service market and strategic behaviour of DBP Telekom, Paper presented at the International Telecommunications Society European Conference, Stenungsbaden Yacht Club.

Enderwick, P. ed., 1989, Multinational service firms (Routledge, London).

Gilroy, B.M., 1993, Networking in multinational enterprises. The importance of strategic alliances (University of South Carolina Press, Columbia).

Noam, E.M., 1989, International telecommunications in transition, in R.W.Crandall and K. Flamm, eds., Changing the rules. Technological change, international competition and regulation in communications (The Brooking Institution, Washington).

Sauvant, K.P., 1986, International transactions in services (Westview Press, Boulder and London).

U.N.C.T.D., 1993, World Investment Report 1993, Transnational corporations and integrated international production, United Nations Programme on Transnational Corporations, New York.

APPENDIX: MAINLY USED ITALIAN ACRONYMS

ASST	Azienda di Stato per i Servizi Telefonici
IRI	Istituto per la Ricostruzione Industriale
IRITEL	Public Telecommunication Carrier
STET	Telecommunication Holding
CSELT	Centro Studi e Laboratori Telecomunicazioni
FINSIEL	Telecommunication Software Company
ILTE	Industria Libraria Tipografica Editrice
ITALCABLE	Servizi Cablografici Radiotelegrafici e Radioelettrici
ITALTEL	Telecommunication Manufacturing Company
SEAT	Division for Publishing and Marketing Services
SIP	Societa' Italiana per l'Esercizio delle Telecomunicazioni
SIRM	Societa' Italiana Radio Marittima
SIRTI	Telecommunication Manufacturing Company
TELESPAZIO	Societa' per Azioni per le Comunicazioni Spaziali
SET	Regional telephone company
STIPEL	Regional telephone company
TELVE	Regional telephone company
TETI	Regional telephone company
TIMO	Regional telephone company

The Race to European Eminence
E. Bohlin and O. Granstrand (Editors)

Internationalisation of Telecommunication Service Provision and the Greek Privatisation Debate

Yannis Caloghirou[a] and Theodore Darmaros[b]

[a]National Technical University of Athens, Laboratory of Industrial and Energy Economics, Department of Chemical Engineering, Division II, Zografou Campus, GR 157 80, Athens, Greece
[b]Planet S.A., Apollon Tower, 64 Louise Riencourt St. GR 115 23, Athens, Greece*

The 1990–93 developments on the Greek telecommunications scene are assessed, focusing on policies to privatise the Greek TO (OTE).

OTE's partial privatisation was initiated by the conservative goverment in order to attract foreign direct investment (FDI) and expertise in an increasingly internationalised environment. For the first time in the history of EC, it included the transfer of management responsibilites to a foreign "strategic investor". The whole process was accelerated by the existence of a large public sector deficit and was nearly completed when the government fell due to, among other reasons, its telecommunication policies including procurement.

Privatisation in Greece failed to be implemented because no political consensus was built around the idea, no adequate regulatory environment was prepared in advance and no clear internationalisation targets were set. The government seemed to sacrifice longer-term strategic opportunities in a rush to obtain short-term financial objectives. Evidently, a more evolutionary and gradual procedure was required, as well as a continuous attempt to intervene openly in the public debate and from a broader consensus in a direction with clear objectives and stages.

Far from a narrow financially-based process, privatisation/internationalisation incorporates strategic alternatives which need to be taken into account for its successful implementation.

1. INTRODUCTION

Telecommunications in Greece, following a decade of waning network expansion and institutional stagnation based on the old analogue-monopoly paradigm, have in the 1990s entered a phase of radical restructuring. In the period 1990–1993, during which the country was ruled by the *New Democracy* ("ND")—conservative—party, the main policy instrument selected to promote the rapid re-organisation of the sector was *privatisation.* Since October 1993, with PASOK (socialists) returning to power, privatisation policies regarding OTE, the Greek Telecommunication Organisation, were put on hold. All signs are that, following political change,

* Financial support for this paper has been provided by Planet SA in the context of the Greek Telecommunications Technology Forecasting Study.

Greek telecommunication policies are about to change yet again. Therefore, a critical assessment of the 1990–93 period seems both timely and useful for the steering of the new course.

With privatisation, the Greek ND government sought to attract foreign direct investment (FDI) and expertise, taking advantage of the current internationalisation trends in telecommunication service provision. The cellular sector was selected first for development by the private sector—not without strong resistance by defenders of state monopoly. Shortly afterwards, plans were announced for the privatisation of the Hellenic Telecommunications Organisation (OTE) itself, 49 percent of which was decided to be sold under a package that included transfer of the Organisation's management responsibilities to a foreign "strategic investor".

Despite the fact that, eventually, policies in the cellular sector proved to be financially successful for the government, the development through FDI of the cellular sector was accompanied by a number of complications. The final controversial outcome of establishing a duopoly with OTE's exclusion from the bidding process was only reached after a series of policy shifts and compromises. With the introduction of service achieved before schedule in the summer of 1993, the resolution of critical regulatory issues has been overshadowed by OTE's own privatisation prospects. This was a much bolder project, characterised by three main features:

First, it represented the only privatisation programme in the European Community so far, which explicitly sought to transfer the TO's management control to a foreign actor. The strategic investor's presence would become instrumental in the internationalisation of the Greek telecommunication market, although one can only speculate over the exact forms this would assume. The prospect of outright foreign control became the chief source of a wide domestic opposition to the programme, leading to the political defeat of the Conservatives in the 1993 election.

Se*cond,* OTE's privatisation was formulated in economic conditions dominated by a high public sector deficit and chronic inflation. The income generated by OTE's privatisation was expected to offer an immediate contribution to the 1993 acute state budget's requirements. For that reason, privatisation had become an urgent priority at the broader level of government economic policies, superimposed on the more specific level of telecommunication policy making.

Third, OTE's privatisation was being promoted *in parallel* with the establishment of a new regulatory regime, rather than *after* its implementation. This introduced a number of procedural complications, as well as an overall uncertainty over the programme's eventual impact on the sector's organisation and operation. In particular, the prospects for OTE privatisation may have been in contradiction with those of promoting competition and therefore attracting FDI in the liberalised market segment by more actors.

Obviously, given the freezing of the privatisation programme, after the October 1993 election result, it will now be impossible to assess its practical implications. Nevertheless, it is still felt that an assessment of the developments leading to its near-finalisation would be useful for two reasons. *First,* because the political debate that has accompanied privatisation initiatives has influenced substantially the thinking of both supporters and opponents of the sector's transfer to private hands; in any case important elements of the ex-privatisation scheme *are* already in place and *will* condition the sector's evolution. *Second,* it is important to place the Greek case in the context of service internationalisation trends, especially as this aspect is largely missing from domestic debates.

2. THE INTERNATIONALISATION OF TELECOMMUNICATION SERVICES MARKETS

Prior to the late 1980s telecommunication operators (TOs) had been predominantly *national*, in the sense that they were solely based on their home markets, for which they usually carried a mandate for *public service provision.* Thus, TOs were highly dependent on the revenues generated from service activities in their home ground. There have even been well-known cases (AT&T before its divestiture, NTT and DBP Telekom) where monopoly TOs were explicitly prevented by national regulation to do business away from their domestic territories.

The internationalisation of service markets that has been initiated in recent years includes attempts, by the more outward-looking national TOs, to provide services away from their home grounds, either in joint ventures with other TOs, or through the cross-border acquisition of foreign telecommunication facilities. Seen from the perspective of smaller or less dynamic operators, this development raises the threat of international aggressive competition inside their home markets or, in a more optimistic scenario, the challenge for international collaboration with incoming players.

It would be a mistake, of course, to view the process of internationalisation in static terms, with two fixed camps of "aggressors" and "defenders". Corporate strategies of telecommunication operators will be determined by a whole array of internal factors, including their size, capital structure, technological sophistication and innovativeness, established relations with business users in qualitative and quantitative terms, relations with other TOs as well as non-TO actors in the telecommunication field etc. Moreover, they will be greatly influenced by the evolution of external institutional and economic conditions.

The restructuring measures that TOs need to implement in order to improve their performance across the above spectrum of criteria can be best understood in the context of the most important factors underlying globalisation trends, presented below.

2.1 The globalisation of the world economy

The growth of global markets requires the provision of business user telecommunication-based services on an international, end-to-end basis. Global economic activities offer particularly exciting opportunities for all those TOs that already provide advanced services—at the national level—to large MN firms in one of their most important market bases. On the other hand, they seem irrelevant to the near-term strategic prospects of public utilities, still pre-occupied with their domestic network expansion and providing a minimal set of services to their business users.

It is often correctly argued that countries lacking a telecommunication service sector that can respond to the requirements of a developing global marketplace are inhibited from participating in globalised economic activities. This is especially true for countries which base their economic performance on traditionally global activities (e.g. shipping, tourism) that certainly require advanced communications support to remain competitive. In such countries, pressure for modernising the telecommunication sector, and the TO itself, is both *internal* and *external.* Indeed, in addition to the indigenous demand by the more outward-oriented advanced user sectors, external pressure is applied by prospective foreign competitors, especially for the provision of international services, possibly through by-passing the local operator. In such an environment, the domestic TO clearly cannot afford to remain *both* dominant *and* uncompetitive.

2.2 The effects of technical change

Revolutionary developments have occurred in virtually all communication technologies, from satellite & fibre-based bandwidth availability, to intelligent switch-based standard interoperability and software defined networks. In general, technical change has lowered the economic and technical barriers for the international provision of services and has multiplied the available service options. In addition, some innovations have favoured the introduction of a new class of advanced services at the domestic level (cellular telephony, CATV, VANs) following a process of international competitive bidding rather than through the commitment of public TO investment. As a result, technical change encourages globalisation through the growth of internationalised activities as well as by increasing opportunities for foreign presence in national developing markets.

A further dimension of technical change, of immediate relevance to the efforts of modernising TOs, concerns its associated forms of *technology learning*. In the past, telecommunication technology learning coincided with the technical learning specific to the engineering practices required to introduce a new system. This has been largely true, even in the context of digital switching introduction, when the central dilemma for technological development was to strike the correct trade-off between technology transfer and indigenous development.

In this era of service globalisation, two important changes are taking place. *First,* technology learning becomes intimately linked with business learning. No technological innovation can successfully be implemented without the development of the matching commercial skills, and vice versa. *Second,* participation in the emerging international networks and clusters of service activities becomes vital for the acquisition of leading-edge technical skills. National operators, even the most traditional ones, are required to broaden their horizons and collaborate with the more advanced players in order to encourage technological diffusion in their domestic base.

2.3 The growing importance of users

It is well known that in striving to evolve into commercially-oriented enterprises, TOs need to become much more user-driven. In global markets, where users are often partners in the development of applications rather than mere clients, user responsiveness is a definite requirement for operators.

"User responsiveness" may include taking up and further developing solutions originally installed and operated by large users themselves. Thus, some business users, operating sophisticated international private networks, find it increasingly difficult and costly to manage and upgrade them alone and require expert assistance, provided in a flexible manner ("outsourcing"). For the TOs that have already faced competition from traffic migration to private networks, the task to win users back cannot be accomplished without a strong international presence. On the other hand, for TOs that have been protected from private network competition the development of virtual private networking skills becomes much more difficult.

On a smaller scale, "user responsiveness" refers to daily tasks like friendly service marketing, effective network management, rising quality of service levels, flexible tariff structures etc. In these mainly *national* activities, there is a definite trend towards the *international* acceptance of a minimum level of formal and informal standards, as well as towards inevitable comparisons, made between operators. As a result, there is already competition between the more advanced TOs on raising such standards and assuming a more user-oriented profile.

This informal race may widen the gap between them and a "second division" of operators, unable to ensure comparable levels of service and still pre-occupied with the utility service approach.

2.4 Continuing liberalisation trends

Some globalisation activities, in particular foreign participation in a national TO, are still prevented by foreign ownership restrictions, imposed by governments for reasons of national security. Virtually all countries, including pro liberal ones (e.g. the US, Japan), have such legislation in place. In less advanced telecommunication environments, state-owned TOs are particularly fervent supporters of national ownership and control, defending the regulatory status quo on national security, national control and domestic development grounds.

However, because of the rise of pro-liberal policies, there has been growing pressure in recent years for the relaxation of foreign ownership rules, particularly evident in some developing and Eastern European countries, seeking to promote rapid FDI and technology transfer. The UK, among advanced liberalised markets, has also encouraged inward foreign investment, to increase competitive pressure on its own dominant TO. Other governments of advanced countries, despite taking advantage of the opening of less developed markets, have been more reluctant to open their own market to foreign presence.

Indeed, as in other service sectors, globalisation of telecommunication service provision has been hampered by asymmetric national regulation and the absence of an international pro-competitive framework. In anticipation of a GATT agreement, most progress at the international level has so far been achieved through regional trading agreements, bilateral agreements, at the ITU level (liberalisation of leased line restrictions) and the European Community.

Regarding the European market, two critical measures currently being studied are the liberalisation of voice telephony and the entry into national markets of pan-European long distance operators. Support for the latter measure in particular seems to be strong and includes larger user organisations, large European and non-European carriers, value added service suppliers, potential new entrants such as water and railway utilities and other organisations operating large private networks and, under certain conditions, parts of the Commission itself. On the other hand, TOs with a more traditional approach are the principal opponents, fearing the loss of valuable international traffic (Thomas, 1992).

If the above developments materialise, there is the distinct possibility that certain "less favoured" parts of the European Community (including Greece) will benefit from an exemption period. This practice has been already followed in previous liberalisation reform measures, including opening of the packet-switched infrastructure and equipment procurement segments. By excluding less favoured parts from *globalisation through competition* the Commission will implicitly back *globalisation through direct foreign acquisition:* indeed, if foreign operators assume the responsibilities granted to national TOs, they will certainly be glad to see a framework protecting them from foreign competition. In that respect, future extension periods will have a very different significance from those already granted if a private monopoly is implemented.

2.5 The state of domestic and international competition

The prospect of competition inside their home markets threatens to reduce monopoly profits and pushes TOs towards new revenue pools in foreign countries. In addition, third party service providers, already operating on an international scale through the implementation of world-wide information networks, are claiming valuable business traffic away from larger national TOs.

Apparently, smaller or less advanced TOs, to the extent that they operate in well-protected markets, seem more secure both from domestic and from international competitors. In reality of course, their immunity to competition is based on protectionist policies which are under considerable pressure to change. If and when the time comes for the removal of protection barriers, such operators will be in very vulnerable positions. This is of course not an argument for further extending their protection, but instead for finding the optimum methods and time scale for removing them.

Is the policy challenge to remove protection from smaller or less advanced national markets compatible with the existing strategic initiatives, by large players, to broaden their international revenue base?

The question essentially refers to the operating conditions imposed on foreign entrants in the opening markets. These however can vary widely: from the transferring—or even the extension—of old state monopoly rights to a new foreign-controlled monopoly, to the exclusion of foreign entrants from key market activities, with only a very gradual removal of protection barriers.

Between the two extremes, which are both based on a sustained monopoly market structure, lie other possibilities of foreign entry in a mixed monopoly-competitive environment. The IFC (1990) report includes, among others, entry into the value-added services segments, the implementation of digital overlay networks, the ownership and operation of specialised networks, the subcontracting of selective installation and operation activities, the implementation of a telecommunication hub, and the diversion of international traffic through the country (Ambrose et al, 1990). A detailed examination of their feasibility depends on factors such as: market size, existing and potential demand structure, regulatory environment, strategies of foreign actors etc.

An additional prime factor determining the answer to the above question is the emerging competition, between countries opening their markets to attract foreign investment on their own home ground. Such competition is reported to be especially acute in the market for TOs under privatisation, following a significant increase in the availability for sale of national TOs in recent years. This development may shift some buyers' strategies from the formation of joint ventures for international service provision—which has seen many aborted efforts in the past—to the direct purchasing of foreign telecom facilities, as a first step that will strengthen their future positions in such ventures.

In general, a buyer will reach a decision only after careful evaluation of a number of factors including price, state of infrastructure, capital structure, monopoly rights (during and after a period of concession) and management control prospects. With more TOs now available on the market, prospective investors may favour polarised "package" solutions, rather than "middle of the road" structures. It is even possible that, in a strong buyer's market, there will be plenty of scope for an old-fashioned type of grouping of national markets to "spheres of influence", rather than fierce competition between buyers. Thus, in drafting their policies and conducting successful negotiations, governments on the receiving end will need to balance carefully their opening-up priorities with the global market situation.

3. A BRIEF REVIEW OF GREEK TELECOMMUNICATIONS

The post-war development of Greek telecommunications can be analysed with the aid of a 3-phase model, summarised in Table 3.1.

3.1 Phase 1: analogue & monopoly-based network expansion, 1949–1981

OTE was founded in 1949 as a state-owned enterprise operating under the law of "Private Greek Firms". The decision for an OTE institutionally independent from the Greek state marked the beginning of a new era in the Greek telecommunication regime. The Greek government entrusted the Organisation with the task to develop the telecommunications sector, which at that time was practically non-existent due to the damage caused by the war during the 1940s. However, the institutional independence of OTE was never accomplished in practise, and the organisation operated for decades as a typical PTT.

OTE's activity during phase 1 concentrated on the expansion of the telephony network (from 1 telephone per 100 inhabitants in 1949 to 31.7 telephones per 100 inhabitants in 1982) based on the monopoly-analogue paradigm. During that period all the necessary investment for the network construction was public and was financed by three sources: state budget, supplier's credit and barter arrangements especially with Eastern European countries, e.g. Czechoslovakia, East Germany, Romania. Under those agreements the Greek state was exporting agricultural products and was importing telecommunications equipment. OTE's procurement policy, which under the scheme of open call for tender or through the barter agreement type of purchasing favoured a variety of sources, resulted in a mosaic of analogue technologies.

OTE's procurement policy had two negative effects: *first*, on the network's operation and *second* on the development of indigenous manufacturing capability in the field of telecommunication equipment. In fact, the only manufacturing activity which was developed in the 1960s and 1970s was limited to assembly operations of analogue equipment by Greek subsidiaries of foreign multinationals in the sector. A considerable manufacturing activity was developed during the 1960s and especially in the 1970s in the cables and wires sector, mainly by subsidiaries of foreign firms, although some significant Greek activity also occurred.

In the mid-seventies new issues entered the telecommunications agenda. First, the digitisation of the basic network which was an early move even by EC standards, and second the decision to create a state-owned company, ELVIL, in which OTE participated with 45% and ETBA (State Bank for Industrial Development) with 55%. ELVIL's aim was the manufacturing of digital switches through a process of transfer of technology. In order to implement this decision an international call for tender was issued in 1980 for the selection of the foreign firm which was to conduct the whole venture in close collaboration with ELVIL. In April 1981, 10 offers were submitted, but the selection process was frozen because of the forthcoming election in October 1981 (Caloghirou, 1991).

Table 3.1
A summary of post-war Greek telecommunication development

	PHASE 1 (1949–1981) Analogue & Monopoly-based Network Expansion	PHASE 2 (1981–1990) Prolonged Transition Crisis	PHASE 3 (1990-present) Digitisation Liberalisation Privatisation
Infrastructure Development	Network Expansion, based on a mosaic of analogue technologies.	Main Line expansion through the exhaustive exploitation of the existing analogue switching infrastructure. On-going selection process for digital switch suppliers.	Network Expansion and Upgrading through Digitisation. Selection process for cellular duopoly (completed early 1992)
Public Investment	Public Investment, financed by national funds, supplier's credits and barter arrangements.	Public Investment and EC Funding (First Community Support Framework, incl. STAR)	Public Investment, EC funding ("Crash Programme")
Private Investment	Analogue equipment manufacturing (assembly)	Digital manufacturing facilities, post-1987 (Intracom, Siemens)	Cellular OTE's privatisation Strategic investor (?) Stock Market (?)
Main Policy Agenda	Digitisation Domestic digital manufacturing Technology transfer	Digitisation OTE internal re-organisation Mobile Telematic Services	Implementation of New Telecoms legislation Privatisation Renewed conflict on digital procurement

3.2 Phase 2: 1981–1990

This period is characterised by an ongoing attempt to finalise the scheme for the procurement of the digital switches necessary for the digitisation of the network. During the period 1982–1985 an attempt was made to increase the local participation and enhance the development effects of the whole venture. Thus, negotiations were conducted with four firms included in the shortlist. This attempt was made in the framework of a successful pro-indigenous procurement policy, which reduced significantly the imported part of OTE's procurement. A significant conflict emerged over the choice of OTE's digital technology supplier. Finally, the initial ELVIL project was abandoned and a decision was made in 1986 to award two contracts to Ericsson and Siemens which were forced to organise local manufacturing through *Intracom* (a Greek-owned firm founded in the late 1970s) and *Siemens Tele* (a subsidiary of Siemens) respectively. The first small tentative contracts were finalised in 1988 and in early 1990 larger contracts were signed, following a decision of the interim all-party coalition government in power at the time.

Due to this long delay because of the on-going selection process for the digital switch suppliers, main line expansion occurred through the extensive exploitation of the existing analogue switching infrastructure. This had a negative effect on the quality of services provided and in addition failed to shorten the waiting list for main line installations.

Because of the long delay to opt for digital suppliers, the 1980s is a period characterised by poor infrastructure investment, illustrated in Table 3.2. As a result, the gap in infrastructure between Greece and many OECD member countries widened, and many problems were accumulated in the sector.

Table 3.2
Greek telecommunication investment, 1981–1990

	Total OTE Investment	% of OECD	% of EC
1981–83	436.99	0.647	1.55
1984–87	349.01	0.487	1.13
1987–89	278.45	0.345	0.81
1990	346.35	0.404	0.90

Note: Investment in US$ million, 1990 constant prices and exchange rates
Sources: OTE (1992), OECD (1993).

In order to reverse declining investment trends, a development plan for OTE was initiated in 1987. This programme was only finalised in January 1989 and included a $2.96 billion investment plan for the next 5-year period, i.e. mainly for the third phase (1988 prices) (OTE, 1989). In the context of this programme, the European Community would contribute approximately 15%, as part of its telecommunication aid to Greece (STAR, Telecommunications Operational Programme, Interreg, et al.). The delay in the finalisation of OTE's investment programme may have caused a 1-year delay in the flow of EC funds but has not affected the total sums.

The Community's contribution reflects its increasingly important role since 1984 concerning the decision-making process and the financing of Greek telecommunication development. The Commission's influence becomes especially relevant following the 1986 discussions for the formulation of the STAR, a regional aid development programme, and after the 1987 Green Paper policy framework. However, the European influence did not always lead to swift domestic reforms. For example, and in view of the forthcoming liberalisation process initiated by the Green Paper, an attempt was made to reorganize OTE internally, a reorganisation which was never completed.

A further uncompleted development of the second phase concerned the emerging mobile telephony sector. Here, an unofficial agreement was reached between OTE and a consortium of Racal (27%), ELVIL (51%), Intracom (11%) and Lambrakis Press Group (11%) in Spring 1989, for the implementation of an analogue TACS service. ELVIL would be restructured as a new services supplier that would represent OTE's interests. This agreement was never finalised because, following the June 1989 elections, a new government was formed.

During the following 10-month period Greece was ruled by short-term coalition governments. After three election battles a New Democracy (conservative) government came into office in April 1990.

3.3 Phase 3: April 1990–October 1993

In November 1990 the Greek Ministry of Transport and Communications, under pressure from the EC, decided to order a study on the development of the telecommunications sector in Greece during the period 1991–2000. After the preparation of the study's terms of reference by a Joint Steering Committee, composed by Greek and EC experts, an international call for tender was issued and the contract was awarded to Coopers & Lybrand. By the end of January 1992 the study was completed and resulted in two plans for action:

i) An urgent *Crash Programme* for the development of telecommunications infrastructure during the period 1992–1993. The proposals contained in that programme focused primarily on improving service quality and reducing the waiting list for main telephone connections (see Table 3.3 for a set of indicators regarding OTE's performance by 1990). Moreover, as a condition for the financing of the Crash Programme by the Commission, the Greek government undertook the responsibility to proceed with all the legislation necessary for the implementation of regulatory reform.

   The Crash Programme included a total investment of ECU 200 mil (about 53 billion drs or $265 mil) which is to be 50% EC funded. In order to speed up technological modernisation of the network, especially in the Athens area, but also in the island of Rhodes and the city of Larissa, the Commission's initial intention was to install an overlay network for business users (*"Attica's Ring"*, in the Athens case). After negotiations OTE succeeded in an agreement to implement the ring so as to serve both its new business and existing residential subscribers, transferring traffic of the existing network to the ring. The contract for Attica's ring was awarded in early 1993 to Intracom, also supplying switches for the main network.

ii) A long-term *Business Plan* (Coopers and Lybrand, 1992) for the development of telecommunications during the period 1992–2000. The scope of this plan was wider than that of the Crash Programme in terms of both the time period and the subjects covered. It aimed at achieving a range of financial objectives and operational targets which will enable OTE to improve the quality and availability of its service offerings so that they are comparable with the standard achieved elsewhere in the EC by the year 2000. In that sense the Business Plan has superseded OTE's own 1989–1993 development plan.

The Business Plan's main objective has been to achieve universal availability of basic telephony throughout Greece at affordable prices and good quality by the year 2000. The plan sets out a range of quality of service targets, as well as quantitative targets for the growth and modernisation of the Greek telecommunications market. On the whole, if implemented, the Plan claimed that it would double the overall value of the Greek market (in 1990 prices) by the year 2000.

The main prerequisite which has been identified by the Plan for a rapid telecommunication development referred to the exclusion of any political or governmental intervention in OTE's operation. Additional detailed suggestions for OTE's restructuring included a market orientation strategy, a programme of commercial development and a structural reorganisation programme.

According to the Business Plan, OTE's true corporatisation could be achieved either by the institutionalisation of strong regulations which prohibit any government intervention in OTE's management or by the transfer of the majority of OTE's shares to the private sector. In any case the study supported the idea of an injection of private capital into OTE up to 49% of its total shares till the end of 1993 in order to finance its investment activities while maintaining a healthy profit.

Finally, the Plan proposed the rationalisation of OTE's tariffs in order to reflect actual costs, especially in the case of local calls which are currently priced at approx. 56% of cost (Technical Chamber of Greece, 1993). Thus, according to the "Business Plan" the tariffs for basic services should be adjusted annually at a rate of 5% above CPI till 1994 and then, till 1999, at a rate of 1 unit below the previous year's CPI at maximum. The adoption of this adjustment mechanism is expected to lead to an annual increase of profits, in real terms, at a rate of 10% till 1995 and 5% till 1999.

The drafting of the Business Plan was, from the outset, regarded by OTE's trade unions and management as an external and unwelcome intervention in the Organisation's affairs. It was claimed that most of the work conducted in the Plan's framework could be performed (or had already been achieved in the context of the 1989–93 Development Programme) by OTE alone. On the other hand, the ND government supported the Plan, which was to become especially influential in the formulation of its privatisation policies. This was despite the fact that ownership status was not the study's major point, neither was privatisation the only policy option contained in that part.

Indeed, shortly after the submission of the Business Plan, the New Democracy government embarked on its *privatisation policies*, which are examined in detail in section 4. By far, the plans for the partial privatisation of OTE represent the main policy initiative of the 1990–93 phase.

In parallel with privatisation preparations, digitisation of the network commenced following the selection of the two suppliers in 1990. In addition to the main digitisation schedule OTE implemented the "Piraeus Telecommunications Park" in order to serve the highly demanding and pressing needs of Greek shipping companies based there. This move was especially welcomed by local business users and improved the Organisation's image in the Piraeus port area (see Table 3.3).

Table 3.3
Main indicators of OTE performance

	OTE (1990)	OECD Average (1990)	OTE (1992–1993)
Main Lines per 100 inhabitants	38.94	42.54	45
Digitisation ratio for local lines	0.5	36.4	14
Revenue per Main line, $US	318.94	1029.29	315
Percentage of revenue from PSTN	85.49	72.66	
Profit per Total Revenue (%)	21.8	12.7	31.6
Revenue per capita, $US	124.2	437.8	142.5
Services market as % of GDP	1.85	2.28	
Investment per main line, $US	87.71	240.08	
Main lines per employee	141	145	173
Revenues per employee, $US	44800	149700	54800
Faults per 100 main lines	55	10.4 (France)	
Waiting time for telephone installation (1990), months	81	3	26

Source: OECD(1993), Technical Chamber of Greece (1993)
Note: 1992–93 calculations are not official figures.

As a result of the growth prospects in the digital systems market, the conflict on digital procurement was renewed and the pressures for the introduction of a third digital system into the network increased, despite Greece's negative experience from the development of the analogue-based network through the use of 14 different systems.

In the Spring of 1993, in order to resolve this conflict a decision was made to issue an international bid for the purchasing of approx. 1 billion digital lines, in order to meet the needs of the period 1994–1996. Five firms, including OTE's two traditional suppliers of digital switches, entered the competition. Initially, the final decision was scheduled to be taken before July 1993 and many options for a reshuffling of the supplier pack seemed to be open. Ultimately however, the ND government decided to allow the new OTEs owners to reach a decision, following the completion of privatisation. The transfer of responsibilities from the state to the private sector was a severe threat to existing established suppliers. The political row that followed as a result became a catalyst for the defeat, in Parliament, of the government, early elections, and the coming to power of the socialist party (PASOK). It is likely that the new government will maintain the status quo in OTE's procurement, although no official declarations were made until November 1993.

4. THE 1990–1993 PRIVATISATION POLICIES

4.1 Privatisation as the main instrument for internationalisation

Greek internationalisation prospects have been, until October 1993, dependent upon the fate of the ND government's *privatisation* policies for the sector. The Greek privatisation programme was designed to attract foreign direct investment and expertise to the Greek market, hence promoting its opening to foreign private (or state-owned, given that some of the candidates are public sector enterprises!) capital. As discussed below, its implementation started in the cellular telephony market and reached its climax in the context of OTE's current privatisation plans, to be stopped short of completion by the October 1993 elections.

Through privatisation, emphasis was placed on the attraction of foreign activities in the Greek market *("inward internationalisation")* rather than the promotion of Greek actor activities in foreign markets *("outward internationalisation").* Outward internationalisation activities of Greek actors, other than OTE, were considered on an independent basis from the privatisation programme. OTE itself was too preoccupied with its own privatisation prospects to pursue any outward internationalisation initiatives. In any case, the formulation of OTE's outward internationalisation plans would logically follow its overdue internal management restructuring. Thus, it was included in the government's proposed privatisation scheme, although the need to restructure the organisation's commercial operations had been recognised even before privatisation became a policy priority.

Although the ND government's privatisation policies explicitly sought to attract FDI and foreign management expertise, it would be a mistake to consider that they had been formulated with OTE's internationalisation business strategy as a priority. As discussed in section 4.4, the main drivers of the introduction of private capital stemmed from the urgent need to improve the current state of infrastructure and services, as well as for wider fiscal requirements. For that purpose, internationalisation became the sole method available, given that most of the private capital and expertise the government sought to attract were simply not available in Greece. Thus, it was left to the future management of OTE to formulate a strategy of outward internationalisation.

4.2 Regulatory change and privatisation in Greece

The prospects for FDI are mainly dependent on the degree of deregulation introduced in the telecommunications market. Two sorts of conditions favour the attraction of higher FDI sums if privatisation is the chosen method: *first,* a guarantee for the long-term protection of the monopoly areas granted to the TO; *second,* a weaker institutionalisation and/or execution of the state's regulatory rights, with respect to the monopolist's exercise of dominant market position. The application however of such conditions discourages FDI through means other than the TO's privatisation, because it prohibits entry into several market areas and handicaps competition in the rest. This basic contradiction underlines the inherent difficulties of combining liberalisation and privatisation initiatives at the same time.

In fact, in Greece the introduction of competition through the liberalisation of the telecommunication regulatory framework was considered neither as a principal motive for privatisation, nor as a necessary step that needs to precede it. Liberalisation has been regarded mainly as an *international obligation,* resulting from Greece's participation in the European Community. In all relevant international debates Greece's position has so far been firmly on the side of the advocates of monopoly rights preservation. This can easily be understood as a combination of three factors: first the level of development of its telecommunications sector; second, the dominant statist tradition; third, OTE's bureaucratic rather than business-oriented corporate culture.

Moreover, in Community negotiations, Greece has succeeded in pushing backwards some of the harmonisation deadlines that apply to more advanced Community members: competition in data networks will be introduced in 1996 instead of 1993 and the opening-up of public procurement in the telecommunication sector in 1998 instead of 1993. Both exemptions have been won in order to protect Greece's efforts to close the infrastructure gap with the rest of the Community. In the context of these efforts, OTE was granted longer adjustment periods to full competition. OTE's Greek-based established equipment suppliers (Intracom—using Ericsson technology, and Siemens' Greek subsidiary) have also benefited from the arrangement, despite strong pressure by their competitors.

The above favourable conditions would be inherited in the post-privatisation environment and the "strategic investor" may well benefit from them. In fact, the conservative ND government obtained a further 5 year extension period (until 2003) for the liberalisation of voice telephony, in common with three more European countries. More important than that is the fact that the new regulatory framework, introduced by Law in July 1992, was left incomplete and inactive, with the "Crash Programme" deadline running out in December 1993. This, as discussed in the following sections, allowed the new managers of OTE and the government to negotiate the right regulatory terms in parallel or immediately after the privatisation process.

The 1992 Greek Telecommunication Law was drafted to comply with the Commission's regulatory model. It separates regulatory from operational activities; caters for the development of competition in the value added services market; and sets the principles for the access and use of public network according to the evolving Open Network requirements.

However, there are key areas in which, pending OTE's sale, there remained a regulatory vacuum. Thus, although the law states that a Regulatory Authority should be established, this was never activated, as the law allowed for a delay period of up to 12 months. The government issued in March 1993 a call for bids for a consultant that will advise it on the detailed necessary steps to be taken in order to implement the law's requirements. Subsequently, this call was superseded by the publication and voting of the new OTE's partial privatisation law, in August 1993.

4.3 The development of cellular telephony through direct foreign investment

Following earlier attempts to introduce an analogue cellular telephony service in 1988 (see section 3) the conservative government decided, shortly after its election into power (April 1990), to review the whole process. The review focused on two main issues: *first*, the selection of the appropriate technology, where it was decided to move from the analogue TACS to the new pan-European digital GSM standard; *second* the implementation of the appropriate regulatory framework for the sector, including the conditions most appropriate for OTE's participation. This was a far more complex issue and although the decision for a duopoly was taken in principle, OTE's involvement was left as an open question.

Following the decision to shift from analogue to GSM, a new call for tenders was initiated in September 1991. Thirty-one foreign service providers, some in joint ventures with Greek companies, applied for the two licences. OTE was not one among them because, according to government thinking at the time, it was to be granted a share of up to 34 percent in *both* winning consortia. This decision was unique, as it undermined the basic competitive principles the cellular duopoly was supposed to promote. Moreover, it effectively excluded OTE from taking part in the bidding process, on its own or in collaboration with its chosen partners, since the company would anyway take part in the network's implementation.

The decision to include OTE in both consortia was in fact taken under pressure from two sides, one internal and one external. In the domestic field, OTE's management and workforce, as well as advocates of state control in telecommunications, demanded that the TO should develop a cellular network on its own, rather than "cream-skimmed" by foreign new entrants. Their position was held throughout the period 1988–1991 and was only abandoned *following* the bidding procedure.

On the other hand, the existing arrangements in most European countries suggested that the TO should be granted one licence under certain conditions, with an open bidding process held for the second. In the Greek case, such conditions would most likely include OTE's requirement to form an alliance with foreign actors in order to achieve the desired transfer of technology and expertise and be competitive with the second consortium.

In an attempt to achieve a compromise between complete OTE monopoly and duopoly, the odd decision was taken that OTE would be included in *both* licensing consortia. The nature of OTE's participation was to be purely on a financial basis, with no effective involvement of OTE in the structure of the bids. However, its precise role in both consortia was never clarified.

Following the above decision, most of the bidders expressed their strong opposition to OTE's compulsory inclusion, fearing it would deprive them of valuable revenues and control. The European Commission may also have communicated their concerns to the Greek government. With the pro-OTE camp also continuing to resist the privately controlled duopoly, the government decided to abandon all attempts at a compromise, wholly excluding OTE from the cellular market. Thus, in the July 1992 Telecommunications Law OTE was explicitly deprived of all rights to operate GSM cellular networks and the two winning consortia were left free to plan their future operations independently of OTE's involvement. OTE was promised to receive 10 percent of the licence revenues as a consolation.

In August 1992 the names of the two winning consortia were announced:

i) Panaphon (Vodaphone 45%, France Telecom 35%, Intracom 10%, Databank 10%).

ii) STET Hellas (a subsidiary of STET, with a 5% stake held by Interamerican, a Greek insurance group with an extensive distribution network).

The Panaphon consortium includes two Greek minority partners: *Intracom*, manufacturing and supplying Ericsson digital switches to OTE's fixed network; *Databank*, a prospective value-added provider, owned by Intracom, Intrasoft (belonging to the Intracom group) and the Lambrakis Publishing Group, a press group with a powerful position in Greek public life. Since both Greek partners were present in the doomed 1988 consortium, Panaphon's choice effectively maintained the balance with important Greek actors, pre-empting domestic criticism. On the other hand, the choice of STET illustrates a recent trend of Italian investment in important Greek industries, also evident in other sectors. STET's confident position is evident in the fact that it was allowed to bid without a domestic partner.

The two winning groups offered $US 163.4 million each for obtaining a licence, to be paid in full to the Greek State by the end of May 1993. In addition, it has been estimated that another $US 560.6 million will be invested by both providers, with two networks targeting 160,000 potential users by 1997. Therefore, on the whole investment in cellular telephony is expected to reach a total of 724 million $US, most of which is expected to be FDI. Employment gain in the sector is estimated to be approximately 7,000, indicating a large positive prospect in a new, technologically-intensive activity. Both groups are planning to use Ericsson cellular systems.

The introduction of cellular service took place in the summer of 1993, approximately 6 months before schedule (Greater Athens area).

4.4 OTE's Privatisation Prospects

Following the opening of cellular telephony, which immediately created a climate of hostility between the government and OTE's trade unions, the Minister of National Economy confirmed, in November 1992, the government's intention to sell 49 percent of its stake in OTE. More specifically, the government initiative included:

i) The transfer of 35% of OTE's shares to a "strategic investor", following an international bidding process. The strategic investor, which could be a single telecommunication firm or a consortium, would also be given the overall responsibility for the management of the organisation during a period of 10–15 years (still not specified), despite its minority stake.

ii) The transfer of an additional 14% to the general public, through the Stock Market. The government based its decision on the following grounds:

   i) In order for OTE to meet the challenges of technical change and market-oriented commercial operation, an "internal shock" in its management structure seemed necessary. This was to be introduced through privatisation, with the strategic investor offering the much-needed managerial and technical expertise.

   ii) OTE's investment needs, estimated at approximately 5.3–7.5 billion ECU (1993 prices) for the next ten-year period, could only be financed with the contribution of private investment sources, offered especially by the strategic investor's credibility. The major government requirements to be met by the strategic investor are summarised in Table 4.1.

Table 4.1
Summary of major government privatisation requirements

A. Main Line Timetable	
1994–1996	1.68 mil. lines (1.08 repl. and 0.6 new)
1997–2000	1.72 mil. lines (1.02 repl. and 0.7 new)
B. Main Line Coverage	
Rural and Island Areas, 1994–1996	0.13 mil.
Rural and Island Areas, 1997–2000	0.045 mil.
C. Transmission Network	
1994–2000	Replacement of cable in 2.1 mil connections
D. Other Civil Requirements	
1994–2000	Network capacity should exceed accumulated demand for main lines by 10–14%
1994–2000	Public payphone penetration of 0.7%
E. Military Requirements	
	A detailed (and confidential) list of requirements, which seeks to extend the present favourable treatment of armed forces in the new regime.

Two additional reasons were recognisable behind the government's initiative:

i) Revenues collected from the OTE's sell-off would contribute to the financing of the budget deficit, currently running at approximately 15.4% of GDP (General Government Net Borrowing Requirement—the EC average is 5.3%). There were many indications that this had been the single most important factor in the New Democracy Minister of National Economy's thinking: for example, the 1993 budget included an item of "income expected from privatisation", valued at approx. $US 1.5 bill.

ii) The entry of OTE shares in the Athens and international stock exchanges was estimated to attract considerable Greek and foreign capital. With privatisation, the ex-government hoped to generate a shorter term capital impulse that would stay in the country and project a longer term "open market image" abroad.

The government's proposals immediately encountered fierce resistance from a broad spectrum of actors. The anti-privatisation camp was led by OTE's trade unions and included: all parties of the opposition as well as prominent members of the ruling "New Democracy" conservative party; influential professional organisations such as the Technical Chamber of Greece (TEE—representing the country's engineers); leading figures from all previous OTE management boards; a number of ex-chiefs of the armed forces, representing the "defence establishment"; last, but not least, the present OTE management, although in a less vociferous manner. All the above presented their criticisms from an essentially "statist" angle, emphasising the need for a continuation of OTE's state ownership and control. In addition, the Confederation of Greek Industries (SEB) expressed its own concerns from a pro-liberal corner.

The "statist" criticism can be summarised along the following lines:

i) OTE is a profitable state corporation with an improving financial situation. It does not pose any burden on the public accounts, being instead a net contributor. In addition, all methods that have been employed to value the organisation underestimate its real asset value, which cannot be approximated in market terms. Therefore OTE should not be sold at all.

ii) OTE's future revenues, complemented by EC funds Greece can claim in the context of the Community Support Framework (CSF) 1994–99 ("Delors II package"), are sufficient to finance its investment requirements. The government is criticised for deliberately excluding OTE from applications for EC financing under CSF 1994–99 in view of its privatisation.

iii) Greek national defence communications are to a large extent dependent on the public telecommunications infrastructure. Transfer of its ownership and/or control to foreign hands is inconceivable and poses great threats to national security.

iv) OTE has a central role to play in technology-based indigenous development efforts. It can only fulfil this by remaining in the state's control and by being incorporated into a broader industrial policy framework. Privatisation would separate OTE from its local suppliers, including its present suppliers of digital switches.

v) The commitment of OTE to universal service would be sacrificed in favour of its profitable activities. In particular, it was expected that a restructuring of OTE tariffs following privatisation will unduly burden residential subscribers.

vi) Despite the ND government's assurances, trade unions expected that privatisation would lead to redundancies and a negative labour relations climate.

In summary, there was unanimous rejection of the prospects of foreign management control and near-unanimous rejection of private sector involvement. There was equally heavy criticism, especially by OTE sources, of the government's involvement in the organisation's affairs. An alternative model was favoured, of a TO owned by the state but at an arm's length from the government and protected from all political or clientist intervention. If such conditions prevailed, advocates argued, OTE would be in a position to perform the necessary management restructuring by itself; following that, it was envisaged that the organisation's productivity would sharply increase, its quality of service be greatly improved and its international image upgraded. Therefore, OTE might even become more active on an international basis, although no specific proposals had been put forward.

A criticism from a radically different perspective was expressed by the Confederation of Greek Industries (SEB), representing both OTE's major business users and its domestic equipment suppliers. SEB is against the substitution of a state-owned monopoly by a private monopoly and emphasised the need for the institution of a liberal regulatory regime that will realise the benefits of competition in the Greek market. To promote such a regime, SEB supports a rapid harmonisation of the Greek regulatory framework to that laid down by the Commission. On the other hand, SEB is also careful not to upset its telecommunication supplier equipment members, favouring *planning agreements* for the benefit of domestic firms active in public procurement. In that respect, SEB also agrees with the relevant exemptions granted to Greece and argues for their implementation.

Following the public announcement of its intentions, the government proceeded at a quick pace, hoping to bypass criticism and complete the whole process before the end of 1993. Thus, in January 1993 a call for tender concerning the strategic investors was issued. Firms that entered the list are presented in Table 4.2.

Table 4.2
Strategic investor candidates for OTE (March 1993)

European Candidates	North American Candidates	Asian Candidates
Cable and Wireless (UK)	Ameritech (US)	Korea Telecom (Kor)
France Telecom (Fr)	AT&T (US)	NTT (Jap)
PTT Netherlands (Ne)	Bell Canada (Can)	
STET (Ita)	GTE (US)	
Telefonica (Sp)	NYNEX (US)	
	Southwestern Bell (US)	

Note: Non-European candidates may be represented by their EC-based subsidiaries. By the end of April 1993 it was reported by Reuters that the shortlisted firms are: France Telecom, GTE, NTT, STET, Telefonica. This however was not officially confirmed.
Source: Greek press (various).

The above table includes a broad spectrum of applicants, with a strong presence by US BOCs. Two Pacific rim TOs are also enlisted, in one of their first attempts to enter the European Community market. NTT, through a subsidiary offering consultancy services, has already offered advice to the government in the cellular selection process and was awarded a similar contract for the evaluation of digital switches corresponding to 1 million lines. Following the NTT's engagement in OTE's privatisation bid, the company declined to accept its evaluation contract for obvious reasons. Of the European candidates, STET and France Telecom are already present in the Greek GSM market. It was envisaged that the buyers list would be shortened to a final shortlist of approximately five candidates (see Table's note).

Regarding OTE's value, widely varying estimates appeared, depending on the method employed and the source. They ranged between 2.8 and 17 billion ECU. The most credible current valuations are presented in Table 4.3.

Table 4.3
Estimates of OTE's total value

SOURCE	METHOD	OTE TOTAL VALUE ($US,bill)	Expected Revenue from "Strategic Investor" (35%) ($US, bill)	Expected Revenue from Stock Exchange Sale (14%) ($US, bill)
Based on C&L Business Plan, 1992	Main Lines at US1300	4.8	1.7	0.7
	Management control premium	0.24	0.24	
	Total	5	1.9	0.7
Reuters 1993		4.5	1.6	0.6
Financial Analysts	P/E ratio	7–7.5	2.5–2.6	1–1.1
OTE 1991–92 Committee	Net Asset Value Estimation	9-10	3.2–3.5	1.3–1.4
OTE Trade Unions	Net Asset Value plus Market Position estimate	17	6	2.4

Sources: Various

Finally, maintaining its fast pace, the ND government presented the law instituting OTE's new structure under the privatisation regime (Law 2167/93) before the summer session of the Parliament. There, the government was forced to accept selective modifications put forward by disagreeing prominent conservative MPs before the law was eventually passed. Most amendments to the original text aimed at achieving a more even balance between the General Manager, representing the foreign strategic investor interests and the Board of Directors, where the state held a majority. When finally a compromise was reached it was not clear to what extent this had been achieved, with both government sides claiming victory.

Thus the *first* phase of the government's partial privatisation plan was completed in controversy and confusion. The *second* phase included the process of nomination of the strategic investor and the finalisation of an agreement between the Greek government and the latter. Finally, the *third* phase referred to the issue of shares through the Stock Market.

However, even before the beginning of the second stage, the main opposition party (socialists) sent a letter to the strategic investor candidates warning them about its intention to scrap the Law. This increased the tension and uncertainty and undermined further the negotiation position of the government. At that stage, although the elections had not yet been declared, it was clear that the government could do very little towards striking an acceptable deal.

In fact, the whole process was never completed. In the middle of September 1993, as already discussed under 3.3 above, the government finally decided not to award the big contract of digital switches (1 million lines) to the established OTE suppliers, despite the organisation's decision following the bid's outcome. The government's choice was made on the grounds that the final decision should be taken by the Strategic Investor and was most likely a last hopeless attempt to accelerate developments.

It was on the day after this decision was made that the government lost its parliamentary majority, due to one of its MPs withdrawing his support. The government, through its affiliated press, linked its fall to the interests of established switch suppliers. The defectant MP included telecommunication policy in the reasons for his decision, without referring however to procurement matters.

The OTE privatisation policy became one of the main issues of the electoral campaign. The government presented the view that if the plan was to be cancelled, there would be an additional burden on the public deficit. However, the socialist party leader made clear that his party was going to reverse this policy, by maintaining state control over OTE's activities, although it would consider selling shares (up to a maximum of 49%) through the Stock Market mechanism. This marked a significant turning point in the party's earlier position, which was 100 percent state ownership.

Following the election results, the socialist PASOK returned to power. One of the first decisions of the new socialist government was to declare that the recent law on the new legal structure of OTE, which was part of the privatisation plan of the former conservative government, would be scrapped. However, at the same time it gave no indication as to what would constitute the new telecommunications policy agenda. Nevertheless, a less bold privatisation is likely. The new Greek government could maintain control over OTE and keep the management of the Organisation, while at the same time issuing shares (e.g. 25%) in the Athens Stock Market in order to finance OTE's huge, by Greek standards, modernisation programme. This is a solution which has also been supported by prominent conservative MPs.

5. CONCLUSIONS: A CRITICAL EVALUATION OF GREEK PRIVATISATION PLANS

Privatisation in the telecommunication sector was the dominant policy direction in the period April 1990–Oct 1993. A summary of main events is presented in Table 5.1.

Table 5.1
Milestones in the Greek privatisation process

April 1990	First announcement of a review of the 1988 cellular market decision.
August 1990	The Prime Minister declares the government intentions to develop the cellular market through a collaboration of foreign actors with OTE.
September 1991	A call for international tenders is issued for two cellular licences. OTE is expected to participate in *both* winning consortia.

Table 5.1 cont.
Milestones in the Greek privatisation process

July 1992	New Telecommunications Law, introducing, but not implementing, regulatory reform measures (Law 2075/1992). In the same law, a special article excluded OTE from any participation in the cellular market.
August 1992	The two winning cellular consortia (Panaphon and STET Hellas) are announced.
November 1992	The government declares its intention to privatise 49% of OTE and transfer management responsibilities to a "strategic investor", who will also own 35% of the shares. The remaining 14% of the shares are to be sold through the stock markets.
January 1993	A call for tenders for the strategic investor is issued.
March 1993	A list of 13 candidates is announced in the press.
August 1993	The privatisation law, defining OTE's new legal structure (Law 2167/1993), is passed by parliament.
September 1993	The government loses its majority in Parliament. A prime justification provided by the defectant MP was the government telecommunication policies.
October 1993	The new socialist government elected repeats its decision to scrap the August law. The government has not yet presented its own plan on the OTE's future regime.

There are four areas which turned out to be critical for the fate of privatisation policies and consequently for the opening up the Greek market to FDI.

5.1 Regulatory inertia

Privatisation developments took place in an ambiguous regulatory regime. In principle, with the introduction of the 1992 Telecommunications Law, OTE's monopoly was restructured along the directions proposed by the European Commission. However, the law was not activated, since the relevant Presidential and Ministerial decrees were not issued. In fact, strictly speaking, OTE operated illegally from 1/1/1993, as it had not been granted a licence setting out its obligations and rights.

The OTE licence was delayed because it was included in the privatisation negotiations with strategic investor candidates. Naturally, during talks, all strategic investors sought the establishment of favourable conditions regarding OTE's monopoly position in voice telephony, tariffs, interconnection obligations and participation in sectors other than the basic infrastructure and telephony. In that context, the extension of the infrastructure and basic service monopoly beyond the Community deadline, achieved by the government, clearly appealed to prospective buyers.

The ex-government policies emphasised private monopoly at the expense of competitive market structure. Policy priorities were for a maximisation of the shorter-term income from OTE's sell-off, rather than the maximisation of the longer term benefits of investment flowing in, due to the creation of an attractive competitive environment. By opting for the development of a private monopoly, the ex-government assumed that the benefits of network expansion and modernisation would automatically flow to the end-users. However, the absence of a suitable pro-competitive regulatory framework could impede this flow and limit the benefits expected from the internationalisation trends at the European level, which could lead to the entry of independent pan-European carriers in profitable Greek market segments, such as international telephony traffic or major business-oriented services.

Arguably, it would be possible to consider action at a later stage, in order to correct structural imbalances created by private monopoly. However, this expectation may underestimate the dominance of the foreign investor in the Greek telecommunications market. Regulatory rules alone, without the aid of competitive pressure, might be too weak to limit the consequences of its monopoly power.

5.2 The nature of OTE's privatisation market

The very large range of estimates of OTE's value, summarised in Table 4.3, produced by diverse accounting methods and hypotheses, reflects the very large diversity in motives behind the privatisation process and the prospects for FDI. Thus, opponents of privatisation supported evaluation methods that overestimated OTE's market price in order to be able to condemn the eventual deal as a scandal. On the other hand, the government may have leaked to the press estimates lower than the actual price it was hoping to achieve, in order to project a success following "tough" negotiations. In reality, however, a major question remained publicly unanswered: was the Greek privatisation taking place in a buyer's or a seller's market ?

In order to answer the above question Greek experience alone is not sufficient, as the full picture of international developments is necessary. However, the fact that the Greek government attempted to sell OTE under pressure from internal political and fiscal problems, in an environment of market openings in a number of other countries, including Eastern European ones, suggests that in the Greek case the *buyer* may be in a favourable position.

The above hypothesis is partially weakened by the large number of candidates. Nevertheless, it is not clear which candidates were seriously interested in the bid and who were mainly participating in order to *learn* from the bidding process, or in order to trade their participation for more favourable terms in other markets. Already by the end of April it had appeared that the shortlist would be narrowed down to five or six, chiefly due to the reluctance of some candidates to proceed with their bids. Therefore, the market purchasing power might have been very unevenly distributed between the various candidates.

Candidates could respond to the Greek market environment in three complementary ways:

First, through linking with Greek actors involved in the telecommunications scene, possibly in the form of a joint venture scheme. This strategy has already been applied, among others, in the procurement of digital switches (Ericsson-Intracom) and in the bids for the cellular network. It has an important political as well as practical impact: the bid appears much less foreign, and the bidding group exploits the accumulated market experience of its indigenous member.

Second, through alliances with other foreign candidates. This could be especially helpful for non-EC actors who can in such a way overcome the unofficial barriers to entry in the Community market.

Third, by exploiting the unofficial influence of foreign governments and/or domestic actors. Telecommunication continues to be a *national* market extremely sensitive to political developments and international relations.

It is important to point out that the above strategic responses to the specific Greek privatisation initiative assume a broader importance, which extends beyond the 1990–93 policies and will be valuable in all future internationalisation developments in the country.

5.3 Future outward internationalisation strategy

As discussed above, Greek privatisation policies emphasised inward internationalisation priorities at the expense of outward internationalisation prospects. The main question however, which remains pending, is the manner in which the outward internationalisation priorities of the strategic investor would have been reconciled with the future outward internationalisation opportunities of OTE.

In opting for a rushed privatisation the government did not include any deal of that sort in the negotiations agenda. For example, no official preference for an EC or non-European candidate was declared. Yet, a decision to opt for a single non-Community candidate could cause strong (unofficial) friction in the relations with Community partners and the Commission. In forming a partnership with a Community TO, OTE would be integrated in European plans for a strong worldwide presence in the industry. On the other hand, in opting for a non-European partner Greece would be encouraging the introduction of external competition in the Community market.

It now appears that future internationalisation must take more account of outward directions, especially in the greater Balkan and Eastern European zones, through the formation of strategic alliances with stronger European TOs and other actors.

5.4 Domestic consensus formulation: political parties, press, Greek actors

The Greek experience suggests that in order for FDI through privatisation to be based in a sufficiently stable environment it is necessary for a broader domestic consensus to be built, at least at the level of accepting in principle the need for foreign involvement. This for example has probably been the case in some Eastern European markets, where there has been more public support for Western investment and de-nationalisation. Greece, however, was and still remains deeply divided over the same issue. In order for privatisation to proceed, this division has to be either bypassed or resolved.

The choice of a speedy progress suggests that the ND government was unable to follow the latter course, resorting to the former. However, even in this case, it would have been necessary to reach a minimum agreement, both within the ruling party and inside OTE's management. Agreement would probably need to include other powerful domestic actors, such as equipment manufacturers, the defense establishment and part of the media.

In summary, in order for internationalisation through privatisation to take place an alliance with strongly-placed Greek actors was required. In fact, the political flexibility of the government in meeting a range of wider requirements related to the telecommunications privatisation procedure proved insufficient.

It is no coincidence that both the first and second stages in the development model presented in section 3 were abruptly terminated due to the external influence of political developments, namely the alternation in power of the two main political parties. It is now clear that the same rule was also followed in the current "stage three" of national telecommunication development. Perhaps, the trends towards the internationalisation, not just of *services* but of telecommunication *policy making processes* may provc stronger than domestic traditions.

A lesson from the Greek case is that in order to proceed effectively with any restructuring process, a more evolutionary and gradual procedure is required, as well as a continuous attempt to intervene openly in the public debate and form the broadest possible consensus in a direction with clear objectives and stages.

REFERENCES

Ambrose, W.W., P.R. Hennemeyer and J.P. Chapon, 1990, Privatizing telecommunications systems: Business opportunities in developing countries, International Finance Corporation, Discussion Paper Number 10, Washington D.C.

British Hellenic Chamber of Commerce and Stanbrook & Hooper Seminar, 1993, EC liberalisation in the telecommunications sector: The impact on Greece, Athens, 19th May.

Caloghirou, Y., 1991, The interlocking between the purchasing power of the state and industrial activity: The case of Greece, Ph.d. thesis (in Greek), National Technical University of Athens.

Constantellou, N., 1993, Liberalising telecommunication markets: Political externalities in the Greek case, to be published in Telecommunications Policy, August 1993.

Coopers and Lybrand, 1992, Greek telecommunications study, ISC 38/91 (sector study) and ISC 42/91 (OTE Business Plan), Athens.

Greek Parliament Proceedings, 1993, Debates on OTE's new legal structure, sessions of August 11–13, Athens.

Kok, B., 1992, Privatisation in telecommunications—empty slogan or strategic tool?, Telecommunications Policy, Vol. 16 Number 9: 699–705.

Mansell, R., B. Holbrook and T. Darmaros, 1993, Telematics services for the less favoured regions of the European Communities: A political economy of development, in B. Mody, ed., Planning communication technology development: alternatives for the periphery (Sage Publications, forthcoming).

OECD, 1993, The 1992/3 communications outlook, Paris.

OTE, 1989, Five-year telecommunication development plan (1989–1993), Athens.

OTE, 1992, Telecommunication statistics in Greece (1982–1991), Athens.

Planet 1993, Greek telecommunications forecasting study, draft progress report, mimeo, Athens.

Technical Chamber of Greece, 1993, OTE—modern telecommunications: a critical review of the government's proposals for OTE's privatisation and their implications, Conference proceedings (in Greek), Athens 3-4 March.

Thomas, G., 1992, Cross-border telecommunication: the prospects for long-distance carriers in Europe, SPRU CICT Report Series No 2.

Vickers J. and G. Yarrow, 1991, Economic perspectives on privatization, Journal of Economic Perspectives, Vol. 5, No. 2, 111–132.

The Race to European Eminence
E. Bohlin and O. Granstrand (Editors)

Internationalization and Foreign Direct Investment in Telecommunications Services in Portugal

Diogo Lucena and Carmo Seabra

Faculdade de Economia, Universidade Nova de Lisboa Travessa Estêvão Pinto, 1000 Lisboa, Portugal

1. INTRODUCTION

The internationalization of the telecommunications services sector seems to be an unstoppable trend, worldwide. The technological progress driving this trend is, if anything, accelerating. The deregulation and liberalization going on in Europe will probably lead to a surge of cross-country investments, alliances and eventually some acquisitions in the EC space. The world-wide battle for positions is already raging, with the American operators moving fast to gain new markets (see Kurisaki, 1993).

The Portuguese market is quite small and is still—like most other European markets—quite closed to foreign competition. But value-added and mobile services are already liberalized and some foreign investment is taking place. With the recent EC ruling about liberalization it became urgent to face the issue squarely. A restructuring of the sector is being decided that will be followed by privatization. It is expected that further significant foreign investment will then take place.

The current organization of the Portuguese sector is under review. Portugal has three operators and there is a strong consensus that a rearrangement is needed, before privatization. Also, the fact that the Portuguese market is quite small and shared among three operators has some importance in explaining the Portuguese investment abroad, in this sector.

The aim of this paper is to give a picture at this date of the foreign direct investment in both directions. Section 2 will give a short description of the Portuguese telecommunications sector. Section 3 will deal with foreign investment in Portugal and section 4 with the Portuguese investments abroad. Conclusions are presented in Section 5, together with our view of the possible evolution in the near future.

2. THE STRUCTURE OF THE TELECOMMUNICATIONS SECTOR

2.1 Suppliers of basic voice services

Unlike most European countries, Portugal has and always has had several telecommunications suppliers. "Telecom de Portugal" supplies local telephony all over the country with the exception of Lisbon and Oporto, long-distance telecommunications and international telephone and telex services to European and North African countries. "Telefones de Lisboa e Porto" (TLP) supplies local telephony in Lisbon and Oporto, the two largest cities in the country, where most business is concentrated. “Companhia Portuguesa Rádio Marconi” (Marconi) has a legal monopoly over submarine cable and satellite transmission. It supplies intercontinental telephone and telex traffic through the two other operators for it can have no direct contact with customers.

Telecom de Portugal, the first entity to operate in the telecommunications field in Portugal, was born as a body of the public administration and transformed into a public company in 1969. Like most European public operators it supplied, until very recently, postal services together with telecommunications.[1] At the end of 1992 post and telecommunications were separated, and the company was transformed from a public into a private company, 100% state owned.[2]

Local telephony in Lisbon and Oporto was formerly a concession to a company owned by British Telecom and in 1967, when the period of the concession was over, the Portuguese Government created TLP, a public company, to supply these services. Although by that time the Government announced that TLP should be integrated in the other public operator, that intention has never turned into reality. However, some steps were taken in this direction and for many years the same board ran TLP and CTT. This hybrid situation remained until recently. In 1989 the legal status of the company was changed and TLP was transformed into a private company completely owned by the state. These alterations in the legal status of TLP and Telecom de Portugal are seen as a needed step towards privatization of at least part of the capital of the operators.

The third operator, Marconi, has always been a private company; currently 51% of its capital is owned by the state.

It must be stressed that the existence of three basic telephony suppliers does not imply and was never meant to imply any kind of competition on the provision of telecommunications services. Each of the operators has a legal monopoly, regionally or functionally defined, that the government seems determined to reinforce at least in the short run. In fact, in April 1992, a holding company, “Comunicações de Portugal” (CN), was created to manage the government's participation in the three operators. According to the Minister of Communications,[3] the main role of CN is to reorganize the sector and may eventually even disappear after that, depending on the specific structure that will be adopted.

2.2 Other telecommunications services

Data transmission services are, since 1987, supplied jointly by a consortium of Telecom de Portugal and TLP, called “Telepac”. In 1992 a second company, “Comnexo”, entered the market of data transmission. This provider is completely owned by a Portuguese private company, Compta. Comnexo has established some commercial agreements with BT.

Analog mobile services are also, since 1990, supplied by a company owned by the two public operators, “Telecomunicações Móveis Nacionais” (TMN).

Value-Added Services (VAS) are totally liberalized since 1990. There are 20 companies supplying these services, most of them integrating a foreign partner, expected to supply technology and marketing expertise. There are no limits on the participation of foreign capital in VAS. The three telephony suppliers own VAS companies.

---

[1] The name Telecom is recent. The company provided both post and telecommunications by the name of Correios e Telecomunicações de Portugal (CTT).

[2] The same happened to CTT which kept the old name but exploits only postal services.

[3] Personal communication by the Ministry, in an interview with the authors on May 24, 1993.

In 1991 two licenses for GSM mobile services were issued. One was attributed to the analog mobile telephone company owned by the public operators, TMN, and the other license was subject to a public tender. In this tender the Portuguese regulator set a limit of 25% on the capital that could be owned by foreign companies. In the consortium that won the license, Telecel, the participation of American companies has been set at the limit, with Pacific Telesis having 23% and LCC 2%. The remaining 75% belong to several Portuguese private groups.

Although the public tender for the GSM licenses was very much disputed by national companies—almost every major Portuguese group participated—European companies did not show any enthusiasm. Formally, the limit on the participation of capital companies was not applicable to EC companies but it is difficult to predict what would have happened if any European operator had tried to have a dominant position in a firm bidding for a license.

Marconi, the intercontinental telephony supplier, intended to participate in one of the consortia applying for the second GSM license, but such participation was not allowed by the ministry. Government then forced the sale to Marconi of 1/3 of the companies formed by TLP and Telecom to supply data transmission services and mobile communications.

Four licenses for the supply of radiopaging services were also issued in 1992. Once again, the Government set a limit of 25% on the share of the capital that could be owned by foreign companies (EC excluded). One of the licenses was given to a joint venture of the three public operators, "Telemensagem", and the remaining three were subject to a public tender. This time, the public operators were allowed to apply for one of the private licenses. The private consortium that won the license for the GSM telephone service was given one of the radiopaging licenses under the name of "Telechamada". The other winning companies are "Contactel", 51% owned by Marconi (which in this way has shares of two competing companies), 15% by Telefonica de Espana, 10% by Matrix Telecommunications (an Australian firm) and the remaining 24% by other private Portuguese firms. The last license was given to "Finacom", a consortium of Portuguese firms, Hutchinson (from Hong-Kong), and Cofira, a French firm owned by Companhie Financière des Eaux.

Two licenses for the supply of trunking systems were also subject to a public tender in 1992. There were only three participants, and only one license was issued. The winning company, "Rádio Móvel", is owned by Finacom, one of the owners of a radiopaging license. A tender for the second license took place in October 1993, and there was only one applicant.

2.3 The structure of regulation

Regulation of telecommunications has always been a direct responsibility of the Ministry of Transports and Communications, although some regulatory functions, like the management of the radioelectric spectrum, were conducted by Telecom de Portugal. Since 1989 regulation is formally conducted by a body specially created for that effect, "Instituto das Comunicações de Portugal" (ICP).[4]

It should be noticed that the effective strength of this newly created regulator is not yet very large. First, because the regulator is partly dependent upon the operators. This body had to recruit its personnel mostly among the employees of the TOs due to the small dimension of the market and it is financially dependent on the operators. True independence takes time to achieve. Second, the government created the regulatory body but did not delegate to it the really relevant regulatory functions. In fact, until recently its responsibilities concerned mainly

---

4 Formally, ICP was created very soon afterwards, in 1981, but was not activated until 1989.

the approval of equipment and the allocation of the radio spectrum, functions that are currently little more than bureaucratic decisions. In turn, the regulatory functions that remained under the direct responsibility of the government—the issue of licenses and tariff approval—are the ones that really determine the competitiveness of the sector.

The recent creation of more competitive market structures, namely in GSM and radio-paging services provision, is changing this picture, as the regulatory body is becoming responsible for the regulation of competitive conditions in these markets. The regulation of tariffs is also becoming more and more a function of ICP. Nevertheless, the Minister continues to be the really relevant regulator.

An important regulatory issue is the position of the government towards Foreign Direct Investment in Portugal. According to the "General Guidelines for the Telecommunications Services Sector", approved by the Portuguese Parliament, foreign companies cannot be allowed to own more than 25% of the country's telecommunications providers. A company is considered as foreign if more than 50% of its capital is held by foreigners. Formally, EC companies are exempt from this type of restriction. Nevertheless, this exemption is probably one of the reasons why the privatization of Portuguese operators is being delayed. The Minister stated clearly as his policy objective the keeping of control of the main operators in national hands.

3. FOREIGN DIRECT INVESTMENT IN TELECOMMUNICATIONS

3.1 A general view

In a situation where the operators are state-owned monopolies, foreign direct investment (FDI) can arise in only two situations: i) when the public operators are privatized and ii) when licenses for the provision of new services are issued. In both cases, restrictions imposed by the government on the acquisition of shares by foreign companies or on the participation of foreign partners in new companies become crucial.

As was mentioned above, privatization of the Portuguese telecommunications suppliers has been postponed, as the government is waiting for the formation of national groups with a sufficient interest and expertise in telecommunications to go ahead with the privatizations. The purpose of this strategy is to avoid the acquisition of Portuguese operators by foreign companies, and in particular by European operators. In fact, these cannot be legally prevented from trying to buy shares of the telecommunications companies when they are put to sale. The only way to eventually avoid foreign acquisition from those sources is to create competition from national firms. The Portuguese Government hopes that Portuguese companies will be able to increase both its international competitiveness, and its expertise in the provision of telecommunications services, so that, when privatization comes, there will be enough know-how and interest in telecommunications to minimize the chances that the Portuguese operators are totally acquired by the large EC companies or even Telefonica.

The chances of success of the government's strategy do not seem to be very large; presently there are only three Portuguese private groups operating in the telecommunications services business: Grupo Espírito Santo and Amorim in GSM and paging,[5] and Interfina in paging. In

[5] In August 1993 these two companies formed a holding company, called Telepri, whose purpose is to buy shares of the Portuguese telecommunications suppliers, if and when they are put to sale. This suggests that they intend to stay in the telecommunications business.

spite of the great interest shown by Portuguese private firms in the public tender for the GSM license it is not clear to anyone (themselves included) whether these investments in telecommunications are anything more than a temporary financial application.[6] Telecommunications is an unknown area of business, requiring very large investments with a very delayed return, and it is possible that they will not become interested in the privatization of the public operators. If this happens, the postponement of the privatization may not bring the desired outcome.

A second reason why privatization has been postponed has to do with the decision to reorganize the sector first. In fact, after private interests are brought in, any restructuring becomes much more complicated.

Very recently "Cabo de Portugal" was formed, as a joint venture of Telecom and TLP. The purpose of the new company is to hold the participations of the two operators in any cabling venture in Portuguese cities, in association with other partners, chosen on a case by case basis. At the moment no Portuguese city has been cabled, and this is a possible venue for the foreign direct investment in the future.

In this situation, the only opportunities for foreign direct investment arose with the issue of licenses for new services, and in particular for mobile services: GSM, paging and trunking.[7]

3.2 Mobile services and VAS

In the GSM private operator, Telecel, Pacific Telesis has 23% and LCC 2% of the capital which in turn amounts to 96.2 million dollars. Until the end of 1993, 88.5 million will have been invested, 22.1 million by the foreign partners. These percentages amount to the allowed maximum. Nevertheless, there are always ways to circumvent legal restrictions on capital. In this case, the escape is the provision by foreign partners of supplements.[8] Hence Pactel is expected to invest an additional 25 million dollars. Of these, 6.7 million have already been called. This makes the total investment by Pacific Telesis amount to 26.9 million dollars, until now. It must be stressed that Pactel is the partner with the technological and commercial expertise, which means its real importance in the consortium does not depend exclusively on the percentage of the capital it owns. In turn LCC has already invested 3.8 million dollars, of which 1.9 are supplements. In Table 3.1, FDI in Portugal is summarized.[9]

The three private suppliers of paging services also have foreign partners. Finacom is a consortium initially formed by Interfina, a Portuguese group, a French company, Cofira (held by Companhie Financière des Eaux) and Hutchinson. Each of these two foreign partners was supposed to have 12.5% of the social capital of the firm. In the meanwhile Hutchinson changed its international strategy and decided to sell its shares in most European companies.[10] Part of its share in the capital of the Portuguese paging firm was acquired by Cofira (7.4%) and the remainder was taken by Finacom. Nevertheless it is expected that the share of Cofira will increase to the allowed 25% as soon as the French legislation applies concerning the

---

6 Personal communication by the President of Finacom, Dr Raúl Junqueiro on May 26, 1993.

7 The same happened in several other European countries. See Elixman and Schnöring (1993).

8 According to the Portuguese law, these supplements are exempt from the restrictions on foreign capital because they do not give voting rights.

9 These data were obtained in personal conversations with the President of Telecel, Dr António Carrapatoso, on May 13, 1993 and with Dra Ana Barreto, from Telecel, in November 1993.

10 The only European business still held by Hutchison is in the UK.

consolidation of the accounts of French firms involved in international business changes.[11] Until now Cofira has invested 1.3 million dollars.

A second paging license was given to Contactel, a joint venture of Marconi, Telefonica de Espana and Matrix Asia. This last company is a joint venture of Marconi itself (22.5%) and Matrix International (77.5%), an Australian telecommunications supplier. In Contactel Marconi has 51% of the capital, while Telefonica holds 15% and Matrix 10%. Social capital amounts to 3.3 million. The foreign 25% thus represents 0.83 million.

The third private paging license was given to Telechamada, a company 100% owned by Telecel, the supplier of GSM. There is no direct foreign investment here, apart from the one involved in Telecel itself.

Three trunking licenses were also recently subject to a public tender. There were three applicants, but only one license was attributed, to Rádio Móvel, a company 60% owned by Finacom, 15% by TLM (a joint venture of TLP and Telecom) and 25% by Protocol Ventures, a UK company. Total investment in this service will amount to 3.2 million dollars during the next two years. Until now total investment was 1.6 million dollars. The foreign direct investment until now was 0.384 million dollars.

TLP and a Spanish company also have a small joint venture in the area of telemarketing, Iberfone. The company is 50% owned by TLP and 50% by Spanish interests, and the total foreign investment is about 0.33 million dollars.

Of the twenty companies that supply VAS, about half have foreign partners. Total investment involved in these companies is small, about 0.2 million dollars each.[12] Nevertheless, in these services, the telecommunications component is usually not very large, the information in itself being the most important part.

Table 3.1
Foreign direct investment in Portugal (million dollars)

Company	Country of origin	Service	Supplier provided	Amount invested	% of capital held
Pacific Telesis	US	GSM	Telecel	26.9	23%
LCC	US	GSM	Telecel	3.8	2%
Cofira	France	Paging	Finacom	1.3	19.9%
Telefonica	Spain	Paging	Contactel	0.495	25%
Matrix Asia	Australia	Paging	Contactel	0.335	15%
Protocol	UK	Trunking	Rádio Móvel	0.384	15%
Ventures	Spain	Telemarketing	Iberfone	0.33	50%

3.3 Financial investment

Marconi is partially in private hands, and is publicly quoted. It is clear that some of the stock is in foreign hands, but it was not possible to reach a figure for that amount. Anyway, the feeling, both in the company and among people involved in the stock market, is that the

[11] Presently, the consolidation is not required for participations smaller than 20%. This is the reason why Cofira acquired 7.4% of Hutchinson's shares, which together with their initial 12.5% makes 19.9%.

[12] Personal information given by Dr José Manuel Ferrari Careto, Director of ICP.

amount is not too important and is mainly held as a financial investment, for diversification purposes, by major institutional investors, especially some US and British pension funds.

4. INTERNATIONALIZATION OF THE PORTUGUESE OPERATORS

4.1. A general view

Internationalization as a strategy to be actively pursued is a very recent concept for the Portuguese domestic telecommunications operators, Telecom and TLP, traditional with their focus on the satisfaction of their internal markets. The relevant international component of the telecommunications business was international telephone traffic, a service that was produced jointly by the TOs. Hence Portugal is, together with Greece[13] and Luxembourg, in the group of European countries less involved in systematic internationalization strategies, in spite of the establishment of several agreements with other European operators, usually concerning the interconnection of services.[14] The existence of three public service providers in Portugal makes internationalization more difficult, as it decreases the relative dimension of each of these already small companies (given the small dimension of the market). Nevertheless, some timid attempts to internationalize operations have been recently made by both suppliers, always directed at Portuguese-speaking countries, particularly African ones. The results of these efforts are summarized below.

Marconi deserves a special word in this setting, as internationalization has been a major objective of the company. The nature of its business—satellite and submarine cable telecommunications—is probably partly responsible for the company's more open attitude, as these facilities are usually provided by consortia of different operators. The fact is that, in spite of its small dimension, Marconi has been trying to diversify geographically.[15] It created a holding, SGPS-Marconi, specially aimed at conducting its internationalization strategy. Its operations are listed below.

The recent creation of the government holding company for the telecommunications sector is having mixed effects on the company's strategy. In a way, a formal association with the other public operators greatly increases the strength with which the company can present itself to public tenders for the supply of telecommunications services outside the country. On the other hand, it is not clear yet whether the holding will sustain a totally decentralized position concerning the strategy of its associates;[16] Marconi may still find out that its degrees of freedom have significantly decreased.

Until now the remaining Portuguese telecommunications suppliers have not been involved in internationalization movements. Telecommunications is still an unknown area of business for Portuguese private firms, and the consolidation of their domestic investments is the priority. However, if they decide to stay in the business as seems probable now, internationalization is seen as a necessary step. Both Telecel and Finacom are interested in the area of mobile communications, in particular in GSM, PCN and paging. To Finacom associations

[13] For more details on the Greek case, see Caloghirou and Darmaros (1993).

[14] For more details on the internationalization strategy of other European operators, see, for instance, Brekke et al. (1993) Karpakka (1993) Henten and Skouby (1993) and Johansson and Granstrand (1993).

[15] Personal communication by the President of Marconi, Dr. Gonçalo Sequeira Braga, on May 26, 1993.

[16] Personal communication by the President of the holding company, Dr. Eurico Cabral da Fonseca, on May 26, 1993.

with the large European private operators would seem the best way to internationalization and Spain could be an ideal location to begin international operations. Telecel is also looking for foreign partners, and Portuguese-speaking countries, in particular Brazil and African countries, are obvious targets.

4.2 Ventures led by Telecom and TLP

TLP is involved directly in only one venture outside Portugal. Together with Telecom they own 50% of Teleserve, a company incorporated in Mozambique (25% each). The other 50% are owned by TDM, the telecommunications operator of Mozambique. This is a small company that commercializes and maintains terminal equipment and an actual monopolist in this market, although not for legal reasons. Teleserve also owns the only license to operate a paging system in Maputo, the capital city. The total amount invested by the Portuguese operators is 0.186 million dollars, 0.093 million dollars each.

The other participations in foreign ventures are made through other Portuguese companies (e.g. Directel), where Marconi has a leading position.

Telecom Portugal has no other direct foreign involvement, beyond the 0.093 million dollars in Teleserve. Again, other involvements are through other Portuguese companies, where Marconi has a dominant position.

Being an international operator, Telecom has participated in several consortia to pay for international submarine cables. Some of these cables are attached in Portugal, and the total investment of Telecom in such systems adds up to 27 million dollars.[17] Others link third countries, and the information Telecom supplied us adds together the investment in such systems and the payment of "rights of way". This item totals 17.5 million dollars.

4.3 The internationalization of Marconi

Marconi is the only one of the three operators with a significant portfolio of foreign investments. Most participations are quite small and its strategy has been to concentrate mainly in the Portuguese-speaking countries, in Africa and Brazil, and in Asia.

Most of these foreign participations are held by a holding company "CPRM-Marconi SGPS", which, in turn, is 100% owned by Marconi itself. A few participations were kept directly in Marconi and we will simply add them together. We organize the information by main "business lines".

The first area is in public telecommunications. Marconi is an important partner in the public operators in Macau, Guiné-Bissau and S. Tomé, the last two being former Portuguese colonies in Africa.

The "Companhia de Telecomunicações de Macau" has the legal monopoly of the public telecommunications in Macau, with the exception of paging services. The company is controlled by Cable and Wireless (51%) and Marconi has a 28% participation. The investment corresponds to about 5 million dollars. Notice that Macau has a special status, for it is a territory under Portuguese administration, although not under Portuguese sovereignty. Anyway the administration will be turned back to China in 1999.

Marconi has a stake of 51% in Guiné Telecom, the legal monopoly for all telecommunications services in Guiné-Bissau, including both domestic and international, fixed and mobile. It

---

[17] These figures include investments since 1971 and underestimate somewhat the true values, for they are not fully corrected for inflation.

has also the sole responsibility for the management of the company. The corresponding investment amounts to 0.5 million dollars.

A similar arrangement is true for S. Tomé, another former Portuguese colony in West Africa. The country is very small and the total investment is also quite small, 0.036 million dollars.

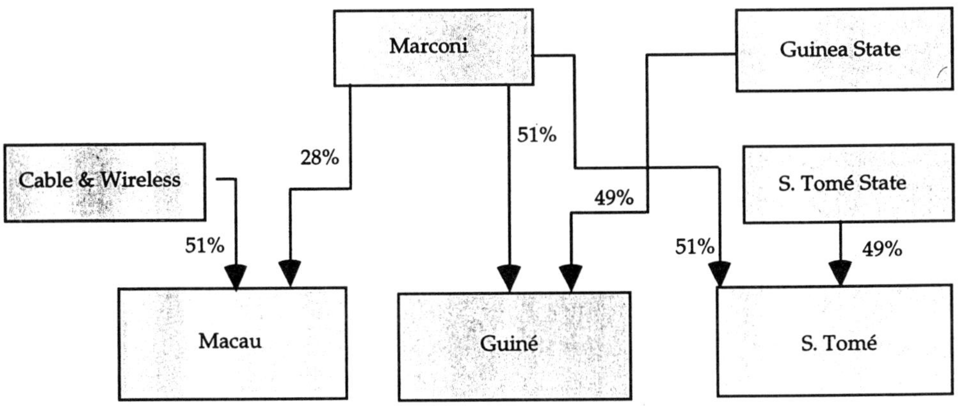

Figure 4.1. Marconi. Foreign investment in public telecommunications.

In the mobile communications area Marconi has interests in both Brazil and Asia. The strategy in all countries has been to make joint ventures with local groups, and Marconi created two holding companies, one for Brazil and another for Asia.

The holding for Brazil is called "CPRM Brasil Telecommunications" and is 100% owned by Marconi. In turn CPRM Brasil owns 49.99% of Mobitel, the operating company which won the tender for one of two national licenses in Brazil for mobile telephony. It is operating already in Rio de Janeiro, S. Paulo and Brasilia and is expanding fast. The Marconi investment until 1993 will be on the order of 2.3 million dollars.

A similar arrangement was made for the investments in mobile communications in Asia. A holding company, CPRM Asia, was created, 100% owned by Marconi. In turn CPRM Asia holds a 22.2% share in another holding company, Matrix Asia, the other 77.8% being held by an Australian company, Matrix International. It is Matrix Asia that holds stakes in three joint ventures, in the Philippines, Thailand and Malaysia. Matrix Asia was formed to bid for tenders for paging licenses in the Asia region. In fact the three companies referred to above hold paging licenses.

"Easy Call Communications Philippines" has one of the two paging licenses in that country. Currently it has 70% of the market, but there will be a third operator soon. The company has over 60,000 pagers and is very profitable. It is looked upon as the "cash-cow" of the group.

"Matrix Thailand" holds one of the 4 paging licenses in Thailand and has about 15% of the market. The local partner—Cellcom—who owns 51% of the company, is a group that wants to enter the telecommunications area, as part of its growth strategy. It is pushing for a very aggressive policy and the company installed 50,000 pagers in 4 months and 120,000 are forecast until the end of the year.

"Matrix Malaysia" also holds one of the four paging licenses for the country. It is a smaller operation and has about 15,000 pagers.

We estimate the total investment of Marconi in this area to be 1.7 million dollars.

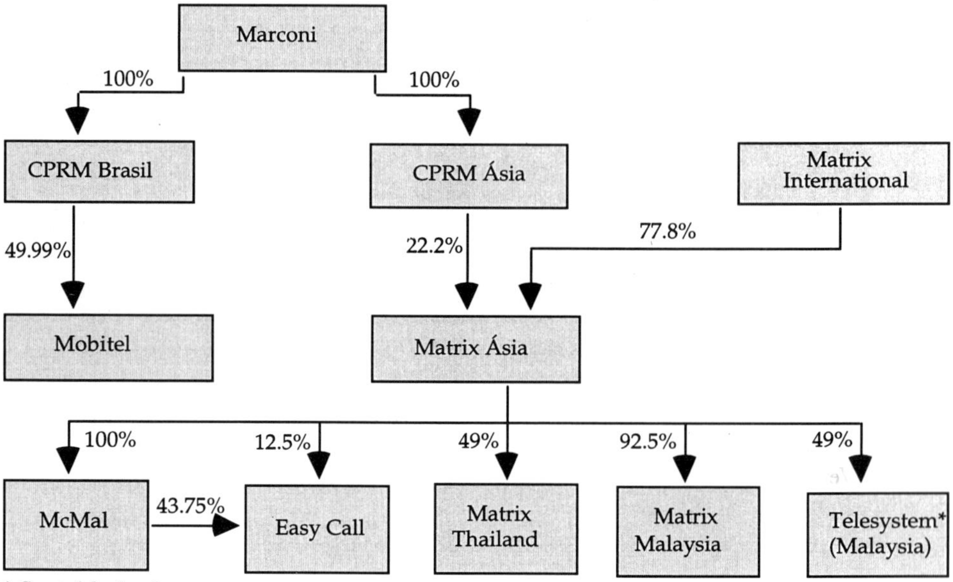

* Created for legal reasons.

Figure 4.2. Marconi. Foreign investment in mobile communications.

A third area where CPRM has some foreign participations is in the value-added services. The more important of these two ventures is a voice-mail company in the UK, "Voice Processing", of which CPRM owns 56.7% and CPRM Asia another 28.2%. As CPRM Asia is 100% owned by Marconi this gives an 85% stake in total. This participation corresponds to 2 million dollars.

The other venture is a data communication firm, operating in Mozambique, "Teledata de Moçambique". Marconi owns 50% of the operation, corresponding to an investment of 0.02 million dollars.

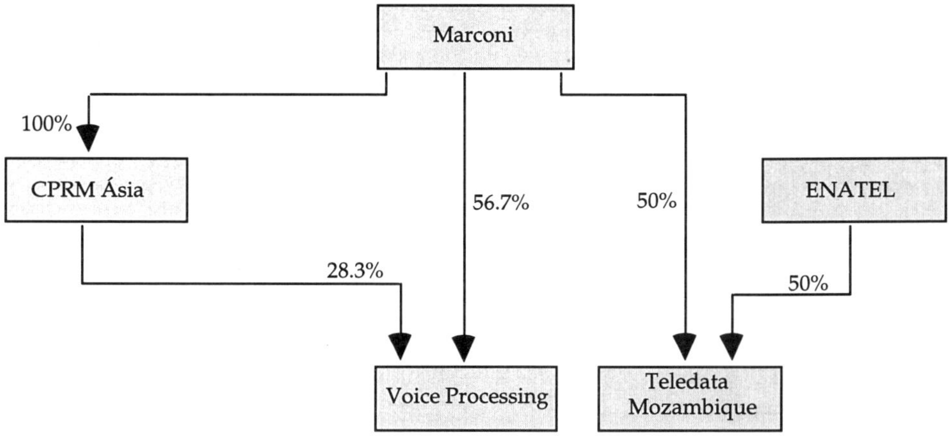

Figure 4.3. Marconi. Foreign investment in value–added services.

Finally, there is a fourth business area where Marconi has some interests abroad, publication of directories. These activities are organized under a Portuguese sub-holding, Directel, owned together by Marconi (60%) and TLP and Telecom (20% each). Directel in turn has stakes in four companies, which do the directory business in Macau, Kenya, Mozambique and Cape Verde. The total Portuguese investment in these four companies amounts to 1.28 million dollars.

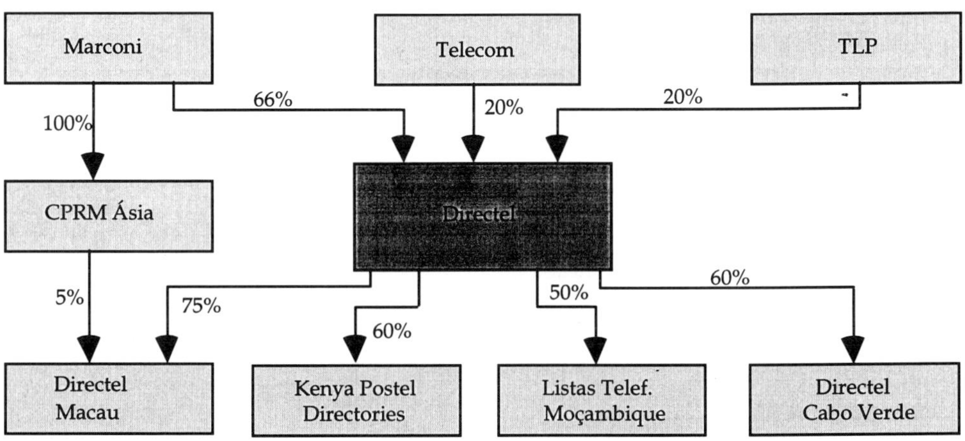

Figure 4.4. Marconi and others. Foreign investment in directories.

Marconi, as an international operator, has participated in several international systems, namely in submarine cables, and has also invested the Portuguese share in the satellite organizations. The following tables summarize these investments.

Table 4.1
Marconi investment in submarine cables (million U.S. dollars)

	Attached in Portugal	Other
Operating	22.0	21.0
In construction	78.6	12.3
Total	100.6	33.3

Table 4.2
Marconi investment in satellite organizations (million U.S. dollars)

Intelsat	10.1
Eutelsat	1.9
Inmarsat	1.6
Total	13.6

* Some of the investments will continue until 1996.

The next table summarizes the Portuguese investment abroad.

Table 4.3
Portuguese investment abroad (million U.S. dollars)

Public Telecommunications	5.54
Mobile and VAS	3.72
Directories and Other	1.47
Sub total	10.73
Participation in International Means	194.00

5. CONCLUSIONS AND FUTURE PERSPECTIVES

We believe that in the very near future there will not be any significant moves in the internationalization of the Portuguese telecommunications.

The new holding company (CN) is very recent and there are strong expectations about the policies it will choose to implement. As referred to in the first part of this paper it completely controls both Telecom Portugal and TLP and has a majority in Marconi. Most people in the sector feel that the current arrangement among the three main operators is quite inadequate and that something must be done to change the "status quo". It is widely felt that the main role of the CN will be to decide and then enforce a restructuring of the sector.

Under those circumstances the existence of private capital in any of the three operators is perceived as an additional restriction that will limit the range of possible rearrangements. The partial privatization of TLP, announced some time ago, was postponed, and we believe this was the main reason. For the moment CN has been trying to define areas or lines of business in which each operator will have a predominance. But probably the rationalization will be much deeper, including the possibility of redefining the scope of each operator's monopoly and even the merger of the new units thus created. For the moment there are no decisions, but the most widely believed speculation, among people in the sector, is that TLP and Telecom will be merged, with Marconi integrating the new company as a second step. The degree to which this integration will be done is still under strong dispute, but it seems clear that it is the new global unit that will be the privatization vehicle. An alternative proposal defends the integration of the domestic part of Telecom with TLP, forming a domestic company, the international traffic of Telecom being handed to Marconi, to form a second, international, operator. Before such moves are made (or before there is a clear announcement that no such moves are being contemplated) it is not possible to consider the privatization of any of the main operators. Actually, the fact that a significant part of Marconi is already in private hands is regarded as a possible source of problems in such a rearrangement.

With such major uncertainties we believe that neither TLP nor Telecom will be in a position to develop a strategy of international alliances or decide on any program of internationalization. Besides, both companies, and especially Telecom, feel that they have their hands full with a major internal reorganization and are not ready to define these strategies.

The Minister of Telecommunications, in a important speech made some time ago, also stated that it is government policy to try to keep as much of the sector as possible in Portuguese hands, even in the context of a future privatization. This is also an objective that runs against an immediate privatization, for the government feels that there are no Portuguese groups with sufficient interest in the sector and has the expectation of promoting the appearance of such groups, through the concession of licenses to the private sector in certain areas like mobiles services (G.S.M., paging, trunking). But this will take some time.

All these timings are conditioned by the dates and scope of E.C. policies for the telecommunications sector. The recent liberalization measures decided by the EC, effective as from 1998, will not apply immediately to Portugal. In fact Portugal was allowed a five-year transition period. The general view at the present moment in Portugal is that this will not be an active constraint in the next few years. It can be assumed that this somewhat relaxed view of the timings is not shared by everyone, especially the people close to Marconi, who feel already that its market power is being challenged by the possibility of (hidden) entrance by others. There is no doubt that the first of the three operators to feel the pressures of competition was Marconi.

Except for a new license to operate a second trunking system, there are no more tenders on the horizon, and so it is not to be expected that new consortia with foreign participation will appear; a possibility sometimes discussed is the privatization of TMN, the mobile company owned by the three operators. But it seems clear now that such a decision will not be taken before the major uncertainties about the organization of the sector are resolved.

To sum up, it looks very improbable that any significant investments will take place, in any of the two directions, in the short run.

Marconi will probably keep pursuing its current policy, trying to get new valuable licenses, especially in the Portuguese-speaking countries and Asia. A major change of policy may happen if, in fact, the European traffic will be transferred from Telecom to Marconi. This would represent a major shift of interests, turning Marconi from a "colonial" firm into a European-centered one.

Portugal has a window of time in which to restructure its telecommunicatons sector and gear it towards a considerably different competitive situation that will develop in the near future. We feel that this window of time is not very big and it is not clear whether the restructuring will be fast enough not to place the new units in a big competitive disadvantage. The simple announcement of the 1988 liberalization will accelerate the rate at which new alliances are being formed and it is crucial for the Portuguese firms that their positioning be made at the early stages of the game.

REFERENCES

Brekke, R., S. Jensen and C. Riis, 1993, Internationalization of the Norwegian telecommunications market, paper presented at the ITS European Conf., June 20–22, held in Sweden.

Caloghirou, Y. and T. Darmaros, 1993, Internationalization of telecommunication service provision and the Greek privatisation debate, paper presented at the ITS Regional Workshop, September 7–8, held in Tel Aviv.

Elixmann D. and T. Schnöring, 1993, Internationalisation of the German telecommunications service market and strategic behavior of DBP Telekom, paper presented at the ITS European Conference, June 20–22, held in Sweden.

Henten, A. and K.E. Skouby, 1993, Inward and outward FDI in Danish telecommunications, paper presented at the ITS European Conference, June 20–22, held in Sweden.

Johansson, O. and O. Granstrand, 1993, Internationalization of the Swedish tele services market, paper presented at the ITS European Conference, June 20–22, held in Sweden.

Karpakka, J., 1993, The Finnish private telcos abroad: Motivations and strategies, paper presented at the ITS European Conference, June 20–22, held in Sweden.

Kurisaki, Y., 1993, Challenge to the globalisation of public telecommunications operators—Toward a new framework of telecommunications policy, paper presented at the ITS European Conference, June 20–22, held in Sweden.

Most of the information required to write this paper was not readily available in published, or even written, form. The following people were very generous with their time, either by giving us long interviews or by supplying the relevant information.

Ana Barreto, Economist, Telecel.
António Carrapatoso, President, Telecel.
Cidália Morgado, Director, Telecom.
Eurico Cabral da Fonseca, Presidente, CN.
Fernando Mendes, President, ICP.
Filipe Paes de Vasconcelos, Member of the Board, SGPS Marconi.
Francisco Murteira Nabo, President, SGPS Marconi.
Francisco Silva, Director, CN.
Gonçalo Sequeira Braga, President, Marconi.
João Mello Franco, President, TLP.
Joaquim Ferreira do Amaral, Minister of Transport and Communications.
José M. Ferrari Careto, Director, ICP.
Júlio Rendeiro, Member of the Board, Finacom.
Lucília Fernandes, Director,Telecom.
Luís Todo-Bom, President, Telecom.
Norberto Fernandes, Director, Marconi.
Raúl Junqueiro, President, Finacom.

The Race to European Eminence
E. Bohlin and O. Granstrand (Editors)

A Note on Internationalization in the Netherlands

Ola Johansson*

Chalmers University of Technology, Department of Industrial Management and Economics, S-412 96 Göteborg, Sweden

1. INTRODUCTION

The Netherlands, with its 15 million inhabitants, is one of Europe's smallest but most densely populated countries. Most of the industry is located in the major area Randstad (10% of the total area) including the industrialized cities Rotterdam and Amsterdam (and the Hague). The mouth of the river Rhine and the city Amsterdam have historically been important communication and transport centers. The Dutch economy is to a large extent based on imports and exports. The Netherlands plays a key role in international goods transport and in the transport of information within Europe. More than 50% of international calls from the Netherlands are made to Germany. The Dutch business customers have traditionally been internationally oriented and are now looking for new solutions independently of the offerings by PTT Telecom BV.

The purpose of this note is to give a brief description of some features of the internationalization of the Dutch telecom services market and of some of the foreign activities by PTT Telecom in order to illustrate the threat to, and the attractiveness of, a highly internationalized, small and densely populated country. The note is not intended to fully cover the internationalization of either the Netherlands or the PTT Telecom. It is rather an illustrative example with scanty data.

The basic argument of the note is that PTT Telecom is highly vulnerable to foreign entry, given the structured characteristics of the Netherlands. Thus, PTT Telecom has had to take several strategic initiatives to prepare itself for internationalization.

2. PTT TELECOM TODAY

The holding company Royal PTT Nederland NV (KPN), who controls both PTT Post BV (the post administration) and PTT Telecom BV, became a liability company in 1989 when a new Telecommunications Act was introduced (some part of the shares is scheduled to be privatized in mid-1994). PTT Telecom has five business units; one is responsible for the fixed network and services, two are responsible for the selling of equipment and services to business and consumers, respectively, one is responsible for the mobile network and services, and one deals with international traffic and services. The business units are responsible for planning and

* The help and guidance of senior consultant Robert Wagenvoort at PTT Telecom is gratefully acknowledged.

investments in networks, while the 13 districts (divided into 32 regions) sell, promote and operate services. There is also a consultancy firm, Nepostel (owned by the Royal PTT), who has activities mainly in developing countries. In addition, there is a research company within the conglomerate which is owned by the Royal PTT and financed mainly by selling services to PTT Telecom. The research company has no major joint research programs with the large manufacturer Philips.

Over the last years PTT Telecom has become more independent of the Royal PTT and is now seeking its own business opportunities on its own conditions. These conditions are changing rather rapidly now and PTT Telecom is facing competition from foreign new entrants. The Netherlands constitutes an attractive market because of its smallness, high population density and internationally oriented industry. Especially the three large cities, with their concentration of the transport and communications industries, form an attractive market segment for foreign entry.

These characteristics may be illustrated by the following figures. The population density in the Netherlands was high in 1990 (366 inhabitants/km^2) in comparison to the average in Europe (101) (United Nations, 1993). Further comparisons reveal that the UK also has a rather high population density (234 inhabitants/km^2 in 1990), while France has a lower figure (102). Turning to international trade figures, the Netherlands had a total import *and* export of 20,000 USD/inhabitant in 1990 in comparison to the average of 6500 USD/inhabitant in Europe (United Nations, 1993). France had a total of 8000, Germany 9500, and the UK had 7500 USD/inhabitant. These simple figures provide an illustrative argument that the Netherlands has a highly internationalized economy and that it is a densely populated country.

3. INTERNATIONALIZATION IN THE NETHERLANDS

This section will describe some of the activities by foreign telcos which have internationalized the telco industry in the Netherlands and, in turn, put pressure on PTT Telecom. Let us first, however, identify some of the drivers of internationalization (Granstrand & Johansson, 1993) in terms of technological development in networks and services, change in customer demand and liberalization of telecom services.

First, in 1991, 90% of the long-distance network and 80% of the local network in the Netherlands were digitized and full digitization of the long-distance network is planned to come into effect in 1995. Switches in the telecom network are supplied mainly by Alcatel, Ericsson and AT&T (through a joint venture with Philips). It seems reasonable to believe that this multi-vendor environment causes difficulties in implementing advanced services that are ahead of standardization. However, the technological development (or rather here the implementation of digital switches) is also creating new possibilities which existing regulations have not considered.

Second, in the Netherlands large customers, such as Philips, have had international business activities over a long period of time, which means that they actively search for new solutions and services which may enhance their international demand for telecommunications. An example is BT who won a contract with IBM in the Netherlands. IBM rents a 2-Mbit leased line from PTT Telecom (who is obliged to provide leased lines upon demand) to the UK and installed equipment in localities which BT (or rather Syncordia) rents. This solution by-passed PTT Telecom for all international accesses of IBM Netherlands.

Third, liberalization in the Netherlands closely follows the development of the increasingly liberalized regulations issued by the EC. The EC has decided to open up for competition in all telecom services in 1998 as proposed by PTT Telecom. The technology development and possibilities arising from new digital networks mean, however, that the race has already started, just as in the case of IBM and BT above. Large users and owners of telecommunication, just as the energy distribution companies, are today still only allowed to use their networks for internal purposes. PTT Telecom has been granted an exclusive concession by the government to provide public services such as voice telephony. Value-added services are, however, possible for anyone to provide without a license and a second cellular (in GSM) telco is scheduled to be authorized within a rather short period of time. Closed mobile networks, inter-connected with the public network, have also been possible to establish.

The changes illustrated above under the three drivers have resulted in several new entrants into the Netherlands. Rather surprisingly, the manufacturing and distributing of switches by AT&T in the Netherlands since the 1980s has attracted the interest of the Italian STET and the Spanish Telefónica. Since 1989, STET holds 19% of this Dutch manufacturing unit and in 1990 Telefónica bought 6% in the same unit.

Bell Atlantic and US West formed a joint holding software company (Atlantic West BV) relatively early (in the late 1980s). Bell Atlantic has also provided network monitoring systems to PTT Telecom.

Turning to value-added network services (VANS), new entrants (e.g. GEIS and IBM) have offered value-added services such as EDI over a period of many years. Sprint was rather early among traditional telcos, when they established an access node (switching center) to their SprintNet in the late 1980s. Since 1991 an increasing number of foreign activities are seen in the Netherlands. AT&T resells leased-line capacity and has installed switches in Rotterdam and Amsterdam. Together with Sprint, AT&T also won an outsourcing contract with Unilever in 1991. AT&T also offers their Accunet international datacom services. Furthermore, both AT&T and MCI have offices in the Netherlands. Ameritech offers voice-mail service over their Tigon network. Cable&Wireless has an access node in the Netherlands for its European MDNS service. BT since May 1993 is "officially" reselling leased lines for VANS and has three access points in the Netherlands for its Tymnet network.

Thus, despite PTT Telecom having been protected up until the early 1990s by regulatory entry barriers, the attractiveness of the Netherlands, with all its new foreign entrants, has created a major threat to them. Moreover, the technological development has made it possible and economically feasible to offer services independently of the regulatory barriers.

The development has thus forced PTT Telecom to change its strategy from being traditionally engaged in international cooperations and standardization bodies (between monopolies), to adopting a more aggressive role. Below, some observations of this change are given.

4. FOREIGN ACTIVITIES BY PTT TELECOM

Over the last five years PTT Telecom has set up offices abroad for foreign representation, e.g. in Tokyo, London, New York, Paris, Brussels and Berlin. However, PTT Telecom concentrates its foreign activities on the European market and has, for instance, cooperations with DBP Telekom (ISDN and a videolink), France Télécom (Minitél (France Télécom's subsidiary Intel Matique owns 16% of the videotex services in the Netherlands), VPN service and ISDN), Belgacom (videolink, ISDN), and BT (VSAT, ISDN and videolink). Moreover, PTT Telecom has had some form of R&D cooperation with NYNEX since the late 1980s.

PTT Telecom seeks to follow its customers and internationalize its operations (through virtual private networks, including leased lines etc.) accordingly. Today, this is being done by the Unisource venture. As an example of this approach, PTT Telecom has established a VPN network between the US, the UK and Hong Kong with interconnections to Japan, Singapore, Australia, Canada and France aimed at companies such as Philips, Nissan, Nedlloyd, KLM and DSM. PTT Telecom also cooperates with DBP Telekom and France Télécom in Transponet to form a European EDI network for the transportation industry. In addition, PTT Telecom established a joint venture (35%) with the American companies American Airlines (AMR) and CSX in 1992 in order to provide an EDI service (Encompass) for the transportation of cargo. PTT Telecom is also active in other international cooperations such as Inmarsat, GEN, Global FON, BoD, Globand etc. In 1990 PTT Telecom invested 5.38% in Infonet. In 1993, PTT Telecom increased that share to 7.17%.

Over the past few years PTT Telecom, however, has complemented its previous strategy, having started up some more aggressive activities such as investments in Eastern Europe (Bulgaria, the Czech Republic and the Ukraine) and the formation of Unisource (with Telia of Sweden and the Swiss PTT). In the Ukraine, the joint venture with AT&T is modernizing and operating services on the long-distance and international network as well as the cellular and paging network. Other investments such as the restructuring of the network in Indonesia are also interesting if they can prove to have an economic value to the PTT.

In the fall of 1991 PTT Telecom and Telia announced that they had created an equally owned holding company called Unisource. Already in the late 1980s PTT Telecom, Telia and British Aerospace cooperated through the joint venture Vesatel (VSAT services), which has now become Unisource Satellite Services. With datacom services and satellite services as a platform, Unisource is developing into switched voice telephony services as well as cellular services. The headquarters of Unisource are located in Amsterdam.

As observed above, PTT Telecom today shows two major directions of its strategy. One is the engagement in Unisource mainly with value-added services (which however is broadened to also include mobile services for instance), and one is the investment in eastern Europe mainly with basic services such as mobile and telephony services.

5. TENTATIVE CONCLUSIONS AND SPECULATIONS

PTT Telecom forms a special case of a traditional monopoly because it has an extremely highly internationalized home market in combination with one of the most dense populations, which in turn is concentrated to three large cities. These characteristics of the economy in the Netherlands have created great incentives for foreign entry into the Netherlands, implying a major competitive threat to PTT Telecom. Although there have been regulatory barriers to entry, entries have been possible and economically feasible because of the technological development of networks (mainly digitization of switches).

PTT Telecom's strategy over the last years reflects a change of direction from a stable situation of cooperation with other PTTs, to a more independent strategy both with respect to the Royal PTT Nederland and other PTTs. PTT Telecom has created close cooperation with Telia and formed the joint company Unisource (with Telia and the Swiss PTT) in order to create a larger platform in Europe. This platform is essential in order to support its highly internationalized customers (often multinational companies) with advanced and international telecom services. Moreover, PTT Telecom has invested in several ventures in Eastern Europe beside the activities in partnership with Telia. In the Ukraine, for example, they are working with AT&T.

PTT Telecom could, thus, be recognized as focusing on two international business areas, partly the European value-added service market and partly basic services in developing countries. A reasonable question that arises is whether PTT Telecom is financially strong enough to simultaneously cope with these two areas. However, there might also be economies which make these two business areas complementary.

Finally, the long-term threat of foreign acquisition of PTT Telecom must be raised. Because of the attractiveness of the Netherlands for foreign entry, the mid-1994 privatization plan has to be carefully monitored. Thus, financially strong and long-term committed shareholders will be essential for PTT Telecom.

REFERENCES

Granstrand, O. and O. Johansson, 1992, Internationalization of Telcos—an early analysis of strategies and behavior, Euro CPR 1992, Sorö Storkro, Denmark, 1992.

Statistical Yearbook, 1993, United Nations, thirty-eighth issue, New York, 1993.

Internationalization in the Nordic Region

The Race to European Eminence
E. Bohlin and O. Granstrand (Editors)

Internationalization of the Swedish Telecom Services Market

Ola Johansson and Ove Granstrand*

Chalmers University of Technology, Department of Industrial Management and Economics, S-412 96 Göteborg, Sweden

CHAPTER SUMMARY

This chapter surveys the internationalization of the Swedish telecom services market according to regulatory developments, growth of market segments, network barriers to entry, foreign entrants, and foreign activities by the domestic companies Telia and Kinnevik. Relevant entries into Sweden are discussed with reference to an earlier hypothesized model of internationalization of telcos, along with some tentative trends and strategic implications.

Even though voice and data services are growing more slowly in Sweden than in some larger European countries, Sweden attracts the interest of large players partly due to the alleged precursory nature of the liberalized free market, which, in combination with its smallness and advanced nature, makes Sweden attractive for trial-and-error, and partly due to the advancement in the Swedish cellular market—one of the most attractive markets. In addition, the Swedish telecommunication network precludes any major network barriers to entry, since it is fairly well digitized and fiberized. In the last decade, there have been new entries in cellular telecom services, public switching telephony and datacom services by both domestic and foreign players.

The FDIs of large telcos, such as Cable & Wireless, Vodafone, Pacific Telesis and France Telecom in Swedish telcos and the FDIs of AT&T, MCI and BT in Swedish offices are also important to note. The initially targeted market segment by these new entrants is the group of large, highly internationalized Swedish MNCs, also an important segment to the incumbent Telia.

Traditionally, Telia has had several cooperation activities, concerning technology and networks, with PTTs in the other Nordic countries as well as with European organizations such as CEPT, ETSI, EMO etc. More recently, however, more commercial cooperation activities have been carrried out partly through the Unisource venture with PTT Telecom and the Swiss PTT and partly through FDIs, as in the Baltic states. Moreover, Millicom International Cellular, in which Kinnevik has a major stake, has made several FDIs in cellular in small developing countries.

* The research support received from Telia is gratefully acknowledged. The research is part of the research program "Economics and Management of Technology" at the Department of Industrial Management and Economics, Chalmers University of Technology, and has been carried out under the auspices of the Institute for Management of Innovation and Technology (IMIT). We also thank our colleagues for valuable comments and support.

1. INTRODUCTION[1]

Customers in Sweden have traditionally been internationally oriented, seeking new solutions and service providers, especially for international services. In spite of the regulatory possibilities and the customers' needs, inward FDIs did not appear in Sweden until the early 1990s because the necessary digital technology had not been installed. However, over the last years an increasing interest in inward FDIs has been evident, which explains why the major revenue base of Telia is threatened, for instance through the appearance of extensive private networks and international calling cards.

Telia predicts (Telia, 1992) a decrease in the total growth of its sales volume from 6% in the 1980s to 2-3% in the mid-1990s, due to the recession and decreased market shares in almost all service segments. This sales figure may be decreased further since an increasing part of Telia's revenues derive from services that are exposed to competition, such as cellular, data services and cable TV (OECD, 1993a).

The new situation reveals a number of complex strategic issues, at both company and government level. One issue is the role of technological evolution both as an element in an internationalization strategy and as an explanatory variable behind market structural change.

The purpose is to describe and analyze the size and orientation of inward and outward FDIs in Sweden with respect to the market development for different services, the requirements of the major customer segments and entry barriers in the regulatory framework, network economies and performance, and price levels.

The focus of this paper is the strategic issues of the telcos and their actual behavior, which have been analyzed on the basis of primary[2] and secondary[3] data.

2. THE SWEDISH MARKET AND BARRIERS TO ENTRY

This chapter outlines some basic barriers to entry of the Swedish telecom services market. In addition, the character and the growth rates of a set of market segments (voice and data services) are analyzed.

2.1 Regulatory conditions in Sweden

This section tries to give a regulatory background to the conditions for and possibilities of inward FDIs in Sweden. The history of Telia's legal structure and status will be discussed, as well as the early and new regulations in Sweden in relation to the new directives from the EC.

Ever since the end of the 19th century, the market structure of telecom services in Sweden has been characterized by the lack of a monopoly grant, while in other EC countries the market structure has been gradually changing from a monopoly status of the national PTT to greater competition in telecom services.

In the EU, the EC recommends further liberalization according to the Rome Treaty, the Service Directive of 1990[4], the ONP Directives in 1990[5] and 1992[6], and the recently proposed

---

[1] This chapter is a part of an ongoing research project which deals with the evolution of international competition in the telco industry in the major OECD countries.

[2] Interviews have been conducted with most players in Sweden during the spring of 1993.

[3] Collection of available statistics from OECD and CEPT as well as some consultancy reports.

[4] Commission directive of 28 June 1990 on competition in the markets for telecommunications services.

Service Review (EC, 1993), through free competition in telecom services in 1998 and a 'Green Paper' for mobile/personal communication before 1994. Thus, the inevitability of full liberalization before the end of this decade is apparently confirmed in the proposed Service Review.

Telia has been a government-owned and government-controlled business agency since its inception. However, in 1981 the legal structure of Telia was changed, when a corporate subsidiary was formed (Teleinvest) to take care of the competitive business not relating to the core telephony services. In 1984 Telia was also allowed to borrow in the capital market through Teleinvest (Bohlin, 1992). In 1989, Swedish Telecom International was formed as a subsidiary of Teleinvest in order to take care of the large Swedish customers going abroad. According to the government bill passed (SFS, 1993) Telia became a corporate company on July 1st, 1993.

Even though there has been a de facto monopoly in Sweden since 1918, Sweden together with New Zealand (OECD, 1993a) has the most liberal regulatory framework concerning telecom services in the OECD. This means that in these two countries competition is formally allowed, although not always realized, in all parts of the public network assigned to telephony, in X.25 data communications and leased lines, as well as in analog and digital cellular telephony, and in paging. Competition is also allowed in other services in Sweden according to the new legislation (SFS, 1993).

In general, regulation takes place partly by restricting the activities of the national PTT (e.g. manufacturing and cable-TV services), partly by setting the structure of the market (e.g. licenses and price levels) and partly by defining the roles and interfaces of the regulatory body (Granstrand & Johansson, 1992). In Sweden there has been a tradition to separate regulatory functions from the government. Those functions have traditionally been included in the responsibility of Telia. However, in 1992 the regulatory functions were transferred to the National Telecommunications Agency (Telestyrelsen), which was set up in 1992[7] with 8 offices as an independent regulatory body responsible for settling disputes between parties on the market and for standardization, representing Sweden in international agreements and supervising the proposed Telecommunication Law (SFS, 1993) which, among other things, upholds and safeguards the national security and the universal service goal.

The liberalization in Sweden has mainly dealt with the interconnection of equipment, i.e. a definition of which equipment was part of the monopoly in 1980, free competition in low speed modems in 1983 and high speed modems in 1988, free competition in telephone handsets in 1985 and in PBX in 1990, along with free competition in the resale of leased lines to third parties and temporary network-to-network interconnect conditions in 1991 (Bohlin, 1992).

The enacted Telecommunication Law (SFS, 1993) states that a license is required for the provision of telecom services, i.e. services enabling transmission of voice, facsimile and data communication with low-speed modems, for the provision of mobile telecom services to a subscriber by using radio access and for the provision of capacity in leased lines to second or third parties. The law is supplemented by a Radio Law and a Tele-terminal Law. Moreover, if

---

[5] Council directive of 28 June 1990 on the establishment of the internal market for telecommunications services through the implementation of open network provision.

[6] Further directives on the application of ONP on leased lines, packet switched data and ISDN.

[7] However, the National Telecommunication Council was formed in 1989, responsible for standardization and frequency allocation, and transformed in1992.

a license is granted, there is an obligation to provide excess capacity to anyone and the interconnect charges should be based on the cost of providing the capacity. It should, however, be noted that the implemented laws give Telia the opportunity to be a dominant actor in a different way than was the case in the UK in the early 1980s, where Mercury was granted a duopoly structure of the market for ten years.

2.2 Performance of the Swedish telecommunications network

Digital networks seem to be a prerequisite for competition and attractive FDIs in telecom services. The sharing of services and telcos in the analog telecommunication network is not successful. The possibilities for a telco to manage a network are significantly improved with a digital network. This means that digitization, fiberization and call failure provide an essential background to the feasibility of FDI. This section is focused on the technical aspects of the Swedish telecommunication network but also tries to give some idea of the economies of the network in terms of the general distribution of main lines and central offices, which in turn may give an indication of the resources required to establish network control.

Over the last few years a major reduction of the cost/performance ratio in telecommunication equipment and provision has led to a major pressure on prices and to several possibilities of by-passing Telia's lines. Striking examples are cellular services and PBX services with leased lines.

Table 2.1
Summary of some key features of the Swedish telecommunication network 1985-1991

	1985	1986	1987	1988	1989	1990	1991
DIGITAL central offices (1000)							
No. of central offices (1000)	735	1014	1206	1624	1829	2300	2800
No. of digital central offices in perc. of total[8]	14.3	18.9	22	29	32	38	47
OECD[9]	n.a	n.a	n.a	n.a	35	42	48
INTER-OFFICE TRANSMISSION							
No. of main lines digital (perc. of total)[10]	n.a	n.a	n.a	39.5	n.a	61	75
OECD[11]	n.a	n.a	n.a	n.a	n.a	n.a	67
FIBERIZATION							
Km. fiber in inter-office transmission	5500	n.a	n.a	n.a	n.a	n.a	10000
Km. fiber in inter-office tr. (perc. of total)	17	27	37	49	58	75	n.a
OECD[12]	15	19	25	32	40	48	n.a
CALL FAILURE							
No. of long distance calls (perc. failure)	2.5	1.8	1.5	1.5	1.4	1.1	1.1
No. of faults per 100 main lines	16.5	17	14.6	14.3	11.8	11.9	10.3
OECD, long-distance failure average	n.a	n.a	n.a	n.a	n.a	2.8	n.a
OECD, average faults/100 main lines	n.a	n.a	n.a	n.a	n.a	29.1	n.a

[8] CEPT (1990).

[9] An average of Norway, Spain, Sweden, the UK, the Netherlands, Germany, Finland, France, Italy, Switzerland, Denmark, Japan and the US (Ameritech, Bell Atlantic, Bell South, Nynex, Pacific Telesis, SW Bell, US West, GTE and United Telecom). Figures from FCC, annual reports and CEPT (1990).

[10] CEPT (1990)); this reference also applies to fiberization and call failure (except the OECD data).

[11] OECD (1993). This reference also applies to the call failures in the OECD (long distance and main lines).

[12] An average of Norway, Sweden, the UK, the Netherlands, Germany, Finland, Italy, Switzerland, and the USA (the networks of Ameritech, AT&T, MCI and Sprint).

As shown in Table 2.1, digitization in Sweden in terms of the number of digital central offices (CO) is about the same as the average in the OECD, i.e. in 1991 47% of the COs in Sweden were digital. Digital main lines in inter-office transmission in Sweden increased to 75% in 1991, just above the OECD average of 67%.

As may be seen in Table 3.1, the time of the introduction of competition in telecom services in Sweden may be defined as the time when Comvik Skyport, AT&T and later BT set up offices in the country, i.e in the mid and late 1980s. At that time digitization reached 20-30% in Sweden, which might be taken as a minimum level of required digitization (on average) for larger inward FDIs. However, this figure is general and has to be analyzed and adjusted further for each service that is considered.

The substitution of optical fibers for coaxial cables, the so-called fiberization, is however not similarly important for FDIs since all cables may be used for digital transmission, even though optical fibers are important because of network capacity. Sweden has had a significantly higher level of fiber in the trunk network over the past years, 75% in 1990 compared to the OECD average of 48%.

Moreover, call failures and technical faults in the Swedish network have been significantly improved over the last decade and the indicators are significantly lower than the average in available OECD statistics (OECD, 1993a). The improvements in these measures may partly be explained by a higher degree of digitization and fiberization, but may also be the result of a more customer-oriented strategy by Telia and of strict requirements by the Swedish government.

The economies of the telecommunication network may be discussed in terms of the distribution of central offices and main lines, measured per inhabitant. These measures may give a general notion of the possibilities of achieving network coverage, e.g. resource requirements in terms of support personnel and number of nodes. In Table 2.2, a much lower number of inhabitants per main line than in the UK, Germany and France is revealed. This may be an effect of the larger area per inhabitant in Sweden, since the number of main lines is dependent on the number of locations of central offices to which the lines are connected. This is also seen in the number of inhabitants per central office, where Sweden had a much lower figure than the other three countries in the early 1980s and probably still has.

It may be hypothesized that the resources required for a high relative degree of network coverage are significantly higher in Sweden than in the other three countries, and it may also be hypothesized that the cost of inter-connection, especially to less populated areas, is higher.

Table 2.2
Inhabitants per main lines and central offices in Sweden, UK, Germany and France 1981-1990
Source: CEPT (1990)

		1981	1982	1983	1984	1985	1986	1987	1988	1989	1990
SWEDEN	Inhabitants/ main line	1.7	1.68	1.66	1.63	1.59	1.56	1.54	1.51	1.49	1.47
	Inhabitants/ central office	1208	1168	1124	n.a	n.a	n.a	n.a	n.a	n.a	n.a
GERMANY	Inhabitants/ main line	2.79	2.68	2.57	2.46	2.36	2.29	2.22	2.15	2.12	2.45
	Inhabitants/ central office	2047	1965	1748	1672	1610	1563	1518	1466	1445	n.a
UK	Inhabitants/ main line	2.93	2.88	2.79	2.7	2.61	2.56	2.49	n.a	n.a	2.31
	Inhabitants/ central office	1973	1936	1910	n.a	n.a	n.a	n.a	n.a	n.a	n.a
FRANCE	Inhabitants/ main line	3.06	2.8	2.61	2.49	2.4	2.31	2.26	2.18	2.1	2.02
	Inhabitants/ central office	2001	1843	1739	1666	1606	n.a	n.a	n.a	n.a	n.a

To sum up, Tables 2.1 and 2.2 give an idea of the network considerations that may be made prior to FDI in Sweden. The degree of digitization in Sweden is on a par with other OECD countries, while the number of installed fibers in the trunk network is significantly higher than the average in the OECD. In addition, the call failures, both in the long-distance network and the average of faults per main line, are significantly lower in Sweden than the average OECD level. The degree of digitization and fiberization in the Swedish telecommunication networks may call for low barriers to entry in terms of technical network considerations, while the distribution of central offices and main lines would imply relatively higher economic network barriers to entry in Sweden.

2.3 The telecom services market in Sweden

Considerations of FDIs in any country involve market potential. As seen in the previous chapter, improved network conditions in Sweden do not seem to be a prerequisite for entry, which implies that it is essential to consider other factors, such as the general market potential, price levels, the market share of the dominant actor and growing services segments before making FDIs in Sweden. This chapter outlines the general characteristics of the Swedish customers, the price levels, the growth patterns of different services, in combination with a discussion on maturity and, finally, the major customer segments and their requirements.

The economy in Sweden is to a large extent affected by the economy in other countries. The heavy international orientation of Swedish industry and, consequently, the more sophisticated demand for reliable, controllable and predictable telecommunications systems, may be explained by the fact that, on average, one third of the gross domestic product is exported and that in Sweden there is a high concentration of large companies.

Telephony charges in Sweden are below average in the OECD (OECD, 1993a) for residential users (42%)[13] as well as for business users (51%)[14]. In addition, there is a substantial cross-subsidisation of local calls from long-distance and international calls. Due to the difference between the charges and the cost structure in the provisioning of telephony services in Sweden, the international and long-distance call charges will decrease while local call charges will increase, even though the Telecommunication Law or the proposed agreement between Telia and the government will state a maximum price level of telecom services at net consumer price index—1%. Cellular prices are lower (36%)[15] than the OECD average, as are the X.25 data communication charges (30%) and the leased-lines charges at different speeds (60-70%[16]).

A result of the new legislation, stating cost-based tariffs, and of the increasing FDIs in Sweden, will be that prices as entry barriers, not the prices as such, will increase, even though the price levels today are relatively lower than the average OECD levels.

Table 2.3 summarizes the market growth in Sweden, France, Germany and the UK, where the average annual growth rates between 1988 and 1991 of voice and data services, respectively, are shown. In Sweden, the fastest-growing voice services are voice messaging (25%) and cellular (12%), while in data services facsimile (45%) and EDI (40%) have the highest growth rates. The only service with a negative growth rate is satellite for voice, which may be explai-

---

[13] For residential users the fixed charges are 20% lower, while the usage charges are 57% lower.

[14] For business users, especially the usage charges on a basket of telephone charges are significantly lower (63%), while the fixed charges are average ones.

[15] On a basket of charges.

[16] This figure is valid as an average when Turkey is excepted, since Turkey has exceptional prices.

ned by the increasing share of fiber in overseas telecom services. However, Sweden seems to have lower growth rates in all services in comparison to France, Germany and the UK.

Table 2.3
Average annual growth rates (%) in voice and data services (1988-1991) in real values for each country

Source: MIRC (1991)

DATA SERVICES

	Facsimile	Online database	Public e-mail	EDI	Mobile data	Satellite (video & data)
Sweden	46%	15%	6%	43%	9%	4%
France	70%	41%	19%	171%	14%	11%
Germany	77%	36%	35%	142%	51%	26%
UK	58%	26%	21%	130%	24%	17%

VOICE SERVICES

	Cellular telephone	Specialized mobile	One-way paging	Voice messaging	Satellite voice	Toll free
Sweden	13%	8%	6%	28%	- 6%	4%
France	16%	10%	5%	14%	- 3%	24%
Germany	32%	20%	16%	52%	- 6%	19%
UK	27%	20%	14%	55%	3%	13%

Potential service segments in Sweden may be those that are growing at a significantly higher pace in other European countries and those which are not already mature in Sweden. The maturity of each service may be measured as the density in terms of accumulated usage per inhabitant. This is, however, not done in this paper, but potential service segments in Sweden are still identified (from Table 2.3). Examples are mobile data, which is growing fast in Germany but not in Sweden in spite of the installed Mobitex[17] system, as well as public e-mail and EDI[18] along with toll-free voice services, which are at least growing at a threefold higher pace in France, Germany and the UK. These services ought to have a potential in Sweden as long as they are not mature (which, for instance, the Mobitex system might be). Cellular, on the other hand, is growing more slowly in Sweden, simply because the penetration is much higher[19] in comparison to the other three countries in Table 2.3.

Finally, the demand and requirements of the Swedish telecom service customers will be analyzed. In terms of usage of telecom services in different industry segments in Sweden, manufacturing has the largest volumes but at the same time one of the lowest growth figures in Swedish industry (on average 25% annually from 1981-1987), while financial/insurance/banking are growing at the highest pace (42%) according to OECD (1993b). Average annual growth rates for other industry segments are for central & local government 13%, for transport 27% and for retail/wholesale 25% over the same period of time. The high growth

[17] A first variant of Mobitex was installed in Sweden in late 1986, while another Mobitex system with higher capacity is now sold to other countries.

[18] The frame relay technology seems to have brought expectations of growth in Sweden in the LAN-LAN segment (not included in Table 2.3) according to the interviewed respondents.

[19] 7.9% in Dec 1992 compared to 2.4% in the UK, 0.8% in France and 1% in Germany (CIT Research, 1992).

rate in financial/insurance/banking is, however, not unique for Sweden, but it may suggest a potential for growth in other segments as well.

According to Thorngren (1993), the initial focus of the telcos entering the Swedish market seems to be pricing, while in the long run a more relevant focus ought to be services at more valuable pricing for the customers and new services. This is in line with decreasing price levels approaching cost levels, which increases the barriers to entry in comparison to the former administrative price levels.

Lindskog (1993) has presented the criteria of the Swedish central & local governments' procurement of data services (approx. 45 MUSD annually) in 1993. The determinant of procurement was price, since all of the offers seemed to have a reasonable technical solution. The prices of BT were 20% higher and those of Megacom[20] 33% higher than the prices offered by Transpac on a basket of services. In addition, it was important for the Swedish central & local governments to have the possibility of determining the agreement within two years as well as measuring its availability and performance. When it comes to the procurement of telephony equipment and services (approx. 185 MUSD annually), Lindskog determined three important customer freedoms, i.e. the possibility of using different PBXs, different telcos and also the possibility of using hybrid solutions of PBX and Centrex[21]. Other relevant customer criteria according to Lindskog are a strong telco/partner, a reliable technical solution, security and a balanced and buyer-friendly agreement.

The procurement criteria of Swedish industry may not be equally focused on price, although it is still important, as it was for the Swedish central & local governments. The possibilities of telecom services developing along with support seem to be at least as important as price for the industry. A striking example of this is the fact that most large companies in Sweden have had to build up their own LAN solutions with their own technical support staff since, according to BT (Sweden), the telcos have not been able to meet the requirements. According to Bäckman (1989), Swedish corporate customers with more than 50 employees ranked price only as the *17th* most important criterion for procurement of data services in 1987. This was however the case when there was still no real competition for customers in Sweden.

At this stage it would have been fruitful to further analyze the customer benefits of the different customer segments for each telecom service in order to provide a framework for entry patterns. Services with high potential growth rates and high customer benefits for a suitable volume of customer usage would generate high levels of FDIs, while services with low potential growth rates and low levels of customer benefits for a low volume of customer usage would be of low foreign interest. An example of the former services may be EDI for the manufacturing industry, while satellite for voice may be an example of the latter. Another service that is identified as having a potential and ought to have high customer benefits is toll-free voice services offered, for example, by the financial/insurance/banking segment to private customers or by parents to their children.

[20] A subsidiary of Telia.

[21] Centrex provides a PBX-functionality built into the network.

3. FOREIGN ENTRIES IN THE SWEDISH MARKET

This section will give an overview of the foreign and new entrants' activities and investments in Sweden as of 1993. Telia will thus not be covered explicitly, even though it will be mentioned when relevant in relation to other actors.

However, a brief comment on the competitive situation of Telia should first be made. As indicated in the previous chapters, Telia's major revenue base in Sweden is threatened.[22] International experience from the UK and the US suggests that Telia should aim at a 60% market share in international calls (it was 90% in 1991) and an 80% market share in long-distance calls (it was 100% in 1991) in 1995 (Thorngren, 1993). Since the early 1980s Telia's market share has dropped to approx. 50% for data services in general and telecommunication equipment such as telephone terminals, small PBXs and modems, while it has a market share of approx. 80% in large PBXs and approx. 60% in cable TV (Cap Gemini, 1992).

These figures give an indication of the volumes of FDIs that have been and will be made in Sweden over the next few years. The FDIs and the corresponding activities will follow different patterns which are important to identify in order to be able to forecast and anticipate future steps by competitors. In this paper, this analysis will be carried out using some generalized options from an earlier proposed model of internationalization of a telco (Granstrand & Johansson, 1992).

In general, no international competition occurred in telecom services in Sweden until the 1980s. Telia was active not only in telephony (fixed and cellular), but also in PBX and terminal manufacturing, packet and switched data services, broadcasting and cable TV etc.

3.1 Cooperation and consultancy activities

Cooperation and consultancy activities occur especially in network cooperation. A natural cooperation alliance is the necessary cooperation between Telia and other telcos concerning leased lines, inter-connect traffic[23] and also concerning some international data services. An important network cooperation agreement is that between telcos and other companies with fiber cables, such as railway and energy distribution companies. Cooperation agreements between cable-TV operators and telcos will also most probably be seen in the Swedish market.

In 1992 the telco Tele2 set up an inter-connect agreement for X.25 with Sprint (to SprintNet). In addition, Tele2 has a long-term strategic agreement with Banverket (the railway company supporting and investing in the railway), who invested in a fiber-optics network (6 pairs of fiber) in 1991. Two of these pairs are partially designated to Tele2 (as well as Comvik GSM and potentially Dotcom, see below). Tele2 also cooperates closely with Sydkraft Telecom, a subsidiary of the energy distribution company in southern Sweden, who also has a fiber-optics network. In addition, agreements have been reached with Svenska Kraftnät and Vattenfall, other large energy distributors, who have placed their power lines at the disposal of Tele2 for the installation of fiber cables. These agreements are an important part of Tele2's strategy to by-pass Telia and a part of their aim to install fiber in the local loop. Tele2 claims that it is cheaper to invest in new networks because of the administrative prices of leased lines from Telia to urban areas, even though they are still using Telia on some parallel routes.

---

[22] The market shares of foreign telcos in Sweden have unfortunately not been available.

[23] Currently priced at 0.65 SEK/min by government decision.

The remaining capacity of the two fiber pairs of Banverket (approx. 84%) has been offered to Dotcom, a Swedish supplier of PBX systems[24] from Alcatel, Siemens, Philips and AT&T along with equipment for e.g. security systems. The option to use the remaining capacity would make it interesting for Dotcom to enter the long-distance and international telephony market and possibly also the data services market. Strategic alliances with, for instance, cellular telcos or CATV operators seem, however, necessary in order to get access to the local customer and in order to by-pass the local network of Telia.

A strategic owner behind several corporations is Kinnevik with strategic ownership in Tele2, Comvik and Kabelvision[25]. These ownership connections, along with the strategic alliances with Banverket and the energy distribution companies, make a powerful conglomerate, as seen in Figure 1.

The cooperation between Telia and PTT Telecom in Unisource is important for Telia's international telecom services in Sweden and consequently for Telia's competitive situation in Sweden.

Important strategic alliances have thus been found between Banverket and the companies Tele2, Comvik GSM and potentially Dotcom. In addition, Tele2 and the energy distribution companies Sydkraft Telecom, Svenska Kraftnät and Vattenfall have powerful agreements. International alliances which may have an effect on Sweden are MCI-BT and Unisource.

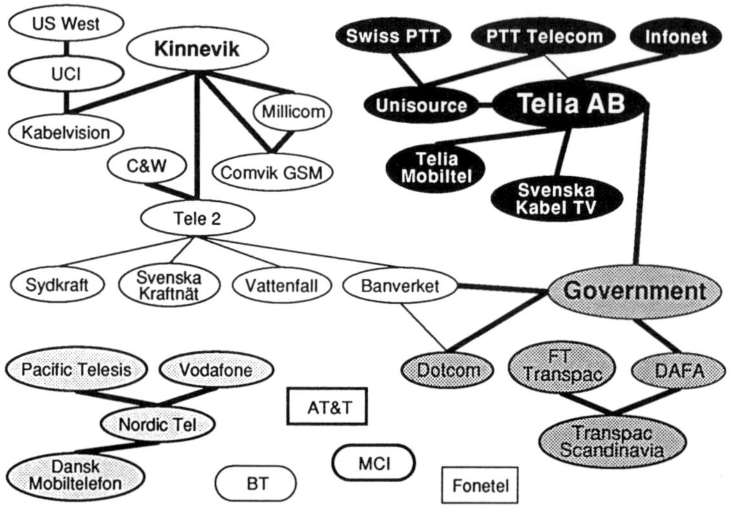

Figure 3.1. Major agreements (thin lines) and major ownership relations (thick lines) on the Swedish market in 1993.

[24] Including the specific solutions of network groups with radio links, leased lines and routers etc.

[25] In addition, Kinnevik has invested in two TV channels in Sweden.

3.2 Special service focus

Interest in FDI in the 1980s was focused on data services, not telephony. In the early 1990s telephony seems, however, to be more interesting for telcos in Sweden and, consequently, a strategic shift from data services to telephony *and* data services is seen in the telecom services industry.

Telcos with an obvious special service focus are actually only VAN operators, cellular telcos and to some extent cable-TV operators in Sweden. Other telcos, such as BT, AT&T, MCI, Transpac Scandinavia and Tele2 provide several services and are also trying to diversify into new services.

One of the operators focusing on VANS is General Electric Information Systems (GEIS). GEIS set up the first datacommunication link over the public switching telephone network in Sweden already in 1968. The business idea was that a centrally located large computer is cheaper for the user than owning a computer. The strategy pursued by GEIS has always been to offer global transmission, processing and storage of business information, including MNS, EDI and e-mail, by using leased lines. The companies Electronic Data Systems and IBM have similar value-added services in Sweden, neither of them in direct competition with Telia.

Another special service worth mentioning is CATV and the 26% interest of United Communications International[26] in Kabelvision,[27] which is the second largest cable-TV company (approx. 25% market share) after Telia's subsidiary Svenska Kabel TV (approx. 60% market share). The largest shareholder in Kabelvision is, however, Kinnevik (37%).

A third special service focus is that seen in the cellular telcos Comvik GSM,[28] Nordic Tel and Telia Mobitel, who have cellular as their main service offer. Telia Mobitel has, however, also mobile services such as land mobile radio, maritime mobile radio and paging. Because cellular has the highest penetration in the World,[29] cellular telcos in Sweden focus on most market segments. They have each invested about 150 MUSD[30] in their existing networks. The area focus in their declared expansion plans for the GSM networks is however somewhat different.[31]

Scandinavian Info Link ought to be mentioned as a special service or application formed in 1988 by some large users[32] in order to establish an electronic-mail system (Memonet). In 1993, telecom services such as facsimile and access to databases are included. The Swedish Post Administration acquired the company recently, and also included a service that automatically converts electronic mail into ordinary letters.

The special service focus discussed above may be summarized as follows. A narrow focus is seen among the VAN operators GEIS, EDS, IBM etc., among the cellular telcos Nordic Tel,

---

26 UCI is partially (45%) owned by US West.

27 Communications Week International, No. 82, pp. 3.

28 Comvik has recently closed their old analog cellular system, which was set up in 1981.

29 CIT Research (1992).

30 According to managers with Comvik GSM and Nordic Tel.

31 Telia Mobitel has their NMT 900 customers who presumably demand the same coverage of GSM, i.e. 94% of the population and 84% of the area. They aim to reach this coverage in 1996 for their GSM network. Comvik GSM and Nordic Tel, however, seem to be satisfied with a coverage of the major parts of the population, i.e. the southern parts of Sweden including major cities and roads in the northern region, with approx. 90% of the population and 60% of the area.

32 Ericsson, Volvo and Scandinavian Airline Systems (SAS). In 1992 approx. 40% of Scandinavia's 100 largest companies were subscribers to the service.

Comvik GSM and Telia Mobitel, among the Swedish Post Administration with e-mail, and among the cable-TV operators, such as Kabelvision and Svenska Kabel TV.

3.3 Special market segment and area focus

A special market segment and area focus may characterize almost all of the telcos that are not cellular telcos, i.e. they have all focused on business customers in large internationally oriented companies, who demand advanced data services and international VPN services. The telcos have followed their customers (MNC) abroad combining this with gaining market shares in the return-to-home-country-market (AT&T and MCI). Transpac and BT, on the other hand, have the ambition to gain larger market shares also in the domestic Swedish market, even though their focus is on large companies.

BT uses the Tymnet network, which they acquired in 1989, to provide international data services together with Syncordia, their outsourcing subsidiary. BT Worldwide (Sweden) states that they have a 25% market share in international data services in Sweden through the former Tymnet. LAN-LAN connections are also established in Sweden through additional leased lines. BT is expanding the original Tymnet network and aims at implementing ATM technology in the network, which will integrate voice, video and data. In addition, BT has focused on video-conferencing services and products. BT will expand their services to include international telephony in Sweden (and the rest of the world) already next year through their Cyclone project, aimed at providing a range of VPN services for international customers at 5-15% lower prices (Telephony, 1992).

BT has a global ambition and sees the Nordic network as an (important) hub in their global network. Large MNCs are penetrated by BT. When BT bought Tymnet, they took over an X.25 network of leased lines in Sweden along with ten technicians for support services in a Stockholm office, which was set up by Tymnet in 1984. In 1993 BT Worldwide has Nordic offices also in Göteborg, Oslo, Copenhagen and Helsinki. Since last year, several qualified sales personnel have been hired, partly from Megacom, a subsidiary of Telia.

France Telecom Transpac focuses initially on data services in general. In Sweden, France Telecom formed Transpac Scandinavia in 1992 through the acquisition of DAFA (a major user of data services with a large leased-line network) in order to get access to a national network with close nodes and an existing customer base. A similar approach has also been adopted by Transpac in the UK, Germany, Switzerland and Italy.

Transpac Scandinavia aims at establishing a multi-domestic base of customers and local networks with close nodes (however connected into a pan-European network within France Telecom Transpac). This implies a strategy towards both international and domestic customers.

Another approach has been adopted by Cable & Wireless (C&W), who together with Kinnevik formed Tele2 out of Comvik Skyport in 1991. The initial focus by Tele2 was affected by Comvik Skyport's data services activities, e.g. national and international X.25 connections through leased lines. However, in March 1993, Tele2 also included international calls (approx. 5% lower price) with direct customer access[33] in their service portfolio, which reflects their ambition to become the second telco with an extensive direct customer access in Sweden. In addition, they intend to launch a calling card, which via transit agreements with

[33] Through a new international prefix (007) instead of 009 with Telia. Their service is available to all cusomters that have an AXE (i.e. a digital connection)-connected terminal (55% of the subscribers in Sweden at the end of 1992).

Finland and Denmark will transfer international calls. Finally, it should also be mentioned that Tele2 has invested in a fiber-ring in Stockholm with a central office in Kista, an industry area with a high MNC concentration outside Stockholm.

Tele2 uses the experiences gained from Mercury in the UK in order to build a new base of customers and networks in Sweden. Tele2 invested approx. 70 MUSD up to 1990 and will invest approx. 15 MUSD annually until 1995.[34] Tele2 focuses on the perceived customer benefits, not on the technical solutions, through professional customer services and additional telecom services, which would mean that Tele2 does not think that new technology is a competitive weapon in the long run.[35]

The approach 'follow your customer's data services abroad' fits well into the activities by AT&T and MCI in Sweden. AT&T Nordics provides international data services. In the late 1980s, AT&T Nordics offered leased lines and private networks on X.25 together with Telia. In 1993, AT&T Nordics also offers switched data services along with VPN for data services (called GSDN). Since their launch of the AT&T calling card, however, international telephony, and especially international toll-free calls, have been included.

AT&T concentrates on America-related customers, i.e. American customers in Sweden and Swedish customers in America. AT&T Nordics established an office in Stockholm in 1987 as a part of their international operations department, with the intention not to interfere with the local telco.[36] However, in 1992 AT&T bought McCaw,[37] which was a clear strategic move towards direct customer access. This calls for a wider view when analyzing AT&T Nordics.

MCI International opened an office in 1992 and invested in an international gateway in Stockholm. The office is a part of their strategy to follow their customers (MNC) abroad and they state that they do not see Telia as an immediate competitor. In contrast to AT&T, though, MCI International Sweden seems to focus on telephony with calling cards and private line services (VPN) along with international toll-free calls. However, they also provide switched international data services in cooperation with Telia. MCI states that they do not have any ambitions towards switched international calls apart from their calling cards.

The debate on area focus by the foreign telcos investing in Sweden must be raised to a European level, since BT, France Telecom Transpac, AT&T, MCI and the GSM telcos (through their owners Vodafone (25% in Nordic Tel), Pacific Telesis (51% in Nordic Tel) and Millicom (20% in Comvik GSM)) are all a part of a European strategy.

The special market segment and area focus is on large MNCs in the major metropolitans in Sweden. AT&T, MCI, BT, Tele2 and Transpac all have this focus. Tele2 and Transpac Scandinavia seem, though, to have broader network coverage ambitions than the other three. The cellular telcos Nordic Tel, Comvik GSM and Telia Mobitel are directing their marketing activities towards the mass market, although with slightly different initial network-coverage ambitions.

---

[34] Communications Week International, April 19th, 1993.

[35] According to managers of Tele2.

[36] According to managers of AT&T Nordics.

[37] The cellular telco in New York.

3.4 Globalized and denationalized telcos

Among the telcos making FDIs in Sweden, it is only AT&T with large international FDI volumes and BT, mostly through their trial-and-error with Syncordia (their outsourcing company), who claim ambitions towards a globalized telco.[38] France Telecom (Transpac) does not have the same expressed strategy to become a globalized telco. Cable & Wireless has also had an extensive global strategy, but Sweden and Tele2 seem to be more a part of a trial-and-error approach in one of the most liberalized countries in the OECD, using the experience gained from Mercury in the UK.

Finally, we have not been able to find any denationalized telcos (telcos with no important home market) and thus not in Sweden either.

3.5 Summary of entry barriers and new entrants

There are no major regulatory barriers to entry. The new Telecommunication Law states only that a license is required for the provision of telecom services. In addition, there is an obligation to provide excess capacity to other actors. However, Telia's dominant position will remain, and the new law will open up increasing opportunities to fully utilize its competitive advantages in Sweden, which may be an important entry barrier within a short period of time for FDIs.

The network barriers to entry seem to be low. Since the end of the 1980s, the quality of the Swedish telecommunication network has been sufficient to attract FDIs. The entry barriers in terms of the economies of the network may, however, be hypothesized to be higher relative to other European countries, if the distribution of central offices and main lines is taken into account.

Since the new legislation states cost-based prices, and since competition in Sweden will increase, prices as entry barriers will increase. Prices will, however, decrease from a lower level relative to the OECD average, even though most prices are administrative prices.

Entry into the Swedish market will most probably be characterized by entries into specific services and market segments, such as the early entrants into international data services, which have followed a service path including domestic leased lines, VPN etc. towards telephony and a broader customer base. The situation in 1993 is, however, rather different, since the digital network allows lower barriers to entry into telephony, which may call for a wider spectrum of potential specific service segments to be entered. The focus on business customers and especially MNCs will, however, most probably remain until the by-pass possibilities to the end-customer have been enhanced, e.g. through cable-TV networks and cellular networks in the local loop.

In Table 3.1 some of the key data for new entrants in Sweden are summarized. The early entries by GEIS, IBM and EDS are, however, not included since their activities in telecom services, i.e. global transmission, are difficult to separate from the other activities of the companies, such as information storage and processing.

In Table 3.1 it may be seen that there are small differences in the number of offices and nodes, but large differences in the number of employees. This may be an indication of the ambitions and level of activity of the telcos.

---

[38] Even though this strategy is now shared with MCI through a joint company including Syncordia.

Table 3.1
Summary of key data for some new entrants. Sources: Annual reports and interviews

Telco	Year of entry	Major owners	Offices (Swe)	Empl. (Swe)	Turnover (1992[39])	Nodes (Swe)	Tot. inv. (mid-93 MUSD)	Entry strategy	Entry areas
AT&T	1987	AT&T (IOD)	1	30	n.a	1	n.a	Cooperation (leased lines)	Int'l dataserv. calling cards
BT	1989	BT	2	40	75 (Nordic)	3	n.a	Cooperation (leased lines)	Data services, Int'l telephony
Comvik GSM	1981 (92)[40]	Korsnäs (100%)	7	170	n.a	2	> 150[41]	Investments in new network	Cellular (GSM)
MCI	1992	MCI	1	2	n.a	1	n.a	Cooperation (leased lines)	Int'l telephony, data services
Nordic Tel	1991	Pacific Tel. (51%) Vodafone (19%)[42]	3	> 150	n.a	3	appr. 150[43]	Investments in new network	Celllular (GSM)
Tele2	1986 (91)[44]	Kinnevik (60%) C&W (40%)	3	130	n.a	6	appr. 70[45]	Coop. (leased lines), inv. in new network	Int'l data. Int'l telephony
Trans-pac Sc.	1992	France Telecom (60%)	8	16[46]	n.a	8	n.a	Acquisition	Data services

Following the pattern described above, Dotcom is an example of a potential new entrant into the international and long-distance call segment for business customers, who initially is required to provide a leased line, or a co-leased line, to the railway station, where Dotcom's potential fiber ends. Dotcom was formed in 1990 out of a part of the former Ericsson Information Systems. An interesting but unknown potential entrant is Fonetel in Malmö, who in 1992 won recognized status to operate network services, similar to Tele2. The initial service focus of Fonetel will be on international telephony and data, probably using satellites and a switching center in Malmö. Fonetel is formed out of Washington-based Veracom International.[47] Other probable potential entrants are the cable-TV operators Kabelvision/Finvision (with a rather high level of network coverage in Sweden) and Stjärn-TV (located in the Stockholm area), who ought to be strategic partners in local networks.

A de facto new entrant is Unisource, jointly set up in 1991, owned by PTT Telecom and Telia (potentially also the Swiss PTT, who owns a part of Unisource Satellite Services since 1992), who recently entered the international data services market. Most probably, Unisource will include international switched telephony in the long run.

[39] Most of the companies had not started full-scale operations in 1992.
[40] Comvik AB started in 1981, Comvik GSM AB in 1992.
[41] According to interviews with Comvik managers.
[42] Three other owners are Spectra-Physics, Volvo and Trelleborg who each have 10% in Nordic Tel Holding, who owns almost all shares in the operating company Nordic Tel (with the brand name Europolitan).
[43] According to interviews with managers at Nordic Tel.
[44] Comvik Skyport started in 1986, Tele2 in 1991.
[45] Communications Week International, April 19th, 1993.
[46] Will increase to approx. 70 beacuse of an order received from the government administration in May.
[47] Communications Week International (1992).

BT is another potential new entrant into the long-distance telephony segment in Sweden, with their Cyclone project, which most probably will be a part of their European telephony strategy to be launched in late 1993 or early 1994.[48] Through VPN services, they ought to be in telephony already, but still not to a large extent.

AT&T and MCI will, though, most probably remain in their existing segment, according to their official European intentions, focusing on the large EC countries in central Europe. Even though Sweden and the Nordic countries in general are an important hub in their global networks, the Nordic countries remain a small market, however with large interesting MNCs. Transpac and France Telecom will also most probably remain in the data services segment, unless obvious opportunities are opened up through technology development, for integrating telephony into their existing Swedish network.

Finally, there is, rather surprisingly, no new entrant in mobile data, which ought to have a potential in Sweden, judging by growth figures in Germany and the UK. Similarly, there is no new entrant in paging.

4. FOREIGN ACTIVITIES AND INVESTMENTS BY TELIA AND KINNEVIK

The foreign investments by Telia and Kinnevik are considered as the only foreign investments made from Sweden in telecom services. Kinnevik owns approximately 49% of Millicom International Cellular (a Luxemburg-based company) and 100% of Comvik GSM through Korsnäs (Kinnevik owns more than 90% of Korsnäs). Thus, the international activities by Millicom International Cellular and Comvik GSM ought to be included in the activities by Kinnevik. Below, the activities are separated into a period of technology and network cooperations and a period of commercial cooperations.

4.1 Technology and network cooperation activities

Kinnevik has not been found to have any of these activities, while Telia certainly has. Until the late 1980s, there was an open-minded and cooperative environment between the PTTs.

Swedtel, a subsidiary of Telia, sells Telia's knowhow outside Sweden. Africa and Eastern Europe are examples of geographical areas which they work in. The foci of the activities are, for example, planning, consultation and operative management.

The Nordic countries have been the major focus of Telia. Already in 1969, the idea of building a Nordic mobile telephony system was accepted at a Nordic telecom conference. In 1975 the specifications of the NMT450 system were decided on and in 1981 the system was introduced. The market development was exponential and the system or its derivates (including the NMT900 system starting in 1986) was later introduced and openly spread in 30 other countries (in addition to the Nordic countries) (Granstrand, 1993). A similar cooperation (Nordtel) between the Nordic telcos was set up in 1980 to coordinate planning and modernization of networks and introduction of new services.

In 1990, Telia further expanded its international datacom network by investing 5.38% in Infonet. Other examples of cooperations in the same manner of network cooperations are: EMO (European Mobile Operations) for the Nordic countries and recently also for PTT Telekom and the Swiss PTT; Nordframe (1991), a regional frame relay public data network

[48] Telephony (1992).

service on leased lines jointly by the Nordic telcos; a European EDI interconnectivity license (1992) to Telefónica Services, Ireland Post GEM and to Infonet; and a European cooperation (in 1993) where the European PTTs will test an ATM network in early 1994.

4.2 Commercial cooperation activities

Millicom International Cellular has received several cellular licenses[49] (either on their own or through a consortium) over the last decade. They have focused on small developing countries, where small networks are implemented with a potential short pay-back period. Millicom has, however, also received PCS licenses in two US cities (1990) and a "wireless to the home"[50] license in the UK (1992). Moreover, Millicom has also owned a service provider in the UK, which they sold to Hutchison in 1991. Kinnevik also owns 33% of the second Norwegian operator Netcom GSM AS (Ameritech and Singapore Telecom are also part-owners of this operator).

Telia has extended its international activities from technology and network cooperations to commercial cooperations in the early 1990s. In 1987, however, a cooperation called STS Telecom was set up by the Nordic PTTs (Telia 48%) in order to supply a high-quality datacom network. However, the commercial cooperation was faced with competition from its domestic networks at the same time as it leased capacity at market price. The cooperation ended its activities in 1991.

Starting with a training program for the Estonian PTT, Telia and the Finnish PTT obtained an agreement to install NMT450, NMT900 and GSM systems. Later on, the telcos were also given the responsibility for reconstructing, partially owning and operating the national and international networks in Estonia, while a similar agreement for the fixed network in Latvia is still pending. Telia has also several interests in GSM ventures, such as a 12.74% stake in a venture in St Petersburg (Russia) and a 14% stake in the Pannon consortium in Hungary, while the GSM license in Italy through the Omnitel venture (Telia 9.8%) is still pending. In mobile data (the Mobitex system), Telia has obtained licenses in France (10% in the TDR venture) and in the UK (5% in the RAM Mobile Data venture).

During the last half of the 1980s Telia started to cooperate with PTT Telecom BV (the Netherlands) and British Aerospace through a VSAT agreement (Vesatel) in order to operate datacom services such as satellite news. They were, however, blocked from services in the UK and later British Aerospace withdrew from the venture.

In October 1991, PTT Telecom BV and Telia announced the joint company Unicom (later changed to Unisource) which today supplies pan-European value-added network services. In late 1993, a mobile subsidiary (as well as three others) was also created. The Swiss PTT[51] joined the satellite business (Unisource Satellite Services) in 1992 and will also (when regulations are set) have an equal share of the holding company. At this stage, value-added datacom services are focused on by Unisource, even though voice services will be offered in early 1994. Unisource has created a network triangle in Europe between Sweden, the Netherlands and Switzerland, which is the base for the world's third largest carrier in terms of outgoing minutes. In addition, Unisource has a service agreement with Sprint's SprintNet.

---

[49] Bolivia (Millicom 46%, Comvik 23%), Chile, Costa Rica, El Salvador, Guatemala, Hong Kong, Mauritius, Mexico, Pakistan, Sri Lanka, Philippines, Latvia, Macau, China (the Guandang province), Russia (Moscow), Ghana along with two PCS licences in the US.

[50] This system is intended to be based on the CDMA modulation technique (which is chosen for PCS in the US).

[51] The Swiss PTT and PTT Telecom BV had already cooperated for two years.

Tele2 (60% owned by Kinnevik and 40% by Cable & Wireless) made a similar agreement with Sprint in 1991, which means that both Unisource Business Networks in Sweden and Tele2 have a datacom service agreement with Sprint.

Unisource became a reality mainly due to close business relationships together with common visions and values between top executives of PTT Telecom and Telia. Both countries (as well as Switzerland) have a high share of multinational companies (20% of the world's 500 largest) which form an important revenue base for the telcos. These factors, in combination with economies of scale, made it possible and maybe also necessary to cooperate.

The cooperation between Telia and PTT Telecom was deepened in mid-1992, when the local telephony districts in Sweden and the Netherlands were linked closer to each other. By benchmarking each other's business and exchange point of views, the two companies aim at increasing efficiency and effectiveness in operations, while at the same time aiming at merging their networks through common equipment and standards.

Another aggressive task has been the application of international voice services to and from the UK. In October 1993, Telia finally got a license to operate this service between the UK and Sweden via leased lines and a separate switch in the UK. In addition, Telia (like AT&T for instance) has started to offer "reverse calls" from Europe, where an international call to Sweden is established in the reverse direction in order to circumvent high call rates from, for instance, Belgium.

Table 4.1
Summary of key data for foreign investments in telecom services by Telia and Kinnevik
Sources: Annual reports and interviews

Telco	Entry market	Entry area	Entry strategy	Year of entry
Telia	Nordic	Datacom	Joint venture (48%)	1987 EXIT 1991
	Global	E-mail, EDI, VPN etc.	Stake Infonet (5.38%)	1990
	Europe	Datacom	Joint venture (50% in Unisource)	1991
	Estonia	Cellular	Joint vent (24.5%)	1991
		Int'l teleph.	Joint vent. (24.5%)	1991
	Latvia	Cellular	Joint vent. (24.5%)	1991 EXIT 1994
	France	Mobile data	Joint venture (10%)	n.a
	UK	Mobile data	Joint venture (5%)	1991
		Int'l teleph.	License	1993
	St Petersburg	Cellular	Joint vent. (12.74%)	1992
	Hungary	Cellular	Joint venture (14%)	1993
Kinnevik	Costa Rica	Cellular	Joint venture (75%)	1989
(Millicom)	Mauritius	Cellular	Joint venture (50%)	1989
	Sri Lanka	Cellular	Joint venture (46%)	1989
	China	Cellular	License	1990
	Guatemala	Cellular	Joint venture (45%)	1990
	Pakistan	Cellular	Joint venture (59%)	1990
	US	Cellular	License	1990
	UK	Cellular	Joint venture (n.a)	1990 EXIT 1990
		Paging	n.a	n.a EXIT 1991
		Service provider	n.a	n.a EXIT 1991
		Wireless home	License	1992
	Bolivia	Cellular	Joint vent. (Millicom 46%, Comvik 23%)	1991
	Chile	Cellular	Joint venture (50%)	1991
	Hong Kong	Cellular	Joint venture (28%)	n.a EXIT 1991
		Telepoint	n.a	1991
	Mexico	Cellular	Joint vent. (37.5%)	1991 EXIT 1993
	Moscow	Cellular	Joint venture (30%)	1991
	Norway	Cellular	Joint venture (33%)	1991
	Philippines	Cellular	Joint venture (40%)	1991
	El Salvador	Cellular	Joint venture (80%)	1992
	Ghana	Cellular	Joint venture (90%)	1992
	Lithuania	Cellular	Joint venture (49%)	1992
	Luxembourg	"Clearing house"	n.a	1992
	Macau	Cellular	License	1992
	Paraguay	Cellular	Joint venture (n.a)	1992

5. DISCERNIBLE PERIODS IN THE INTERNATIONALIZATION OF THE SWEDISH MARKET

Are there any corresponding patterns in the way that former nationalized and mostly monopolized *markets* gradually become internationalized through FDIs? This question can naturally not be answered on the basis of a single case like the Swedish telecom services market alone. However, a few summary features could be extracted from the observations so far and tentatively arranged in periods or stages in the evolution of the Swedish market, in order to facilitate further comparative analysis with corresponding market evolutions in other countries.

A few current features that distinguish the Swedish market deserve to be summarized first. The Swedish market in 1993 is small, not particularly fast growing, fairly advanced in terms of technology and customers (with a sizeable segment of MNCs), sparsely populated with medium-sized rural areas, highly liberalized, on average low-priced with a dominant and efficient telco (Telia) in an international comparison, having a long-standing close supplier relation to a large, internationalized equipment manufacturer (Ericsson). Sweden attracts a significant degree of foreign interest as a trial market, an interest that has been reinforced to the extent that it has been considered a precursory market in many regulatory, technological and customer developments.

In the century-plus history of the Swedish telecom service market the following features, sometimes constituting periods or stages, can be discerned. *First*, when the radical new technology of telephony came along there was a fairly short period of mixed international and national competition between the Bell Company and local telephone companies.[52]

Second, the market then became nationalized, i.e. without foreign players. A service monopoly (in 1993 named Telia) became established as a state company/agency hybrid with a preferred supplier (Ericsson). Telia partially integrated backwards into manufacturing in order to gain cost control and bargaining power against its preferred supplier. National geographic expansion and integration of local networks became significant, as did automation later, while service differentiation outside basic telephony remained insignificant. Postal services were traditionally organized as a separate state company/agency. A few user networks were built but of minor importance on the whole, although they could be considered as niche entries by domestic entrants.

A *third* period, one of isolated new niche entries, began in 1967 when an entry by a foreign entrant, GEIS, was made in a new special service, datacom, based on a new technology (mainframe computers).

A second isolated new niche entry was made in 1981 by a mainly domestic entrant, Comvik, in a new special service, land mobile public telephony, based on new technology (cellular radio). Like GEIS, Comvik was not a traditional telco viewed from their parent company's point of view, as they were diversifying into telecom services. However, in contrast to the entry by GEIS, the entry by Comvik was fiercely challenged by the incumbent Telia, who had been a prime mover of the new service and the new technology.

A *fourth* period, one of multiple and sequential trial entries, began in the second half of the 1980s with a first real, although small, wave of new entries and inward FDIs on the Swedish market. In this period several of the major PTTs entered the Swedish market (AT&T, BT, France Telecom, Pacific Telesis) plus a few other telcos (Vodafone, Cable & Wireless, MCI). The leading entrants were almost all foreign, with an Anglo-Saxon dominance. DBP Telekom and NTT have not yet (1993) entered the Swedish market.

The PTTs made it mostly on their own, while the other telcos mainly allied themselves (at least initially) with domestic majority partners, primarily large users (so far equipment suppliers had refrained from integrating forwards).

---

[52] The Bell Company missed taking out telephone patents in Sweden.

The initial entry market segment was the large user group, especially domestic and foreign MNCs, and the infrastructure firms,[53] especially the railway company and the energy distribution companies. The preferred entry service area was datacom, being a weak point of the incumbent Telia in the mid-1980s with capacity and quality problems. The preferred entry mode was "soft and low key"—not to compete head on right away and/or unnecessarily trigger retaliatory actions by the incumbent and/or regulatory bodies.

The entry was mostly just a part of what appeared as a sequential entry strategy with steps to be taken later when opportunities arose towards attractive parts of the incumbent's core business (e.g. international calls) from some market position in adjacent business areas (e.g. datacom) and/or from some advantageous technology position (e.g. calling cards and their handling). Conceivably subsequent steps could also be taken towards entirely new businesses once a customer base was secured.

One may note here that Telia's strong position in analog mobile telephony most likely has deterred foreign entry (although an alliance between a foreign entrant and Comvik would have been a possibility). Given the by-pass possibility and other possibilities of wireless (e.g. land mobile, satellite and cordless) communication, it appears to be highly important for the future evolution of the Swedish telecom service market as well as for the future evolution of Telia how it manages transitions to new mobile communication technologies (e.g. digital).

This fourth period signifies substantial internationalization of the market in terms of FDIs. To some extent the FDI pattern has followed the simple 'establishment chain model', predicting an FDI progression from agents, to sales subsidiaries, to production and possibly also R&D operations. That is, what are essentially foreign sales office establishments and leased-lines arrangements have been followed by increased establishment of network ownership and control. Note in this context that the traditional arrangements in international telephone traffic could quite reasonably be viewed as export activities carried out by a web of PTTs operating essentially as each other's agents, which makes the establishment chain model quite fitting. The same model then also predicts that in the future some foreign telecom R&D will be established in Sweden, something which is quite conceivable in some form, e.g. an R&D consortium.

Another observation regarding the FDI pattern in telecom services in general (i.e. not only Sweden) is that it may have followed, rather than led, FDIs in manufacturing industries, as has been observed by Dunning (1991) for FDIs in service industries in general, but then perhaps only to some extent initially. As for Sweden, with a generally small flow of inward FDIs since long ago in manufacturing industries, her telecom inward FDIs seem to have been caused mostly by other factors.

What will happen after the fourth period must largely be speculated about. Patterns from other industries suggest a period of continuing entries, possibly crowding, shake-out, specialization, consolidation and temporary stabilization. The alliance propensity, which for several reasons is high in telecom, may lead to somewhat other period characteristics, however. Alliance formation will dilute national identities, perhaps to the point of denationalization (i.e. loss of a single national identity) of a player, the mere prospect of which would probably provoke regulatory action. It is also possible that the regulatory pendulum, possibly influenced by the incumbent, will swing back, at least to some level of competition perceived as acceptable and controllable (Bohlin and Granstrand 1991). Table 5.1 summarizes features and periods in the past, as well as hypothesizing some of the future. The transitions to a 5th and

---

[53] The term infrastructure firm is here used to mean firms with some kind of a network, e.g. energy distribution companies, gas companies, water supply companies, railway companies etc.

possibly a 6th period are gradual. What especially marks the transition from a 4th to a 5th period is the change from defensive to offensive moves by Telia, which is a fairly natural chain of reactions by a challenged incumbent.

Table 5.1
Actual and hypothetical stages in the internationalization of the Swedish national telecom service market[1)]

Period	Characteristics	Illustrative starting years for the case of Sweden[3)]
1st	Mixed international and national competition	1880s
2nd	Nationalization. Local monopolization	1890s
3rd	Isolated new niche entries by domestic and/or foreign entrants through greenfield operations or acquisitions (Entries typically made in areas with weak, unclear or absent regulation[4)])	1960s
4th	Multiple and sequential trial entries (Entries typically made in response to liberalization) Domestic entrants being complemented by foreign entrants Defensive moves by Telia	1980s
5th[2)]	Restructuring of Telia and shift towards offensive moves (Typically internationalization and subsequent service specialization, plus alliance formation Also e.g. change of legal status, reorganization and privatization) Limited market invasion and restructuring by domestic and foreign competition	1990s
6th[2)]	Appearance of major proprietary service innovations and subsequent but limited creative destruction Telecom service branding Integration with foreign markets Network disintegration and reintegration Challenge of dominant firm positions Emerging denationalization Integration with other information service markets Restructuring of supplier relations (possibly increasing quasi-integration)	?

1) Variations in pace and scope for different service areas are understated. A steady pace of liberalization without major, persistent anti-competitive regulation is assumed.
2) Hypothetical stages. To some extent the 5th stage has been entered.
3) The time division in periods is approximate, especially since some of the characteristics refer to processes with gradual change rather than to events.
4) The absence of regulation was in some cases due to a narrow conceptualization of telecom industry and services on behalf of regulators.

What especially marks the possible transition to a 6th period is the appearance of entrepreneurial entries made on the basis of proprietary service innovations, thereby fostering Schumpeterian competition, based on proprietary new technologies. The role of R&D and close relations to advanced users and suppliers thereby increases. At the same time various integration among actors, markets, networks and other services is likely to proceed. In this period the real challenges to a domestic, dominant firm will develop.

Finally, it is largely an open question how much competition is, to use a typical Swedish term, "lagom" (i.e. moderate, "middle way", appropriate). This question has not least been reopened by the new international trade theorists, advocating strategic trade policies, rather than "pure" free trade. On the other hand, the general answer 'as much as possible' has been received for a long time from the free trade liberalists with Sweden as one of their strongholds, but then certain areas have been excluded, e.g. service areas like telecom.

6. CONCLUSIONS

The Swedish telecom service market became subjected to international competition already in the early days of telephony. The 1890s signified the beginning of a 2nd period in which by and large a national monopoly/duopsony became established on a mixture of regulatory and de facto grounds. The first foreign entry was then made in the 1960s outside the core regulatory areas, signifying a 3rd period of isolated new niche entries, followed by a 4th period in the 1980s, in which multiple foreign entries were made.

Significant de facto competition in telecom services in Sweden was established when Comvik Skyport, AT&T and later on BT set up offices in the mid and late 1980s. At that time digitization reached 20-30%, which could be seen as a minimum level of required installed digital central offices in order to attract larger volumes of FDI. Today, the degree of digitization (and fiberization) calls for low technical network barriers to entry, while the wide distribution of main lines in Sweden calls for relatively higher economic network barriers to entry. Moreover, cost-based pricing of telecom services will yield even higher entry barriers to Sweden (where prices already are relatively lower).

Entries into the Swedish telecom services market may be separated into those who have acquired a network with customer access and those who gradually have invested in a new network. The acquisitions by France Télécom in European datacom networks illustrate the trend towards increasing local network control (multi-domestic) which gives direct customer access. Another example of this is AT&T who bought McCaw, the cellular telco in New York, obviously in order to get direct customer access. In Sweden AT&T, MCI and Tele2 have (although to various extents) invested in new networks or just used leased lines connected to their own international switches (gateways).

It seems difficult to enter a new market without the *fast* establishment of a customer base in the targeted market. This has not really been proved yet, since most new entrants have been protected by regulations, such as Mercury in the UK. In Sweden, however, there is a case of an early (in 1981) second cellular operator, Comvik GSM, who did not succeed and who gradually only got some small percent of the cellular market. They were simply not aggressive enough. Since the needs of the business customers seem to be more diverse and difficult to fulfil with a single service (or even a single supplier of services), services have to be more customized, which makes it more interesting for customers to look for a new supplier of services, implying lower barriers to existing customers for new entrants. This will also call for a closer look at customer segmentation by telcos. There are no longer only business customers and residential customers. In addition, there is a trend towards a separation of service providers (telcos) and teleshops selling equipment and subscriptions.[54]

Apart from the large MNCs in Sweden, who will be monitored by the telcos with international coverage, it is difficult to identify distinct market segments apart from the traditional business-residential ones. It seems instead clear that different telcos will concentrate on different segments to a larger extent than in 1993. There are some examples, however, such as home-country related customers (MCI and AT&T seem to focus on American firms in Sweden for instance), large domestic companies and central & local governments (BT and Transpac) and metropolitan customers in general (this is obvious for all new entrants, whether they provide telephony services or datacom services).

There are some different service focuses among the telcos on the Swedish market. There are a group of value-added network service providers (e.g. EDS, GEIS, IBM), a group of cable-TV operators (e.g. Svenska Kabel TV, Kabelvision etc.), cellular telcos (Telia Mobitel, Nordic Tel, Comvik GSM) and EDI/E-mail providers (e.g. Scandinavian Infolink). These telcos have, thus, a low scope of activities, even though Telia Mobitel is a part of the Telia group. Furthermore, potential fast-growing services segments in Sweden ought to be mobile data, e-mail, EDI and toll-free voice services.

It seems reasonable to believe that there will most probably be new and different kinds of telcos in Sweden. There appear to be telcos concentrating on international network coverage without direct customer access and telcos concentrating on domestic coverage with direct customer access. In Sweden it seems reasonable to forecast that BT, MCI, AT&T and Unisource will belong to the former group (where Sweden or the Nordic countries may form an important hub in a larger international network) while Telia, Transpac (and probably also BT) and the cellular telcos presumably will belong to the latter group. In addition, we believe that there will probably be a third domestic group of telcos, providing by-pass opportunities to Telia's network. An example of such a telco is the potential entrant Dotcom, the telecommunication subsidiaries of the energy distribution companies and cable-TV operators. Finally, strategic alliances (network cooperation) in order to get geographic coverage (and direct customer access) are anticipated. Examples are the cable-TV providers and the cellular telcos cooperating with each other or with a telco with by-pass capacity, such as Tele2 potentially cooperating with Kabelvision.

Turning to outward foreign investments, Telia has changed its early phase of technology and network cooperation activities, and turned into a phase of commercial cooperation activities. They have started a close cooperation with PTT Telecom through the formation of Unisource and joined several joint ventures, mainly in cellular but also international switched telephony. Examples are cellular ventures in Estonia and Hungary. Kinnevik, on the other hand, has joined a large number of joint ventures through the company Millicom. They have focused their activities on small developing countries, where small networks are implemented with a potential short pay-back period.

Finally, it seems clear that the telcos in Sweden providing data services in 1993 will, within a short period of time, also provide telephony services, when their networks and their marketing departments are ready. In addition, it may be hypothesized that the relatively higher competition in cellular along with the relatively more advanced technology used in the GSM networks will make the implementation of personal numbers through calling cards (including

---

[54] This is seen in the UK and is forecasted e.g. by GEAB, a large Swedish distributor of mobile phones.

the SIM-card) faster in cellular networks in Sweden. This may be a major threat for Telia since the traffic through the cellular networks will increase substantially with new personal calls going through the cellular network apart from the calls to and from the cellular terminals.

REFERENCES

Bohlin, E., 1992, Telecommunications liberalization in Sweden 1980-1993: An overview, in M. Christoffersen and A. Henten, eds., The proceedings of the 7th Conference on European Communications Policy Research, EURO CPR'92.

Bohlin, E. and O. Granstrand, 1991, Strategic options for national monopolies in transition. The case of Swedish Telecom, Telecommunications Policy, October, 1991, pp. 453-476.

Bäckman, B., 1989, Är marknaden beredd?, Affärsutveckling med tele- och datakommunikation (Is the market prepared? Business development using tele and data communications), Idébok, Televerket, 1989.

Cap Gemini, 1992, Swedish telecommunication services—promoting an efficient market, Swedish Ministry of Transport and Communications, Stockholm, 1992.

CEPT, 1990, Statistiques des Télécommunications.

Communications Week, 1992, Sweden to get third operator, Communications Week International, New York, USA, No. 97, 1992, p. 7.

Dunning, J.H., 1991, The internationalization of the production of services: Some general and specific explanations, GSM Working Paper, 91-11, Graduate School of Management, Rutgers University, New Jersey, 1991.

EC (European Commission of the European Communities) (1993), Communication to the Council and European Parliament in the Consultation on the review of the situation in the Telecommunications Services Sector, COM (93) 159 Final, Brussels, 1993.

Granstrand, O., 1993, The evolution of Nordic mobile telephony, International Telecommunication Society's conference, Stenungsbaden, Sweden, June 20-22, 1993.

Granstrand, O. and O. Johansson, 1992, Internationalization of Telcos—An early analysis of strategies and behavior, in M. Christoffersen and A. Henten, eds., The proceedings of the 7th Conference on European Communications Policy Research, EURO CPR'92.

CIT, Research, Mobile Communications in Europe 1993, Communications and Information Technology Research, London, 1992.

Lindskog, H., 1993, Oral presentation of the procurement criteria of the Stattel delegation, STF Ingenjörsutbildning, Telekomdagarna 1993, Stockholm.

MIRC (Market Intelligence Research Company), 1991, World telecommunications services factbook 1991, Brussels.

Nilsson, A-M., 1993, Oral presentation of the procurement criteria of the Stattel delegation, STF Ingenjörsutbildning, Telekomdagarna 1993, Stockholm.

OECD, 1993a, Communications Outlook 1993, Information computer and communications policy, OECD, Paris, 1993.

OECD, 1993b, Usage indicators, Information computer and communications policy, OECD, Paris, 1993.

SFS, 1993, Telelag och en förändrad verksamhetsform för Televerket mm. (Telecommunications act and a changed form of operations for Televerket and others), 1993:597, CE Fritzes, Stockholm.

Telephony, 1992, BT whips up a storm with cyclone supercarrier plans, Telephony, August 24, 1992.

Telia, 1992, Treårsplan 1993-1995 (Three-year plan 1993-1995), Farsta, Stockholm, 1992.

Thorngren, B., 1993, Oral presentation of future customer competition, STF Ingenjörsutbildning, Telekomdagarna 1993, Stockholm.

The Race to European Eminence
E. Bohlin and O. Granstrand (Editors)

Internationalisation of the Norwegian Telecommunication Market*

Ragna Brekke[a], Sissel Jensen[a] and Christian Riis[b]

[a]Norwegian Telecom Research, Instituttveien 23, 2007 Kjeller, Norway,
[b]Foundation for Research in Economics and Business Administration, Gaustadalleen 21, 0371 Oslo, Norway**

Within a European context, the Norwegian telecommunications policy is characterised by a quite liberal position with respect to market deregulation. Telecommunications traffic is clearly internationally oriented. This is reflected in Norway's status as a small nation with an economy directed towards export, opening up interesting opportunities for foreign operators. Furthermore, the entry process is relatively easy to accomplish since the telecommunication infrastructure is largely digitised and as fibre is widely introduced.

1. INTRODUCTION

Within a European context, the Norwegian telecommunication policy is characterised by a quite liberal position with respect to market deregulation. The markets for terminal equipment and value-added services were deregulated in the late eighties, resale of leased lines is now allowed, and competition in mobile communication is introduced. However, as regards the organisational structure of the telecommunication administration, the policy is clearly more conservative. Norwegian Telecom remains a public corporation. But it seems clear that, over a period of time, the process of internationalisation will force a separation between the State and Norwegian Telecom.

The potential for competition in the Norwegian market is improving with time. Technical change is modifying the scale problems of a "small country". The traffic is clearly internationally oriented, reflecting the country's status as a nation with an economy directed towards export. Independent networks exist (for instance cable television and transmission capacity connected to the railway, and within energy distribution). Furthermore, significant capacity based on leased lines is accumulated in private hands. Finally, several private suppliers of various services are already established within the liberalised market segments.

---

* We thank the organizers of the ITS Conference and anonymous referees for valuable comments.
** This work is a part of a proje ct financed by Norwegian Telecom.

2. BACKGROUND AND STRUCTURE OF THE NORWEGIAN TELECOMMUNICATION MARKET

The unique topography of Norway together with special communications demands have been a challenge in building national communications networks. The decades following the Second World War were a period marked by appeals to the authorities for more funding for necessary investments. In the early 80s a more generous line on the part of the authorities enabled Norwegian Telecom to increase the investment level and this removed the waiting lists in a few years. Today, there are around 11,500 kilometres of optical fibre in the national network, 61% of which is in the trunk network. Fibre is used in about 2/3 of total channel km (64 kbit/s circuits). According to plans, the trunk network will be 100 per cent digitised in 1995, and the subscriber network likewise by the year 2000.

Norwegian Telecom International (NTI) was established in 1991 with the overall purpose to enable Norwegian Telecom to better meet competition in the international arena, protect Norwegian Telecom's home market and support Norwegian industries engaged in business abroad.

To some extent the Norwegian telecommunications market is concentrated in smaller geographic areas, reflecting the degree of concentration in business and also in population. 47% of the total number of business subscribers are located in the urban areas of Oslo, Stavanger and Bergen, while 38% of the residential subscribers belong to these areas. These main cities generate 46% of the total operating revenues from the public switched telephone network (PSTN) of Norwegian Telecom. PSTN revenues are about 65% of total revenues. The public telecommunications service market is about 2.5% of national GDP, the average for the OECD countries being nearly 2.3%.

Norway has a modern telecommunication infrastructure as indicated in Table 2.1. However, the fault rate is clearly high compared to other countries:

Table 2.1
Features of the Norwegian telecommunications market, compared with Sweden, France and the United Kingdom[1]

	Norway	Sweden	France	UK
Digitised switching capacity of total exchange lines (1991)	45.0%	56.0%	70.0%	45.0%[2]
Telecommunications gross investment per capita (1987–1989)[3]	147.52	148.91	99.36	77.10
Telephone main lines per 100 inhabitants (1991)	51.4[4]	68.4[5]	51.0[6]	44.6[6]
Mobile phones per 1000 inhabitants (1993)[7]	79.35	88.57	8.67	29.26
CATV connections per 100 households (1991)	33.3	29.4	1.5	1.4
International outgoing traffic per main line per minute (1991)	144.7	112.8	81.7	106.3
Faults per 100 main lines (1990)	21.4	12.0	10.4	16.0
Percentage failure on long distance calls (1990)	2.2	1.1	0.7	0.6/0.7[8]

1. Source: OECD, Communication Outlook 1993, except lines three and four.
2. Estimate for British Telecom; for Mercury the estimate is 100%.
3. In USD at constant prices and exchange rates, a three-year moving average.
4. According to Norwegian Telecom, 1993.
5. According to Swedish Telecom, 1993.
6. Financial Times, Oct. 18, 1993.
7. Source: Mobile Communications, Financial Times, Sept, 23, 1993.
8. The estimate is 0.6 in the subscriber network, and 0.7 in the trunk network.

Norway is a small country with an open economy. Exports as a percentage of GDP are more than 43%; the percentage for France and UK is about 24%.

2.1 Concentration

Statistics show concentration in both outgoing and incoming telephone traffic. The "Top 5" destinations account for 63% and the "Top 10" destinations account for 78% of the total outgoing traffic. Incoming traffic is even more concentrated. For Norway, outgoing voice traffic is about 15% higher than incoming traffic. Regarding bilateral traffic relationships, traffic imbalances exist with more and more countries.

International outgoing telephone traffic accounts for 20% of operating revenues, and for a substantial part of total surplus from voice traffic. In addition there are accounting revenues from incoming international calls. Considering the revenue created by international telecommunications, it is estimated that business calls are responsible for 67% (while business calls account for about 50% of the total revenue). Taking into account a relatively high degree of concentration, the market for business communication in the main city areas is of vital importance in the internationalisation process. Estimates show that the 200 largest business subscribers account for 30% of Norwegian Telecom's operating revenues.[1]

With relatively small investments Norwegian Telecom's competitors can establish nodes and give business customers access to their existing global networks. This is a situation which will have a significant impact on Norwegian Telecom's operating revenues and profit.

3. ENTRY CONDITIONS AND REGULATIONS

At present, Norwegian Telecom has the legal monopoly rights to the basic infrastructure network and to voice telephony. Within mobile communication, duopolistic competition takes place.

3.1 The process of liberalisation in Norway

In 1987, the Norwegian Telecommunication Regulatory Authority was established, to separate the regulatory matters (type approval of equipment, licensing, standardisation, frequency administration, etc.) from Norwegian Telecom. The markets for all types of terminal equipment (incl. internal networks) were liberalised in 1988. One year later, competition in value-added services on leased lines was introduced. In the market for digital mobile communication, a licence to a private competitor, NetCom, was given in 1991. At present, the duopolistic competition between Tele-mobil (owned by Norwegian Telecom) and NetCom is in the opening phase (presupposed to cover the whole country by 1997 at the latest). From 1993, resale of leased capacity is allowed; however, resale of leased capacity for voice telephony is prohibited. Satellite communication is still regulated. A liberalisation in this market is expected to take place in 1995.

3.2 Types of regulation

Norwegian Telecom remains a public corporation. Consequently, the charges of monopolised services are determined by the Government. Within the liberalised market segments private

[1] According to Ministry of Transport and Communications, Finansavisen 30.09.93.

competitors have in principle access to the public network on non-discriminatory conditions. To avoid cross-subsidising between the monopolised and competitive sectors of the Norwegian Telecom Group, limited companies (and in some cases, separate accounting units) have been established in the competitive areas.

The aim of the EC is to achieve liberalisation of all public voice telephony services by 1 January, 1998. Norwegian Telecom's board of directors want deregulation according to EC ideas and want to become a shareholder company. The Norwegian Ministry of Transport and Communications will evaluate the organisation structure in 1994. It is expected that Norwegian Telecom will become a state limited company.

3.3 Conditions for foreign investments in Norway

As Norwegian Telecom is a public corporation, foreign ownership is prohibited. Within the telecommunications sector in general, there are no restrictions on foreign ownership, except one. The exception is cable television, where the present regulation states that the networks are under Norwegian control. The regulation is based on a statute on foreign ownership within the manufacturing sector. To get a licence for cable TV, the company must have Norwegian owners. For a limited company, this means that at least two thirds of voting shares have to be owned by Norwegian subjects.[1] In addition, the majority of the board including the chairman must be citizens of Norway. The restriction is incompatible with the European Economic Area agreement and will be changed.

4. PRIVATE TRANSMISSION CAPACITY

In Norway, as in other Nordic countries, the use of telecommunications and local area network (LANs) is extensive (Svendsen, 1993). An increasing number of companies have already installed or are planning to install LANs. The use of the "high speed" data services, (2 Mbit/s and 64 kbit/s leased line and 64 kbit/s X.25 service) is concentrated on a few major companies with a large share in the total number of subscriptions. This is shown in Table 4.1.

Table 4.1
Relationship between subscribers and subscription for the data services in 1992 in Norway

Data service	% share in number of subscribers	% share in number of lines
2 Mbit/s leased line	5	51
64 kbit/s leased line	3	44
64 kbit/s X25 service	9.5	47

Table 4.1 shows that 5% of the subscribers of 2 Mbit/s leased lines have a 51% share of the total number of line subscriptions. This implies a trend towards a few companies leasing a lot of lines.

[1] The cable-television company Janco is granted exemption from this regulation.

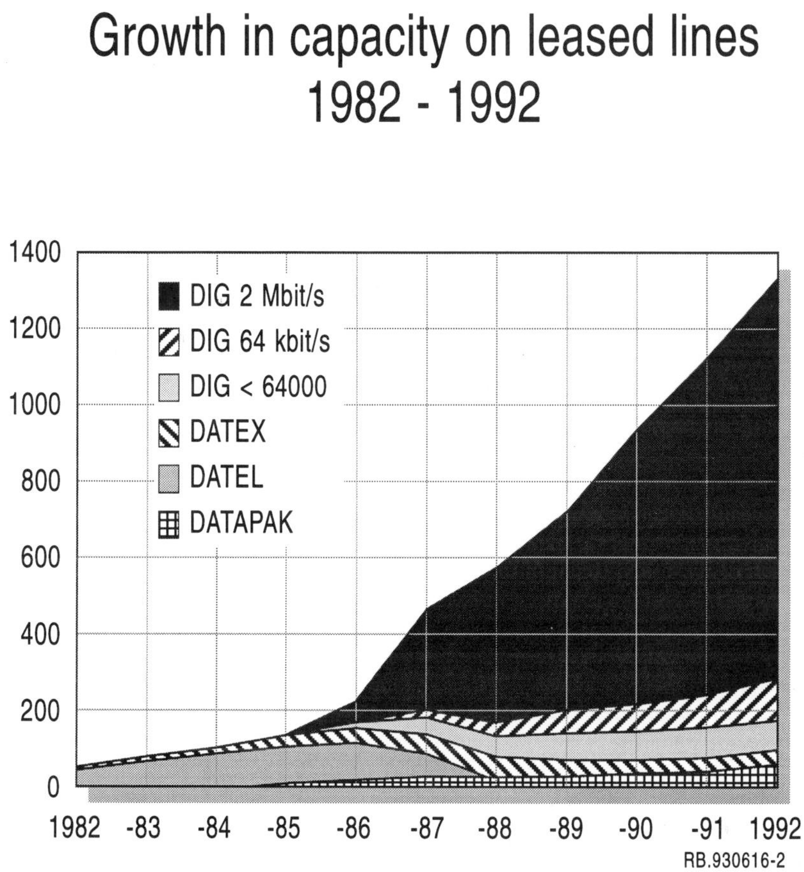

Figure 4.1. Growth in capacity on various data communication services 1982–1992. (Capacity is measured in mill. bit/sec.)

The total capacity of data networks is growing fast in Norway. Figure 4.1 shows the increasing demand for capacity on various data communication services during the last decade. The figure shows the number of subscriptions multiplied by the capacity of each line. This gives the total volume of data lines for domestic and foreign use. Capacity is measured as mill. bit/sec.

The average growth in capacity on different types of data lines per year has been 8.8% in the period 1986–1992. Since we know that a few companies account for a large amount of these lines and for capacity, there is every reason to believe that there are large private networks in Norway. Bypass of the public switched network is already notable. The networks can also supply access to the Norwegian market for foreign telecom competitors. The challenge for Norwegian Telecom is to keep the marked share when customers have a choice of supplier.

4.1 Cable-TV connection in Norway

According to Table 2.1, 33.3% of Norwegian households are connected to various kinds of cable-TV systems. There are three main companies; TBK, Janco and Norkabel. These companies cover 86% of all CATV subscriptions.

Table 4.2
Details about the cable-TV connection in Norway

Cable company	Owners (share)	Subscribers	Network coverage[1]	Main area
TBK	Norw. Tele. 100%	185,000	270,000	All main areas
JANCO	Sanoma corp. 100% (Finnish)	160,000	n.a.	Oslo
NORKABEL	Jan Helen[2] (70%) UCI[3] (30%)	140,000	210,000	The southern parts of Norway

1. Number of households which are or may easily be connected to the cable company.
2. The numbers refer to voting shares. UCI owns 86% of the non-voting shares and Jan Helen 14%.
3. United Communication International, which is owned by UHS (8%) and Telewest (92%), which in turn is owned by TCI (50%) and US West (50%).

The cable TV networks cover a relatively large part of Norwegian households. Considerable capacity for telecommunications is available within the three large operators. Except for Janco, which covers the main areas of Oslo, the cable networks are split into several local networks. The cable companies may establish a complete parallel network to Norwegian Telecom, either by connecting the local networks wireless or by building/leasing line capacity. One interesting aspect of the cable-television companies is that, except for TBK, they have connections with international telecommunication companies through foreign ownership. It seems that Norwegian Telecom (TBK) is facing this competition by buying up independent cable companies.

5. INWARD FDI—THE PRESENT LEVEL OF INTERNATIONAL COMPANIES' ACTIVITIES IN NORWAY

Norwegian Telecom has almost 100% of the Norwegian market share when it comes to international telephony. The competition from foreign carriers has not been a big threat to Norwegian Telecom until now. As a consequence of deregulation, this situation will change in the future. Many companies have supplied value-added services, for example EDI, for a long period of time. However, the competition from foreign telcos is not yet very strong.

Table 5.1 gives key data for some of the new entrants in the Norwegian telecommunications market, even though GEIS cannot be considered as a new entrant.

Table 5.1
Summary of key data for some new entrants in the Norwegian telecommunications market.

Telecom Company	Year entry	Major owners	Offices Norway	Empl. in Norway	Turnov[1] 1992	Nodes Norway	Investm.mid-93	Entry strategy	Entry areas
AT&T[2]	1980	ATT	Yes[3]	No	n.a.	1	n.a.	America-related customers	X25/SNA Calling Cards
BT[2]	1990 1991	BT	No	No	n.a.	2	n.a.	Large customers Int'l serv.	X25/SNA Int'l voice
MCI[2]	1992	MCI	No	No	n.a.	No	n.a.	Large customers America-related customers	Data services Telephony
Unisource	1993	Telia PTT-Netherlands Swiss Telec.	1	3	0	1	n.a.	Int'l customers outsourcing Managed Data Services	X25 Int'l data services
NetCom	1993	Ameritec, Singapore Telec, Orkla	1	90	0	n.a.	n.a.[4]	Price competition VAS	Cellular GSM
Sprint	1984	Sprint	1	8	5.6	Yes[5]	n.a.	Hardware Joint vent. MNCs	X25/SNA Calling Cards Resale
GEIS	1970	General Electric	1	7	2.8	Yes[6]	n.a.	Shipping Transport MNCs	Int'l data services EDI
EDS	1990 1991	General Motors	No	No	n.a.	1	n.a.	Car industry	Int'l data services
Telegroup	1992	n.a.	1[7]	160	2.2	No	n.a.	Call Back services	Int'l telephony

Source: Reports and questions.

1. In USD.
2. AT&T, BT and MCI have offices in Stockholm covering the Nordic area.
3. Through NCR, National Cash Register.
4. Estimated investments of 265 mill. USD within 5 years.
5. Sprint shares its network with Fellesdata A/S and has interests in IF Assistor and Finansnett A/S.
6. Cooperation with NIT A/S (Norsk Informasjonsteknologi A/S).
7. Through an agent Teleteam A/S.

5.1 Access to international data networks

The share of leased lines out of Norway was low in 1992, but the number is growing. An increasing number of international firms are leasing lines out of Norway: Geis, IBM IN, Reuter, Sprint, BT GNS, Scitor (SITA), among others. Several international telecom operators have installed nodes for international traffic in Norway, such as AT&T, BT, Sprint, and Tele2 (Swedish).

The total use of Datapak to foreign countries was doubled in 1992. Most of the traffic went to Sweden, with UK, Denmark, Germany, and USA, in descending order. The share of subscribers of leased digital lines out of Norway was about 2% of the total amount and less than 1% of the analogue leased lines. The main traffic in the Nordic Datex network was domestic. This means that the share of subscription only intended for international data communication is still small in Norway. This does not necessarily mean that the traffic is low. One high capacity leased line out of Norway is enough to take a lot of traffic out of the public switched network.

The present level of activities in data communications by international firms seems to be low. But the deregulation of the leased line market and Norway's open economy will probably increase the demand for leased lines out of Norway in the future. Norwegian Telecom has to cope with tough challenges in the new market situation.

5.2 Country direct and public operators charge cards

Charge cards and Country Direct services let users make calls abroad and have them charged to their home telephone account or billed to a credit card. From Norway most of this traffic terminates in the USA. At the moment this accounts for more than 12% of the traffic to the USA. For other countries, the traffic is insignificant. Outgoing telephone traffic is nearly 6 times larger than incoming traffic on international Free Phone numbers (mainly Country Direct).

5.3 Call-back operators

Call-back operators provide arbitrage on the basis of national differences in international rates on PSTN. The services may be called dial back, call reverse, or code calling. Many of them are carried out over international free phone lines. Almost all operators are US-based.

Tariff comparison in Norway shows some differences on a bilateral basis. On a peak rate international telephone call between the OECD Member countries, there is no relation where Norway is the most expensive (taxes excluded)[2] . Compared with calling from the US, Norway is slightly more expensive in some traffic relations. Because the margins are small the USD exchange rate is important. Norway is ranked third in an OECD basket of international telephone charges after Japan and Finland. The basket uses tariff data for a four-minute direct dialled call. Tariff reduction for the basket in the period 1989-1992 (January) has been 42% in real terms, the OECD average in the period being 10%.

This implies little saving on bilateral reverse dialling and may explain why International Discount Telecommunication (IDT), which operated in Norway for a short period (during 1991/1992), left the market. Another US-based company has recently announced call-back services in Norway. The relatively short period that this service has been available suggests that this service has no great business potential in Norway.

It is quite clear that the international accounting rate system[3] plays an important role in a global market characterised by rapid changes in ownership, structure, and innovation. The practice has supported a monopolistic form of market organisation, and is not designed to work efficiently in more competitive environments.

---

[2] OECD tariff comparison model, OECD (1993).

[3] The convention used by telecommunications carriers to settle net balances of account arising from bilateral international traffic flows.

5.4 Resale business in Norway

With regard to voice communication, leased lines in Norway may be connected to the public switched network at one end only. Corporate users can route their own traffic across leased lines to public networks in other parts of the world, but they cannot resell it as a public service. Resale companies are presently testing the limit of international resale regulations. Resale of data communications on leased lines was made legal from 1993. Even though resale of voice on leased lines is prohibited, it is impossible to control.

The effect of competition from resale will force traditional carriers to balance tariffs and renegotiate accounting rates closer to costs. Norway has for a long time initiated renegotiations, but has not always succeeded.

Sweden, our neighbouring country, allows simple resale from the United Kingdom. The new international operator in Sweden is 60% owned by Kinnevik and 40% by Cable & Wireless. This can give Norwegian Telecom a reason to fear transit refile.

5.5 Mobile communications

The two analogue generations of mobile systems, NMT 450 and NMT 900, were both quite successful in the Norwegian telecommunications market. The market penetration rate has been among the highest (see table 2.1). Norwegian Telecom has played an active part in formulating the basis for the CEPT (now ETSI) specifications for global systems for mobile communications (GSM). During 1993 GSM mobile telecommunication was introduced in Norway. Two companies have been granted licence to operate: Tele-mobil (a subsidiary of Norwegian Telecom) and, after a competitive bidding, NetCom. This company is mainly owned by foreigners and is a joint venture confederating Singapore Telecom, and the American Ameritec International which owns 49% and the Norwegian company Orkla, which owns 49.9%. The Swedish company Kinnevik has non-voting shares in NetCom.

As a precondition for the licence both operators have to expand their networks to cover all regions within the next six years. At present the operators are constructing separate networks also in remote areas of the country, clearly on the cost of economies of scale. (However, there is a possible benefit connected to the monitoring of performance as separate infrastructures imply less need for coordination.) In addition, the systems are based on incompatible roaming standards.

A possible source of inefficiency relates to an asymmetry in market position, as Tele-mobil is granted the monopoly rights of the analogue mobile markets, NMT 450 and NMT 900. This enables Tele-mobil to skim the monopolised analogue market as long as they could possibly do so. Additionally, due to the monopoly rights to the analogue market, Tele-mobil will profit from any delays in the construction of the digital networks. Furthermore, Tele-mobil has a strategic advantage in the digital market as it may cross-subsidise the GSM activities. There are no types of conduct regulations either in the digital or in the analogue market.

5.6 The IT industry in Norway

The IT industry in Norway is modest. Total production in 1990 amounted to 2.6 billion USD, while total export was 1 billion. The industry is mainly owned by foreigners (Alcatel, ABB Nera, Ericsson, Siemens). The Norwegian companies are rather small; among the most important are Tandberg (total sales 22 mill. USD) and Stentofon (50 mill. USD).

The potential for Norwegian international expansion is probably connected to mobile and satellite communication, both areas where Norwegian firms have developed new products and new technological solutions, and where basic competence and know-how have been built up.

The Norwegian telecom industry has seen a growing importance in the international satellite equipment market. For example, ABB Nera is in the leading position of Inmarsat traffic and control stations and mobile earth stations.

6. OUTWARD FDI—NORWEGIAN TELECOM'S INTERNATIONAL ACTIVITIES

6.1 International operation

Norwegian Telecom is a signatory of INTELSAT (share of 0.5%), EUTELSAT (share of 0.7%), INMARSAT (share of 10.7%). In addition to this, Norwegian Telecom has become a significant operator of satellite communication services. This makes Norwegian Telecom an international network operator.

Norwegian Telecom operates two of the earth stations employed by INMARSAT. One is directed towards the Atlantic Ocean and one towards the Indian Ocean. The Eik station will now expand to cover satellite communication with aeroplanes (Inmarsat Aero). This facility supports the new public air-to-ground communications network provided jointly by BT, Norwegian Telecom, and Singapore Telecom under the name Skyphone. Norwegian Telecom also established an earth station at Nittedal in 1987 for European data communication via the EUTELSAT system, as well as a traffic control station for the NORSAT-B system. The NORSAT-B system offers connections between unmanned earth stations in Europe. NORSAT-B is a digital system providing the opportunity for switched high speed transmissions. The system was developed in Norway. Satellite communication is also used by Norwegian Telecom as part of its domestic telecommunication network, connecting the oil installations in the North Sea, the fleet of Norwegian ships in national and international seas and the Arctic islands of Spitsbergen (the NORSAT-A system) to the domestic public network in Norway. Norwegian Telecom has recently bought the TV satellite Marco Polo II from British Sky Broadcasting. The satellite has been renamed THOR and is intended to supply five different TV channels to the Nordic countries.

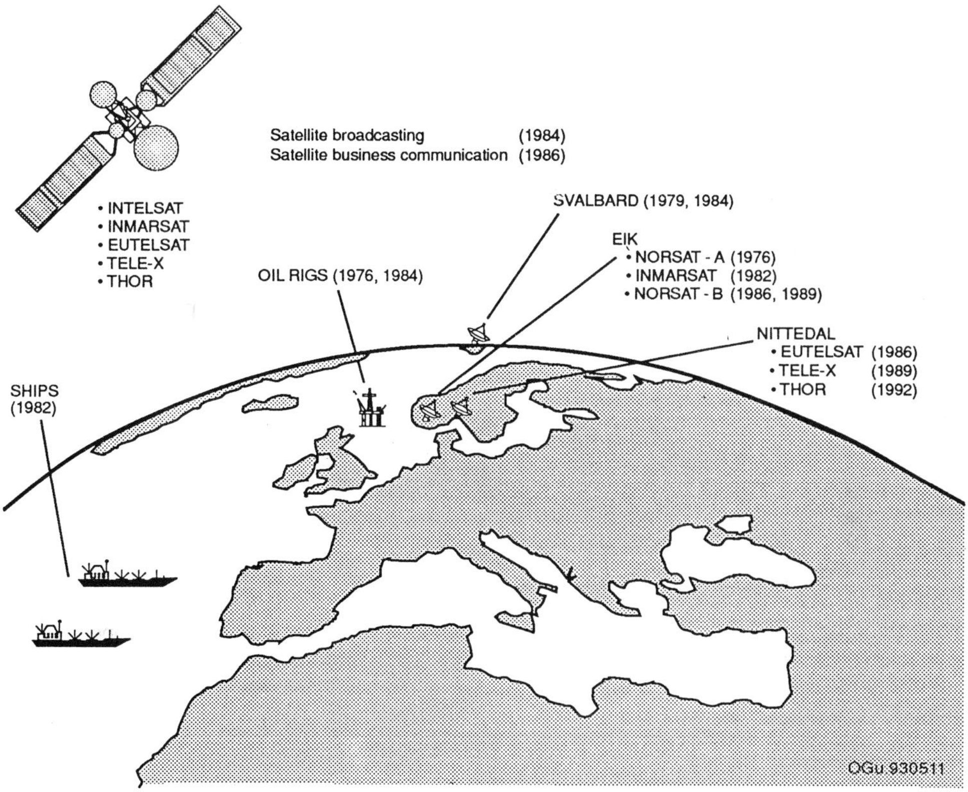

Figure 6.1. The Norwegian satellite involvement; main earth stations and utilised satellites.

6.2 International investments

Norwegian Telecom also participates in international activities, some of which involve investments outside Norway. Generally speaking, Norwegian Telecom is a newcomer in this field, but is starting to internationalise its commercial investments. Traditionally, FDI by PTOs started in those countries where the levels of liberalisation were high, and where the headquarters of multinational enterprises were located. This has not been the situation in Norway. The financial position and the needs for national investments, as pointed out in section 2, have also restricted the possibilities for international investments. Norwegian Telecom International was established as late as May, 1991.

A couple of years ago Norwegian Telecom established an international consulting unit called Nortelcon. Nortelcon has completed various analyses and projects in the fields of training, management development, and personnel planning in Africa and Eastern Europe. Among its projects, Nortelcon has also been responsible for putting into service a set of mobile switches in Albania.

A Norwegian-Russian joint venture company called Kolatelecom Ltd. has been established for cooperation in Murmansk. This is 50% owned by Norwegian Telecom. Another, called Severtelcom Ltd., is being set up for the purpose of cooperating with the telecom authorities in Archangel. The Kola project has been awarded a licence to route international traffic via Norway, and the second project is expected to obtain a similar licence. Investments so far carried out on the Kola peninsula are estimated at approximately USD 2.2 million. In return for these investments Norwegian Telecom will gain traffic rights and parts of the revenues from the circuits.

In 1992 there was a call for tender for 12 regional GSM licences all over the former Soviet Union. In a joint venture with Swedish Telecom International, Telecom Finland, St Petersburg Trunk-Line Exchange, and Leningrad Oblast, Norwegian Telecom formed a consortium which was granted the licence to establish and operate a digital cellular mobile network in St Petersburg. The total network costs are estimated at USD 160 million. The network will be put into service towards the end of 1993 with a limited number of subscribers and will initially cover the central parts of St Petersburg and the main highways leading into the city. A total of 7.5 million people live within the planned coverage area, and the companies' market share is expected to grow fast. The number of Western businesses in St Petersburg has increased considerably over the past years. The licence is for 10 years, and no other licences will be granted during this period.

6.3 International cooperation

For centuries, Telecommunication administrations have cooperated in setting standards and linking international networks. Most of this has focused on the public services and the switched data communications networks, telex, etc. Cooperation between PTOs is increasing as a major means to serve international customers, while supplementing geographical coverage of PTOs with one another.

Large carriers have formed alliances and strengthened their market position by supporting global network solutions like Eunetcom, Unisource, and Infonet. There will be both short-term and long-term advantages from forming alliances.

Norwegian Telecom has not yet formed strong alliances. Continued negotiations with several different partners are taking place. Unisource is one partner that has been under consideration for a long time, but has not been found interesting enough so far.

It is not clear which advantages Norwegian Telecom may obtain by establishing close links with international operators. In the future, international telecommunication markets will be characterised by large volumes, but probably low profit margins. In its international efforts Norwegian Telecom should probably concentrate on satellite and mobile communications, where it has comparative advantages, rather than forming alliances that aspire to achieve a position as a large global telecommunication operator. The latter strategy would involve a level of investments which probably cannot be justified within a small and relatively indebted operator. However, forming alliances to satisfy the Norwegian customers' needs for international communications, and also to be able to face the liberalisation and increased competition in the Norwegian telecommunication market, can be justified.

7. SOME POSSIBLE SCENARIOS

The alternative policies are clearly limited by the fact that the liberalisation and deregulation process continues internationally. Accordingly, from a national point of view, a main point of interest is how to secure an efficient supply of telecommunication services in Norway within an internationally integrated market. A second point of interest is which role one may expect Norwegian Telecom to play within the future market.

As the degree of competition increases, the present organisational structure of Norwegian Telecom is too inflexible and clearly too much biased towards the political sphere, to be able to compete efficiently. Historically, the Government has intervened in major price and investment decisions in conflict with the firm's own view. Furthermore, as the employees are civil servants, layoffs are restricted. Since the technological development has considerably reduced the labour requirement, a large proportion of the employees are now more or less redundant. However, due to the restrictions on layoffs, a new division, Norwegian Telecom Resource Pool, has been created.

Within a few years the formal status of Norwegian Telecom will change, probably into a public-owned limited company. Also, the question of privatisation has been raised.

As the market liberalisation process continues, Norwegian Telecom may meet the competition with an aggressive strategy, with the aim of maintaining the position as a dominant player in the main segments of the Norwegian telecommunications market. However from the moment that the basic markets are opened, it is not obvious that Norwegian Telecom's market dominance is feasible in the long run, unless formal alliances with some of the larger international operators are created.

As an alternative to the aggressive strategy, one may consider a more accommodative, niche-oriented strategy. Norwegian Telecom and also the Norwegian telecommunications industry have their main competitive advantages within mobile and satellite communications. Within these market segments Norwegian firms may become major players internationally.

8. CONCLUDING REMARKS

As the infrastructure is largely digitised and as fibre is widely used, entry into a liberalised Norwegian telecommunications market is easy to accomplish. The fact that the traffic is more internationally oriented than the OECD average makes the Norwegian market even more interesting.

However, there is a long political tradition in Norway of demanding the same level of infrastructure services in remote areas of the country as in the central districts, and it remains to be seen what consequences this may have for the entry process. It may in fact encourage entry as the potential for cream-skimming urban areas improves. However, if the Norwegian government, in order to get rid of cream-skimming incentive, extends the obligation of universal service to all network operators, entry is clearly discouraged. As the cost of creating networks to cover the whole country is extremely high in Norway, a principle of universal service obligation for all network operators will perhaps inhibit all entry.

In the long run, a complete liberalisation of the European market will take place. From a Nordic point of view one might gain from liberalising the Nordic markets much earlier. Considering our geographical structure, one may expect significant gains of network integration in regarding Scandinavia as one market. As in the other Nordic countries, large investments are made in a digitised trunk network, which at present has a high degree of idle capacity. Certainly, an effective competition in the trunk network can be created by letting the Scandinavian Telecoms compete. Furthermore, the operators may gain from experiencing competition in the market for basic services.

REFERENCES

OECD, 1993, Communications outlook, information computer communications policy.

Svendsen, B., 1993, High speed data services for LAN interconnection in Norway, Proceeding from IBCN & S, April 19-23, Copenhagen.

Televerket Statistikk (annual), Norwegian Telecom, Oslo.

The Race to European Eminence
E. Bohlin and O. Granstrand (Editors)

Inward and Outward FDI in Danish Telecommunications

Anders Henten and Knud Erik Skouby

Institute of Social Sciences, Bldg. 301, Technical University of Denmark,
DK-2800 Lyngby, Denmark

1. HISTORY AND INTRODUCTION

With the recent development in outward FDI trends, Danish telecommunications, in a sense, seems to be back where it all started.

Telecommunications was from its inception a highly internationalized enterprise. The first "real" telephone company in Denmark was established in 1881 by The International Bell Telephone Company.[1] Already in 1882, the Bell company was taken over by the Danish company Kjøbenhavns Telefon-Selskab, which later changed its name to KTAS.

The formation of the Great Northern Telegraph Company in 1869 and the visions of the company founder, C.F. Tietgen, concerning the position of Denmark as a centre for telegraphic connections in the Northern hemisphere, furthermore, emphasize the highly international character of telecommunications at that time. The idea was to make Denmark a transit area for telegraphic messages between East and West.

C.F. Tietgen had in 1868 obtained a licence from the Danish and the Russian governments for building a telegraphic cable between Denmark and Russia, and in 1869 the cable running via the Danish islands of Møn and Bornholm to Libau (in Latvia) was completed. This cable was planned to link a Trans-Siberian cable with a North Atlantic cable connection. Unfortunately for C.F. Tietgen, the North Atlantic cable did not come into existence, because it was defeated by a cable at a lower latitude between Ireland and Newfoundland. The Ireland-Newfoundland cable also stopped a cable project across the Bering Straits between North-East Russia and Alaska that caused the Trans-Siberian line not to be extended further than Kiachta in Mid-Siberia.

The idea behind Tietgen's plan was to take advantage of the favourable geographical position of Denmark in terms of bridging East and West and of the effort of the Russian government to build a Trans-Siberian line. Exactly the same idea forms the background for the deployment of the new fibre optic telecommunications cable in the Baltic Sea between Denmark and Russia which was opened in the Spring of 1993. This indicates that even today geographical position tends to have some importance for the strategy of telecom operators.[2] Furthermore, the Great

[1] In 1879, Kjøbenhavns By- og Hustelegraf (City and House Telegraph of Copenhagen) was established offering a combination of telegraph and telephone services.

[2] "A large part of Telecom Denmark's international activities is based on Denmark's central geographical location in

Northern Telegraph Company is once again active on the international telecommunications arena, now under the name GN. The company is a partner, e.g., in the new Denmark-Russia connection and in the cabling project between Russia and Japan/Korea. An extensive cooperation between Telecom Denmark and the Great Northern company constitutes an axis of Danish international telecommunications expansion.

Tietgen's activities thus illustrate that from the beginning, in the second half of the 19th century, telecommunications operators as well as operations had a very international character. The new technology was diffused by a limited number of firms mastering it, and it was perfectly suited to enhance the efficiency of international communications that were fast expanding along with international trade, which gained its momentum from new technologies as steamships and railways. After this 'international period' a long period of almost 100 years followed where telecommunications became a mostly nationally based activity.

The origin of the national, monopolistic regimes in telecommunications was essentially political. It is evident from the political discussions that a central driving force in creating the regimes generally was an intention to consolidate the economic and political advantages known from the control of the postal services.[3] Furthermore, the strategic military value of the telecommunications technology was evident and the telegraph network was developed in close contact with military authorities. Telephony gradually took over the strategic functions of the telegraph and with wars and protectionism in international relations, the scene was set for the monopolistic regimes.

From the late 1980s, telecommunication seems to have resumed its internationalism. The legitimacy of the old regime has been overthrown by a combination of new technology and conflict of interests between large users and the old regime—the challenge starting out in the US. The guiding principles of the emerging regime seem to be international competition and free trade.[4]

2. ISSUES AND TYPES OF INTERNATIONALIZATION

Against this background, an empirical account of a limited aspect of the discussion is presented in the following: Inward and outward foreign direct investments (FDI) in relation to Danish telecommunications—in line with the discussion above, confined to network operators and excluding equipment producers.

Based on this empirical material, we will discuss two issues: 1) The internationalization strategy of Danish telecommunications operators, 2) the question whether the internationalization of telecommunications operators has special features.

The internationalization of telecommunications has two different foundations. The first is

the Baltic Sea. Using this strategic location as a starting point, Telecom Denmark has established an extensive telecommunications network of cables, radio and satellite links that makes Denmark a natural gateway for international telecommunication between West and East". This is an excerpt from a brochure from Telecom Denmark, September 1992.

[3] This is documented as well for Denmark in "P&Ts historie 1850–1927" (The History of P&T) (1992), pp. 253-60, 367-71 as internationally in Peter Cowhey (1990).

[4] The guiding principles of the old regime correspondingly were 'natural monopoly' and 'universal service'.

based on the hitherto national monopolistic character of telecommunications and, therefore, deals with international interconnection of national operators. International submarine cables and satellites are examples of this type of internationalization. The second foundation is based on the upcoming international competitive regime and, consequently, involves types of internationalization, e.g. joint ventures and strategic alliances found in traditionally competitive business areas. From the following description of Danish outward FDI in telecommunications, it will be apparent that Tele Danmark is involved both in the traditional consortium type of cooperation in relation to international cables and satellites and in forming joint ventures in specific areas with changing alliances. More strategic alliances on a broader international scale have not yet been formed.

3. INWARD FDI

In Denmark—as in other EC countries—the telecommunication services sector is organized in three regulatory areas: the monopoly area, the special rights area and the competition area. As permitted by EC regulations, telephony and network infrastructures are currently in monopoly. GSM is the only service in the special rights area and VANS and data communications are in the competitive area. In spite of these differences, all three areas are or could be subject to FDI.

Tele Danmark holds the concession in the monopoly area. Tele Danmark is a limited company—where the state has the right and obligation to own a majority of shares. By the end of 1993, after the recent reshuffling of the sector, the state still owns a vast majority of shares, but the aim is to come down to just above 50%, and there is no law against foreigners owning Tele Danmark shares. The possibility for minority foreign ownership thus exists, although Tele Danmark shares are not very attractive because of the so-called rebuying clause: The state has the right to rebuy Tele Danmark shares at a 125 quotation.

In the special rights area, two operators have a licence: Tele Danmark and Sonofon. Tele Danmark Mobil is a 100% subsidiary of Tele Danmark while Sonofon is owned by the Great Northern company (36%), Bell South (29%), NordicTel (20%) and Incentive (15%). Sonofon is, therefore, a genuine case of inward FDI. So far, the GSM system is still in the build-up phase[5] and GSM investments are, therefore, not profitable at present. But they presumably will be and, furthermore, important experiences will be drawn by the companies involved. Bell South attaches great importance to its GSM activities in Denmark. Experiences will be used in other regions of the world.

In the competitive area, there is obviously scope for activities of foreign companies. But at present, there is not much foreign investment. IVANS have some activities. GEIS, for instance, has an office in Copenhagen. But activities are at a low level and the nationally oriented VANS are owned by Danish capital. Furthermore, BT, AT&T, Sprint, France Telecom and a few others have opened offices in Denmark, approaching business customers and preparing for new possibilities in relation to further liberalizations.

---

[5] The two NMT systems (450 and 900 MHz) have well over 200,000 subscribers. The NMT system is in the monopoly area and, therefore, a part of the concession to Tele Danmark.

The most striking effect, at present, of the question of inward FDI in Denmark, in reality, seems to be the big scare that clearly formed an important background for the establishment of Tele Danmark in 1991. Previously, concessions had been given on a regional basis and there were fears in Denmark that this fragmented structure could lead to an inexpedient rivalry between the regional companies which would open up the field to foreign competitors.

4. OUTWARD FDI

In telecommunication services operation, outward FDI from Denmark is more extensive than inward FDI into Denmark. This type of balance is traditionally the position of large and strong economies, and though Denmark in the hierarchy of developed countries is far from the top measured by GDP or powerful industrial companies, the position in telecommunications is relatively advanced.[6] Furthermore, Denmark has a strong standing when compared with developing countries and former socialist countries in Eastern Europe. This relationship takes up a central feature in the current internationalization of telecommunications. Still, developed countries have great difficulties in getting into the telecommunications markets of other developed countries. But markets in developing and Eastern European countries are fertile soil.

Outward Danish FDI can be grouped in five categories: Mobile joint ventures, investment companies, backbone cables/links in foreign countries and submarine cables, VANS and international satellites.[7] The two main actors are, as mentioned, Telecom Denmark (which is responsible for international activities of Tele Danmark) and the Great Northern company.

4.1 Mobile joint ventures:

- UMC (Ukrainian Mobile Communications) owned by the Ministry of Communications of Ukraine (51%), Telecom Denmark ($16^{1}/_{3}$%), Netherlands PTT Telecom ($16^{1}/_{3}$%) and DBP Telekom ($16^{1}/_{3}$%).
- Comliet UAB (establishment of mobile telephone networks in Lithuania) owned by Lithuanian Telecom (41%), Millicom (29%), Telecom Denmark (20%) and Antene UAB (10%).
- Pannon (GSM mobile telephone network in Hungary) owned by Telecom Denmark (16%), Telia, Televerket/Norge, Telecom Finland, Netherland PTT Telecom and three Hungarian companies.

---

6 The two NMT systems (450 and 900 MHz), which are in the monopoly area and, therefore, a part of the concession to Tele Danmark, have 250,000 subscribers. By the end of 1992 there were only a few thousand subscribers on the GSM system. Vigorous price competition including subsidies to the buyers during the last month of 1993, however, drove this number above 100,000 by the end of 1993.

7 Two additional types of international activities should be mentioned: Systems/equipment exports and consultancy. Danish Telecom International (DTI, owned by the four regional subsidiaries of Tele Danmark) takes care of systems/equipment exports, while Danish TeleConsult International's (DTCI) field of activity is consultancy assistance. DTCI is also owned by the four regional subsidiaries of Tele Danmark. Activities of DTCI may be important for Danish telecom operators in terms of preparing the ground for further involvement. This is, for instance, the case in Lithuania.

- It should, furthermore, be mentioned that Telecom Denmark together with Bell Canada and DBP Telekom have been involved in negotiating the conditions of a mobile licence in the Moscow region.

4.2 Investment companies:

- DPTG (Danish Polish Telecommunications Group) owned by Telecom Denmark (25%) and Great Northern, GN (75%).
- DRTG (Danish Russian Telecommunications Group) owned by Telecom Denmark (50%) and GN (50%).

* DBTG (Danish Baltic Telecommunications Group) owned by Telecom Denmark (50%) and GN (50%).

4.3 Cables/microwave links:

Telecom Denmark is a co-owner of a number of submarine cables,[8] mostly transatlantic and inter-Scandinavian cables and cables to Great Britain. Five cables will be mentioned here because they clearly illustrate the international strategy of Telecom Denmark:

- CANTAT 3: Together with 20 other network operators, Telecom Denmark in October 1992 concluded an agreement concerning the installation of an optical fibre cable between Europe and Canada.
- Together with GN, Telecom Denmark has established a North-South Link in Poland (1,500 km from the Baltic Sea to the Czech Republic).
- In Lithuania, an optical cable has been completed connecting Kaunas and Vilnius.[9]
- A submarine optical cable between Copenhagen and Kingisepp in Russia was opened in April 1993 and this link has been extended to St. Petersburg and Moscow by way of a radio relay system. The Copenhagen-Moscow link has been established by Telecom Denmark and GN in cooperation with the Russian Intertelecom.
- In the Far East, Telecom Denmark and GN are participating in the deployment of a fibre optic cable between Naoetsu (Japan), Pusan (South Korea) and Nakhodka/Khabarovsk in Russia. So far, it has not been decided finally how to link Khabarovsk and Moscow.

---

[8] Some of these sea cables should probably go into the category consortiums, as they have many operators as co-owners. But other cables clearly fall under the category joint ventures. For the reason of not separating cables, they are all included here.

[9] Both the Poland cable and the Lithuania cable have been financed by the above-mentioned investment companies and, thereafter, transferred to the local telecom operators.

4.4 Value added network services:

- Temanet is an alliance between Telecom Denmark (50%) and Maersk Data (50%). Temanet is an international network operator, offering both transmission services and value added services to business customers. Temanet has taken over (and digitalized) the international network of Maersk Data that operated international communications for A.P. Møller (shipping). A.P. Møller is by far the biggest customer of Temanet and the primary targets of Temanet are Danish firms going abroad.
- Another example of a VANS in which Telecom Denmark participates is Infonet. In 1992, Infonet reached an agreement with the Danish pharmaceutical producer Novo Nordisk, and The East Asiatic Company (shipping and manufacturing) is also an Infonet customer.
- Furthermore, a 100% subsidiary of Tele Sønderjylland (one of the four regional operators of Tele Danmark), Tele Schleswig GmbH in Gemany, should be mentioned. Tele Schleswig is active in selling, installing and servicing integrated communications solutions.

4.5 Satellites:

In international satellite transmission, the consortium type of cooperation is common. Such consortiums are mostly concerned with pure transmission arrangements. Telecom Denmark is a member of:

- INTELSAT
- EUTELSAT
- INMARSAT

5. TELECOM DENMARK/GREAT NORTHERN TELEGRAPH COMPANY

The main axis of Danish international telecommunications expansion is centred around cooperation projects between Telecom Denmark and the Great Northern company. There is cooperation in the cable field and in the investment field. In the mobile field, there is no cooperation. On Danish ground, GN and Tele Danmark are competing in the GSM area.

Figure 5.1 illustrates the cooperation initiatives between Telecom Denmark and Great Northern.

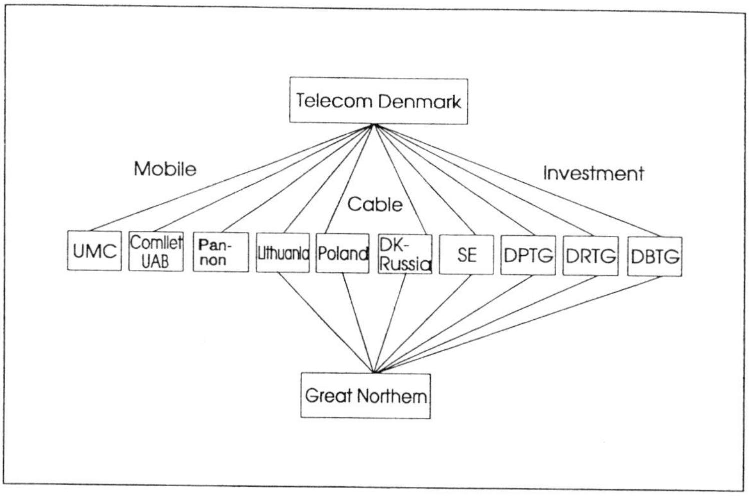

Figure 5.1. Cooperation between Telecom Denmark and Great Northern, 1993.

6. THE DANISH STRATEGY

The strategy of Tele Danmark and its subsidiary Telecom Denmark for expanding internationally has 4 building blocks: Cables, mobile licences, VANS and investment funding. The strategy related to cables, mobiles and investment funding seems relatively clear while the VANS strategy is somewhat muddled with the involvement in both Temanet and Infonet.

Tele Danmark is focusing on making Copenhagen a telecommunications hub in the Baltic area by attracting East-West transit traffic. The completion of the Copenhagen-Moscow link; the involvement in Poland and the Baltic states; the cable project in the Far East; and the involvement in CANTAT 3 all evoke memories of the telegraph visions of C.F. Tietgen in the 19th century. There are also signs that the relatively good results achieved by Tele Danmark in this area are very much due to the relations and results established during the first part of the century as to geographical position.[10]

Telecom Denmark has until recently considered participating in projects in Africa but has decided not to go ahead. There is not enough Africa experience in the organization and the market is not evaluated as promising enough to justify development of this experience. South-East Asia and Latin America, on the other hand, are becoming a target area for Telecom Denmark. South-East Asia is an economic growth zone and Telecom Denmark will seek to build on its engagement in the Russia-Japan-South Korea cable as a stepping stone to South-East Asia. Telecom Denmark and GN have been assigned by the Russian Intertelecom to provide for connections

[10] Great Northern upheld a representation in Leningrad/St.Petersburg until the 1960s.

out of Russia to South-East Asia.

Concerning the regional aspect, the strategy of Telecom Denmark, currently, seems to focus on East-West connections and the former socialist countries of Eastern Europe, while the technology/services aspect is concentrated around specialized areas (electronic highways, mobile and VANS). The strategy appears to be rather ambitious for a relatively small operator, but until now it has been relatively succesful. Four elements can be pointed out as important for the development:

- traditional and well developed relations with respect to telecommunications to Russia and the Baltic states
- the telecommunications market in Denmark has for a relatively long period been advanced considering technology and penetration based on a high GDP per capita
- the conventional monopolistic regime was modified by a type of competition or rivalry between the regional companies
- the organizational structure of Tele Danmark, where the international operator (Telecom Denmark) is acting separately, seems to be both flexible and efficient.

In a future international competition, only the last element is likely to be of some relevance—the first one will still be an advantage on the specific markets, whereas Nos. 2 and 3 are likely to lose or have lost their significance.[11]

7. CONCLUSION

It is, obviously, very difficult to draw any general conclusions from the description of the strategy of one operator, which is even one of the smaller ones. But the core features identified in the Danish strategy seem to be common: Focus on weaker regions with apparent growth potentials and on specialized services for particular segments in society (businesses and rich residential customers). Two structural arguments inherent in telecommunications are suggested to account for these common features.

1) It is very difficult to penetrate telecommunications markets in foreign developed countries and, furthermore, the establishment of cable-based generalized telecommunications systems (e.g. PSTN) is extremely expensive. There can be no doubt that telecommunications has entered a period of internationalization, but this internationalization primarily affects "peripheral" areas both in a geographical and a technical sense. The vast majority of telecommunications operators are tied to their local network as their primary business area.

2) Trade in services has been widely discussed lately, but in comparison with most other services, telecommunications require a physical infrastructure. This forms the background

[11] It is highly possible that Telecom Denmark very soon will be abolished in connection with a general restructuring of Tele Danmark.

for the specifics of the internationalization of telecommunication services. By definition, trade is taking place when a call is made from one country to another. The country receiving a call gets paid for its part of the transmission service and is, therefore, the exporter. In addition, more genuine types of telecommunications export can be found, for instance in the case of Telecom Denmark providing a satellite link for Lithuania and Moldavia to the outside world. But generally, the internationalization of telecommunications requires foreign direct investments (FDI) which, as described, lead to very high initial costs in the case of cable-based systems.

REFERENCES

GN Store Nord, 1993, Beretning og regnskab 1992, Copenhagen.

Andersen, M.M., 1987, Dansk telehistorie og det statslige tilsyns placering)Poul Selvig, Copenhagen).

Lange, O., 1978, Finansmænd, stråmænd og mandariner (Gyldendal, Copenhagen).

Granstrand, O. and O. Johansson, 1992, Internationalization of telcos—an early analysis of behavior and strategies, in M. Christoffersen and A. Henten eds., Telecommunication—Limits to Deregulation (IOS Press, Amsterdam).

P&Ts historie 1850–1927, 1992, P&T, Copenhagen.

Cowhey, P., 1990, The international telecommunincations regime: The political roots of regimes for high technology, International Organization, Vol. 44, No. 2.

Tcle Danmark, 1993, Annual Report 1992, Copenhagen.

Telecom Denmark, 1993, Annual Report 1992, Copenhagen.

Telecom Denmark, 1992, Update, As at 1992, Copenhagen.

Telecom Markets no. 233, Oct. 28, 1993.

The Race to European Eminence
E. Bohlin and O. Granstrand (Editors)

The Internationalization of Finnish Telecommunications: Motivations, Strategies and Activities

Jari Karpakka

Helsinki Telephone Company, P.O. Box 148, 00131 Helsinki

In Finland, the area of telecommunications is being opened up for full competition between the local telcos, known as TeleGroup of Finland, and the former state agency, Telecom Finland. The telcos have been expanding from providing local traffic to national traffic and are now expanding their operations to include also international traffic, whereas Telecom Finland, which has traditionlly been handling long-distance and international traffic, is seeking to expand its share of the local market. Both have been active in marketing their knowhow abroad, particularly in local network operations and cellular services, with considerable success. Finland's geographical position next to neighboring Russia and particularly the St Petersburg area, combined with the long-term Finnish trading experience with the Eastern neighbor, offer Finnish operators a unique opportunity to become involved in providing services to that area on a local and international level. Several international operators have already extended their global services to Finland mainly in cooperation with the domestic players, but without establishing any extensive own facilities. As the international services are about to be opened up for competition as well, there is now a growing interest among international operators to seek the opportunities of this liberalization.

1. AN INTRODUCTION TO FINNISH TELECOMMUNICATIONS

Finnish telephony has traditionally been handled by private local telcos besides the national PTT. The telcos have handled local telephony to most of the subscribers in the country, while the PTT has handled the remaining lines (about 28% in 1993) and the long-distance and international services as well as the mobile services. This division of responsibilities was characteristic before the new legislation in 1987, although there was also some earlier competition between the local telcos and the PTT, especially in data services.

The PTT as a government agency was responsible for the regulatory affairs as well until the Telecommunications Act of 1987 and its later revisions, after which it has become a concessionary network operator similar to the telcos. Since that time, competition has gradually been introduced also into mobile and voice services, mainly between the telcos and the PTT. However, licenses have been granted to other, smaller operators in the market as well. From the beginning of 1994 all domestic Finnish telecommunications, and from July 1st, 1994, also all international traffic will be opened up to full competition, mainly between the two initially equally-sized players.

The relationship between the telcos and the PTT has been characterized by both cooperation and rivalry. On the one hand they have for decades cooperated to make this multioperator (and multivendor) environment function with its variety of signalling, billing and accounting systems. As a result the Finns enjoy top quality telecommunications with one of the highest penetrations of telephony and the lowest tariffs within the OECD countries. On the other hand, there are parallel competing networks for both voice and data services, and for almost any telecommunication service there are at least two suppliers available. Even the residential customers have a choice; the wireless networks are becoming a true alternative to a wireline connection.

Finland's local telcos were until the late 1980s focusing their operations only on domestic affairs. During recent years they have changed from fully separate local telephone companies to a closely cooperating nationwide business federation. While most of the activities take place on the local level where the customers are, there is also a need for nationwide coordination in providing services. The telcos' response to this has been jointly owned affiliates. There are currently 47 local telcos ranging in size from two thousand to seven hundred thousand main lines. The nationwide and regional affiliates have telecommunication licenses for specific services, but they primarily utilize the local telcos' infrastructure and manpower.

Led by Helsinki Telephone Company, the local telcos and their affiliates for nationwide services, known as the TeleGroup of Finland, are competing in all telecommunications services with the state PTT. On the international front the PTT has, until recently, enjoyed a monopoly status. Competition has now been adopted besides all domestic services also in international data services and voice services between neighboring countries. Some of the telcos have in 1993 opened direct international voice traffic to Estonia, Sweden and Karelia in Russia. A subsidiary of TeleGroup of Finland, Finnet International Ltd., applied for a full international license on May 7th, 1993. The government has recently granted this license, allowing Finnet to begin international traffic from July 1st, 1994. This is expected to lead to full competition between the two groups in international voice services as well.

The PTT, which now is a public enterprise, and which has renamed itself Telecom Finland, is actively trying to increase its market share in local services to offset the obvious losses in long-distance and international services to the TeleGroup of Finland affiliates. It has even unsuccessfully tried to bid for some of the local telcos.

The biggest business of Telecom Finland is actually its mobile services, which bring twice the revenues of its domestic long-distance services. Practically all of this is due to the successful NMT networks. Hence the speed of GSM take-up is a serious threat to Telecom Finland, as the rivalling Radiolinja is bound to take a significant market share in the digital system (owned by TeleGroup Finland, private and business subscribers, and institutional investors less than 5%). The other side of the coin is that the existing, well-entrenched analog NMT service with over 400,000 customers and (in international comparison) low tariff levels makes the life of a new GSM operator more difficult than in countries with less developed services. A particularly Finnish feature is that both the telcos with their affiliate Radiolinja and Telecom Finland have the potential and freedom of integrating their mobile and fixed telephone networks and services to create new services and repackage their existing services, as long as they do not misuse the monopoly position.

In addition to the TeleGroup of Finland companies and Telecom Finland, there are other network operators in the Finnish market such as the national railways and Telivo, a subsidiary of a Finnish state-owned electricity company. Telivo has both cooperated and competed with the two major players. It leases capacity from the fiber-optic transmission network built into its power-grid. It's also providing this capacity directly to business customers, as well as offering voice and data network services. Telivo has had activities in St Petersburg, to which it has an optical fiber transmission link. Telivo has since the end of 1992 had a license for domestic long-distance and international services to Sweden, Norway, the Baltic countries, Russia and the UK. The international license was recently expanded to cover the whole world. Telivo's license is, however, limited to a maximum 5 percent of the annual turnover of the total business for both domestic long-distance and international services, respectively.

There are also separate cable-TV companies, which in principle could start competing in telecommunications services. According to the legislation, the cable-TV companies can either build their own networks or lease the capacity from others. In practice most cable-TV operators are either directly controlled by the telcos (or Telecom Finland), or lease substantial parts of their transmission network from them. However, the integration of telecommunication and television services would rather be encouraged by the present set-up, as soon as there are markets and technologies for it to make it commercially viable. In addition to cable services, there are activities in the broadcasting industry, which might have impacts on telecommunication services. A datacast service was recently launched jointly by the national broadcasting company and Datatie, a TeleGroup of Finland affiliate. The new digital radio system could offer new possibilities for both local and national radio stations.

2. DRIVERS AND MOTIVES OF INTERNATIONALIZATION

As technologies and markets develop in Europe, the inevitable trends lead to increased internationalization of European telecommunications services. No network operator can expect to remain protected in a home market. This evolution has long been acknowledged in Finland by the regulator as well as the operators. Actually, much of what is now taking place in Europe has had similarities to the domestic development in Finland. The need for end-to-end control is a primary driving force, as the emphasis moves from traffic to services. To remain competitive the Finnish network operators have slimmed their organizations, restructured their businesses, cooperated with others, repackaged their services, created new, flexible tariff structures and increased their customer responsiveness in general.

Most European PTTs have responded to the internationalization of telecommunications services by following their customers and by setting up offices abroad. The TeleGroup of Finland does not, however, intend to establish its own offices and international networks to compete with the large international operators. Instead, the Finnish telcos are determined to remain the best in the business in Finland. In this way they are the natural choice as a business partner and provider of services in Finland for those international operators, who need to establish a link to provide services to their worldwide customers in Finland. Consequently, the telcos can become the distributor of those international services to their own customers.

Telecom Finland has traditionally had the monopoly in international services. It has long experience and established relationhips, which it will obviously try to utilize in maintaining its market share high for as long as possible. It provides transiting services to Russia and the Baltic countries, and provides international value added services such as Finland direct and program tranmission. Telecom Finland has also been active in becoming involved as a public network operator, particularly in central and eastern European countries through joint ventures.

Finnish knowhow as an export product is clearly an important driver of internationalization for all Finnish operators. In this way they can utilize their competitive skills in other countries to enhance the profitability of domestic operations by creating economies of scale. This will also give their staff new, interesting challenges and an opportunity to learn new things. Also, it is very important to actively monitor affairs abroad to obtain new information and ideas for application at home.

3. STRATEGY OF INTERNATIONALIZATION

3.1 TeleGroup of Finland

Finnet International Ltd., the new international network operator, is an affiliated company of the TeleGroup of Finland companies. Finnet has recently been granted a full international license to offer services from June 1st, 1994. Finnet is currently establishing direct interconnection agreements with those countries to which there is most international traffic from Finland.

In providing services to international customers, Helsinki Telephone Company and the other telcos are cooperating with big international service providers rather than establishing own networks and offices abroad. This enables them to serve the needs of Finnish customers as well as the needs of global players in Finland, who need to provide services to their customers. Examples of these cooperation agreements are those with Sprint International and Unisource Business Networks.

Keeping the eyes open for the latest knowhow and testing it in the Finnish market, learning from experiences of other operators, and working closely with the leading suppliers of technology are seen as essential by the telcos. Through the Association of Telephone Companies in Finland (ATC Finland) the telcos participate actively in international organizations such as ETSI, ETNO and EURESCOM. The latter one, which contracts its members into pre-competitive research, has placed several projects with Omnitele, Helsinki Telephone Company and other members of TeleGroup of Finland. There has been active collaboration and discussions between the TeleGroup of Finland companies and operators and suppliers in both Europe and the USA.

The international activities not directly related to services to Finnish customers have thus been targeted carefully to countries where there is interest in local telco operations, and to other areas where there have been particular competencies to offer. The establishment of GSM networks has clearly become an export product for Helsinki Telephone Company and Omnitele, the engineering and R&D arm of the telcos. There are also particular competencies in the data and value added services in general, as these have been developed in an environment with the longest tradition of true competition in the whole of Europe. This competition started between the telcos and the PTT as early as in 1969 in data modem transmission and has led to parallel fiber-optic back-bone networks, packet-switched services, managed bandwidth services, frame-relay and TCP/IP services supplied not only by the two major groups but also by other companies.

3.2 Telecom Finland

Telecom Finland's strategy so far in its FDI activities can be described in the following way. Its main effort has been to utilize Finland's geographical position next to the emerging new

Russian and Baltic markets together with the experiences acquired from operating in the advanced and competitive Finnish marketplace. It has several both fixed and mobile network joint ventures in the region together with local and other Western, mainly Nordic operators. It has also offered some "over-the-border services" to the regions adjacent to Finland.

Another, clearly discernible part of Telecom Finland's FDI strategy has been utilizing its mobile telephony experience in foreign investments without any gateway or area strategy. Examples of these are memberships in the Hungarian and Turkish mobile network consortia.

Telecom Finland is internationally regarded as a pioneering provider of new corporate network services. It has not, however, been very quick in utilizing these product strengths in foreign markets. Currently it is mainly targeted on Finnish customers with subsidiaries abroad. For that purpose it has frame-relay nodes in Stockholm and Brussels. In Belgium it has the license to offer value added data services, while in Sweden this is not required. So far it has not entered into any serious joint ventures abroad in this product area.

A different type of FDI for Telecom Finland is the small minority share holding in the UK operator Ionica. The investment has been made mainly to acquire the rights to Ionica's radio fixed local loop technology, which Telecom Finland aims to use in the local network competition with the private telcos in Finland.

4. TECHNOLOGIES, SERVICES AND COMPETENCIES OF PRIMARY IMPORTANCE FOR FDIS

For all Finnish international operators their position as neighbor to Russia and particularly St Petersburg with five million inhabitants in the city itself and more in the surrounding areas is very strategic. The city is only about 150 kilometers from the border, and there is already one fiber-optic cable route available between Finland and St Petersburg with more under consideration.

4.1 TeleGroup of Finland

There is no country in Europe to match Finland in mastering local telco operations. This has led to a unique expertise in running telecommunications in a multi-operator environment now facing most European countries, if not in the form of new local operators then at least in the form of new data and mobile network operators. Besides multiple operators there are also multiple supplier technologies on the Finnish market. The major suppliers in Finland have traditionally included Alcatel, Ericsson, Nokia and Siemens. Hence, there are expertise, models and supporting software available for the complicated environments of competitive telecommunications.

Another area where the telcos have particular expertise is in setting up GSM cellular operations. As challengers to an existing, successful provider in analog services, the telcos and their consulting and engineering company Omnitele have succeeded in getting several international contracts for GSM feasibility study, quality audit, tender evaluation or network planning from governments and operators.

Also, turn-key deliveries of telephone and data networks to both existing and new operators have been successful because of the high level of knowhow and efficiency of the staff. As an example, Helsinki Telephone Company employees have been subcontracted to build local telephone networks in the eastern parts of Germany, e.g. in Dresden.

4.2 Telecom Finland

Telecom Finland has used its domestic experience of operating NMT networks to expand these services to the Baltic countries and Karelia. It has also participated in international consortia in setting up NMT and GSM mobile networks in several eastern and central European countries.

5. REGULATIONS AFFECTING FDI

5.1 Regulations affecting foreign operators in Finland

No telecommunication license is required for domestic or foreign operators to provide switched data or value added services in Finland. The provision of switched data services requires, however, a notification to the Ministry of Transport and Communications from a company or a subsidiary registered in Finland. After that it is subject to the same basic requirements as if they had a concession, namely to publish tariffs and provide their services in a non-discriminatory way to all customers. Fixed-link services and telephone switching still require an appropriate concession subject to political decision-making, even though the legislation as such does not bar foreign operators from entering the market.

5.2 Regulations affecting Finnish operators abroad

There are no regulatory restrictions barring Finnish telcos or the PTT from foreign operations.

For the telcos the unique user-ownership concept, which has led to an effective self-regulation of tariffs, and hence has kept the organizations productive, is also a factor affecting foreign direct investments. It requires of the foreign activities to make sense in business terms, and to invest only where their own competences can contribute to the achievement of successful results.

6. MEASURES OF FDI AND DEGREE OF INTERNATIONALIZATION

6.1 International operators in Finland

International telecommunications operators have established themselves in Finland mainly in cooperation with either Helsinki Telephone Company or Telecom Finland. During 1992–1993 British Telecom (Tymnet), Sprint International and Unisource Business Networks have located their international nodes in Helsinki. Their nodes are hosted either by Helsinki Telephone Company or Telecom Finland, and none of the three has any technical staff of their own in Finland. More international operators are expected to bring in their nodes towards the end of 1993 and in 1994, in particular, as also the international voice services are to be opened up for competition. In addition, Telecom Finland owns 90 percent of Infonet Finland, and provides Infonet services to its customers.

6.2 TeleGroup of Finland abroad

Compared to the size of the private telcos (1992 staff: 9000, turnover 750 millon USD) the international activities of TeleGroup of Finland still comprise a relatively small portion of the total. However, when taking into account that so far the private telcos have been barred from handling international traffic directly, and the very recent nature of their international activities, the achievements so far are quite remarkable.

The number of employees abroad in 1992–1993 has varied from 50 to 200, depending on projects. There are experts abroad in several countries on long-term assignments. Direct foreign investments total only a few million USD, mainly in a local telephone company in Hungary. The total revenues from abroad, excluding international traffic, however, are well over 20 million USD. There are no permanent offices abroad, only on project basis.

6.3 Telecom Finland abroad

Telecom Finland has traditionally handled international services, the revenues of which in 1992 were approximately 850 million FIM (=USD 150 million), while its total turnover was 5005 million FIM.

Telecom Finland has direct offices abroad in Brussels and St Petersburg. In addition to that, Telecom Finland participates in joint ventures in Estonia, Latvia, Murmansk, Turkey and Hungary. The total number of people involved in these operations is probably about twenty and the committed investments so far about one hundred million USD.

7. ORGANIZATION AND STRUCTURE OF INTERNATIONAL OPERATION

7.1 TeleGroup of Finland

Most foreign activities take place either through Omnitele, Helsinki Telephone Company or its subsidiaries. Helsinki Telephone Company is directly involved in Germany as subcontractor building turn-key telephone networks, and in the UK and Sweden as provider of technical consulting and assistance to new telecom operators.

Helsinki Telephone Company owns, together with a Finnish institutional investor, Telcofounding Ltd, which has invested in First Pest-City Telephone Company (EPT) in Hungary. Helsinki Telephone Company also provides long-term consulting to EPT.

Helsinki Telephone Company's subsidiary Comptel employs 80 data processing professionals, and it has long experience in providing various software systems for telephone companies. Internationally it provides software systems for local telco (TelePro) environments, such as for EPT in Hungary, and billing systems for cellular networks, e.g. for NMT450 in Thailand and for GSM in Hong Kong. Comptel has also provided a directory assistance system for KTAS together with IBM Denmark.

Omnitele, owned by all the telcos, provides consulting and engineering services for demanding telecommunications projects in all telecommunications related areas. Its core staff is about thirty experts, who have gained experience in carrying out joint projects for the TeleGroup of Finland companies, and who can also subcontract the resources in telcos where appropriate. Omnitele's international projects now comprise more than a fifth of its business.

7.2 Telecom Finland

Currently Telecom Finland has two SBUs responsible for the foreign business ventures, Telecom Finland International and Telecom Finland Mobile International. Starting January 1st, 1994, all foreign business ventures will be centralized into one business group, Telecom Finland Ventures.

At the time of writing (1993) Telecom Finland International has a subsidiary and representative office in Brussels and branch offices in Tallinn and St Petersburg.

8. ALLIANCES AND JOINT VENTURES

8.1 TeleGroup of Finland

Helsinki Telephone Company has signed distributorship agreements with both Sprint International and Unisource Business Networks for providing their international services to Finnish customers throughout Finland. In this context, Helsinki Telephone Compny also operates the necessary nodes and equipment in Finland for the services.

TeleGroup of Finland companies have an extensive list of international clients on project basis. Regarding long-term activities in other countries Helsinki Telephone Company has made joint venture and cooperation agreements in Hungary and in the UK.

8.2 Telecom Finland

Most joint ventures are for mobile networks and include the following NMT networks, with Telecom Finland's ownership indicated in brackets: Estonian Mobile Telephone Co (24.5%) with Telia and Telecom Estonia, Latvian Mobile Telephone Co (24.5%) with Telia and three Latvian operators, TeleNord in Murmansk (50%) with a Russian operator. Telecom Finland also participates in three GSM consortia: North-West GSM in St Petersburg with Telia, Norwegian Telecom and two Russian operators, Turkcell in Turkey with three Turkish companies and Ericsson, and Pannon in Hungary with Telia, Tele Danmark, Norwegian Telecom, Dutch PTT and three Hungarian companies. Besides the mobile networks, Telecom Finland owns 49% of Lenfincom, an international operator in St Petersburg, and 10% of Infocom, which markets Infonet services in Moscow.

9. THE FINNISH MANUFACTURING INDUSTRY

The flagship of the Finnish telecommunications manufacturing is undoubtedly Nokia corporation, which in the 1970s acquired the state-owned electromechanical telephone switch manufacturing. Nokia Electronics diversified during the 1980s also into consumer electronics and computers, the latter of which was sold in 1991. Nokia has developed digital switching products based on its DX200 family for public, mobile and special networks, as well as transmission equipment, cables and mobile phones. Nokia has succeeded in securing about forty percent of the domestic market in competition with bigger international rivals. In the 1980s the Soviet Union was its most important foreign market, but in the 1990s it has truly become a global player, particularly in mobile switching and mobile telephones. Nokia has clearly benefited from the liberalization and deregulation in Europe, as this has meant new potential customers and new technologies, as well as more open procurement even by the incumbent operators. Nokia is now present not only on most major European markets, but in the USA and Far East as well. Its telecommunications turnover grew by more than eighty percent in the first eight months of 1993, mainly related to the boom in the cellular business worldwide, where it is the world's second largest manufacturer of both systems and mobile phones.

The other Finnish manufacturers are far smaller, yet very much oriented towards the international markets, as typically most of their production is sold abroad. Another mobile phone manufacturer, Benefon, a spinoff from Nokia, has been growing fast and provides NMT phones. Martis provides digital networking systems and base-band modems. Tecnomen provides paging systems and Teleste provides cable-TV and satellite master antenna equipment as market leader in Scandinavia and with a subsidiary in the UK.

Several big European manufacturers have long been established in Finland. Ericsson and Siemens were even owners of several telcos in Finland in the 1930s. Ericsson, ITT and Siemens were exporting to the Soviet Union through Finland in the 1970s and 1980s, and even presently Ericsson's subsidiary in Finland is exporting to Russia and the Baltic countries besides its extensive R&D for the Ericsson group.

In total, telecommunications manufacturing has been one of the fastest-growing segments in Finnish industry, and taking into account the size of the five-million-inhabitant domestic market, Finland is relatively the biggest exporter of telecommunications products in the world!

10. SCENARIOS FOR THE FUTURE

TeleGroup of Finland, because of its existing infrastructure and strong presence at customer interface, is likely to capture substantial portions of both the domestic long-distance and the international traffic already during its first year of operations. The two competing groups, TeleGroup of Finland and Telecom Finland, will represent the major part of operations in Finland for both domestic and international services. Because tariffs are already more cost-oriented than anywhere else in Europe, competition can work fairly smoothly without too much political and regulatory intervention. In addition to these two players, there will be several small operators providing services ranging from simple resale to value added services, partly cooperating, partly competing with the two major groups.

As the full competition is about to begin in 1994, it can be expected that the reactions of the major players will initially be quite strong. There will be abrupt changes in market shares and even cutthroat pricing might occur. The Ministry of Transport and Communications has, however, announced that it will monitor the activities in the market very closely and, where appropriate, take necessary action in order to prevent unfair practices and risks of too severe price competition. Although open competition is encouraged, it is also considered to be very important that operators can still afford reinvestments and development of new networks and services.

It is a generally accepted view in Finland that close monitoring and action by the regulator is needed so long as there are unfair monopoly positions, but that the ultimate goal is to let normal business rules apply in a not too distant future.

Regarding internationalization, there will be competition in international services, the Finnish operators will have extensive business activities abroad, and there will be close cooperation between Finnish and international operators. Regarding FDI to Finland there is a differentiating feature in Finnish telecommunications compared to most European countries. Foreign investments in other countries are a catalyst in creating competition through new market entrants. In Finland there are already well-established, domestic operators in competition, and foreign investors would need to offer other advantages than simply the injection of capital (unless the existing operators exhaust themselves through price-wars in mutual competition). However, a potential scenario is that there will be new ownership structures among current operators involving also foreign operators.

11. SUMMARY

The Finnish telecommunications marketplace is like no other in the world. There are two equally sized groups providing telecommunications with initially divided responsibilities but now bound to enter into each other's businesses in full competition, the one group as the traditional master of local networks, and the other as long-distance and international services provider. The former needs end-to-end control and the latter needs customer access.

The TeleGroup of Finland companies have only recently started their internationalization. With this in mind, the achievements so far have been remarkable, particularly as international operations of the TeleGroup have been a business in their own right, not merely investments abroad with uncertainty about future returns. Becoming a full-scale international operator the telcos through their subsidiary Finnet are about to take the next, significant step in their internationalization and in the forms of cooperation with foreign partners. Telecom Finland has also been successful in selling its knowhow of NMT and cellular operations, and it has been active in participating in international consortia in setting up mobile networks.

The international players have so far not invested in Finland, even though the market is from the regulatory point of view becoming open to foreign companies. The international service providers have brought in their nodes and are expected to partner with the Finnish companies as currently in data services and in the future in voice services as they also will be opened up for competition from 1994 onwards.

REFERENCES

Liikenneministeriö, 1993, (Finnish ministry of transport and communications), Kansainvälisen teletoiminnan kilpailu (Competition in international telecommunications), Liikenneministeriön julkaisuja 38/93, Helsinki.

Liikenneministeriö, 1993, (Finnish ministry of transport and communications), Telecommunications Statistics, 1992, Helsinki, Valtion painatuskeskus.

Various sources inside Finnish telecommunications operators.

Investment Strategies in Eastern and Central Europe

The Race to European Eminence
E. Bohlin and O. Granstrand (Editors)

Transformation of Poland's Telecommunications and Foreign Investment

Tomasz Jarmoszko

Indiana University School of Business, Department of Decision and Information Systems, Bloomington, IN 47405, USA

1. INTRODUCTION

The perception of a growing technological gap to the developed world, and the inability of the Communist system to arrest this growth, played a central role in the demise of the Soviet Bloc (Geipel and Jarmoszko, 1991). In Poland, that perception was especially acute, especially vis-à-vis telecommunications. In virtually every analysis of reform prospects the issue of a grossly inadequate telephone system was raised. The popular press in the late 1980s lamented that throughout Europe only Albania had a system worse than Poland's. In a somewhat overstated call to action, the Polish media repeated what Roman Polanski, during a visit to his former homeland, was reported to have said. When asked about the possibility of filming in Poland he replied that it was easier to make a movie without a camera, than without a telephone.

In response, since mid-1989, the leadership of the Republic of Poland, prodded by international organizations such as the World Bank, has undertaken significant reform of the country's telecommunications. The 1990 Act on Posts and Telecommunications liberalized all markets for telecommunication services except those for international telephony. The government-accepted program to improve the nation's telephone network, made public in March 1990, projected growth in telephone penetration from 8.3% in 1989 to 30% by the year 2000. Attracted by credit guarantees from Polish and western governments, all major producers of telecommunication equipment have tried to enter the Polish market and by early 1993 equipment deals worth some $700 million were signed (see Table 1.1). With nearly $1.5 billion committed to telecommunication projects, penetration in the public network grew to 9.8% in 1991 and 10.2% in 1992. In the otherwise dark picture of the Polish economy, the telecommunication sector has been a bright spot. This paper addresses the principal issues affecting foreign investment in Poland's telecommunications.

Table 1.1
Funding for selected projects in telecommunications, 1989–1991

Lender	Amount in $mil.	Grace Period	Repayment Period	Comments
World Bank	120.0	5 years	12 years	Interest tied to Libor
European Investment Bank	75.0	5 years	12 years	Interest tied to Libor
Government of Spain and Alcatel SESA	120.5	10 years		Each lender provides half the total amount; Interest: 1.5% per annum
Government of Korea	50.0	5 years	20 years	Interest: 3.5% per year
AT&T	63.0			
AT&T-NSI	2.5		48 payments	Leasing arrangement
Banque Société Générale and Alcatel CIT	70.0	3.5 years	5 years	
Siemens AG	7.8		5 years	
Siemens AG	38.2		8.5 years	
Israeli Source	1.4	0.5	5 years	
Danish Source	26.0		15 years	
TOTAL	584.9			

Based on: Iwona Ryniewicz, "Przyszlosc na kredyt" (Future on Credit), Gazeta Bankowa (Warsaw) (July 14-20, 1991), pp 28-29.

2. ECONOMY: FROM COMMUNISM TO CAPITALISM

The system of a centrally planned economy from which Poland and the rest of the European CMEA (Council on Mutual Economic Assistance)[1] have tried to emerge is a system that was defined by Stalin in the 1930s and transplanted to East-Central Europe after World War II. A Stalinist Centrally Planned Economy (SCPE) is based on three crucial precepts: (1) that the state must directly control the majority of the means of production, including foreign trade; (2) that industrial concentration, as a way to derive the benefits of economies of scale, is the best development policy; and (3) that heavy industry is an essential component of any modern economy. Until the early 1960s, the system seemed to function adequately. Central control over resource allocation facilitated a recovery from the ravages of World War II, maintained high growth rates of national income, and produced islands of technological sophistication, such as the Soviet space program. These and other achievements, in the eyes of the Third World, established the SCPE as a real alternative to the development models provided by the West.

[1] The Council on Mutual Economic Assistance was created to promote economic integration within the Soviet Bloc.

By the early 1960s, however, systemic shortcomings of the SCPE became clear to many leaders within the Soviet Bloc. Growth rates that had been achieved by encouraging labor migration from the countryside to the cities, or by adding to the available pool of physical resources, began to falter. Since the SCPE did not absorb innovation very well and the machinery stock began to age quickly, the CMEA share in the world trade stopped growing, and even declined. The world economy seemed to have taken a turn somewhere, leaving the CMEA countries behind.

Despite the repeated attempts at reform, the basic structure of an SCPE remained unchanged. Isolated from world trends, Poland's economy continued to rely on heavy industry and agriculture to a degree disproportionate to the developed, and even the developing, world. For example, in 1960, the role of agriculture, manufacturing, and services in Spain's economy was more or less on par with Poland's economy. The two countries are similar in size, population, and to a limited degree in history and culture. Some 30 years later, the role of services—a sector of particular importance to telecommunications—has grown and the role of agriculture has declined to a much greater extent in Spain than in Poland (see Table 2.1). In another diverging trend, the average size of a Polish firm has remained much larger than the average developed economies. In 1989, three-fourths of Poland's state enterprises employed at least 400 people—more than four times the average in the West.

Table 2.1
Labor force employed in services, manufacturing and agriculture in Poland and Spain

	Services	Manufacturing	Agriculture
Poland	38%	37%	25%
Spain	53%	33%	14%

Derived from: The CIA World Factbook, Washington DC, 1989 Edition.

2.1 Shock therapy

The basic elements of the reform package introduced in Poland on January 1, 1990 were a reaction to the principles of an SCPE. In the words of Leszek Balcerowicz—the Minister of Finance in the Mazowiecki and Bielecki governments and the principal author of the reform—a well-functioning, modern economy is characterized by:

- a stable, freely convertible currency,
- predominantly private ownership of property, and
- openness to competition from foreign firms.

To achieve these goals a number of measures were adapted to achieve: price deregulation and an end to administrative allocation of goods and services; balanced budget through a sharp reduction in subsidies and spending; a tightened money supply and elimination of easy credit; limit on wage growth to 20-30% of inflation; an end of full-employment policy; a new international trade regime based on a convertible zloty (Zl), low import-duty tariffs, and elimination of income-tax credits for export-oriented firms; privatization of state enterprises; liberal joint-venture regulations, including freedom to expatriate profits.

From the vantage point of mid-1993 policies based on the Balcerowicz Plan must be given a mixed rating. Certainly the goal of economic stabilization has been a qualified success. Inflation was cut from 600% per year in 1989, to 70% in 1991 and 45% in 1992. Public confidence in the zloty has surged, as a result of internal convertibility, fairly stable Zl/$ exchange rates, and increased savings in zloty accounts. The lowering of import duties and the lifting of restrictions on private trade, as well as price liberalization, have virtually eliminated chronic supply shortages for consumer goods. By the end of 1992 some 1.7 million private enterprises were registered and an estimated 45% of the Gross Domestic Product (GDP) came from the private sector. Following the elimination of many subsidies, a surplus in the 1990 state budget was recorded. Similarly, a positive trade balance was achieved. In 1990, hard currency exports grew by 50%, in relation to 1989, whereas imports grew only by 4%.

However, not all news has been good news. Poland's GDP fell 14% in 1990, 7% in 1991 and registered an anemic 1% growth in 1992. By the end of 1992 unemployment rose to 13.5%. The 1990 budget surplus turned into a deficit in 1991—a result of reduced tax revenues from the contracting economy and lower-than-expected income from privatization of state property. The 1991 exports fell and a trade deficit was recorded. The restructuring of the economy had not taken place as quickly as was expected, particularly in the realm of converting large, inefficient enterprises into smaller, economically viable units. In August 1991, the credit rating of 1.9 thousand enterprises was too low to obtain any loans at all. Several highly publicized financial scandals have shaken the public's confidence in the banking system.

2.2 Privatization

One element of the Balcerowicz Plan that has provoked widespread criticism is privatization. All political forces within Poland agree that the success of Poland's transformation is highly dependent on the extent to which the country's industrial base can be privatized. Janos Kornai expressed this sentiment better than anyone when he wrote that state ownership is inherently inefficient because it leads to bureaucratization and a soft budget constraint.[2] Unfortunately, beyond the initial acceptance that privatization is good, no consensus on the best way to sell off state assets has emerged.

The beginning was auspicious. Armed with support from various strata of society, the first post-Communist government of Tadeusz Mazowiecki set out to convert as much as possible of the state property into private property, as quickly as possible. Within the first several months of 1990, most small state retail shops, service points, and restaurants were either auctioned off or leased. The percentage of private retail sales, as part of all sales, rose from 20% in 1989 to 80% in 1991.

In July of 1990, the Parliament passed the Privatization Act, outlining two scenarios for privatizing state enterprises. In one scenario, called "large" or "capital" privatization, a state enterprise is transformed into a joint-stock company owned entirely by the state treasury. In the vernacular of privatization, the enterprise is "commercialized," which for practical purposes has meant that laws applicable to state enterprises—for example those on worker representation in management—could be ignored. Thus cleansed from employee pressures,

[2] Soft budget constraint is a condition in which management relies on ultimate help from the government and therefore can never impose efficiency. See Kornai (1990), for an eloquent elaboration of this point of view.

the management can then prepare for sale of shares to investors by assessing enterprises' assets and liabilities and by restructuring. The sale of shares can occur via auction, public offer, or invited private offer, and is limited only by two conditions:

- that 20% of shares be made available to workers of the enterprise at 50% of the floatation price;
- that foreign investors obtain the approval of the Agency for Foreign Investments if they intend to purchase more than 10% of shares.

The second privatization scenario provided for by the Privatization Act involves liquidation of assets. The objective of this scenario is to end the existence of a state enterprise as a separate legal entity and to dispose of its property. Article 37 of this Act outlines three liquidation methods: 1) the sale of an enterprise, or of its constituent parts, to independent entrepreneurs; 2) the transfer of enterprise property to one or more other enterprises; or 3) the leasing of property, in whole or in part, to a cooperative established by more than half of the employees of the liquidated enterprise (Cetwinska, 1990).

Despite the relative success of "small" privatization (retail shops), the selling of large enterprises, the bulk of Poland's economy, has been painfully slow. While 1,258 state enterprises were instructed, between August 1990 and January 1992, to undertake privatization measures, only 26 were privatized through a "capital" sale to either foreign or domestic investors and 198 through a liquidation procedure. In another 752 enterprises, privatization through liquidation was initiated, but not completed. In other words, despite pronouncements of urgency by all decision-makers, a mere 3% of the 7,800 state enterprises eligible for privatization (224 enterprises) were in fact privatized, and in another 10% privatization measures were in the process of being implemented.

Needless to say these results have satisfied neither the public nor any of the organized political groups. At the end of 1991, the prevailing opinion seemed to favor a mass privatization plan which attempts to balance the political need for fair distribution of state property, with economic efficiency considerations of restructuring. The plan would establish about a dozen National Wealth Management Funds (ZMN)—a type of mutual funds, in which the capital would be entire industries. Western and Polish consulting firms would be hired to manage the funds and would be paid in a scheme that ties ZMN performance to compensation. All adult citizens of Poland, residing within the country, would be assigned shares within each fund and the state treasury would retain a percentage of shares. The fund managers, with minimal state supervision, would be free to restructure or liquidate the enterprises within their ZMN, basing their judgement on profit potential and trends in the world economy. After about two years, all those who own ZMN shares[3] would be free to trade them, thereby unleashing the market forces of resource allocation.

In April 1993, the Parliament approved the ZMN concept by passing the Law on national Investment Funds and Their Privatization.

2.3 The international dimension

The collapse of the Soviet Bloc transformed Poland's economic relations with other countries. During some 45 years of Communism, the principal trading partner for all of CMEA was the Soviet Union, the source of critical raw material and energy supplies. Although trade within

---

[3] In the interest of fairness, initially, all citizens would be given shares in EACH fund.

CMEA never lived up to its potential, and in fact stagnated throughout the 1980s, it nevertheless was important for all countries in the Bloc, including Poland. In some categories of products, such as the broadly defined electrical-machinery sector, CMEA members depended on each other almost entirely. For example, in 1989, nearly 90% of all telecommunication equipment made in Poland was exported to other CMEA countries.

The principal reason for intra-CMEA dependence was CoCom. Since the onset of the Cold War, the 17-nation Coordinating Committee for Multilateral Export Controls has set limitations on the selling of sophisticated technologies to countries deemed dangerous to the established international order. In general, despite many instances of circumventing its restrictions, CoCom has been successful not only in limiting access to targeted technologies, but also in forcing expensive development programs on the CMEA countries (Jarmoszko, Geipel and Goodman, 1988). In 1990–1991, CMEA and CoCom, the two pillars of the established international order in East-Central Europe, tumbled. Both the CMEA organization and the CMEA trading system of bilateral accounts in transferable rubles, supplemented by barter and limited hard-currency transactions, were dissolved. With Poland, Hungary and Czecho-Slovakia interested more in EC membership than in reforming the old system, no replacement framework for regional economic cooperation was created, and instead, a series of bilateral trade agreements favoring hard-currency transactions was signed. Predictably, a sharp drop in intra-CMEA trade resulted. For example, in 1991 Poland's trade with other CMEA countries fell on the average by half, and trade with the former Soviet Union plummeted 60%.

CoCom restrictions on exports to Poland have been substantially eased. In an effort to help the fledgling democracies of East-Central Europe, and as part of a new policy of building higher fences around fewer products, CoCom members agreed to revise the International Industrial List: the enumeration of military-civilian dual-use items subject to controls. Following a series of high-level debates, a number of advanced technologies were decontrolled, or else directives were issued to give favorable consideration for export to Poland, Hungary and Czecho-Slovakia, and to other post-Communist countries. The removal of restrictions has contributed to the opening of Poland's economy to the West, particularly the EC countries, and to the increasing of investment from abroad.[4] The change in the composition of Poland's trade is shown in Figure 3.1.

By mid-1993 much was unclear about Poland's transition to capitalism. Large parts of the Balcerowicz Program remained not implemented, privatization was moving at a pace that satisfied no one, international economic relations were far from stable with both the former CMEA countries and the West. The uncertainty in the economy at large carried over into the realm of telecommunication regulation.

3. REGULATION: FROM MONOPOLY TO COMPETITION

Before 1989, the authority of the Communist Party on issues in telecommunication regulation was unchallenged. Even though in theory the constitution of the Polish People's Republic provided for independent legislative, executive and judicial branches of government, in practice all important decisions were made by the Communists in a highly centralized manner. The

[4] To be sure, the degree of the opening to the West did not satisfy Poland's decision-makers. Exports to the EC countries were hampered by trade restrictions on agricultural goods and textiles. Similarly, following an increase in Polish exports to the US in 1990 to $408 million, in 1991 a drop to $357 million was recorded.

party-state apparatus—led by the First Secretary of the Communist Party and the Prime Minister—set policy and all branches of the government accepted edicts passed from the top down. The Minister of Posts and Telecommunications, his deputies, and all other officials with any significant power had to satisfy the requirements of *nomenklatura*—an unofficial list of individuals from which all decision-making positions were filled. The content of television and radio programming—and of other mass media—was monitored by cells of the Communist Party and the office of the censor.

Poland's regulation regime was defined by the Act on Posts and Telecommunications of November 1984, and by its amendment October 1987. In the European tradition of telecommunication monopoly, the state was granted the right to monitor and control activities related to the relaying of electro-magnetic signals. With the exception of the private networks of the Ministries of Defense and Internal Affairs, and the Polish State Railways, the only providers of telecommunication services were the Polish Post Telephone and Telegraph (PPTT) and the Polish Radio and Television (PRiTV). Stiff laws prohibited usurping of monopoly powers.

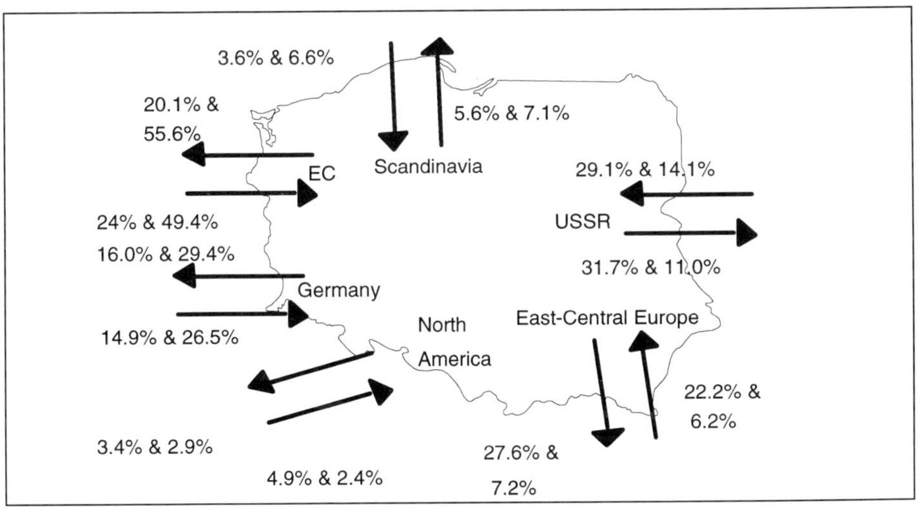

Figure 3.1. Composition of Poland's 1977 and 1991 (respectively) exports and imports.

3.1 Institutional and legal framework

The political changes of 1989–1991 began the process of redefining the regulation of telecommunications, a process that was by no means completed by the end of 1991. Following the elections of May 1989 and October 1991, the Parliament, consisting of the lower chamber, the *Sejm*, and the upper chamber, the *Senat*, assumed the role of a legislative institution in the emerging system of parliamentary democracy. Primarily through the work of the *Sejm* Committee on Trade and Services, the new supreme law-making organization of the land passed the 1990 Act on Posts and Telecommunications—the basic telecommunication law. Interrupted by elections and political squabbling, the Copyright Act and the Act on Radio and Television were still in the legislative pipeline in the early part of 1992.

The *1990 Act on Posts and Telecommunications* (APT) introduced the brave new world into Poland's telecommunications. Almost one year in the making, the Act eliminated the state monopoly on the provision of telecommunication services and facilities; reorganized the Polish PTT (PPTT) into Polish Post and a joint-stock company, Polish Telecommunications (TP SA); and redefined the rules for foreign investment. The general thrust of issues addressed in the Act was the introduction of market forces into an environment that had long been monopolized.

The *Act on Radio and Television* created the most controversy within the Parliament on issues ranging from regulation of programming to the role of private television and radio stations. When it was finally approved in early 1993, its provisions preserved the existing structure of two public television channels: one with national and one with regional focus. Regional television, however, was allowed to operate fairly independently of the center, particularly in the realm of programming and finances. The Ministry of Posts and Telecommunications (MPT) and the National Council on Radio and Television (KRRT) were given the responsibility of licensing private stations. The MPT manages frequency allocation, and the KRRT regulates programming content.

In the new telecommunication regime, the MPT and the KRRT are essentially new organizations. Before APT became law—on January 15, 1991—the MPT was both the country's principal regulatory body, and the organization responsible for the PPTT—the provider of most services. In a typical Communist-system arrangement, the MPT policed itself. This, in theory, is not true any more, for the MPT no longer has any administrative links to the successors of PPTT, the Polish Post and the TP SA. The new MPT, with a much reduced staff, is supposed to become Poland's supreme regulatory body, overseeing all providers of telecommunication services.

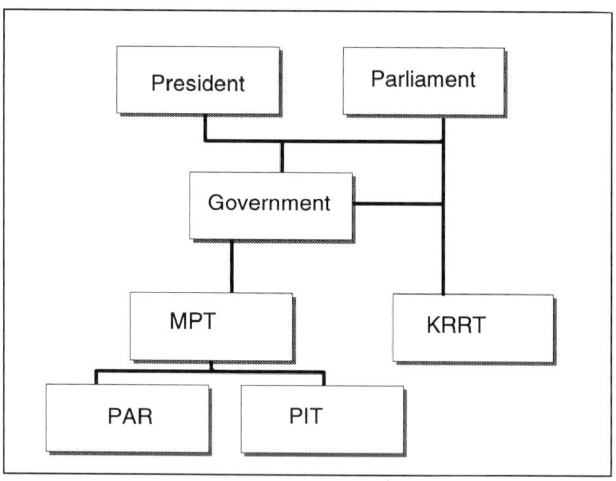

Figure 3.2. Institutions of Poland's new telecommunication regime.

Among others, the functions of KRRT entail: setting of subscriber fees for public broadcast services; defining of broadcast standards; licensing of private radio and television stations; and formulating, in cooperation with the MPT, of national objectives for radio and television broadcasting. The last function has been a point of controversy over the degree to which codified law should outline broadcast standards and national objectives.

A new organization, which is supposed to collaborate with the KRRT on matters of frequency allocation, is the State Radio Agency (PAR). PAR and another new organization, PIT, the State Telecommunication Inspectorate, were created by the APT to be the two "executive arms" of MPT. Their principal areas of competence stem from the two distinct transmission media: public airwaves and wired lines (see Figure 3.2).

Another law of relevance to the new telecommunication regime, unfinished as of early 1993, is the *Copyright Act*. The impetus for this legislative measure stemmed from the agreement on cooperation between Poland and the US, signed by the Bush administration and the Mazowiecki government. The agreement stipulates that Poland pass laws to protect the rights of software creators, analogous to laws on authors' rights. The version of the Copyright Act deliberated in early 1992 follows the spirit of similar laws in the US and Western Europe, in that: a) the original shape of software is protected and not its content or function; b) it is permissible to make two backup copies of legally acquired programs. The delay in the passing of the Act was caused by a debate on potential costs and benefits of enforcing a law of this type in what still is a fairly underdeveloped economy.

3.2 Regulatory tools

While passage of the Copyright Act and the Act on Radio and Television will introduce new requirements for service providers, the Act on Posts and Telecommunications already has. The APT requirements are enforced through powers assigned to decision-making institutions and the following is a discussion of these powers, of regulatory tools, in the emerging regulatory framework.

3.2.1 Market entry

One of the principal goals of the reformed telecommunication regime is introduction of competition. Article 4 of the APT stipulates that, in addition to TP SA and the ministries of National Defense and Internal Affairs, telecommunications operations can be carried out by "entities granted telecommunications license." The MPT Minister grants, revokes or refuses to grant licenses (Article 13) based on conditions specified in Article 18. The latter Article lists five, fairly specific, reasons for which a license application can be refused: national security, contradiction with international agreements, technical unfeasibility, lack of frequency/numbering capacity, or previous license revocation. A refused application can be appealed to one of the administrative courts.

APT proscribed market competition in only one area: Article 16, paragraph 1 states that no permit can be issued for international telecommunication services. Since the only economic agent needing no license to operate on the Polish market is the successor to the Polish PTT, the TP SA (Article 4 paragraph 1), monopoly for public international services has been thereby retained. Similarly, Article 16, paragraph 3, in prohibiting foreign entities from directly connecting to foreign networks, necessitates the leasing of circuits from TP SA .[5]

[5] "Licenses shall not be granted: ...to a foreign entity or a corporation with part of the stock held by foreign entities for the purpose of operating telecommunication equipment, lines and networks directly connected, by cable radio or through international satellite systems with lines, equipment or networks located outside the territory of the Republic of Poland." (Article 16 paragraph 3.)

Prior to the passage of APT, the issue of foreign capital in Poland's telecommunications was widely debated in the Parliament. Legislators wanted to encourage foreign investment in Poland's infrastructure, while reserving the most lucrative market segments (e.g. international services) for Polish firms. The issue of foreign influence over the country's airwaves was also a consideration. The final version of APT limited the role of foreign capital to 33% in entities providing public broadcasting services, and to 49% in entities providing inter-city (long-distance) telecommunication services (Article 16 paragraphs 2 and 4). There are no restrictions of this type on ventures providing local services.

The issue of foreign capital has come up in every approved license application, and in many pending applications. By the end of 1991, the Ministry of Posts and Telecommunications issued licenses to:

a) BTT—a venture between the National Bank of Poland and the American firm EDS to provide computer-networking services;
b) TESA—a telecommunication firm with Polish and Austrian capital—to provide a host of electronic communication services including e-mail, telex, telefax;
c) Polpager—a joint venture between TP SA and a Belgium-based Polish emigre—to provide paging services;
d) Centertel—a joint venture between TP SA, Ameritech, and France Telecom—to provide cellular telephony services in all of Poland.

While the Act on Radio and Television (ART) was in the legislative pipeline, temporary licenses for broadcasting services were issued to private stations. In this category were three private radio stations—Radio Z and Radio Solidarity in Warsaw, Radio Fun in Cracow—and one television station *Echo* in Wroclaw. In early 1992, there were 631 pending applications to allocate frequency for private radio stations, and 208 applications for television frequencies.

3.2.2 Tariffs

Article 39 of APT defines the setting of tariffs for public telecommunication services. Prices are set by network operators unless the Minister of MPT deems it necessary to impose a maximum charge (cap) on specific public services. Although, according to the letter of the law, only tariffs on international services must be set in consultation with the Minister (paragraph 4), the position of the government has affected TP SA pricing. This is likely to continue as long as the state treasury remains a significant stockholder.

Following the adoption of APT prices for all telecommunication services rose quickly. In 1990, when inflation was 220%, the average price for a telephone connection, use fees (meter pulse) and a monthly subscription fee increased 122%. In 1991, inflation was reduced to 70%; the same price average, however, increased 230%. The trend continued through mid-1993.

3.2.3 Interconnection

Article 38, paragraph 1, of APT mandates that an operator of public telecommunication services may not refuse to connect to its network the network of another operator. Although paragraph 2 of the same Article and Article 39, paragraph 2, stipulate that charges for an interconnection must be agreed on, contractually, between the parties in question, nothing in the APT provides a mechanism for mediating disputes. TESA, Polpager, and Telbank connected their networks to the network of TP SA without major disagreements because, with the exception of data messaging—which TP SA and TESA both provide—there was no real service competition. Because this situation is unlikely to continue, some form of regulatory mechanism for settling future disputes must be introduced.

3.2.4 Technical standards and equipment certification

Articles 7–9 of APT codified some de facto rules for installing equipment and systems, including cabling, in the public network. All such devices must have a "homologation certificate" issued by the MPT Minister and must comply with technical standards as defined by the Minister.

The legal restrictions notwithstanding, the events of 1990–91 have introduced a variety of equipment brands and types into Poland's telecommunication system. Attracted by its potential size, many manufacturers tried to enter the Polish market by donating equipment, reducing prices, and offering favorable financing terms. MPT initially attempted to limit the number of different technologies in the network by declaring that only four systems could be connected: System-12 (Alcatel SESA), E-10 (Alcatel CIT), EWSD (Siemens), and DMS (Northern Telecom/Kapsch). Subsequent offers by AT&T, ITALTEL, Ericsson and Samsung dissipated the initial resolve.

By late 1990, because of the perceived necessity to limit the number of systems in the network, and in an attempt to attract foreign investment, MPT had begun to require that equipment sold to the TP SA be, at least in part, manufactured in Poland. To satisfy this requirement, the principal foreign competitors set up joint ventures with Polish firms: Northern Telecom/Kapsch with Elwro, Siemens with ZWUT, Alcatel SESA with PZT, Alcatel CIT with Teletra, and AT&T with Telfa. To attract additional capital, Poland's authorities made it clear that investment in the country's manufacturing base would be looked upon favorably when orders for equipment are made. Consequently AT&T purchased the majority of Telfa stock, the two Alcatels collaborated in buying the PZT and Teletra and Siemens acquired the controlling share in ZWUT and Elwro.

3.2.5 TP SA

By far, the principal purchaser of telecommunication equipment and provider of telecommunication services is TP SA. As prescribed by Article 76 of APT, on January 1, 1992, the PPTT was split into two constituent parts: TP SA, a joint-stock company owned by the state treasury and Polish Post, a public-utility company. The difference in the organizational type of the two PPTT successors stems from the announced intention to privatize TP SA,[6] following an assessment and upgrade of assets. Exactly when this privatization will be initiated has been left unclear.

---

[6] Article 82 reserves the right of TP SA employees to buy shares, with the same privileges as privatized state enterprises.

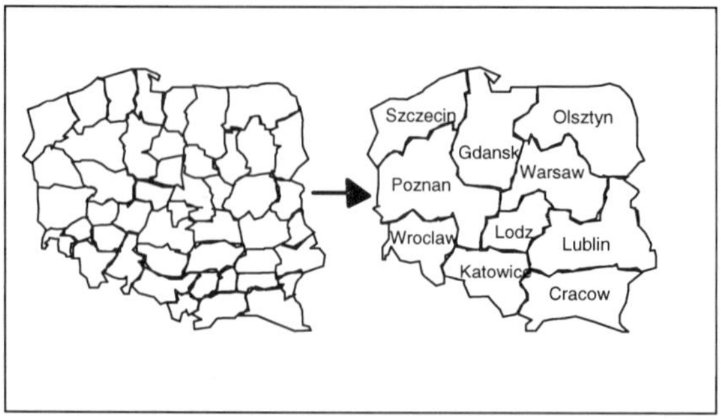

Figure 3.3. Regional Reorganization of PPTT.

The divorce of the postal and telecommunication services, recommended if not required by the World Bank, was a consequence of economic trends. While postal services have always been labor-intensive, beginning in the 1980s, the trend toward telecommunication services becoming much more technology-intensive has produced a conflict. That is, any firm which tries to provide both telecommunication and postal services, while maintaining common employment and investment policies, encounters problems (Saunders, Warford and Wellenius, 1983). This conflict was particularly evident in the case of PPTT, which for years failed to attract high-caliber employees because of low wages and poor working conditions.

TP SA, at its birth, also received a new territorial and management structure. The forty-nine provincial units of PPTT were consolidated into ten much larger Regional Operating Companies of TP SA. The resemblance to the Bell system was not coincidental. The PPTT reformers initially wanted to create only five ROCs, thus utilizing the experience of the American firm to define the TP SA corporate structure. The final decision on the number of regions was a compromise between technical necessity and local interests.

The management structure of TP SA attempts to combine the advantages of central coordination with regional independence in operating the network and establishing service priorities. TP SA is governed by a management board, consisting of "national" members—those whose primary interests lie in the firm as a whole—and directors of ROCs. The board makes decisions on major investments in the network infrastructure, on technical standards and quality-of-service targets. Each ROC is free to decide which services it will provide and how operational resources are to be expanded. Profits are divided equally between center and each region.

4. POLICY: FROM NEGLECT TO PRIORITY

The poor condition of Poland's telecommunication network at the end of 1991 was a consequence of long-term inadequate investment and neglect. It is sometimes asserted that this condition can be explained by the Communist policy of inhibiting uncontrolled communication among citizens. This is a simplistic view. Throughout the 1980s evidence can be found that decision-makers recognized the dire need to improve telecommunication infrastructure. The reforms undertaken by the last Communist government of Mieczyslaw Rakowski are the case in point. These, however, were half-hearted administrative measures aimed more at appeasing the public than at fixing the infrastructural bottlenecks. All reforms notwithstanding, sufficient resources to expand, modernize or even maintain the existing network were never allocated throughout the 45 years of Communist rule.

While the Communist commitment to improving the telecommunication infrastructure could be questioned, under the Mazowiecki and the Bielecki governments, the issue gained prominence. Prodded by the World Bank, officials of both governments stressed that telecommunications had become one of the top priorities. In March 1990, the Economic Committee of the Council of Ministers adapted a telecommunication development program to the year 2000. In mid-1989, to underscore the importance of the issue, a cabinet-level position was created for telecommunications, following the dismemberment of the Ministry of Transport, Marine Economy, Posts and Telecommunications.[7] The new Ministry of Posts and Telecommunications (MPT) played an important role in drafting the legal acts delineated above, in expediting the legislative process, and in defining the strategy for network development.

4.1 Network development and penetration goals

Telecommunication services can be thought of as consisting of three hierarchical layers: international, long-distance, and local. The strategy to improve the infrastructure, which emerged in 1989–1991, approached the three layers in a top-down fashion. Because eliminating the bottleneck in international services promised the greatest pay-off relative to investment, Poland's strategy began there. Rebuilding the local networks, on the other hand, by far the costliest and the least profitable, was expected to be gradual and span many years.

Related to the top-down approach is the plan to digitize the network. At the end of 1991, the great majority of installed equipment was based on outdated analog technology. Instead of modernizing the existing equipment piecemeal, the strategy called for the building of a parallel network, a digital overlay, and for gradual migration from the first to the second network. The technique makes it possible to introduce ISDN services in those areas where digital technology is available. To phase out analog systems, the strategy called for no new analog equipment to be introduced at any layer of the network after 1994. A complete transition to an all-digital infrastructure was expected by 2010.

The year 2010 was also set as target for reaching the same level of infrastructure development advanced economies exhibited in the early 1990s. In 1989, with 3.121 million subscribers in Poland's network, the DEL penetration stood at 8.29 per 100 population, one of the lowest in Europe. The announced growth goals are:

- to add 1.8 million subscribers by 1995 and increase the network size to 5.1 million;
- to build a network for 10–12 million subscribers and reach penetration of 30 DELs per 100 population by the year 2000;
- to reach penetration of 50 DELs per 100 population in a network of 22 million subscribers by the year 2010 (Geborys Lucjan, 1991).

To accomplish these ambitious goals, the number of subscribers would have to grow by 9% per year between 1991–1995, 14% between 1996–2000, and 8% in the 10 years between 2001 and 2010. Even though in 1990 the network grew by a mere 5.5%, in 1991 the growth rate jumped to 8%. This tendency continued in the early part of 1992, as the number

---

[7] A thrust of SCPE reforms in the late 1980s was to eliminate layers of bureaucracy by consolidating ministries.

of subscribers increased 9.6% (annual rate) during the first quarter of 1992. To fulfill these plans a number of development projects were announced and some implemented.

4.2 Projects on international and long-distance networks

The following are network projects that began in 1990–1991:

- Improving Poland's links to the outside world is top priority for telecommunication reformers. In September 1990, PPTT, in cooperation with AT&T NSI, began offering international telephony services via *Komertel*, a dedicated network connecting an ESS exchange in Warsaw with telecommunication centers in Frankfurt, London, and Pittsburgh via satellite. Since Komertel does not interface with the public telephone network, within Poland subscribers can call only other Komertel subscribers. In 1990 it cost Zl.10 million ($1000) to get a Komertel line, and Zl.1 million ($100) per month to maintain a subscription. Even at these prices, and fairly high use fees, the 2000-line capacity of the network was nearly exhausted by the end of 1991.
- The success of Komertel was at least in part due to satellite links. In October 1990, AT&T NSI signed a contract with PPTT to digitize Poland's INTELSAT connections via the earth station in Psary and to build a digital radio line, Warsaw-Psary-Katowice. In May 1992, Satellite Transmission Systems, an American firm, was contracted to build another Psary earth station, this one to work in the EUTELSAT system. The latter earth station will provide improved connectivity with 18 countries.
- To expand the public network's connection with the outside world, TP SA signed contracts with AT&T NSI, Siemens and Alcatel CIT for three new exchanges to handle international traffic. In early April 1992, a 5ESS exchange, delivered and installed by AT&T in Warsaw, provided 3.5 thousand new international lines, doubling capacity at that time. By the end of 1992, a Siemens ESWD and Alcatel E-10B exchanges for international traffic were scheduled to be installed in the cities of Katowice and Poznan, respectively.
- The three international exchanges interface with the long-distance network layer, which will be based on twelve exchanges to be installed in Gdansk, Bydgoszcz, Poznan, Bialystok, Warsaw, Lódz, Radom, Lublin, Wroclaw, Kielce, Walbrzch, and Cracow. In January 1992, Siemens won a TP SA contract, financed by the World Bank, to deliver and install these exchanges. The imlementation schedule calls for completion of the project by the end of 1993.

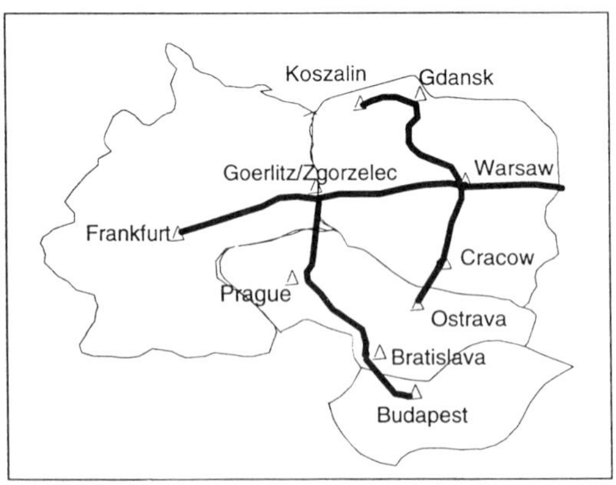

Figure 4.1. North-South and Trans Europe Line.

• Plans call for fiber and digital radio transmission in the digital overlay. In February 1992, TP SA signed a contract with the Japanese firm Mitsui to provide the equipment for and to install 16 digital radio lines. In March 1992, a $26 million contract was signed with AT&T to transfer the necessary technology and to install 875 miles of fiber lines. The fiber cable will be manufactured by OTO Lublin, a Polish company. Both contracts are financed by loans from the World Bank.

• Danish companies NKT and Northern Telegraph Company, Swedish Telecom, P&T Finland, and TP SA are collaborating on the installation of a North-South Fiber Line. In December 1990, the Danish island of Bornholm was connected with the Polish city Koszalin by a 140 Mbps system. By December 1992, the northern cities Gdansk and Bydgoszcz were expected to be connected, via Warsaw, with Cracow, Katowice and Cieszyn in the south. Sixty percent of project costs is financed by the Danish participants.

• In December 1990, Bundespost Telekom, TP SA, and the PTTs of Czecho-Slovakia and Hungary signed an agreement to build a fiber Trans Europe Line (TEL). TEL runs through Frankfurt, Zgorzelec/Görlitz, Warsaw, and on to Poland's border with Belarus. In its initial conception the line would continue on to Moscow, and may still do so if appropriate agreements are signed. In Zgorzelec/Görlitz, TEL branches off toward Prague, Bratislava and Budapest. Once the line is extended to Moscow, its total length will be 3200 km. Poland's component is expected to be financed by a loan from the World Bank ($30 million).

• Many firms and individuals in Poland still depend on telex and telegraph services. Even though these networks are likely to be replaced by telefax or a form of electronic mail, the existing facilities have been modernized. PPTT signed a contract with Siemens for a new telex exchange to be installed in Warsaw. At the end of 1991, the Swiss firm *Ascom Hasler* was negotiating to sell telex exchanges for Poznan, Wroclaw, Katowice and Lódz. The same Swiss firm had already delivered telegraph exchanges for Lublin and Cracow. The financing for both deals was guaranteed by the government of Switzerland.

4.3 Projects on local networks

The development strategy for local networks is designed to utilize a variety of capital sources. In addition to the monies collected by social committees for telephonization, the principal sources of funding for projects on local networks have been foreign governments, which—through government-guaranteed loans—have tried to promote exports. The PPTT and TP SA have asked foreign companies trying to sell equipment on the Polish market to make proposals for network upgrades in particular provinces of the country. A number of such proposals have been made and accepted and the following is a synopsis of the available data on the subject.

• Alcatel SESA, the Spanish division of Alcatel and the producer of System 12 exchanges, was invited to submit proposals to improve the network in the Warsaw Province. As a result, PPTT bought eight transit exchanges, six subscriber exchanges for 75 thousand lines, fiber optic cable and other equipment for the local network. The contract, worth $78.6 million, was financed by a Spanish-government guaranteed loan. Subsequent negotiations resulted in a deal to improve the infrastructure of the Cracow Province, and of 18 other provinces.

• The Katowice Province is the domain of Siemens. The German equipment manufacturer signed contracts with the PPTT to install five transit-subscriber exchanges in Bytom, Gliwice, Katowice, Sosnowiec and Tychy. The contract, worth approximately DM65 million, is expected to expand the local network by 171 thousand subscriber lines.

- The French division of Alcatel has entered the most provinces. PPTT and Alcatel CIT signed a contract for exchanges of the E-10B type with capacity of 314.5 thousand lines. The exchanges are planned to be installed within three years (1991–1994) in ten provinces, including Poznan, Czestochowa, Legnica, Lublin and Lódz. The contract, worth $70 million, is financed by *Banque Societe Generale* and guaranteed by COFACE.
- In June 1991, AT&T Network Systems International signed a contract with PPTT to deliver switching equipment for 128 thousand lines. The equipment worth $63 million is to be installed in the networks of Szczecin, Siedlce, Wloclawek and Plock provinces.
- The South Korean Samsung is providing equipment for the Opole Province. In 1989, the government of South Korea offered to the Polish government a $50 million loan to buy Korean technology. The loan was made available to the PPTT which devised a plan to install 113 thousand telephone lines in the province with one of the lowest penetration levels in the country.
- Northern Telecom (Canada), with its Polish partner Elwro, signed a contract to deliver DMS exchanges for the Kielce, Wroclaw, Rzeszów and Bielsko-Biala provinces. The Bielsko-Biala contract calls for the installation of 100 thousand lines by 1995 and the Kielce proposal projects an installation of 45,000 new lines by the same year.

Following up on the announced goals for network growth, researchers at the Institute of Posts and Telecommunications in Miedzylesie prepared a plan for network development to the year 1995, based on the existing infrastructure, and modernization plans. According to this plan, the telecommunication network will be organized into 56 numbering areas, with one long-distance exchange serving each numbering area. The lines of the digital overlay will interconnect all areas, and will provide the means for connecting local networks.

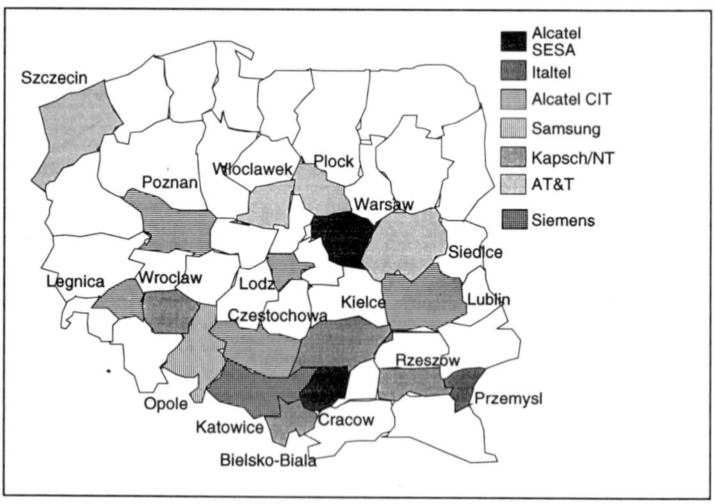

Figure 4.2. Foreign equipment in local networks.

5. TELECOMMUNICATION EQUIPMENT: FROM AUTARKY TO INTEGRATION

One dominant feature of an SCPE was autarky. The fathers of Communism in East-Central Europe saw themselves as surrounded by a hostile capitalist world, lurking to subvert their system by any means, including economic subjugation. This attitude, a derivative of the Bolshevik fear of encirclement, prodded development policies oriented toward CMEA self-sufficiency in all industries. Trade with the capitalists was viewed only as a necessary stop-gap measure intended to satisfy current needs until the same products could be made in one of the CMEA countries. Although in the 1960s the self-sufficiency policy was recognized as erroneous and increased ties to the developed world were pursued, the quality and the range of CMEA products did not attract many buyers in the developed world.

While limited trade with the capitalist world was, at least initially, a result of a conscious policy, a systemic deficiency caused the relatively weak economic links among CMEA countries. Without horizontal information channels between producers and consumers and without price mechanisms to communicate information about opportunity costs, economic interaction was determined by the preferences of planning organizations, and not by consumers' needs. Any planning "mistake" simply added to the stockpiles of products no one really wanted. Some of these products were traded either because the Soviet Union saw it in its political interest to promote economic interaction with its allies, or because similar products were embargoed by the West. As the pace of technological innovation increased, planning "mistakes" occurred more often, further weakening CMEA trade ties. Instead of ushering multi-lateral integration, CMEA developed into little more than a bureaucracy to manage the barter of Soviet raw materials for East-European consumer goods.

5.1 Changing industry structure

With limited trade ties to either the West or the East, most industries in the CMEA countries veered toward vertical integration of domestic firms—from input components to finished products. In the name of efficiency, industry associations were created to coordinate this vertical integration, thus removing any remaining incentive for competition. The telecommunication equipment industry in Poland structured itself into an association called *Telkom*, which produced everything from electronic components to sophisticated digital switches. Whenever needed, Telkom interfaced with the two other broadly defined "electronics" associations: *Mera* for computer equipment and *Unitra* for consumer electronics. Despite repeated attempts at reform, the association structure proved to be remarkably resilient. Following the last round of SCPE reforms in late 1980s, enterprises of the three electronics associations found themselves pulled by the centripetal force of *Elpol*—yet another form of industry association.

In the 1980s, domestic monopolies were tempered by the emergence of private computer firms, numbering at one time as many as several hundred. Aware of the on-going computer revolution and unable to rely on domestic industry, Poland's authorities made the decision to encourage private importation of microcomputers and microcomputer components. The lowering of import duty and toleration of a tax loophole on computer-related transactions made it possible for a real market in microcomputers to emerge. As the number of competitors increased prices declined and quality and post-sale service improved.

Most of the private computer firms, however, did not have sufficient capital to buy the more advanced telecommunication equipment. The preferential tax treatment was for microcomputers, and not transmission or switching gear, and the existence of CoCom restrictions added to transaction costs. The state telecommunication industry did not feel sufficient pressure to reform and its structure remained more or less intact. Manufacturing was still performed by the same firms, and although distribution channels were expanded somewhat, there was never any real competition for the customer.

The radical character of the Balcerowicz Program stemmed, at least in part, from frustration with previous, failed, attempts at reform. "Shock therapy" was designed to confront the economy with the reality of competition and technological sophistication needed to function in the 1990s. The drastic devaluation of the zloty increased the costs of component imports. The lifting of many tariffs on foreign products—between July 1990 and July 1991—put pressure on domestic manufacturers. The elimination of subsidies and various preference programs faced the state industry with real costs. The ultimate goal was to restructure the economy by forcing firms to modernize, and if they did not to let them go bust.

While the objectives of the Balcerowicz Program were never in question, implementation proved to be much more complex. The process of restructuring did not progress quickly enough; privatization hobbled along; and allowing large enterprises to go bankrupt proved to be politically untenable. Instead of choosing between modernizing or closing shop, many enterprises chose to linger on by reducing employment and refusing to pay bills—including taxes and accounts payable. As GDP and living standards plummeted, voices were raised in favor of an industrial policy—or government support for selected industries.

Aspects of such a policy have indeed emerged and the telecommunication equipment industry was directly affected. First, tariffs of 15–20% on telecommunication equipment were reintroduced. Second, in a number of policy statements and in negotiations with foreign manufacturers trying to enter the Polish market, the MTP made clear that it preferred deals that included manufacturing in cooperation with Polish firms. As the economy worsened, emphasis on such cooperation increased, culminating in a decision that after June 30, 1992 no equipment with less than 50% domestic content could be installed in the TP SA network.

5.2 Manufacturing joint ventures

With MPT preferences clearly stated, foreign equipment providers and Poland's telecommunication firms scrambled to find each other. ZWUT, the largest producer of step-by-step Strowger and crossbar Pentaconta exchanges in Poland, has negotiated with Alcatel SESA, before finding Siemens. A joint venture between the two firms, CEWIS, is owned 51% by the Polish manufacturer, and 49% by the German manufacturer of ESWD switches. Following TP SA's decision to award a contract for 12 long-distance tandem exchanges to Siemens, CEWIS is likely to benefit from the deal, and perhaps emerge as one of the principal producers of large switching equipment on the Polish market. The firm has announced its intention to eventually produce one million exchange lines a year.

For a company whose primary markets collapsed only two years before, ZWUT has positioned itself fairly well in the new conditions. Its strategy has been to turn itself into a holding company of joint ventures with partners able to transfer technology. In early 1992, ZWUT signed joint venture agreements with Arnica, an American firm, and with AMT, a British firm. The first joint venture, DTS-ZWUT, with 33% Polish ownership, will employ 400 of ZWUT's workers to make small-capacity digital exchanges. The other joint venture, with 49% ZWUT capital, will also make small switches and key telephone systems (KTS).

In their strategic plans, ZWUT and its foreign partners hope to sell on the markets to the East. Before 1991, 40% of company's production was sold to the former Soviet Union, and many of its Pentaconta exchanges are working in the networks of the Ukraine and Belarus. ZWUT has maintained those ties and in one recent barter agreement, an exchange was sold to Belarus for 16 train-loads of potassium nitrite.

With 60% of its production going to the CMEA countries, Wroclaw's Elwro was also hit hard by the collapse of trade within the Soviet Bloc. Elwro used to make mainframe computers (ES-1032 and ES-1034), IBM-clone microcomputers, communication processors (ES-8371), calculators and some household products such as tea-pots. The company's answer to the loss of markets has been to shed half of its employees—from five to two and half thousand—and to leave the computer and enter the telecommunication equipment market. After negotiations with Siemens and the Austrian Kapsch, Elwro set up a joint venture with the Canadian manufacturer Northern Telecom—Northern Telecom Elwro. NTE will manufacture the NT switches, DMS-10 and DMS-100 for installation in the Bielsko-Bia_a, Wroc_aw, Rzeszów, and Kielce provinces and "Elektron" telephone sets. The expected production volumes are 350 thousand lines and 520 thousand telephone sets by the end of 1993.

Northern Telecom's greatest competitor in the U.S., AT&T, has also entered the fray for the Polish market. The U.S. company, through its Dutch subsidiary AT&T Network Systems International, has signed a cooperation agreement with Telfa of Bydgoszcz, a manufacturer of Polish-design small exchanges for rural areas—ECW-B and U-10. The arrangement seemed to be one of necessity for both sides. At one time, Telfa staked its future on Ericsson and even went to court to force the MPT to certify the AXE-10 system of the Swedish manufacturer. Telfa won the case but Ericsson failed to convince the TP SA to buy its equipment[8] and the idea of a joint venture between the two soured. AT&T, on the other hand, was much more successful in its dealings with TP SA. After securing contracts for Warsaw's international exchange, the American company was asked to submit bids to supply its ESS exchanges for the Szczecin, Siedlce, and Plock provinces. With Telfa looking for a foreign partner and AT&T interested to satisfy the domestic-content requirement, the two firms found each other.

Another major competitor, particularly in the switching market is SETEL, a joint venture between the Spanish Alcatel SESA (51%) and two Polish firms: PZT, a producer of transmission equipment (34%), and Elektrim, a trading company (15%). SETEL has benefited from favorable financing offered to TP SA by the government of Spain for the explicit purpose of purchasing Spanish equipment. Contracts to supply SESA's System 12 switches for the Warsaw and Cracow provinces have been signed and SETEL has announced its intention to produce 100 thousand lines in 1992.

Alactel CIT Polska is a joint venture between the French Alcatel CIT (51%), Elektrim (14%) and Teletra (35%)—the only CMEA manufacturer of digital exchanges. In 1972 Poland bought a license to manufacture the then state-of-art Thompson-Alcatel E-10 switch and Poznan Teletra was assigned to set up production. Between 1972 and 1987 a mere 300 thousand lines were manufactured in Poznan—a number that dwarfs in comparison to the one million lines a year that CEWIS is projected to make. In the last two decades, the French Alcatel modernized the E-10 design but these innovations, for the most part, did not find their way to Teletra's products. Alcatel CIT signed a contract with TP SA to deliver 314 thousand

[8] In an attempt to secure for itself the Cracow province, Ericsson offered to pay for the renovation of the Cracow Symphony building—a cost of about $1 million—if the local TP SA bought its equipment. The deal did not go through and the contract was awarded to Alcatel SESA.

lines and Alcatel CIT Polska will benefit from that contract.

In a somewhat different field is Thomson Polkolor, a producer of color-picture tubes for television sets. The French Thomson, one of the largest makers of television sets in the world, owns 51% of the firm and the Polish firm Polkolor the other 49%. The venture plans to produce three million picture tubes a year by 1995, and to sell 20% of that volume to the two manufacturers of color television sets in Poland—Unimor of Gdansk and Elemis of Warsaw. The remaining 80% are expected to be exported, in equal proportions, to the West and former Soviet republics.

6. CONCLUDING REMARKS

Transformation of Poland's telecommunications has not run its course and many issues still remain unresolved. The privatization and restructuring components of the Balcerowicz Plan were clearly not completed. The legal framework was still being formulated, with the Copyright Act awaiting legislative approval, and the 1990 Act on Posts and Telecommunications considered to be lacking. As the new constitution was being prepared, the executive, the legislative and the judicial branches of government were positioning themselves to exert greater weight in the future balance of power. In sum, an air of uncertainty persisted, despite the apparent consensus on the need to continue the general direction of reforms in place.

However, in the early 1990s, many countries could not describe with certainty their telecommunication environments. In the US, the Federal Communications Commission persisted in its campaign to introduce greater competition among different types of service providers. Telephone companies were allowed to enter the markets for some video services, and expected changes are likely to remove restrictions on cable TV firms to offer telephony. In France, the postal services were separated from telecommunications and the rules for providing value-added services were liberalized. In Germany the tight, constitutionally guaranteed monopoly of Bundespost was eased, as authorities licensed several providers of paging and cellular services, and the Kohl government announced its intention to privatize Telekom, Bundespost's telecommunication division. In Great Britain, cable-TV companies, many of which are controlled by the Baby Bells, have been allowed to offer local telephone services. The list can go on.

Despite the general state of flux, some trends can be recognized in the environment of Poland's telecommunications. The following sections describe four such trends, or agents of change: institutional pluralism, market competition, shortage alleviation and market restructuring. The final section raises questions about development models in telecommunications.

6.1 Institutional pluralism

The emergence of the new telecommunication regime in Poland has multiplied the number of actors affecting the shaping of the telecommunication environment. Although the government through the MPT still plays the dominant role in the formulation and execution of policy, other institutions have been aggressive in asserting their roles. In contrast to the past when the legislative bodies rubber-stamped the decisions of the executive organs, both houses of the Parliament, the Senate and the *Sejm*, have played active, if not activist, roles in drafting telecommunications-related legislation. Similarly, during the debate on the Act on Radio and Television (ART), Lech Walesa, as President of the Republic, which was formerly a ceremonial post, has indicated a strong interest in having influence over the setting of broadcast standards.

The debate on ART has contributed to political fragmentation that has characterized Polish politics since the Solidarity coalition disintegrated in mid-1990. In the European political tradition, public television and radio have been perceived more as propaganda tools for the dominant political forces, than as ideologically neutral media that disseminate information. In contrast to Western Europe where the emergence of private broadcast and cable networks has lessened the political importance of public radio and TV, in the post-Communist Poland—where airwaves were abused for so long—the public's sensitivity to real or imagined programming biases has not diminished. Parliament's failure to pass the Act on Radio and Television two years after it was first introduced is a reflection of this sensitivity. As the dominant political institutions in the country—the President, both houses of the Parliament, and the government—fought to have a voice in the shaping of broadcasting policy, the reform of public networks and the development of private networks have been delayed.

The relative paralysis at the top on ART did not mean cessation of activity at the bottom. Over 400 applications for radio and television licenses were pending in April 1992 and a number of station owners have started to broadcast without proper permits. Following the success of social committees for satellite television, some two thousand social committees for telephonization, often in rural areas, have been created to lobby for and take part in infrastructure improvement projects. Umbrella organizations such as Polish Telephones or the Private Radio and Television Industry Association have coordinated the local efforts. The telecommunication regime can no longer be described as centralized and immune to pressures from below.

6.2 Market competition

A central component of the Balcerowicz Plan, and telecommunication regulation, has been the introduction of competition in the economy in general, and telecommunication markets in particular. Laws on joint ventures with foreign capital and a liberal policy on import tariffs were designed to encourage competition between state monopolies and new foreign and domestic market entrants. Both APT and ART devoted much space to licensing issues, an indication of the value placed on introducing competitors to the TP SA and PRiTV. Idealizing competition in the early stages of transformation, the reformers perceived it as critical in the campaign to force domestic monopolies to adapt to new market realities or to go bankrupt.

The introduction of competition in markets for telecommunication equipment has multiplied the number of options available to network builders. TP SA—which had to buy equipment one to two generations behind global trends from mostly CMEA suppliers—can now select its gear from a wide variety of top brands. Similarly, telecommunication managers and subscribers in setting up their private networks are now limited much more by resource constraints than by problems of access to the latest technology. For all intensive purposes, the chronic shortages so characteristic of an SCPE have been eliminated.

As expected, the competition resulting from the opening of Poland's economy to imports has hurt domestic manufacturers. Although in the two years from 1989 to 1991 there were no major bankruptcies among the former Telekom Association enterprises, most firms have drastically cut costs by reducing employment and refusing to honor debts. Responding to market pressures, many have tried to form joint ventures with established foreign firms, for the purpose of transferring technology. Those who succeeded in finding a partner stood a much better chance in securing TP SA contracts than those who did not. As unemployment rose and standards of living continued to decline, calls for government support for domestic producers became more and more widespread.

As the initial fascination with the touted "invisible hand" faded, a much more cautious approach to market benefits emerged. Following a series of equipment purchases in which favorable financing was the most important consideration, the issue of how domestic manufacturers would be affected began to play a greater role. In an attempt to protect domestic suppliers, the MPT Minister issued a decree that all equipment installed in the TP SA network must have at least 50% domestic content. In a similar trend toward protection of domestic providers, the MPT has not allowed any real competitors to TP SA to emerge, even though APT permits competition for all but international services.

Given current trends, the degree of competition in service markets in the future might depend on the decisions made by the judicial branch. Thus far the courts have played virtually no part in the shaping of the telecommunication environment. Since APT is very explicit in outlining when a license can be denied, legal action may eventually force access to markets. A debacle between the Anti-Monopoly Administration and the TP SA Lublin, reported in the popular media, might be a precursor to a more activist role for the courts.

There is no question that competition will have a role in the emerging telecommunication environment, both in the equipment and service markets. Its extent, however, will be limited by what is politically feasible and by perceptions of the public good. Even though the government has tried to extricate itself from micro-managing the economy, most Poles still believe that ultimately the political leadership is responsible not only for the condition of the economy as a whole but also for the economic well-being of each firm, and by extension each citizen. The building of a capitalist system and markets will depend, at least in part, on the degree to which the philosophy of self-reliance and individual initiative takes root in the public psyche.

6.3 Shortage alleviation

The most visible change of the preceding two years has been the alleviation of shortages for virtually all products, including telecommunication equipment and services. The rapid increase in the number of international lines has removed the bottleneck of communication with the outside world. An application for telephone service in Warsaw is now likely to be filled within two years, and other major urban areas are not too far behind. The issue is no longer whether or not something is available, but how much it costs. Although there is still a sizable waiting list for installation of telephone services, particularly in the rural areas, the expensive option of becoming a subscriber to cellular telephony is now available in Poland's principal cities.

Goodman (Goodman, 1987) argues that all technologies with universal appeal go through five evolutionary stages:

1) an experimental rarity, often an entrepreneurial or laboratory discovery;
2) an exotic tool or toy used by a small group of experts;
3) products that are well known and manufactured in modest quantity, but direct use is limited to a small fraction of industrial or other institutional environments;
4) widespread production and availability, with direct use requiring little or modest training in a broad domain by a sizeable minority of the population;
5) the technology has become part of the fabric and infrastructure of daily life, and its absence is often more noticeable than its presence.

With the exception of terrestrial radio and television, most telecommunication services in Poland are between stages two and four. The new information services were in stage two, and paging was arguably approaching stage three. Telephone services had not yet reached stage four, but were already past stage three.

Associated with stages four and five is technology user culture. For the telephone to become part of daily life's fabric, it must be used not only for occasional calls across town, but also for communicating long-distance with friends, for dealing with various bureaucracies, for mail-order shopping, or for accessing information services. TP SA claims that in the Warsaw network more than half the subscribers do not generate enough traffic to pay the costs of installation and maintenance. To such subscribers, the telephone is more a passive device to be used in emergencies than an active communication tool.

6.4 Market restructuring

Yet another agent of change has been the restructuring of subscriber markets. Even though privatization has not taken place as quickly as planned, progress has been made. By the end of 1991, most small retail shops were privately owned, and the process of breaking up large state enterprises into smaller saleable units was slowly edging forward. Even though the political dangers of worker displacement and unemployment have been apparent, the issue was never whether to privatize, but what the sale price should be, who should be the buyer, and how quickly.

Although the final results of privatization are still years away, one likely outcome is the reduction in average enterprises' size in terms of both capital and labor. Similarly, the businesses being created are more likely to be small, family-run operations, often in the service sector, than large manufacturing firms. Because the range of telecommunication services needed is related to how big the enterprise is and in which sector it is active in, a provision matrix is likely to evolve. For example, instead of large PBXes, smaller customer premises switches or key telephone systems might be in greater demand and because of lower start-up costs the attraction of Centrex services might be greater. Furthermore, the emergence of the mixed residential-professional market segment is likely to increase traffic volume per subscriber.

If the experience of the more advanced countries is considered, private networks' development patterns are likely to carry over to the public networks. In Germany and the US, the rapid growth in private facilities has pushed the process of liberalization of data-carrier services. Because private networks reduce traffic on public networks, in general, public operators try to discourage enterprises from installing their own facilities by lowering tariffs. By mid-1993 tariffs for most TP SA services were fairly high. As the experience of US long-distance operators shows, a price reduction can lead to greater profits in the long run due to a greater traffic volume.

6.5 Mobile networks and telecommunication development models

A clear trend in telecommunication services is the movement toward mobility. In cellular telephony or in paging, the important feature is the capability of transmitting or receiving the electromagnetic signal without being constrained to a wired, stationary terminal. The decisions on frequency management made at the World Radio Conference point to a likely emergence of worldwide mobile telecommunication services. Similarly, European standardization organizations have promoted the use of compatible technologies and frequencies for cellular and data services, in an attempt to build continent-wide networks. The functional advantage of mobility, reinforced by declining technology costs, raises the possibility of alternative, to the traditional, models of developing telecommunication infrastructures.

Could countries with inadequate telecommunications substitute for, or at the very least complement, the fixed telephony networks with mobile services in the vein of personal communication networks? Although such a theory is plausible, it remains to be tested in Poland. While paging networks have proliferated, penetration indicators for cellular telephony have not been overwhelming. In June 1992, *Centertel*, a joint venture between TP SA, Ameritech, and France Telecom, began offering cellular telephony in Warsaw and the surrounding areas. With the installation cost $500, a monthly subscription fee $25, usage fees of about 25 cents per minute and equipment cost of about $1000, the service could have been afforded only by the very rich. By mid-1993 a few thousand have signed on.

The potential of mobile services, particularly in the realm of cellular telephony, is clearly greater. This is because costs of providing cellular technology have been on the decline and because foreign investment could play a more significant role in local loop development. The 1990 Act on Posts and Telecommunications permits 100% foreign ownership of local networks. Given more liberal licensing policy on the part of MPT, foreign capital could be utilized effectively to the benefit of the foreign investor and Poland's telecommunication infrastructure alike.

REFERENCES

Cetwinska, J., 1990, Prywatyzacja, komercjalizacja, likwidacja przedsiebiorstw panstwowych (Privatization, commercialization and liquidation of state enterprises) (Wektory Gospodarki S.A., Warsaw, October 1990).

Geborys, L., 1991, Charakterystyka stanu istniejacego i koncepcja technicznego rozwoju sieci telekomunikacyjnej w Polsce w latach 1991–1995 (Characterization of current state and a technical conception of developing the telecommunication network in Poland in the years 1991-1995), Przeglad Telekomunikacyjny (Warsaw).

Geipel, G.L., A.T. Jarmoszko and S.E. Goodman, 1991, The information technologies and East European societies, East European politics and societies, Vol. 5, No. 3.

Goodman, S.E., 1987, The information technologies and Soviet society: Problems and prospects, IEEE Transactions on systems, man, and cybernetics, Vol. SMC–17, No. 4, July–August, pp. 529–552.

Jarmoszko, A.T., G.L., Geipel and S.E., Goodman, 1988, Computing technologies in Eastern Europe: The impact of reform in John P. Hardt and Richard F. Kaufman, eds., Pressures for reform in the East European economies (Government Printing Office, Washington).

Kornai J., 1990, The road to a free economy (W.W Norton & Company, New York, London).

Saunders, R.J., J.J., Warford and B. Wellenius, 1983, Telecommunications and economic development (The John Hopkins University Press, Baltimore and London).

The Race to European Eminence
E. Bohlin and O. Granstrand (Editors)

Foreign Investment in Hungarian Telecommunications

Ferenc Kiss

Chief Economic Advisor, Hungarian Telecommunications Corporation,
55 Krisztina krt., Suite 438, 1013 Budapest, Hungary

The paper analyzes the legal-regulatory framework for telecommunications in post-communist Hungary, especially from the point of view of foreign investment, the country's Telecommunications Act and telecommunications policy as well as regulatory rules and procedures. It summarizes existing foreign telecommunications investment and offers some detail and elaboration on investments and opportunities in the privatization of the national carrier Hungarian Telecom (HTC), local service supply, cellular and paging services, and professional services.

Great pent-up demand for telecommunications services, good long-run economic growth potential, a well educated labor force and political stability created an attractive telecommunications investment climate at the beginning of Hungary's post-communist era. However, actual foreign investment has been small in volume. This is attributed to the failure of political institutions and processes to deliver the industry's legal-regulatory framework, without which foreign investment in the voluminous telecommunications core activities could not take place and foreigners were discouraged from pursuing investment opportunities. With the emergence of the legal-regulatory framework, HTC's privatization, wired local and wireless service concessions, foreign investment will increase and the investment climate will improve. Further improvement is needed with respect to the legal-regulatory framework.

1. INTRODUCTION

The main objective of this paper is to analyze opportunities and conditions for foreign investment in the Hungarian telecommunications service sector as they exist at the end of 1993.

The paper consists of 11 sections. Section 2 explains the attractions of post-communist Hungary for foreign investors. Section 3 deals with the slowly emerging legal-regulatory framework for telecommunications. Sections 4 to 9 elaborate on foreign investment. A brief summary of the existing situation is followed by some elaboration on investment opportunities in HTC's privatization, local network development, cellular services, paging and professional services. Section 10 examines some more recent discouraging phenomena. Section 11 offers a summary and conclusions.

2. AN IRRESISTIBLE INVESTMENT OPPORTUNITY?

Foreign investors have been showing significant interest in the Hungarian telecommunications service sector since the great wave of fundamental political change swept away the communist regimes of Central Europe in 1989-90. The attraction has been quite irresistible. Huge pent-up demand by the country's poorly served 10.5 million inhabitants, combined with a prediction of strong economic growth in the long run, promised very rapid domestic residential and business market development. The growth potential in international traffic appeared even more favorable, due to quickly expanding political, cultural and economic ties with Western Europe, North America and the rest of the world. To all this, Hungary's geographic location in the heart of Europe added the promise of becoming an East-West (and, to some extent, North-South) telecommunications hub—with the associated high growth in transit traffic. Favorable geographic, climatic and demographic conditions together with the presence of a well educated and rather cheap labor force seemed to guarantee that network construction and service development could be achieved quickly and at an attractively low cost. The outlook for technological progress was also highly favorable. Hungary had an elaborate and efficient educational system, capable of producing highly qualified professionals in sufficiently large numbers. R&D traditions, organizations and activities were also rather strong in telecommunications. It seemed that there were no significant obstacles to rapid technology transfer in order to catch up with West European levels of technology.

Hungarians appeared anxious to introduce modern, efficient and ubiquitous telecommunications services. They readily understood the grossly inadequate nature of their telecommunications infrastructure—not only in comparison with Western European countries but also relative to Hungary's own per capita GDP level. They understood that inadequate telecommunications generated very large economic losses and prevented the country from attracting foreign capital. Perfecting loss estimation became the favored pastime of a number of researchers. Various declarations of key telecommunications policy objectives included market equilibrium, universal service, modernization, rapid development, efficiency, restructuring, regulation and privatization. The long overdue structural changes to the telecommunications sector were introduced early and without hesitation. As a result of divestiture from the Post Office, the Hungarian Telecommunications Company (HTC)[1] was formed at the beginning of 1990. HTC quickly assembled a 10-year strategic development plan and a 3-year program to build the network infrastructure. At the core of the new infrastructure was a digital overlay backbone network with primarily Ericsson and Siemens electronic switches and fiber optic (plus some digital radio) transmission facilities. Plans were created for the development of network intelligence and the deployment of various classes of advanced services. Information services (videotex) were introduced. Data communications were planned to develop and expand significantly. The public packet switched (X.25) network of HTC was scheduled for (and indeed underwent) a major expansion, and the choice of data private line services was also to expand significantly.

The rapid introduction of competition was one of the major policy goals. The transformation of terminal equipment and value added service markets began very early and progressed at a high rate. At the same time, various ideas surfaced and were discussed about redefining the boundary between monopoly and competition in basic service markets. In preparation for incoming competition, an elaborate organizational separation was carried out within HTC.

[1] In Hungarian: Magyar Távközlési Vállalat Rt. or Matáv.

Entry into existing and newly emerging competitive service markets was carried out in part by a series of specialized subsidiaries. For example, Comex Ltd. and Centrophone Ltd. offered PABX services, Please Ltd. sold data services (packet switched and private line), Secotel Ltd. did repairs and servicing of computing and telecommunications equipment, Tel-ex Ltd. provided technical support for data services, Sat-net Ltd. entered the satellite services business, Videotex Ltd. did what its name implies, and Westel Ltd. became the country's first cellular service provider. Anticipating limited competition in local service markets, local network development companies were also formed as HTC subsidiaries and associates.

HTC was to be privatized. By the end of 1991, the 100 percent state-owned HTC was transformed into a joint-stock company. A large percentage of its equity was earmarked to be owned by a large foreign strategic partner. From the foreign investor, HTC expected not only money but also technological and management expertise. Privatization was expected to take place very quickly. The financing plan of the 3-year development program relied on proceedings from the 1992 privatization. Conditions seemed highly favorable, and the success of privatization seemed assured.

The general social and political conditions also appeared favorable. Since the 1960s, Hungary had been regarded as the most liberal and most market-oriented of the communist countries of Central Europe. Her leading role in the region's move toward fundamental political and economic change was greatly reinforced in 1988–90. Hungary has begun to attract a lot of foreign investment. During the first two years of the post-communist era, Hungary alone attracted more than 50 percent of total foreign investment in former Warsaw Pact countries.

3. LEGAL-REGULATORY FRAMEWORK

With political change came the urgent need to revamp the entire legal system of Hungary—all the way from her constitution down to the last ministerial decree. The new Parliament elected in 1990 had a gargantuan task on its hands. Having decided what laws were possible to preserve with changes, what laws had to be replaced and what new laws had to be created, Parliament began an almost industrial mass production of laws.

At a time when the country was having no workable constitution, criminal code, civil code, etc., the legal framework for telecommunications was understandably not the most important piece of legislature for Parliament to create. Nevertheless, despite its lesser importance, work on a new Telecom Act took a very early start and at least two major drafts were produced and discussed before the end of 1989. As one drafter explained to the author of this paper in March 1990, "the Telecom Act is now ready and waiting for the elections and the formation of the new Parliament; it will be passed in a few months." Unfortunately, reality failed to live up to such optimistic expectations.

3.1 The Telecom Act

The Telecom Act is the centerpiece of the legal-regulatory framework of Hungarian telecommunications. At the end of a lengthy debate and after considerable changes to its originally proposed contents, the Hungarian Parliament finally passed "*Act 72(1992) on Telecommunications*" on November 23, 1992. However, the Act did not come into force with its passing by Parliament. The following legal conditions were necessary for the Telecom Act to become effective and operable: (1) the passing by Parliament of the Media Act, (2) the passing by

Parliament of the Frequency Act (on frequency distribution), and (3) the issuance of a series of governmental and ministerial decrees as specified in the text of the Telecom Act. Legislation set the deadline. The Telecom Act was to take effect no later than July 1, 1993. The first requirement has been eliminated by a subsequent decision not to formulate a Media Act at this time. The Frequency Act was substantially delayed and finally passed by Parliament on April 27, 1993. Section 27(1) of the Frequency Act put the Telecom Act into force as of July 1, 1993. However, the governmental and ministerial decrees were still absent; thus the Act remained inoperable. Some decrees gradually emerged during the second half of 1993, others are still being prepared at the time of the writing of this paper (November, 1993).

At this point, the reader's attention is drawn to an interesting phenomenon, namely that the Act is undoubtedly somewhat short on substance. Some basic characteristics of the Hungarian legal system have made this phenomenon possible. Acts passed by Parliament, like the Telecom Act, are not supposed to be capable of functioning all by themselves. A series of governmental and ministerial decrees provide the detailed rules that are necessary for the law to function properly. This structure is supposed to guarantee increased flexibility, since government, and especially ministerial, decrees can be tailored to changing circumstances in a much more sensitive fashion than acts by Parliament. Increased flexibility, however, has a price. In this structure, it is possible that issues on which consensus cannot be achieved during debate are simply taken out of the text and referred to the governmental and ministerial decrees that are to follow the act. Hungarians sometimes refer to this phenomenon as the "emptying" of the law.

From the point of view of foreign investment, it is important to note that the Telecom Act has set up a system of telecommunications service concessions. In Hungary, only a concessionaire is allowed to provide the following services: (1) public telephone services, (2) public mobile radio telephone services, (3) nationwide public paging services, and (4) national and regional distribution and broadcasting of public radio and television programs. Other services (*e.g.,* public switched data transmission) only require a service license, while private non-radio networks within property limits require no license. Concessions are to be issued by the Minister of Transport, Communication and Water Management ("the Minister"). The Act sets the major procedural rules and refers to a ministerial decree for more detailed rulemaking on concession procedures and fees.

The Act empowers municipalities to initiate concession tender procedures. Any municipality may, at any time, request the Minister to issue a concession tender for the primary district to which the municipality belongs. The Minister has an obligation to examine the merits of the request and decide within 90 days whether to issue a tender. However, "if more than one half of the municipalities within a primary district propose the issuance of a tender and such municipalities undertake to reimburse the costs of an unsuccessful tender, the Minister shall have an obligation to issue a concession tender within 90 days after the proposal." (Section 4(3))

It is also of importance that the Act sets up a Telecommunications Fund to be used, among other things, "for the purposes set forth in Section 4 (3)" (*i.e.,* to cover the expenses of concession tenders) as well as for some other purposes, including the partial financing of local operators whose operations are not profitable due to unfavorable market conditions. (Section 36(2))

3.2 Telecom policy

Government submitted a *"Telecommunications and Broadcasting Distribution Policy Statement"* to Parliament for discussion and approval in March 1993. Since that time, the document has been in limbo. Parliament never discussed it (though it was discussed and apparently accepted by a parliamentary subcommittee) and it is unclear whether there are definitive plans to include it on Parliament's agenda. In fact, it is not even clear whether it is still being discussed or pursued by Parliament or whether the government has informally withdrawn it.

The policy document contains a great deal more than is usual for policy statements. The main reason for this is that it attempts to fill several gaps which are left by the (1) non-existence of policy research, (2) insufficient coverage of the Telecom Act, (3) absence of governmental and ministerial decrees. The policy document also tries to shape subsequent work on the decrees.[2]

The Telecom Policy specifies major telecom policy objectives in view of the text of the Telecom Act. It calls for the acceleration of development and declares that goals that have been set for 2000 should be achieved by 1997. It declares that priority must be given to the development of "profitable areas." At the same time, it also calls for cross-subsidization in order to aid the development of "underdeveloped regions."

Two markets, cellular voice and paging, are identified in the Telecom Policy as areas of limited competition. Limited competition will be achieved via the issuance by the Minister of a limited number of concessions. For either area, the Policy specifies that in addition to the already existing single supplier, two additional concessions will be issued in the near future for Pan-European systems. The two concessions will be awarded to two winners of a single tender.

Fully competitive markets are listed as follows: (1) other mobile services linked to the public telephone network; (2) satellite services (operation of ground stations, VSAT services, data network links, etc.); (3) public switched data transmission services (circuit or packet switched); (4) value added services; (5) leased line services; (6) operation of "special purpose" (dedicated) networks; (7) operation of "closed loop" networks; and (8) design and construction of licensed networks.

With respect to HTC, the Telecom Policy makes a series of important declarations. It says that the Policy is based on the assumption that HTC will remain the dominant carrier in Hungary in the long run. It identifies HTC's protected monopoly on long distance and international markets as an important policy objective. It declares that such a monopoly should be maintained until 1999. Cross-subsidization of uneconomic local network development is identified as the reason for this. It declares that "primary national interests" are attached to the successful privatization of HTC. Privatization is understood as a means not only to inject more capital into network development but also to obtain "professional expertise" necessary for HTC to become an efficient carrier.

The policy statement specifies and expands the general framework that is set up for local service provisioning in the Telecom Act. First, it states that, on account of economies of scale, local networks must not be smaller than a primary area or a designated local network area in Budapest. It does not say why the firm size based on a primary area would be sufficiently large for local carriers. Second, it allows for competition in local service provisioning, wher-

---

[2] For example, it would not normally be the task of a policy statement to assemble a complete list of governmental and ministerial decrees to be issued or to quantify specific annual service goals for telecom development for the period 1993 to 1996.

ever entry is judged economic. HTC would have a National Concession, which would allow HTC to provide local services *anywhere* in the country. Competitive entry by non-HTC suppliers would take place via the Minister issuing "second concessions." *De facto* monopolies may be formed but they would not be legally protected. Third, it declares that any *de facto* monopoly supplier has an obligation to serve (last resort supplier). Wherever there is competition between HTC and a non-HTC concessionaire, the latter would have last resort supplier obligations. Wherever there is no other supplier, HTC would be the last resort supplier. HTC, however, would have no obligation to serve the uneconomic parts of service areas in which there is a non-HTC supplier. Furthermore, HTC may freely withdraw from any area after entry by another concessionaire. Fourth, "second concessionaire" local suppliers might be interconnected with one another exclusively through the HTC long distance network, and must not provide long distance or international services. Bypass is not allowed.

The Telecom Policy is Janus-faced. It has a progressive face which expresses significant evolution in Hungarian thinking about industry structure (especially monopoly and competition). It also has a retrograde face in that its underlying philosophy still bears the stamp of central planning.

Most importantly, the Policy fails to rely on market (as opposed to administrative-bureaucratic) solutions or to understand how a market economy is supposed to work in telecommunications. There is no sign of a recognition that suppliers are to be run in the future as normal business enterprises. Running them in such a manner poses certain requirements for entrepreneurial freedom which have not been fulfilled, and the creation of freedom requires policy action. Expressions such as "profit," "profit incentive" or "profit maximization" are studiously avoided throughout the text. In contrast, it is mentioned as desirable that authorities should be able to redistribute "the greatest possible portion" of corporate profits. A careful reading of the Policy reveals that carriers are supposed to be severely limited in their freedom to make some of the most important types of business decision. Development targets are to be set by the authority (in the old-fashioned quantitative way; *e.g.,* as annual increases in the number of main lines) and will be written into concession contracts. Pricing decisions which, under normal circumstances, form and reflect corporate strategy and market behavior are not mentioned at all. Regulation is mentioned; however, even the most fundamental regulatory issues (type of agency, type of regulatory constraint(s), incentives, major principles, rules and procedures, etc.) remain undetermined. The philosophy that seems to lie beneath all this is that corporate freedom to make fundamental business decisions is not a right but a privilege, which is created solely via being granted by the state authority.

Unfortunately, the Telecom Policy does not deal with foreign investment in Hungarian telecommunications in an organized and systematic manner. The unavoidable conclusion is that, as of now, Hungary has no explicit policy in this regard. Various references to foreign investment do appear throughout the policy document. These are useful in revealing some of the current Hungarian thinking, and will likely remain the only guidance for interested foreign investors for some time.

Ministerial action subsequent to the assembly of the Telecom Policy has made changes with respect to the nature and structure of both HTC and local service markets. These changes are dealt with in Sections 5 and 6 below.

3.3 Regulatory institutions and procedures

The Telecom Policy contains a description of the system of regulatory institutions. It consists of Parliament, the government, the Minister, and the Telecommunications Inspectorate. Parliament passes laws with respect to telecommunications. Its telecommunications and privatization subcommittees may discuss issues of regulation. The government issues decrees with respect to telecommunications. Its Economic Cabinet may discuss issues of regulation. The regulator of Hungarian telecommunications is the Minister of Transport, Communication and Water Management. The Minister issues decrees and concessions. The creation of a politically independent or semi-independent regulatory agency is not considered in Hungary at this time. The Telecommunications Inspectorate has largely technical regulatory functions such as licensing or frequency distribution.

The work of regulatory institutions is aided by a number of regular and *ad hoc* committees. These carry out a wide variety of functions such as arbitration, consumer representation and even policy research.

3.4 Ministerial and government decrees

The Telecom Policy specifies eighteen governmental and ministerial decrees to be issued in order to bring the Telecom Act into full force. Government decrees are to be issued on closed networks (defense and the like); network interconnection; subscription contracts; telegraph service; the Telecommunications Inspectorate; service licenses; penalties; and information supply obligations. Ministerial decrees are due on local service concession procedures and fees; quality control; subscriber protection; technical specifications and standards; equipment and network licenses; penalties; rights of way; information supply obligations and fees; financing; entry fees (contributions); and prices (including regulation). The Telecom Fund and a few other matters require further legislative action.

3.5 Price cap regulation

The regulatory regime is described in a ministerial decree on tariffs and tariff regulation. It is a pure price cap regime in which three disaggregated and an aggregate price cap are formulated. Initial tariffs are set in a tariff table for January 1, 1994. Beginning in 1995, for any calendar year (t), the *ex ante* aggregate annual price cap is equal to a lagged (from June (t-2) to June (t-1)) value of the observed proportionate change in an aggregate domestic producer price index (PPI), as reported by the Central Statistical Bureau. There is no explicit productivity offset in the formula. The price cap also allows the carrier to pass on to their customers (in the form of tariff increases) some of the cost impact of extraordinarily large devaluations of the forint, the Hungarian currency.

Disaggregated price caps are set for (a) subscription fees, (b) local and short-haul (Zone I) domestic long distance tariffs, (c) all other regulated service (long-haul domestic (Zones II and III) and all international) tariffs. For a number of years, the disaggregated price caps for subscription fees, and for local and Zone I tariffs, are allowed to be higher than the aggregate price cap (by multipliers of 1.05 and 1.07, respectively). At the same time, the third disaggregated price cap is constrained to be lower than the aggregate price cap by a factor of 0.96. The multiplicative factors are determined in the above described manner in order to regulate the speed of rate rebalancing.

Carriers are allowed to alter their tariffs twice a year: on January 1 and at an unspecified date. Compliance with the price cap is to be verified once a year with the aid of Laspeyres price indices. It is not entirely clear what happens if the carrier is found to have violated the price cap.

The price cap decree is a major step forward in the regulation of Hungarian telecommunications. It represents a departure from the 40-year practice of individual price determination by the state authority during the communist era. It paves the way towards the creation of normal market conditions by allowing service prices to carry market signals. It gives regulated carriers a certain degree of freedom in setting service prices. First, the carrier is entirely free to set prices as long as no price cap is violated. Second, it is allowed to postpone rate increases to the following year. Third, some (though perhaps insufficient) "room" is created for rate rebalancing. Fourth, the carrier may alter not only the nominal price but also the price form. For example, time-of-the-day rate periods may be altered and optional pricing schemes may be introduced.

Significant weaknesses exist in the following forms: (1) low initial rates and insufficient allowed rate increases for the next few years, (2) unnecessary restrictions of corporate freedom to price, (3) "white spots," gaps and inconsistencies, (4) a general lack of detail and elaboration with respect to procedures, (5) some technical errors with respect to the calculation and use of index numbers. The initial tariffs for 1994 appear low in the sense that they do not allow HTC to earn sufficient profits to finance network and service development according to its existing development program. The PPI is seriously downward-biased. It does not represent the actual factor inputs and, thus, does not reflect the true inflationary pressures on HTC. Labor inputs are not represented at all, capital inputs are poorly represented because imports are voluminous, while materials and other inputs (*e.g.*, energy) are represented but with distorted weights. Observations and forecasts suggest that the PPI is systematically 4 to 6 percentage points lower than the consumer price index (CPI), hence the non-representation of labor is the source of a significant downward bias. The exclusion of the productivity offset from the formula may compensate for the bias in later years but not during the next several years, when HTC's massive development program will temporarily strongly repress productivity gains. Besides, it is generally inappropriate to represent expected productivity gains by an index bias, since nothing seems to ensure their equality.

Corporate freedom to price is unnecessarily restricted by: (a) *ex ante*, as opposed to *ex post*, price caps, (b) too many price caps, (c) limited opportunity to alter prices. Pre-setting price caps for entire calendar years brings no identifiable benefit but raises difficulties due to the presence of forecasting errors in price caps. *Ex post* verification of compliance after the end of the year, based on actual index values, would be simple, efficient, and adequate. Four price caps are too many. The aggregate price cap is entirely unnecessary and will cause difficulties because, due to unexpected weights, it will be possible for the carrier to meet all disaggregated price caps and violate the aggregate cap at the same time. Since, for various reasons, quick rate rebalancing is in the interest of the carrier, the disaggregated price caps will slow down rather than force or accelerate rebalancing. The multiplier for the price cap of subscription fees (1.05) is lower than that for the price cap of local and Zone I calls (1.07). This relationship will slow down rebalancing by preventing local carriers and HTC from raising the relative price of access to its economically desirable level. It will also prevent carriers from increasing the currently very low volumes of traffic per subscriber by keeping prices of usage low. The multipliers do not seem to take into account technology-induced changes in the cost structure in the future. For this reason, they do not deviate sufficiently from 1.0. It may well happen that in the next 6 to 8 years the proportions of directly attributable long-run incremental costs per unit of output will change as much, or nearly as much, as the maximum allowed change in relative prices. The creation of roughly cost-based prices may not be achieved by the time competition emerges in Hungarian telecommunications. Uneconomic entry may result in long distance and international service markets. It would be adequate to deploy two price caps: one for subscription fees, local and Zone I long distance tariffs, and another one for all other regulated service tariffs.

The discussion of other significant weaknesses is beyond the scope of this paper. It is important to mention, however, that the price cap regime will be imposed by the regulator; *i.e.*, the social contract nature of the price cap will not be allowed to shine through the regulator-regulated relationship.

3.6 The concession agreement as a regulatory tool

The Telecom Policy states that Hungarian telecommunications will be regulated by the Minister, principally with the aid of (1) decrees and (2) concession agreements. This implies that, in addition to their usual role in specifying the rights and obligations of the state and the concessionaire, concession agreements will play a regulatory role in Hungarian telecommunications.

This role may create some difficulties because concession agreements cannot fulfill regulatory functions very well. Concession agreements set the signing parties' rights and obligations for a very long time. Regulation, on the other hand, requires continuous monitoring of the regulated firm and corrective actions (*e.g.* price changes) when, to the extent, and in the form, such actions are necessitated by the ever changing conditions under which the regulated firm operates. Even though some automatic trigger and adjustment mechanisms can be built into concession agreements, the adjustment period cannot possibly be long enough to cover the concession period; thus, repeated renegotiations of adjustments would be required. Another problem is that pre-setting triggers and adjustments raises the need for concession agreements to contain rules with respect to regulatory procedures. This has two undesirable consequences. First, it may create overlaps, increasing the potential for inconsistency with ministerial decrees. Second, interdependence between non-overlapping rules creates interdependence between decrees and concession agreements. The concession agreement may alter the nature of regulation as defined in a decree. For example, a decree may introduce pure price cap regulation which, then, may be turned into rate of return regulation by the concession agreement if the latter introduces a rate of return based trigger mechanism for renegotiating the price cap. Another likely result is that the rather inflexible concession agreements become an obstacle for making desirable changes in the more malleable ministerial decree. For example, with a given rate of return trigger it may be impossible to change the incentive mechanism of the price cap (rules on how profit changes due to productivity gains are distributed between firm and customer) towards giving a greater efficiency incentive for the firm because such a change would "pull" the trigger. Draft concession agreements give an indication that many of the above described problems may well become reality in Hungary.

In their traditional role of setting specific service volume and quality targets, the concession agreements represent a development of the Telecom Policy in three aspects. First, arbitrary quantitative service targets are to be set only for the first few years of the 25-year concession period. Second, service volume targets are subject to the existence of sufficient market demand for the targeted volumes. Third, a real market target is introduced for access volumes: the degree of satisfaction of market demand.

4. SUMMARY OF FOREIGN INVESTMENT

There is a variety of small foreign investment projects in markets which have been declared fully competitive. Interested foreign investors range from small equipment manufacturers through small and medium sized firms which are allied with large manufacturers and service providers to giants such as Hughes, GTE or AT&T. Large and very large firms may be involved directly or indirectly (through subsidiaries and business associates). Regardless of their size and profile, few foreign investors attempt to invest in Hungary without first securing the cooperation and partnership of existing Hungarian firms. The Hungarian partners may be equipment manufacturers (such as Microsystems or Müszertechnika), service providers (HTC) or potential clients (for example, banks or insurance companies). Existing, emerging and potential Hungarian private data network owners represent the bulk of the targeted market. VSAT services (especially for banks and insurance companies) and larger private data networks have attracted the largest amounts of foreign investment so far. The number of seriously interested but not yet actually involved foreign investors is very large. This is a good indication of a significant potential for rapid growth in foreign investment in these markets.

In markets with limited competition (cellular and paging), each of the two or three competing suppliers have, or are expected to have, foreign partners. The analog cellular voice service provider Westel is a Hungarian-American joint venture. There is also foreign involvement in existing paging service provisioning. Future digital cellular as well as paging service providers will be partnerships between large individual foreign partners or consortia on the one hand and Hungarian service providers, broadcasting signal distributors or equipment manufacturers on the other.

Local network development is a particularly hot area for foreign investment. It is highly investment-intensive and, since local networks are underdeveloped and technologically obsolete, the country obviously needs a very large program of local network development. Both HTC and non-HTC groups of existing and potential developers have secured foreign participation in a variety of projects. Local network development is discussed in detail in Section 6 below.

The greatest and the only truly large-scale foreign investment opportunity is represented by the privatization of HTC.

5. PRIVATIZING HTC

The idea of finding a major strategic partner for the Hungarian national telecommunications carrier through privatization had existed as early as 1988; *i.e.,* well before the first post-communist government was formed in 1990. The new government wasted no time in embracing the idea and encouraging efforts to carry it out. As mentioned above, HTC was transformed into a temporarily 100 percent state-owned joint-stock company at the end of 1991. Its first designated state owner was the State Property Agency (SPA). For a while, this agency was exercising ownership rights on behalf of the state in conjunction with all state-owned enterprises. With a subsequent division of functions, companies which were to be fully privatized stayed with the SPA, while companies which were to remain in full or partial (majority or minority) state ownership in the long run were handed over to the newly formed State Holding

Corporation (SHC).[3] There were 162 such companies and HTC was one of them.

The privatization of HTC has been designed to take place via selling a "significant minority shareholding" (30 to 35 percent) to a foreign strategic investor. Foreign ownership is expected to result from a combination of (1) purchase of existing SHC-owned shares and (2) issuance and sale of new shares. The remaining (15 to 20 percent) equity capital would be owned by employees (via an Employee Share Ownership Plan or ESOP) and others such as municipalities, foreign and Hungarian institutional investors and Hungarian private citizens.

The privatization effort was kicked off in April 1993 (approximately two years later than originally planned) by the joint issuance by the Minister of Transport, Communications and Water Management and the President of the State Holding Corporation of the "*Concession and Sale Tender Memorandum and Notice*." As the title implies, the privatization of HTC has been tied together with the process of giving it a telecommunications service concession. There are legal reasons for this concession-cum-privatization process. According to Hungarian law concessions can only be created through awarding them to winners of concession tenders. The national concession, however, is supposed to be given to the existing national carrier, which is HTC. To allow other potential nationwide carriers to compete with HTC for the national concession would be wasteful and inefficient. The problem caused by a lack of competition is resolved within the framework of privatization. Privatization will provide the required competition and the winner will be awarded a national concession via ownership in HTC.

Echoing objectives that have been subject to much discussion during the last four or five years, the principal sectoral goals defined in the Memorandum are as follows:

- Create a modern, efficient telecommunications infrastructure that will support the economic development of Hungary.
- Improve the quality and quantity of service and provide for basic telephony services throughout Hungary within as short a time as possible.
- Support Hungary's development as a regional telecommunications center.

According to the Memorandum, privatization would serve the above mentioned goals by providing investment, technology and management expertise. More specifically:

- Through privatization, HTC will acquire additional financial resources to support the company's capital expenditure program.
- The incoming strategic partner will assist in the development and operation of the company, and support further implementation of a customer-oriented corporate culture.
- Privatization will help facilitate compliance with the European Community's and other international bodies' policies and technology standards as they relate to the telecommunications sector.

The Memorandum was sent to a fairly large number of potential foreign investors. A total of 13 (six European, four American and three Asian) investors signaled interest in investing in HTC.

---

[3] In Hungarian: Ållami Vagyonkezelö Rt. or ÅVRt.

On August 13, 1993, the Minister and the SHC jointly issued an *"Invitation to Tender"* for the national concession and to acquire significant minority ownership of at least 30 percent of HTC. The national concession is to include international, domestic long distance and local services—the latter in 29 geographic areas of Hungary, which are specified by name. This differs from the Telecom Policy, which called for an HTC concession to provide local services anywhere in the country. Other concessionaires would have been allowed to enter through the concession process and create competition to HTC in any area. The current idea is that the Minister awards exclusive concessions to HTC as well as non-HTC concessionaires, thereby creating a string of legally protected local service monopolies.

The Invitation repeats or redefines the main features of the concession. Duration is set at 25 years, 8 years of which will be exclusive; *i.e.,* HTC will be a legally protected monopoly supplier.[4] There is an option to extend the concession by 12.5 years. It is stated that HTC will remain in majority state ownership. The main financial conditions are specified. An annual concession fee of 0.1 percent of revenue will be paid by the concessionaire. The first 40 billion forints of the proceeds (the sum of the purchase price of existing and new shares, plus a lump-sum concession fee) will be reinvested in HTC. There are other requirements such as Hungarian content in the construction program. Development targets are identified as (1) at least 15.5 percent annual growth in the total number of main lines for at least the first six years of the concession period;[5] and (2) beginning in 1997, at least 90 percent of declared demand for access must be satisfied within 6 months and 100 percent within one year from application. The Invitation also sets the bid evaluation criteria and the requirements that investors must satisfy in order to become eligible to bid. Apparently, the eligible bidder must be (alone or in consortium with) a telecommunications carrier with total annual revenue of at least $1 billion (US), at least 1 million subscribers, and substantial recent experience in telecommunications infrastructural development.

In September 1993, interested investors signed a confidentiality agreement and purchased an *Information Memorandum* which contains a large volume of information on HTC and Hungarian telecommunications in general. *Due diligence* took place in the same month. The investors' inquiries were facilitated by a series of presentations by HTC management, ministerial and SHC officials as well as by consultations with HTC experts and the use of a "data room." Preliminary bids were submitted on November 5, 1993. There were two single-firm bidders (STET of Italy and Telefonica of Spain) and two consortia (France Telecom teamed up with US West, while Deutsche Bundespost Telekom, Ameritech and Cable & Wireless formed another consortium).

An *Evaluation Committee* has been formed by the Minister and the SHC. This committee will evaluate the preliminary bids, form a short list of bidders (who then present their final bids), evaluate the final bids, and make recommendations to the Minister and the SHC as to who the winner should be. The Minister and the Board of Directors of the SHC make the decision, the Minister awards the concession, and the SHC signs the appropriate agreements concerning the sale of shares.

According to a highly ambitious timetable, an exchange of contracts between the winner and the Hungarian government would take place before December 31, 1993. Certain details

[4] Implying that competition would be introduced in Hungary at an earlier date than in the less developed countries of the European Community.

[5] This ambitious target may prevent the rebalancing of rates between access and usage if, in order to prevent market demand for access from shrinking, the regulator keeps the price cap of access low.

would be ironed out later, but the Minister would award the national concession to the privatized HTC before April 30, 1994—the legal deadline.

The privatization of HTC does not bring investment opportunities to an end. Further—Hungarian or foreign—investments may take place in the future via the purchase of existing or newly issued HTC shares. Future purchases require "room" in the equity structure of HTC. Such room is likely to be created by changing the current legal requirement of majority state ownership. Bi-annual reviews of the ownership structure are scheduled. It is reasonable to expect the quick elimination of the majority state ownership requirement, and the reduction of state ownership to 25 percent plus one share. In the long run, the Hungarian state may well decide to leave the telecommunications services sector entirely and ensure the observance of national interests by means other than ownership.

6. INVESTING IN BASIC LOCAL SERVICE SUPPLY

Historically, a single nationwide supplier provided local telecommunications services in Hungary. This supplier was the Post Office, then, after divestiture, HTC. With competitive franchise bidding for specific local service areas, and eventually perhaps some *de facto* competition among territorially overlapping suppliers, multiple suppliers are expected to emerge. This makes investment conditions particularly complex in this field.

6.1 Structure, concessions and geography

The Telecom Act stipulates that basic local services can only be provided in Hungary by concessionaires. Concessions are awarded to winning bidders of concession tenders. As mentioned earlier, winning a tender is the only way for a concession to be created.

The geographic unit for local service concessions is the primary area—the area served by a primary switch. A primary switch is the customer-side end-node of the public network. There are 54 primary areas. One, Budapest, deserves special consideration for its size (20 percent of the population), customer density and market structure (high demand for business and international services). Budapest is part of HTC's planned national concession. Local service concessions may contain multiple primary areas.

Low penetration makes local service markets small in Hungary—even in Budapest. It will take substantial growth in most geographical areas for the local service markets to become large enough to support two or more service suppliers. In some parts of the country, the market does not seem large enough to support even one. Small markets and very high costs of entry will make monopoly the dominant local service market form in Hungary for quite some time to come. The future, however, will differ from the past in some important ways. First, the wired network monopoly will have to endure competition from not one but at least three wireless suppliers. Second, the single HTC monopoly of the past will be replaced with a mosaic of geographic monopolies which will be owned by HTC as well as by a number of other concessionaires. Third, the new local service monopolies will be stand-alone local operations. The only exception is integrated HTC itself, where accounting separation will be introduced in order to resolve issues of interconnection and cross-subsidization. Fourth, a variety of corporate forms may come into existence. HTC as well as non-HTC concessionaires may form joint ventures with municipalities and various other domestic and foreign investors in their effort to increase capital availability. Development may be separated from operations.

When this is the case, newly constructed equipment may be handed over to the operator via various arrangements (purchase, rental or leasing). Return on investment may also be generated in a variety of ways, including revenue sharing arrangements. Fifth, the new local monopoly will be protected for a limited time only. Competition from wireless operators and the increasing threat of entry by wireline service providers are expected to have an efficiency-increasing impact on incumbent operators.

One feature which could increase the growth rate of local service supply and enhance the market structure is HTC private franchises. The scheme whereby HTC would be given the right to create a chain of HTC local franchises via competitive bidding was considered a few years ago. Eventually, it was discarded.

6.2 Local competition vs. HTC privatization

Due to the expected emergence of non-HTC local suppliers, HTC's role in the supply of local services and, of course, the geographical boundaries of its serving area became temporarily indeterminate. This indeterminacy has been a source of considerable difficulties in the process of HTC's privatization because it prevented the Hungarian state from offering a precise definition of the business it intends to privatize.

One solution would be the postponement of the privatization of HTC until a new local market structure develops. This, however, would limit HTC's chances to obtain local concessions, since less funds would be available for bidding before than after privatization. Alternatively, the tender process for local concessions may be postponed until HTC's privatization is completed. This solution would be advantageous for the local tender process, since the privatized national concessionaire would be in a position to bid for a maximum number of local concessions, thereby increasing the degree of competition in the process. Unfortunately, there is no time for either solution, since the Telecom Act has set a very tight deadline (April 30, 1994) for awarding all telecommunications concessions. Thus, the government has been seeking a solution which would allow for the approximately simultaneous creation of concessions for HTC and non-HTC local service providers.

In April 1993, the Minister requested municipal initiation of concession tender procedures for exclusive concessions. More than two dozen municipalities (mainly from Budapest's less developed suburbs, the city's immediate vicinity and the country's less developed areas) responded to the Minister's request. In July 1993, the Minister announced that he intended to issue a total of 25 tenders for exclusive local service concessions. The 25 primary areas were identified by name. Subsequently, tenders were issued. Interested investors must submit their bids by January 3, 1994.

From the point of view of local concessions, the near-simultaneity of the two processes is likely to result in a radical reduction of the degree of competition.

From the point of view of HTC's privatization, four fundamental uncertainties remain in the scheme. First, the fate of HTC personnel and facilities in the 25 primary areas is currently unclear—and will remain so beyond the end of HTC's privatization process. Second, HTC may bid for and win some of the 25 concessions. If these are not considered in HTC's definition the foreign investor will obtain a share in their ownership at zero price. In order to define what is to be privatized, it must be known not only where HTC intends to bid but also where HTC will, or is likely to, win the concession. Third, areas for which there are no bidders or acceptable bids will revert to HTC. The identity of such areas will not be known until after the conclusion of HTC's privatization. From the point of view of the foreign investor, these areas will be "free." However, at least some of them may be uneconomic service areas.

Potential competition may lead to another complication. Unless all primary areas become exclusive concessions, competing local service suppliers may emerge and reduce the incumbent's market share from 100 percent to some lower level. Given the primary areas' small size (in terms of subscribers and traffic), this may reduce demand for investment to a significant extent. Even if all concessions are declared exclusive, some doubt remains whether the Minister has the right to issue exclusive local concessions at all. Some legal experts argue that the Telecom Act makes it compulsory for the Minister to issue a second tender if the majority of an area's municipalities request it, regardless of whether the first tender is qualified as exclusive.

6.3 Demand for and supply of investment

Who are the potential foreign investors? The following categories come easily to mind:

1. small and medium sized local carriers
2. international and long distance carriers
3. large equipment manufacturers
4. international construction companies
5. specialized telecommunications investment groups

The most obvious investment opportunity is to invest in specific individual local telephone companies. The investor may want to become the developer and/or operator of a single telephone company or multiple telephone companies. Various classes of investors may be interested in this kind of opportunity for a number of possible reasons. Small and medium sized foreign local telephone companies may view Hungarian local network development and operation as activities in which they have significant competitive advantages due to their existing expertise and experience (use of existing organization, employees, processes, software, etc.). Such advantages are not available to other classes of investors. International long distance carriers may pursue Hungarian local network investment opportunities to improve their geographical positioning. Interest by U.S. Sprint International in Hungarian local network development in the Szeged region is an illustration of this point. Large equipment manufacturers of any profile may want to establish a market presence in Hungary or close to the potentially large Eastern European markets to increase their sales of equipment. Construction companies may pursue network development projects very much for the same reasons. When traditional construction contracts are not available but equity participation or revenue sharing promises an attractive return on investment, construction companies may well decide to expand the range of their business arrangements in new directions. Telecommunications investment groups may be interested in investing in Hungarian local networks for a great variety of reasons.

It is likely that significant economies of scale and the potential for a variety of attractive business strategies will provide a strong investor incentive for multiple telco construction and/or operation. Scale economies and business strategies may also play a significant role in investment in ancillary activities such as consulting, financing, planning and design, management services, etc. The following kinds of support agencies may be formed: (1) financial agencies; (2) consulting agencies; (3) planning and design bureaus; (4) construction companies (main and subcontractors); and (5) integrated investment agencies.

Financial agencies deserve some elaboration. They would finance multiple local network development efforts. They could take advantage of scale and scope economies in the process of finding and distributing funds. Financial agencies would raise debt and equity from domestic and foreign sources. Foreign sources would include multinational banks, investor groups as well as the above listed five categories of potential foreign investors. Domestic sources might include banks, other institutional investors, the above listed consulting agencies, planning and design institutions, construction companies, or even income from revenue sharing arrangements (B.O.T. or B.O.O.T. schemes).[6] A full range of financing arrangements may be used from straightforward loans to project financing in various arrangements. With integrated construction activities, agency ownership of the newly constructed facilities would make a variety of arrangements (sale, rental, leasing, equity contribution, operation by contract, etc.) possible. The main advantages of relying on financing agencies are: high growth potential; easy and quick entry for concessionaires; various equity options; tax advantages; risk distribution; loan guarantee by government or municipalities; exit strategy options (*e.g.,* transfer to HTC); and several expansion paths to further investment (HTC and non-HTC development projects, non-local projects, non-telecommunications projects).

Consulting, planning and design as well as construction activities could be carried out by small or medium sized foreign telecommunications carriers, equipment manufacturers, construction companies or their consortia.

Finally, some or all ancillary activities might be combined into a single integrated development agency. Full integration would make one-stop shopping available for local service concessionaires in need of development capability.

The formation of large and sophisticated support agencies would be beneficial for local network development activities in Hungary. Such agencies could do much to remedy the relative technological backwardness that is expected by some to result from heavy municipal and other local participation in the development process. It is thus unfortunate that the current absence of a legal-regulatory system at least temporarily prevents support agencies from being formed.

[6] B.O.T.= Build-Own-Transfer; B.O.O.T.= Build-Own-Operate-Transfer. The builder of facilities transfers ownership to an operator in exchange for a certain share of operating or net revenues for a certain period of time.

Table 6.1
Some actual and potential (interested) foreign investors[7]

Investor	Invested or interested in: Company	Location	Number of lines
Alcatel Austria	Dunatel	Dunaújváros	90-135,000
FTC (Fail Tele-communication Corp.)	see IFC		
Giltek Telecommu-nications, Israel	Digitel 2001	Fót	40-49,000 (plus CATV)
HTCC (Hungarian Telephone & Cable Corporation)	Elsö Hazai	Nagykovácsi	600-1,000 (CATV: 600)
	Pilistáv	Solymár	6-10,000
	Rába-com	Sárvár	15-40,000
IFC (International Finance Corp.)	Biharcom	Berettyóújfalu	25-55,000
Swisscom	Jászcom	Jászberény	15-20,000
US Telecom East	Budatáv	Budapest	90-150,000
	Valitáv	Bicske	65-130,000
	Bakonytel	Zirc	25-30,000

The Telecom Act allows foreign investors to own majority interests in local telecommunications companies in Hungary. The opportunity of majority ownership has created additional interest among foreign investors to invest in local network development and service provisioning. However, most foreign investors have so far been unable to take advantage of the opportunity. Minority foreign ownership is far more frequent than majority foreign ownership. Instead of forming new ventures, foreign investors typically try to join existing development companies, founded by Hungarian investors (typically with heavy municipal participation). A new foreign partner may buy up existing shares of Hungarian partners or may appear with additional capital.

[7] The following information may be quickly outdated. Any use of information contained in this section should be made with utmost care, including further verification.

7. FOREIGN INVESTMENT IN CELLULAR SERVICES

The current market structure is simple. Westel Ltd. operates as a monopoly supplier of cellular services. Competition will, however, occur with the emergence of two digital cellular service providers in 1994.

7.1 Westel Ltd.

Westel was established in December 1989 to offer analog cellular voice services (NMT 450) and also to sell cellular equipment and accessories. It is a joint venture of HTC (51 percent ownership) and U.S. West (49 percent ownership). It began operations in February 1990. The parties have invested a total of $85 million in the venture so far. The system has capacity for 50,000 to 100,000 subscribers (depending on traffic characteristics). There are currently approximately 30,000 subscribers. Westel's geographical coverage of Hungary will be 100 percent by the end of 1993. The service is expected to develop further in the future, even after the introduction of digital cellular services, as prices drop. Cellular service is already used rather extensively to serve underdeveloped areas, where wireline service is currently unavailable. Service ideas such as card-operated mobile public phones in booths are being investigated.

7.2 Digital services (GSM)

A tender was issued by the Minister of Transport, Communications and Water Management late in 1992 for the so-called Pan-European or GSM system of digital cellular service. The tender generated considerable interest among potential investors. The pre-bid conference was attended by 50 participants, 34 of whom purchased the tender document. The following five tenders were opened on May 28, 1993: (1) the Vodaphone group; (2) the Mobile Phone Hungaria group, consisting of Cable & Wireless, Singapore Telecom and Mannesmann Eurocom; (3) the Pannon GSM group, consisting of the Danish, Dutch, Finnish, Norwegian and Swedish telephone companies and MOL, the Hungarian oil and gas company; (4) HTC and U.S. West; (5) the DBFH group, consisting of Deutsche Bundespost Telekom, British Telecom, France Telecom and Antenna Hungaria, the Hungarian television signal distributor. Two winners were announced on August 26, 1993. They are the Pannon GSM group and the HTC-US West duo. According to the latest announcement, no additional permit will be issued to cellular service providers during the next fifteen years. The GSM concessionaires are currently expected to achieve 90 percent geographical coverage within one year from the start of operations.

With the emergence of GSM service providers, the Hungarian cellular market will undergo several fundamental changes. First, the current monopoly will give way to a competitive market structure in which one analog and two digital cellular service providers will compete with each other and with wireline service for subscription and traffic. Second, the market will undergo rather radical expansion. Total required investment into a fully built Hungarian digital cellular system is expected to be between $100 million and $300 million, putting the break-even point in the neighborhood of 100,000 subscribers. Each of the two GSM winners is expected to invest anywhere between $100 and $200 million into the business. The survival of three competitors does not appear possible, unless the total number of cellular subscribers exceeds 250,000. The eventual total system capacity may be as high as 2 million subscribers. To understand the significance of these figures, one should recall that in mid-1993, the total number of telephone lines in Hungary was 1.2 to 1.3 million. By the turn of the century, cellular services will have a very high market share. Third, GSM will open the door to a wide range of data services and significant cellular voice service enhancement in the future.

8. FOREIGN INVESTMENT IN PAGING

There is 41 percent foreign (French) ownership in *Operator Ltd.*, which is the existing paging venture of Antenna Hungaria. It uses the USW (Ultra Short Wave) band of Radio Petofi (one of the nationwide radio stations). It has good geographical coverage (approximately 80 percent of the country) and more than 10,000 subscribers. Prices are generally judged to be very reasonable. Future growth potential in the presence of competition is judged to be moderate.

Work is currently under way on a tender for nationwide paging services to use the Pan-European ERMES (European Radio MEssage System). The European frequency allocation scheme gives Hungary a total of four frequency bands to use for paging. Two of these will be used by two competing nationwide service networks, while the other two will serve a multitude of regional and local networks.

The tender procedure for nationwide paging has been modeled after the GSM tender. The bid evaluation criteria are very similar in nature. The most important criteria are low service prices, rapid introduction, large geographical coverage, maximum use of Hungarian-made equipment, and transparent data transmission capability. Each supplier is expected to cover at least 75 percent of the country's population at the end of the second service year.

The paging tender is to be issued in November 1993. Bidders are requested to submit their bids in the spring of 1994. Heavy bidding is expected by foreign investors. Evaluation will be carried out in one or two phases, depending on the number of bids. Two winners are scheduled to be announced in June 1994.

Local and regional paging service is now being offered by *Radiocontact Ltd.* One of its owners is HTC. Further local and regional operators will emerge when concessions are issued at later, hitherto undetermined, dates. Local and regional paging is expected to attract some foreign investment.

9. FOREIGN INVESTMENT IN PROFESSIONAL SERVICES

There are several areas of unregulated professional service activities which are wide open to foreign entry, either in the form of consulting contracts or as joint ventures to provide professional services. These have been briefly mentioned in connection with local network development in Section 6. The most promising ones are: (1) financial services, (2) consulting services, (3) network planning and design.

HTC was the first Hungarian corporation to go beyond foreign consulting contracts by setting up a joint venture to deliver professional services. It was hoped that a permanent organizational arrangement with clear strategic objectives would do more to increase the volume and, more importantly, the quality of its financial analysis and financing. After extensive search and bargaining, *Investel Ltd.* was founded in 1991 by HTC, Irish Telecommunications plc (ITI) and three Hungarian banks. HTC owns 75 percent and Irish Telecom owns 20 percent of its total equity. Investel has three main functions: (1) raise money for HTC; (2) manage the debts of HTC; (3) do various types of financial and other analyses for HTC in support of HTC's own effort to evaluate investment opportunities. In connection with all three, Investel channels in Irish expertise in financial analysis and financing makes it available to HTC, carries out analytical and financing tasks as required, transfers its own expertise to the HTC organization, and continuously develops new and more advanced methods and analyses, including project evaluation and various other engineering economic analyses.

10. ENCOURAGED OR DISCOURAGED?

Having read the above account of social, political and economic attractions, it may come as a surprise to the reader that, as of the end of 1993, the total value of foreign direct investment in Hungarian telecommunications is rather small. The great Hungarian promise has so far not been fulfilled in telecommunications. Some argue that the promise itself has diminished to a considerable degree.

Foreign interest in investing in Hungarian telecommunications has undoubtedly diminished to a certain degree during the last year or so. There are several reasons for this.

First, there are external (non-Hungarian) reasons which are related to worldwide phenomena in the economy in general or specifically in telecommunications. Worldwide recession has reduced the cashflow of many potential telecommunications investors; thus, the funds to invest have diminished. At the same time, what appeared just a few years ago as an exciting but somewhat remote and futuristic domestic investment opportunity in future network and service developments (advanced high speed switched data, image, video and multimedia services, advanced intelligent networks, broadband ISDN) has quickly become a must for large carriers for survival in the long run. National "network of the future" programs such as President Clinton's electronic superhighway or the equivalent Japanese program are capable of consuming tens, or even hundreds, of billions of dollars—far more than what even the largest carriers such as AT&T and NTT can generate in the near future. The network of the future with its unprecedented appetite for funds now begins to compete with privatization for the available scarce funds. The international privatization scene itself has also changed substantially in the last year or two. A series of attractive competing telecommunications privatization efforts emerged all over the world. They all compete with one another. The supply of telco privatization begins to exceed the demand for it.

Second, there are internal Hungarian, or at least specifically Central European, reasons for diminishing foreign interest in investing into Hungarian telecommunications. It is important that the initial euphoria about Central Europe gave way to sober consideration of real business opportunities. Furthermore, not only the investors' approach but also the Central European economic reality has changed as the national economies of the region have become increasingly seriously plagued by problems such as foreign debt, state budget deficit, declining domestic production, disappearing export opportunities, bankruptcies, rapidly growing unemployment, and high inflation.

Written and verbal declarations by representatives of foreign corporations, newspaper articles, the closing of some Budapest offices of foreign corporations and other evidence seem to indicate rather convincingly that the above mentioned negative conditions and events have, indeed, resulted in a somewhat diminished foreign investor interest.

Fortunately, unlike in some other countries of the region, the economic difficulties were not accompanied by an unattractive political climate in Hungary. Government has been stable (though decreasingly so) and the evolution of democratic political institutions has proceeded in a fundamentally healthy and orderly way. Even though her long-held advantage seems to be fading somewhat, Hungary still compares favorably to other countries of the Central European region.

The current low level of foreign investment in Hungarian telecommunications cannot be explained by diminished investor interest alone. On the contrary, it can be concluded from responses to the most recent GSM and HTC tenders that substantial foreign interest still exists. This interest, however, finds it difficult to materialize in the form of actual investments because delays and other imperfections in the Hungarian political process have prevented the creation of an adequate legal-regulatory framework that would have allowed foreign investments to take place. The failure of Hungarian political institutions and processes to deliver the legal-regulatory framework for telecommunications in time and of an acceptably good quality is the single most important reason for the current low level of foreign investment in the telecommunications industry of Hungary. In the absence of a legal framework, entry into the voluminous core activities of telecommunications has in fact been barred. Only the smaller, though rather important, non-core markets such as analog cellular voice or VSAT data services have been open to entry by foreign investors.

This situation will undergo fundamental changes in 1994 with the successful completion of the GSM tender, the privatization of HTC, and the awarding of local service concessions in 25 local service areas.

11. SUMMARY AND CONCLUSIONS

With the collapse of the communist regime in 1989-90, Hungary became an attractive country for foreign investment in telecommunications as well as in other industries. Geographic, demographic, economic and political conditions were generally attractive in Hungary, comparing favorably with other countries in the Central and East European region.

Several important but smaller foreign investment projects have taken place in areas such as cellular voice or satellite data services. However, until the end of 1993, the core activities of telecommunications remained a forbidden area for foreign investment because the legal-regulatory framework was very slow to emerge.

The Telecom Act was passed by Parliament approximately two years behind schedule in November 1992. The Act introduces the institution of service concessions into telecommunications, specifies the services that can be supplied only by concessionaires, and defines the role of municipalities in concession tender procedures. The Act does not offer a complete legal framework and leaves important areas fully or partially unregulated. The Telecom Act is made operable with the aid of a large number of government and ministerial decrees, whose issuance has been very slow.

The Telecom Policy defines the structure of telecommunications service markets. The Policy is strongly in favor of competition. It would allow monopoly to survive, for a few years, only in the provisioning of long-haul domestic and international calls. The Policy also has retrograde features in that it does not grant adequate economic freedom to carriers and focuses on administrative-bureaucratic, as opposed to market-based, solutions.

Regulation is defined in a ministerial decree. Price cap regulation is introduced. It represents a departure from authoritative individual price setting by state authorities during the communist regime. It has several attractive features but, at the same time, it is plagued by a series of weaknesses and inconsistencies.

The emergence of the legal-regulatory framework has made it possible to initiate the privatization of HTC. A tender process is under way, and expected to come to a conclusion before the end of 1993. The privatized HTC will have a national concession to offer long-haul domestic and international services in the entire territory of Hungary, and local and short-haul domestic services in 29 pre-specified local calling areas. Also under way is the tender process to award further local service concessions in 25 local service areas of Hungary. Two concessions were awarded in 1993 for digital cellular (GSM) services and two concessions will be handed out in 1994 for nationwide ERMES paging services.

Foreign interest in these investment opportunities has been fairly high. Incoming foreign investment is expected to open up further investment opportunities in the future. Despite significant expected inflow of foreign capital, the investment climate needs improvement in Hungary. Foreign interest has diminished somewhat in the last year or so as worldwide recession has reduced available cashflows. Several international privatization projects have emerged and "networks of the future" have begun to compete for the scarce resources. Hungary's ability to attract and accommodate foreign investment could be greatly enhanced by further improvements in the telecommunications sector's legal-regulatory framework.

The Race to European Eminence
E. Bohlin and O. Granstrand (Editors)

Options for the Reduction of Deficiencies in Telecommunications in the Czech Republic

Johann Cas

Technology Assessment Unit, Austrian Academy of Sciences
Postgasse 7/4/3, A-1010 Vienna, Austria

This paper summarises the preliminary results of a current research project undertaken by the Austrian Academy of Sciences in collaboration with Czech partners which analyses options for the rebuilding of the Czech telecommunications infrastructure. It provides a discussion of the interrelationships between regulatory issues and possibilities of financing investments in infrastructure, including foreign direct investments. It also aims to assess the appropriateness and long-term consequences for the Czech Republic of different measures proposed.

1. INTRODUCTION

The rebuilding of the telecommunications sector will play a key role in economic restructuring and development of the Central and Eastern European Countries. The success of the transition from centrally planned decision making to a market economy will strongly depend on the availability of efficient means to transmit and exchange information. Apart from the enabling features of telecommunications for a modern economy, several studies show that there is a close relationship between economic development and the level of telecommunications infrastructure. Economic development is virtually impossible without an adequate telecommunications infrastructure and telecommunication services that meet the requirements of day-to-day business. The social and political importance of telecommunication is of equal weight; it enables better contacts between people, it provides better access to information of all kinds, and it improves the access to and efficiency of public and private services in general.

Accordingly, the rebuilding of telecommunications is of high priority. This finds its expression in the objective of the C&EE countries to reach penetration rates of 40 main lines per hundred inhabitants within the next 10 to 15 years, a rate roughly equivalent to the current OECD average. A very complex set of decisions and measures in various areas is required to achieve this goal.

This paper is based on a current comprehensive research project that the Technology Assessment Unit undertakes jointly with Czech partners. The purpose of this project is to analyse feasible options for the rebuilding of the telecommunications sector in C&EE, taking the Czech Republic as a case study.

The Czech Republic faces the problem of rapid modernisation and extension of telecommunications under severe economic restrictions. At the same time it also needs to find a new regulatory and organisational frame for telecommunications. The problem of modernisation and closing of technological gaps is not restricted to infrastructure. It has to take place also in the national telecommunications industry if substantial losses in jobs and an almost complete dependence on foreign suppliers are to be avoided. Although a considerable gap in the level of technology exists between the Czech Republic and Western countries, the telecommunications industry is not the worst industry in this respect. A large domestic demand and a base of skilled labour force make the telecommunications industry a potential aspirant for establishing a high-tech industry with spill-over effects for the whole economy. The effectiveness of the CoCom restrictions in the past may also contribute to the wish for more independence from foreign producers.

The on-going study includes the rebuilding of the telecommunications infrastructure, the development of a regulatory framework as well as some aspects of telecommunications industry. It concentrates on the various dependencies within and among these three spheres, aiming at the identification of consistent strategies of development. In a second phase of this project, which is planned to start at the beginning of next year, the results will be implemented in a computerised decision support model, the aim of this model being to support decision makers in their choice between strategies according to their preferences.

This paper summarises some of the preliminary results of the above-mentioned research project. It aims at providing a discussion on the interrelationships between regulatory issues and possibilities of financing investments in infrastructure, including foreign direct investments. It also aims to assess the appropriateness and long-term consequences for the Czech Republic of different measures proposed.

2. OUTLINING THE BACKGROUND

2.1 Economic development

Czechoslovakia was once the economic backbone of the Habsburg Empire. Before World War II it belonged to the 15 most developed countries in the world. It survived the war in relatively good condition and was by 1948 almost back to pre-war production levels. Four decades of central economic planning opened a considerable gap in the level of infrastructure and of economic performance.

With a total GDP of 44.5 billion US$ as of 1990, the GDP per capita (approximately 2,800 US$) in the CSFR amounted to about one sixth of the average value of the OECD. After substantial decreases in GDP in recent years (-2% from 1989 to 1990, -14% from 1990 to 1991, and -7% on average in 1992) the economy in the Czech Republic stabilised in the second half of 1992. Whether a continuous recovery can be maintained in the near future is, however, still very uncertain. The effects of privatisation on the economy as well as the consequences of the division of CSFR will at least in the short term lead to further falls in output and increases in unemployment. In addition, the current recession in Western economies makes it more difficult to get access to new markets that could compensate for the breakdown of COMECON. Therefore the prospects for the economy of C&EE countries to reach Western levels are expected to take decades. However, the economic development in the past as well as the current achievements in economic stabilisation place the Czech Republic among the most promising candidates for a relatively quick recovery.

2.2 Telecommunications infrastructure

In 1992 a penetration rate of 16.74 main lines per 100 inhabitants was reached, a rate which means a top position in C&EE (except Bulgaria), but is still far below the average of OECD countries with about 43 for 1990. The state of telecommunications is far from sufficient to fulfill the requirements of the economy, of the public administration, and of the inhabitants. The penetration rate gives only an incomplete view of the need for network modernisation and expansion. With a few exceptions analogue technology is employed. The transmission systems have a very low transfer rate compared to current technological standards. As a consequence very high rates of unsuccessful attempts to make long distance or international calls are the rule. In addition, high noise and distortion ratios prevail in the network, limiting the possibilities to connect fax and data transmission equipment to the network. A similar situation concerning switching systems exists, most of the exchanges being electromechanical and more than 20 years old. The average figures do not reflect the situation of rural regions as main lines are often concentrated in urban centres; e.g. in Prague the penetration rate exceeds 30% whereas it amounts to less than 5% in most rural regions.

In the following table some key data for the CSFR and comparisons with other C&EE countries on the one hand and the OECD on the other hand are presented.

Table 2.1
Comparison of key figures for 1990/91

	CSFR	Poland	Hungary	OECD
GDP per capita in US$	2841	1661	3068	19346
Main lines in thousands	2335	3293	996	357178
Main lines / 100 inhabitants	14.9	8.6	9.4	42.9
Revenues in MUS $ [1)]	542	520	523	367848
Revenues in % of GDP	1.7	0.7	1.25	2.27
Revenues per main line in US$ [1)]	220.5	157.9	463.4	1,030
Revenues per inhabitant in US$ [1)]	34.5	13.6	49.3	438.0
Annual growth rate of revenues 1980-1990 in %	2.97	-0.74	4.33	4.8
Investments in MUS $	125	169	211	89545
Investments as % of GDP	0.28	0.27	0.65	0.69
Investments per inhabitant	8.0	4.4	19.9	106.7
Investments in % of revenues	16.9	32.5	59.4	24.4

Sources: ITU, OECD, own research
[1)]1991 data for CSFR and Hungary

2.3 Regulation

A substantial change in the telecommunications sector has recently taken place. After the dissolution of the Czech and Slovak Federation the responsibility for telecommunications moved from the former Federal Ministry of Communications to the Czech Ministry of Economy. Within the Ministry of Economy the Czech Telecommunications Office was established, acting as a regulatory body. The separation of telecommunication and post services took place in January 1993; the SPT Prague was split into two companies, the Czech Post and SPT Telecom.

With respect to privatisation the Czech Government has stated that it intends to privatise 49% of the equity of SPT Telecom, 51% remaining in state ownership at this stage. A tender is planned to involve foreign operators in privatisation, the share of equity that will be open to foreign capital being restricted to a certain percentage rate. A major objective of inviting foreign participation is to gain access to modern management skills and to increase overall efficiency.

The former PTT Research Institute Prague has been converted into TESTCOM, the Technical Centre for Telecommunications and Posts, which is also responsible to the Czech Ministry of Economy but independent of the operators on the Czech territory.

Provision of basic telephone service will not be opened to competition for a certain period, SPT Telecom remaining the monopoly operator. Mobile services and a public data network are operated by EuroTel Ltd. Prague, a joint venture of SPT Telecom and the US association Atlantic-West. A paging service is operated by Radiokontakt Ltd., a joint venture of SR Prague, the operator for broadcasting and microwave links, and TéléDiffusion de France. Apart from these main services several licences were granted to the private sector for limited operation of telecommunication facilities outside the public network, for VSAT operation, and for private mobile networks.

A new telecommunications regulatory law is currently in the process of preparation. The passing of this new law is expected in 1994. As this new law is also expected to be more in line with EC regulations it will increase the possibilities of foreign participation in the telecommunications sector. It remains, however, open, to what extent concerns about national sovereignty will be affected, allowing, for instance, majority foreign ownership of service providers or network operators. The relatively good performance of the Czech economy would allow the rebuilding of telecommunications without losing national control.

2.4 Current status of telecommunications in the Czech Republic and goals of development

The current status (as of 1991) of the Czech telephone network is outlined in the following table.

Table 2.2
Telephone network in the Czech Republic (1991)

Main lines (in thousands)	1,707
Main lines /100 inhabitants	16.00
Waiting list (in thousands)	381
Quality of network (failures in local call attempts) (CSFR)	30%

Sources: Czech and Slovak PTT; Annual Report 1991

In 1992 the telephone density increased to 17.6 main lines in the Czech Republic. The corresponding increase in unsatisfied demand, however, exceeds the increase in main lines by far. Average growth rates in main lines of 3.5% in the last decade have been accompanied by an exploding demand. As of spring 1993 the waiting list amounts to about 500,000 in the Czech Republic alone. The average waiting time is about 4 years; 16% of the new subscribers had to wait for more than 8 years for their telephone.

In both republics together there were 6,400 data terminals and 29,400 fax machines connected to the public network. Some 2,000 subscribers used the paging service. By mid-1993 about 5,000 subscribers to the cellular mobile telephone network had been registered. This figure indicates that mobile services have so far not played an important role as regards substitution for the public switched telephone network, although immediate connection to the network is offered. However, as connection charges, equipment costs, and usage tariffs are very high compared to the basic network, mobile services are not affordable for the majority of potential private and business users. As the operator of mobile services, EuroTel Praha (see section 2.5), was granted a licence for 20 years with exclusivity for 5 years at the 450 MHz band, the situation may not change considerably in the near future.

In order to rebuild the telecommunications infrastructure in a structured way a programme for the period up to the end of this century has been developed with the assistance of foreign consultancy. This "First Telecommunications Project" includes detailed plans for the extension and modernisation of the network with respect to technical realisation and financing. In terms of main lines the overall goals are shown in the following chart.

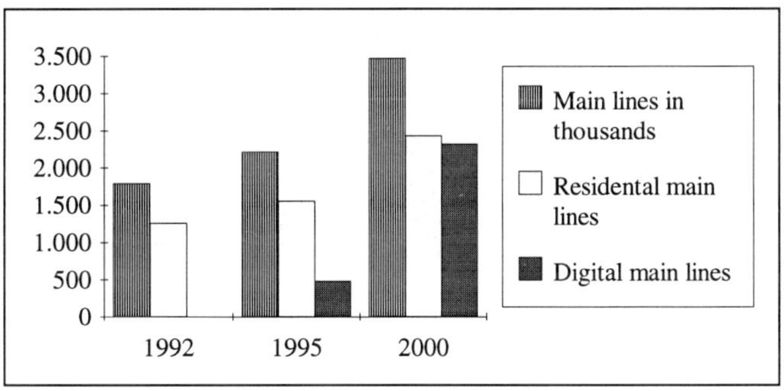

Figure 2.1. Network expansion and digitisation plans in the Czech Republic.

The full implementation of this plan has been delayed to some extent as a result of the division of CSFR and the consequent difficulties to provide the state guarantees which were required for financing. Apart from this delay, the need to create bodies to take over responsibility from the former federal government arose. On the level of network operation, the division did not cause important changes as there were already separate operators working in both republics of the former CSFR. The finality of the division in telecommunications was further recognized with the mutual introduction of international tariffs for calls between both republics.

In the first phase of this project (until 1995) emphasis is on the construction of a Digital Overlay Network (DON) covering the whole territory of the Czech Republic. This DON should remove the bottlenecks in long distance and international transmission and hence create the preconditions for a massive network expansion and quality improvements in the second phase (1996–2000). The installation of digital local exchanges will be restricted in the first phase to areas of high concentration of business demand. For the increase of main lines in general, analogue technology will still be employed.

The main share of expansion is planned to take place in the second phase, after the principal links of DON have been completed. In this period it is also intended to replace old equipment to a considerable extent by digital technology. The aim is to reach a penetration rate of 33 main lines at the end of the century, two thirds of the main lines being digitised.

Total investments required for this project will amount to about 4 to 5 billion US$. According to current plans less than 25% of the total investments will be financed by external resources, the main contributions coming from international loans by the World Bank, the European Bank for Reconstruction and Development (EBRD), and the European Investment Bank on the one hand and from national loans on the other hand with about 10% respectively. Only minor contributions to overall requirements are planned from export and supplier credits. The international loans will be utilised in the first phase, accounting for about one third of investments. The national loans will mainly be used for the network expansion in the second phase, contributing little more than 10% to total investment in this period, which is expected to be chiefly financed by revenues.

The main prerequisites for the viability of the project are the continuation of the monopoly in basic services for a certain period, flexibility in tariffing, and the implementation of efficient management methods by privatisation or by participation of foreign operators.

2.5 Telecommunication services joint ventures in the CSFR

In the telecommunication services sector joint ventures between the US operators US West and Bell Atlantic on the one hand and the SPT Prague and the SPT Bratislava, EuroTel Praha Ltd. and EuroTel Bratislava Ltd., have started to implement a mobile telephony and a data communications network in both republics in 1991. The share of equity held by the CSFR partners is 51%. Paging services are offered by Radiokontakt Operator, Ltd., a joint venture between the French enterprise Télédiffusion France and the SR Praha and Bratislava. A further joint venture, Telecomspol Ltd., has been established between SPT Praha and the Dutch operator PTT Telecom Holland. The main purpose of this company is the development of application software for telecommunication systems and services.

2.6 Telecommunications industry joint ventures in the CSFR

A number of joint ventures between domestic and foreign manufacturers have been established in the CSFR to maintain national production of telecommunications equipment.

Unlike the service providers who have a protected status in their domestic markets, the telecommunication industry is open to direct competition from other operators on the global market, especially on its export markets. The CSFR telecommunications equipment sector, for instance, created a trade surplus of more than 200 mil US$ in 1989. Competition will also grow in their domestic markets as with increasing liberalisation the operators will be free to purchase wherever they get best value for money. To prevent further decrease, the industry has to acquire quickly up-to-date technological know-how as well as management and marketing expertise.

Table 2.3
Foreign investments in telecommunications industry

Czech partner	Foreign partner	Origin	Established JV
Tesla Karlin, Sb.	Siemens AG, München	Germany	Tescom GmbH, Prag
Tesla Lanskroun AG	JCC Engineering Co,LTD, Akigawa Tokyo	Japan	Naseng AG, Lanskroun
Tesla Lanskroun AG	Far East Engineering Co,LTD, Nagano	Japan	Nascon AG, Lanskroun
Tesla Liberec Sb.	Ivo Kosina, Saint Galen	Switzerland	Teko AG, Liberec
Tesla Intes Sb.	Elpol Warszawa Roznov p. Radhostem	Poland	Tespol AG
Tesla Kolin, Sb.	Melewar Cor., Kuala Lumpur	Malaysia	Testech Kolin, AG

Source: Zahranicni Kapital, vyhoda ci riziko, Ekonom 20/1992, pages 11–13

3. OPTIONS FOR THE REBUILDING OF TELECOMMUNICATIONS

The main focus of this section is to outline some important relationships between options of telecommunications sector development on the one hand and the fulfilment of general objectives as well as requirements and possibilities of financing on the other hand.

3.1 Options for infrastructure development

Several options exist for how to develop the infrastructure with respect to speed, technology, setting priorities, or the shaping of single elements of the network. Most of them have an impact on the amount and shaping of the resources required. First of all, the speed of expansion and modernisation obviously determines the amount of external resources needed. Keeping in mind the role of telecommunications for economic and social development and the explicit goals to close the gap between Western and Eastern countries, a step by step strategy based entirely on internal revenues, as followed by Western nations in the past, is practically impossible. It would signify that the achievement of sufficient penetration rates would be delayed for several years with significant growth rates in network not appearing in this century, and thus the depressing effect on the economy would persist for an unacceptably long period. The speed does not only affect the amount of external funds required. In combination with the technology chosen it also shapes the share of equipment that has to be imported from Western countries and thus the share of hard currencies. Keeping in mind the backwardness of domestic industry, it will take some time before national suppliers will be able to provide state-of-the-art digital equipment. Further, taking into account the social costs and long-term disadvantages of closing down the national telecommunications industry, it can be quite reasonable to maintain installment of analogue components for some period, even though they are more expensive and less powerful than imported equipment.

Regarding the setting of priorities two principal options can be distinguished: the DON strategy and the island strategy.

The DON strategy, which has been chosen by the Czech Republic, can be characterised as follows: As a first step the DON acts as a parallel network to the existing analogue one. It links the main centres and provides connections to the foreign countries. In following steps less important cities are equipped with digital exchanges and connected to the DON until the whole country is covered and the old analogue network is fully replaced.

In this way the most congested parts of the existing network, the long distance and international connections are unblocked in an early phase and a solid base for future network expansion is created. Large subscribers can directly be connected to the DON, and the improved quality can also justify higher tariffs than for the analogue network. The incomes from large users and from increased international traffic can contribute to expansion of the entire network. This strategy involves heavy investments in the initial stage, which makes backing by international financing inevitable. As also state-of-the-art technology is necessary, the share of investments that require hard currencies is high. In the first years residential users will benefit from this approach only as far as business users move to the DON, thus freeing capacities in the old network.

In the case of the island strategy, emphasis is on the full digitisation of chosen priority areas down to the level of local exchange. As these islands must be connected to each other, investments in a backbone network (DON) are also involved but to a lesser extent as it is not aimed to cover the whole territory initially. This strategy can increase the penetration rates more quickly. It is, however, uncertain to what extent the self-financing of investments is increased compared to the DON approach. As the possibilities of price discrimination and of targeting large business users are limited, the outcome depends strongly on the composition of residential and business users in the chosen priority area. For transmission and switching equipment digital technology and hence hard currencies are necessary immediately. However, as less complex technology is involved in lower levels of the network, it might be easier for local suppliers to adopt the required technologies. Additionally, a considerable share of total investments, the local loop, does not at all require foreign assistance. The concentration of investments in certain regions would further increase existing regional disparities within the country.

Apart from the principal decisions on infrastructure development also the shaping of technology in single parts of the network can have a considerable impact on financing. Although external resources are inevitable for a quick expansion, the main source has to be revenues generated by users. To link the user to the network is the most expensive task as placing the cables into the earth involves a lot of construction work. The employment of wireless access lines in the local loop can provide a cost-effective means to connect additional users wherever free capacities in the local exchange are available. However, funding for the cable laying operation is not yet available. With the expansion of the wired network the wireless access line equipment can be moved easily and thus be used several times at different locations. As the installation of a new exchange only takes some weeks whereas building up the associated local access lines takes several months, this technology could help to improve overall utilisation of capacity.

3.2 Options for funding

The estimated costs of 4 to 5 billion US$ for realisation of the First Telecommunications Project represent the minimum amount of finance required in the Czech Republic up to the end of this century. Prior to full digitisation of the network, additional investments will be necessary in the next century as the process of catching up does not end with a density of 33 main lines. Furthermore, the question whether the plans outlined in this project are sufficient is still under discussion in the Czech Republic.

Obviously such amounts of investments can only be initiated with backing by external resources. It is, however, just as obvious that loans alone will not be sufficient to develop the infrastructure. Only a small proportion can be expected to come from multilateral sources. The main share of investments will have to be self-generated within the country. Internal generation of revenues from the market will necessarily be the main source of funds.

3.2.1 Internal funding

Existing tariffs in the Czech Republic remained largely unchanged from 1979 to the middle of 1993. As a result the profit per main line decreased by about one third between 1986 and 1991. Tariffs were also very much out of line with costs; particularly the charges for line rental are much too low (less than 2 US$ per month), whereas the tariffs for long distance and international calls are too high. Keeping in mind that equipment has to be purchased at world market prices, whereas current revenues per main line are only about 20% of the OECD average, the adjustment and rebalancing of tariffs forms an essential precondition for achieving long-term financial viability. A major step to increase internal funding was taken in autumn 1993 by increasing usage-related tariffs by 50% on average. A further increase in revenues can also be expected from the introduction of international tariffs for calls to the Slovak Republic.

In the short run, however, the effects of tariff adjustments may be limited. Rapid increases may not be accompanied by matching increases in revenues as, even if tariffs are adjusted to costs, the purchasing power to provide sufficient revenues will not exist in the near future. Increases would be acceptable for business users, but would seriously endanger the affordability for residential customers. Apart from social policy also economic stability is concerned as inflation would be accelerated. Technical constraints of the current network also limit the adoption of modern tariff schemes such as the differentiation between peak and off-peak hours.

A smooth adjustment and increase of tariffs is inevitable. It will not only increase the self-financing possibilities; it will also improve the access to external funds. A particular opportunity to increase incomes will be offered by the DON as the higher quality of services justifies higher charges. A considerable increase of total revenues can be expected from network expansion and modernisation in general. With the expansion of the network and quality improvements not only the number of users but also the usage per main line can be expected to increase. Additionally the utilisation of capacities can be optimised by new tariff schemes.

Further possibilities to enlarge profits result from efficiency gains and cost reductions. As improvements in management and organisation take time, and significant reductions in employees occur due to digitisation, only few immediate effects are to be expected.

3.2.2 External funding

It will not be feasible to realise the plans of accelerated rebuilding without some form of external financing. Various sources of funding are potentially available: Initial investments can be financed by international institutions like the World Bank, EIB, EBRD, bilateral government grants, private financing institutions, credits from suppliers, private sector participation, etc. New schemes of financing like B.O.T. (Build-Operate-Transfer) can help to overcome initial requirements without overburdening the financial resources of C&EE operators.

In this section some important features of different options will be outlined in brief. It is a difficult task to find an appropriate mix of financial sources. Each source will be linked to conditions. The conditions may differ, but they all come down to one basic condition, that the financing is, and must remain, repayable. Ultimately, the funds can only come from subscribers.

Loans from international institutions (the WB, the EBRD and the EIB) play a central role in financing in C&EE countries. They do not only offer favourable terms of repayment on their own; as they possess a status of preferred creditor that can be extended to co-financiers, they can also attract additional sources of finance that otherwise would be reluctant to face the risk. However, due to the scarcity of available resources and to rules that limit the share in single projects, funding does not normally fully cover the investment. In addition to a satisfactory profile as regards economic return and technical viability, there are specific conditions linked to these sources that restrict their feasibility. Loans from the WB require for instance international tenders. They are therefore suitable for clearly defined projects such as the construction of a DON, but not for the extension of the network in general. Additionally, international tenders leave almost no chance for domestic manufacturers to participate, at least in the short run, as it will take some time until world standards in quality can be obtained. A principal requirement of EBRD loans is a government guarantee to permit tariff increases to keep pace with inflation, limiting the scope of regulatory options, e.g., the incorporation of social objectives.

So far lending from foreign commercial banks has not occurred to a significant extent. This is not very surprising if the risks of loans to C&EE countries are taken into account. The risks of political instability, of uncertainties about the economic development, and of unstable currencies are included in the lending conditions, making such loans unfavourably expensive. Revenue sharing models, such as B.O.T., B.O.O.T. (Build-Own-Operate-Transfer), or B.L.T. (Build-Lease-Transfer), have not yet played an important role. These schemes require projects with clear boundaries, such as the expansion of local networks, the improvement of long distance networks, or cellular mobile telephony. They diverge mainly in the sharing of risks and the complexity of the required agreements, B.L.T. involving the least complex relationships. The impact of such agreements on the long-term ability of the network operator to expand the network needs also to be taken into consideration, especially in the case of long distance and international networks. Credits from suppliers directly or from an export bank of the exporting country are a further source of funding investments. They can, however, only contribute minor shares to total investments. To use these sources extensively would imply the co-existence of several systems in one network as they are restricted to single suppliers and exporting countries. Several different systems would limit the efficiency gains associated with modernisation as they also require several teams for operation and maintenance.

The possibilities of loans from the national capital market are usually limited in C&EE countries as mature capital markets have not yet been established. As an alternative, loans could be directly raised in the form of subscriber bonds, offering a priority of getting access in exchange. The realisation of this priority might, however, be difficult in practice as there is no easy way to link subscriber bonds to network planning. A general obligation for new subscribers to provide loans would, on the other hand, bring about unequal treatment of old and new subscribers and reduce the affordability for residential users.

3.2.3 Privatisation and telecommunications sector reform

Privatisation and telecommunications sector reforms are more indirect ways to raise funds for the development of telecommunications. Incomes from privatisation are usually not transferred from the government to the privatised operator. Privatisation is rather regarded as a means to increase the efficiency of the network operator, thus offering better access to and more flexibility of financing. Telecommunications sector reforms set limits to the established operator's monopoly provision of services. This offers possibilities of the participation of foreign operators or the private sector in the provision of advanced services for the general public or for their own purposes. Licensing of joint ventures between the established operators and foreign companies allowed the quick introduction of mobile telephony, a packet switched data network, and of paging services in the Czech Republic. Licensing has so far not been linked to the introduction of competition. On the contrary, taking into consideration the risk associated with investments in the Czech Republic, extensive rights of monopoly provision have been granted to the joint ventures. Although obvious advantages with respect to the supply of telecommunication services exist, some difficult questions concerning the consequences of privatisation and sector reform arise, especially if the Czech plans to adopt EC regulations are taken into consideration.

Even developed industrial countries may have difficulties in mobilising the funds to acquire state-owned companies within their country. A solid capital market is required to achieve privatisation without problems of fairness. Too early and too extensive privatisation in C&EE countries carries the risk of not finding sufficient demand or of being bought up by foreign capital. The danger is that one of the most important sectors of and for the economy will become controlled by foreign capital. Another issue is how to avoid cream-skimming. Capital coming from Western countries is profit-oriented. Profitable returns comparable to investments made in alternative destinations are required as soon as possible from the investments ploughed into the eastern telecommunications. Only the most profitable segments of eastern markets are likely to fulfill these requirements. To open these segments could seriously reduce the capability to develop the basic network. A possible strategy to lessen this risk is to offer packages only, as these packages contain both rich and poor segments. It is, however, questionable to what extent such packages remain attractive enough for foreign capital.

To separate the market into a profitable and a non-profitable one would cause irreversible damage to telecommunications in these countries. To open the most profitable segments of the market, that is, business demand in agglomerations, can help to overcome shortages in serving businesses and thus contribute to overall economic development. In the long run, however, it bears the risk that the most profitable markets are run by private operators, the revenues produced in the first market being drained away and not available anymore for expansion of the network. The public network operator, or the state, would have to allocate more resources than necessary for the development of the basic network.

To conform with EC rules all forms of cross-subsidising effects should be removed and cost-based tariffs be introduced. It can, however, be questioned whether this principle should apply to C&EE at this stage of development. The current level of universality of telecommunications in Western countries has been reached by long periods of using revenues from business customers, long distance and international traffic to expand network coverage to rural and less profitable residential customers. There is no theoretical reasoning why the same phasing as for the EC should apply to C&EE, keeping in mind the different stages of development. The full adoption of EC telecommunication rules will further require the opening of the provision of VAS (Value Added Services) to competition. This requirement may be highly appropriate to increase choice and to reduce cost for users in countries with a developed infrastructure. For countries in which the infrastructure needs to be built up first, competition could seriously reduce the financial ability of the operators to fulfill this task. Taking into consideration the low levels of purchasing power and the sometimes unfulfilled expectations of non-basic services providers in some C&EE countries, it is also doubtful that there would be any foreign operator willing to compete in these markets. On the contrary, protection against competition might be a prerequisite for investment.

3.3 Options of development of the telecommunications industry

Due to the technological gap of domestic manufacturers the modernisation of the telecommunications infrastructure involves heavy hard currency requirements. The share of hard currencies in total investment is expected to increase up to 50% in the next few years. Whereas these initial requirements can largely be covered by loans from international financing institutions, the whole telecommunications project is endangered without considerable increases in domestic provision of equipment. The substantial effects on the balance of payments, which are linked to equipment imports, raise the question of the economy's ability to compensate for the deficit without the need to raise real interest rates to an extent that could have a negative impact on economic activity in general.

To overcome the backwardness of local industry with respect to technology and management, contracts with foreign suppliers are often linked with an obligation to establish joint ventures with domestic manufacturers. To secure long-term viability, this strategy has to be accompanied by at least two measures. As value-added is more and more embodied in integrated circuit design and software development, in addition to or instead of mere assembling, these components should form an essential part of the joint venture. The building of production capacities in each single C&EE country involves the danger of excess capacities and production at unfavourable scales. Therefore these joint ventures should be incorporated into a regional strategy for C&EE on the whole. However, the appearance of such a regional strategy is not very likely within the next few years. The experiences of four decades of forced trade within the COMECON caused a strong aversion to renewal in most of the former "Eastern Bloc" countries. As also many Western companies are willing to establish local production facilities in exchange for market access, regardless of their long-term viability, co-ordination between these companies would be necessary.

4. CONCLUSIONS

Any decision maker in a C&EE country has to be aware that there is an evident trade-off between short-term satisfaction of urgent demand and long-term interests of network expansion. Cross-subsidy from long distance and international traffic has been used in most Western countries to finance network expansion to help enable the provision of the obligatory universal service. The telecommunications industry has strongly benefited from national procurement policies. Thus a cautious approach regarding the early application of EC telecommunications rules is required in order to limit their potential disadvantages for countries with an underdeveloped telecommunications infrastructure.

There is, however, no single path, no single set of measures to resolve the conflicting goals and interests involved in the development of telecommunications. There are different preferences, different ways to achieve the same goal, different situations prevailing in different countries, and different stages of development that may have an impact on the set of policies feasible. Every country must choose the specific mix that best fits its own requirements.

REFERENCES

Allione, M., 1992, Bridging the telecommunications' gap in the new Europe, ITU World Telecommunication Forum, Europe Telecom 92, Policy Forum, Budapest, 81–84.

ITU, 1992, European telecommunications. Indicators (ITU, Geneva).

Laidlaw, B.H. and A.F. Odeh, 1992, Potential for private sector participation in the development of telecommunications services, Background paper, ITU, Geneva.

Müller, J., 1991, Closing the capacity and technology gap in East European Countries, in: EU-RDC Working Documents, International Telecommunications Union, Geneva, 1992.

Nulty, T.E., 1990, Telecommunications markets in Eastern Europe, in World Telecommunications, speakers' papers, Financial Times Conferences, London, 5.1–5.7.

OECD, 1993, Communications Outlook, Paris.

PART C: INTERNATIONALIZATION ISSUES

Challenges for Policy and Regulation

The Race to European Eminence
E. Bohlin and O. Granstrand (Editors)

The Race to New International Mobile Satellite Services*

Patricia Bonanzinga[a], Francesco Castelli[a] and Claudio Leporelli[b]

[a]Fondazione Ugo Bordoni, via B. Castiglione, 59, 00142 Roma, Italy
[b]Universita' degli Studi di Roma "La Sapienza" Dipartimento di Informatica e Sistemistica, via Buonarroti 12, 00185 Roma, Italy

Technological innovation and the growing demand for global personal communication services have opened opportunity windows, in space and time, for new commercial applications of satellites. This paper characterizes the competitive setting of the preparadigmatic stage of this new industry as a time-based competition among coalitions of supply subjects who operate in different segments of the value system. Regulatory, R&D and industrial policy agencies play a crucial role in this game, pursuing a number of possibly conflicting national goals, such as the competitiveness of domestic industry and the efficiency of resource allocation. This requires new regulatory approaches at an international and national level (in the spectrum allocation, operating licence granting, standard setting, and tariff control) to explicitly take into account the effects of regulatory decisions on the allocation of the benefits of innovation among countries.

1. INTRODUCTION

International satellite telecommunications service markets have been historically non-competitive and characterized by the cooperation of national agencies within consortia at a worldwide (INMARSAT and INTELSAT) or regional (EUTELSAT) level. Consequently, almost everywhere, Public Telecommunications Operators (PTOs), as signatories of the international consortia, have monopolized the national access to regulated satellite services. The demand for these services has until now derived almost exclusively from the telecommunication (TLC) carriers, as suppliers of complementary terrestrial services, or from broadcasters (Commission of the European Communities, 1990) (OECD, 1992).

Moreover, public R&D and operational civil and defence projects, at national and regional levels, such as those sponsored by NASA and the European Space Agency (ESA), have anticipated and paralleled commercial projects. This public demand, on the one hand, determined the growth of a private space manufacturing industry, first in the USA, and then, to a

* This paper has been carried within the framework of the agreement between the Italian PT Administration and the Fondazione U. Bordoni. Moreover, support was provided by the Italian National Research Council (CNR) Progetto Strategico "Trasferimento delle tecnologie dei progetti finalizzati". The research project originated from an ESA contract to the Fondazione U. Bordoni. The authors gratefully acknowledge suggestions by A. Agostini and A. Pinglier of ESA D/TEL in Paris.

lesser extent, also in Europe. On the other hand, it gave rise to positive indirect externalities (through scale, scope, and learning effects) on the commercial demand.

In recent years, the satellite scenario has been characterized by the supply of a set of new services, provided through the available space infrastructure (such as those provided by private VSAT networks and the vehicular low capacity mobile services), and by the proposal of innovative systems and services (such as the new non-geostationary,[1] non-GEO, configurations for personal voice communications services with global coverage and quality comparable to that of terrestrial cellular systems). These latter services will hereafter be referred to as *Mobile Satellite Services* (MSSs).

TLC markets have also been characterized by deregulation and an opening to competition and internationalization, which originated in the USA, the United Kingdom and Japan and is now moving to Europe (Noam, 1992).

Technological, regulatory, and demand trends have had synergy effects on the evolution of the satellite market organization. The main real or potential consequences have been: the emergence of privately owned satellite networks for the provision of international services (at times in competition with the existing international public consortia); the possibility to compete with the national PTO, directly reaching the end user via satellite (e.g., in the VSAT or in personal mobile communication networks) which in principle is open, not only to private service providers, but also to international public consortia.

It is interesting to note that all the proposals regarding non-GEO systems, except Project 21 by INMARSAT, come from private United States-led consortia. This can be seen as a consequence both of the experience gained in a national deregulated service market and of the existence of a skilled and competitive manufacturing industry looking for new markets after the decline of military demand.

In the new competitive framework of the satellite markets, the race among new systems in competition among themselves and with the terrestrial cellular systems has begun. This is due to the fact that given the high sunk development and capacity costs and the scarcity of spectrum and orbital resources, only a few systems are foreseen to coexist and break even (Commission of the European Communities, 1992) (Pelton, 1993).

The shortage of natural resources has created new allocation and usage regulation problems, at both an international and national level, that strictly interact with the national and regional industrial and technological policies (Commission of the European Communities, 1990) (Commission of the European Communities, 1992). Obviously, attitudes and decisions of regulatory and industrial policy agencies will depend on the role of satellite systems, as either a competitive or complementary infrastructure variously interconnected to the terrestrial network (Stehman, 1992). Moreover, technological evolution requires an analysis of the optimal satellite network industry structure, from the point of view of incentives to the innovative R&D efforts and the new features of indivisibility and service specificity of non-GEO constellations, rather than from the point of view of cost subadditivity. Furthermore, the supranational dimension of the break-even markets implies a decisive role for the national regulatory authorities in determining the feasibility and in limiting or protecting from competition the profitability of a new project.

The objective of this paper is the characterization: of the competitive framework of the new markets for global mobile satellite services, which are still in the preparadigmatic phase where

[1] A geostationary satellite is a satellite which due to its circular orbit with a ray of 36,000 km remains in a practically fixed position with respect to the Earth.

the competition among firms is a competition among different projects; and of the problems posed by innovative technology and new market organization in the development of regulatory and industrial policy procedures.

The paper is organized as follows. In section 2, the main evolutionary features of the *Satellite Communications Industry* (SCI) are briefly discussed, pointing out the effects on the organization of the international markets of the new technological opportunities and of the competitive pressure of deregulation.

In section 3, the competitive setting of the satellite manufacturing and service provision markets is analysed to propose an interpretative model of the opportunities and the incentives to horizontal and vertical cooperation among firms who are in charge of different elementary activities in the product life-cycle. The advantages to be had in being the first to enter a new market give rise to a race among potential competitors for the control of specialized assets (scarce frequency spectrum and orbit position, national distribution channels) and for the pre-emption of the market (through the commitment constituted by sunk investments and the constitution of a client base).

In section 4, reasons, objectives and effects on the competitive setting of public intervention are discussed and the necessity of new regulatory procedures which take into account the time dimension of the allocative process is stressed. In those cases in which there is room only for one or a limited number of supply subjects, artificial entry barriers (besides those deriving from sunk costs) could be created for late comers and time-based competition could cause a wasteful duplication of investments.

In section 5, opportunities and options of a common European industrial and regulatory policy with respect to the new non-GEO satellite systems are summarily discussed.

Lastly, section 6, gives conclusions and guidelines for public policy.

2. THE EVOLUTION OF THE SATELLITE INDUSTRY

In this section the evolutionary trends of the SCI, as determined by technological change, deregulation, and demand requirements for global personal communications, are discussed.

In recent years, satellite technology has been characterized by a series of incremental innovations in the traditional geostationary communication, which has allowed bi-directional end user terminals for direct reception and transmission to be offered on the market. More importantly, a radical innovation has been proposed: non-GEO orbit satellite constellations which support systems for global personal communication granting the possibility to use the service wherever and whenever (*global international roaming and availability*).

The distinctive features of this new satellite technology and of its cost structure, both in the manufacturing and operation of the infrastructures, have changed considerably with respect to previous technology. This calls for a re-examination of the problem of the optimal industry structure and, consequently, of the determination of public intervention strategies.

2.1 New satellite technology features

Specific features of the established satellite communication technology have restricted commercial uses of space capacity to long distance point-to-(multi)point carriage services complementary to terrestrial network services: to carry international traffic (a declining market after the introduction of optic fiber cables (OECD, 1988)), to grant spare capacity and permanent backup in transoceanic long-haul systems (Howes, 1992), and to serve the needs of broadcasters (which represented 58% in the USA and 75% in Europe of the total satellite service revenues in 1991 (OECD, 1992)).[2]

The continuous incremental technological innovations in GEO satellite communications have had, as their most relevant market effect, the supply of low-cost terminal equipment which has been used both to extend terrestrial networks where terrestrial alternatives are inefficient or under-developed, and to allow direct links with end users for the provision of fixed services or mobile services which still need vehicular or briefcase terminals.

However, the available space infrastructure does not allow the provision of the most promising new telecommunication services: mobile voice communication with handheld terminals and full coverage (global or regional).[3] Even if the new services are still a commercially unexplored opportunity, in the USA, the most advanced satellite service market, the already available land mobile satellite services, accounted for about 80% of total MSSs revenues in 1992 (OECD, 1992). In this market segment satellite systems have a competitive advantage over terrestrial cellular systems because of their cost in widely dispersed areas with low residential density or their shorter lead time in deployment or coverage homogeneity. This demand opportunity, along with the push coming from both the terminal industry, which could benefit from skills and cost complementarities deriving from their terrestrial cellular production, and the space industry, which is suffering from the decline of military and international carriers' demand, has led to the proposal of many new satellite system projects.

The relevant trade-offs in satellite communication technology, available power on earth being equal, are "orbit ray vs antenna size" and "footprint area vs terminal size". In addition, smaller footprints allow for higher frequency reuse (cellular-like coverage obtained through a large number of small spot beams per satellite). Consequently, in order to provide personal communication services via satellite, both innovative GEO systems with very complex satellites, large antennas and cellular coverage, such as TRITIUM by Hughes Aircraft, and systems based on satellites in medium altitude orbits (*Medium Earth Orbit*—MEO), such as Odyssey by TRW, or in low altitude orbits (LEO), e.g. IRIDIUM by Motorola and GLOBALSTAR by Loral, have been proposed.[4] This paper deals for the most part with LEO systems in so much as they encompass most of the projects, even if many of the concepts considered can be extended to MEO systems.

These new projects are very ambitious from a technological point of view, having also to face serious market and regulatory problems. In particular, they have to break even against the (potential) competition of other satellite or terrestrial systems, and therefore they are forced to

---

[2] Traditional services still account for 95% of the total satellite service revenue, which in OECD countries is less than 0.5% of the total TLC services revenue: i.e. commercial applications of satellite technology represent only a small segment of the overall TLC service market.

[3] Currently available mobile voice services are globally supplied through the GEO satellites INMARSAT-2. The similar systems INMARSAT-3 (that will provide 4800 voice channels worldwide using four GEO satellites) and MSAT (which is private and will provide 3000 voice channels over North America, using two satellites) will both become operational by the mid-90s, but will still require vehicle-mounted or briefcase-size terminals due to amplifier, antenna, and power supply requirements.

[4] Also systems based on satellites in Highly Elliptical Orbits (HEO), such as the experimental system Archimedes by ESA, have been designed to overcome low elevation angle problems typical of both LEO and, at northern latitudes, of GEO systems, which can reduce visibility and, therefore, service quality, in particular in an urban environment, where the complementarity with terrestrial cellular systems could prove to be necessary.

meet strict quality, cost, capacity and time-to-market requirements, and to justify themselves as global or regional users of the scarce frequency spectrum.

The new LEO configuration concept tries to benefit from completely different trade-offs among the system cost elements and resource utilization, with respect to GEO technology (Maral, De Ridder, Evans and Richharia, 1991).

- Access to the already congested geostationary resource is not required (LEO satellites are deployed in circular orbits at altitudes ranging from 500 km to 1500 km), even if problems of spectrum shortage are exacerbated (Poskett, McDougal, 1990).
- Low altitude and small *footprint* allow handheld terminals with low power requirements and, in turn, low satellite mass. The unit launch costs decrease with decreasing mass and altitude, and multiple launches become possible, but total deployment costs are very high because of the large number of satellites.
- Constellations of LEO satellites are made up of a large number of *minisatellites* (50–500 kg). This is because non-GEO orbit satellites move with respect to the Earth at a relative speed which increases as the orbit ray decreases. Therefore, in each area the connection to the system is granted by different satellites at different times. The number of required satellites increases along with the elevation angle and the temporal availability requirements, and decreases together with the orbit altitude. It also increases with the extension of the service area: for example an initial deployment of 24 satellites will allow GLOBALSTAR coverage of the USA, and subsequently 48 satellites will be necessary for global coverage. The demand for a large number of small satellites (easily designed and launched) allows cost reductions through line production, reduced spare capacity and limited reliability requirements, given the redundancy and smooth decay of dense constellations, and the low cost, low lead time replacement of damaged satellites, which also affects the insurance costs.
- The redundancy and low vulnerability of the system, with respect to a GEO configuration, is also of interest for military communications.

The disadvantages of LEO configurations, for the most part, are due to the non-stationarity of the position/coverage of a LEO satellite which implies the more difficult operation and orbit keeping of a constellation of many satellites, an increase of the Doppler effect and the short average visibility period of a satellite (5–10 minutes) which requires frequent *handover* procedures (shift of an earth surface area from one satellite *footprint* to another). Furthermore, it is impossible, with LEO systems, to confine the coverage to a prescribed area, as is done with GEO satellites, and, therefore, a large portion of the voice channel capacity will be wasted over oceans and desert land masses.

There are considerable differences in the technology embodied in different LEO system projects, as occurs in the preparadigmatic phase of an industry before the appearance of a dominant project. For example, IRIDIUM proposes the use of techniques never before experimented in GEO commercial systems such as *Inter Satellite Link* (ISL) and *On Board Processing* (OBP) for switching and call routing, requiring no more than 20 gateways to the terrestrial networks worldwide. Instead, GLOBALSTAR is a transparent system whose architecture assigns most of the intelligent functions to earth stations, with satellites in orbit acting as simple repeaters. As a consequence, it needs a much larger number of earth stations, both for the interface with Land Mobile Networks and Public Switched Telephone Networks, and as Network Coordination Centres for processing and system control functions (at least one gateway is needed within the footprint of each active satellite). However, it has the advantage of being more flexible and able to integrate with other networks. Note, therefore, an important relationship between system architecture and possible market organizations and vertical relationships among LEO system companies and terrestrial network operators (Ciarniello, 1992). It is presumable that PTOs will be favourable to the diffusion of complementary systems such as GLOBALSTAR, designed to be able to route and bill its services through existing carriers, without bypass ability. On the contrary, they will be competing against systems such as IRIDIUM, designed to also serve areas already served by terrestrial cellular systems and to bypass local and long distance terrestrial networks. This will affect user charge sharing among carriers: GLOBALSTAR will charge its tariff to regional suppliers adding their services and (probably higher) charges to end users; IRIDIUM, being an interlink network, will have a relatively short tail connection and therefore modest tail charges.

In addition, different proposed systems use different modulation and multiple access techniques which, together with frequency reuse, determine their efficiency in the use of the spectrum, i.e. the number of voice channels they provide in a given bandwidth. The possible coexistence of different systems must be assessed, and the fact that the chosen multiple access techniques make efficient spectrum sharing possible must be proven. In particular, the FCC,[5] which is currently engaged in frequency assignment among the major LEO projects for voice mobile service, has to examine the conflicting proposals of Motorola and of the other applicants and rule accordingly. Motorola's system uses the *Time Division Multiple Access* (TDMA) technique, while all the others plan systems that use *Code Division Multiple Access* (CDMA) approach. Any number of different systems can dynamically share the entire allocated spectrum under the CDMA approach; on the contrary, only one system can use TDMA in a given frequency band, and therefore permanent bandwidth segmentation would be required. Motorola's design is accused of being inefficient and in conflict with international standards. The company replies that the rival design is flawed because of the risk of frequency interference (Seitz, 1993). In the beginning Motorola proposed to segment the band allocated for LEO satellite communications: part for its TDMA approach and the rest for CDMA systems, even if CDMA operators and the FCC tend towards full-band sharing. Recently (October, 1993), Motorola has reached an agreement with its largest competitor, Loral Qualcomm (GLOBALSTAR), and they have jointly proposed a licensing scheme, based on the first-come first-served principle, which would assign the entire band to the first system entering the market and subsequent equitable divisions of the available spectrum when additional systems become operational.

LEO technology allows a choice among a set of architectural options in order to meet the service and system requirements deriving from market choices. Each competitor in the MSS market chooses the architectural option that, given the institutional and natural resources constraints and the competitive setting, maximizes his benefit (defined according to suitable criteria). It is important to note at this point the following, which will be further discussed in subsequent sections:

---

[5] Federal Communications Commission, the US government agency which regulates telecommunication domestic markets and foreign trade.

- service and system requirements settings are primary drivers of the system cost; they derive from the competitive strategy and may involve choices that negatively affect the social welfare; in particular, the entry timing may be chosen to erect entry barriers and may involve negative cost-time trade-offs;
- the use of scarce natural resources is not reflected in private costs, which may lead to inefficient spectrum and orbit utilization for a potentially unlimited time horizon;
- the private cost of knowledge assets may differ from social ones because of the multiple uses of those assets (indirect externalities). This could justify industrial policy measures which favour only determined technological options.

2.2 Technology and industry structure

Technological innovation has modified the parameters of the theoretical discussion regarding the optimal structure of the SCI. In the past, the emphasis was on the cost subadditivity of the space infrastructure, while at present it has shifted to the minimal optimal dimension of satellite systems. The proposal of global LEO systems imposes an assessment of the efficiency of various market structures (monopolistic or oligopolistic), and therefore licencing decisions within a dynamic international scenario.

The traditional arguments which justify the monopolistic and public operation of the terrestrial network, i.e., on the demand side, the need to guarantee service universality, and, on the supply side, the network cost subadditivity, have also been used to support the worldwide cooperative satellite service provision through consortia of national monopolistic operators.

The natural monopoly assumption was based on the relevance of fixed costs and indivisibilities of deployment and operation of the space capacity: the given, large capacity of each geostationary satellite determines *fill economies;*[6] scale economies derive from the reduced unit capacity cost of large satellites and from the reduced incidence of spare capacity with the increase of the supplied capacity; scope economies derive from the usage of a common spacecraft for multiple dedicated payloads and from the joint usage of flexible capacity (Snow, 1986).

More recently, it has been noted that most of the previous arguments merely demonstrate the existence of a minimum optimal dimension of the space capacity and that, given the technology modularity, separate systems could be optimally operated by different subjects (Roller and Waverman, 1991) (Rochhia, 1990). Furthermore, the more scale-sensitive activities, such as orbit monitoring, can be jointly supplied by specialised firms. Finally, a satellite system which operates independently of monopolistic terrestrial networks could separately supply differentiated services without additional social costs (Stehman, 1992).

The presence of both *static* scale economies (deriving from large fixed and sunk investment costs), and *dynamic* scale (*learning by doing*) economies (that shift downward firm cost functions with the increase of the cumulative output (Roller and Waverman, 1991)) in space and terrestrial equipment manufacturing should not be considered an obstacle to competition in services. In fact, even a competitive satellite service provision industry could enjoy indirect externalities derived from manufacturing industry scale economies. In other words, a competitive service provision does not imply the duplication of R&D and manufacturing fixed costs if the competitors adopt the same established technology.

---

[6] The increase of the usage level of a given satellite capacity, within the indivisibility interval, implies a decrease of the unit cost per circuit.

The innovative LEO configurations require a reconsideration of the previous arguments from a number of viewpoints.

- *The indivisibilities of the space infrastructure arise at the configuration level rather than at the single satellite level*: while a single large GEO satellite can independently provide services, LEO configurations make use of many small satellites, which are individually useless; therefore the whole constellation becomes indivisible. An increase in the supplied capacity, in terms of the covered area, or of the number of channels available over a given region, can be obtained by coexistent separate constellations (based on the same or on different technology) or by an enlargement of a single constellation. The limited modularity of capacity expansions with different systems together with the current dimension of the global mobile service market render questionable the ability of a LEO system to compete and break even.
- *LEO systems are designed for a rather specific class of mobile services*: voice and data communications to handheld units capable of being used in the entire covered region. To meet their main service requirements, they are quite complex and too limited in capacity and bandwidth to pose a commercial threat to other markets served by GEO satellites. Therefore they can be considered to have much less flexible space capacity than GEO systems comparable to single multiproduct plants (Maral, DeRidder, Evans and Richharia, 1991), their flexibility and ease of integration with other systems decreasing along with their complexity and on-board intelligence.
- *LEO systems require new, irreversible investments*, whose specificity or possible future reuses have still to be assessed, while, on the contrary, the traditional GEO systems have very low R&D relevant costs, in so much as they make use of past irreversible investments in public production inputs (most sub-system designs and components are readily available). The risk of wasteful duplication of competitive efforts in R&D should be confronted with the social benefit of competition in R&D, manufacturing, and operation.

2.3 The competitive pressures on the international cooperative market structure

In those international telecommunication service markets that are the most profitable, PTOs have always pursued cooperation policies that go well beyond technical coordination needs. Satellite communications are no exception.

With the creation of INTELSAT in 1964, the international civil satellite communication market for telephony, point-to-point video and data transmission, and television has been organized as a coordination of local monopolies for the access to a centralized service provision. In fact, each member state has designated a Signatory, a carrier in charge of managing incoming and outcoming INTELSAT traffic (INTELSAT cannot sell services to end users). In many countries the national PTO, e.g. France Telecom or DBT in Germany, was chosen as the INTELSAT Signatory; in other countries a specialized state agency or a private company was designated, e.g. Telespazio in Italy or Comsat in the USA. In the latter case, the Signatory is a carrier's carrier, because it cannot reach the end user but is only an intermediary between INTELSAT and one national carrier.

This Organization prevents national governments from authorizing or participating in other satellite systems without (1) the *technical coordination* with INTELSAT, in order to avoid risks of radio interferences, and (2) the demonstration of not causing significant *economic harm* to INTELSAT (Commission of the European Communities, 1990).

Subsequently, in 1979, from a similar international agreement, INMARSAT, the organization for terrestrial, aeronautical, and maritime MSSs, was instituted.

In 1975, the creation of ESA initiated a joint European space program aimed at reducing the technological gap between Europe and the USA, even if some countries, such as France, Great Britain, and Germany, have continued to devote a relevant share of their space investment to autonomous national programs (Noam, 1992). In 1977, as a consequence of an ESA resolution, many European PTOs collaborated in creating EUTELSAT, with the aim of providing fixed satellite services among European countries. EUTELSAT underwent the INTELSAT coordination procedure and demonstrated that its market entry did not cause economic harm to INTELSAT (Convention Establishing the European Telecommunications Satellite organization, "Eutelsat" 1984).

Both INMARSAT and EUTELSAT adopted the INTELSAT rules in defining the national Signatories' coordination duties.

Since *International Satellite Organizations* (ISOs) are nominally non-profit organizations, their governing bodies set for their Signatories space segment tariffs whose sole aim is to cover operating and depreciation costs and to obtain a fair rate of return on invested capital, fixed at about 14% (Pelton and Snow, 1977). However, the effective tariffs, which are essentially compensation for internal transactions and not published, have allowed INTELSAT and INMARSAT to exceed their return goals (between 1973 and 1983, an average of 16.1% for INTELSAT, in 1991 31.5% for INMARSAT, but only 8.5% for EUTELSAT in 1990, because of the rather slow growth of its market).[7]

Two peculiar characteristics of the international market cooperative structure are sources of inefficiency.

- The Signatories' role of exclusive retailers of the ISOs' space capacity introduces a chain of price distortion due to the fact that each monopolistic intermediary adds its mark-up at each stage of the service provision chain, as shown in Fig. 1. The inefficiency deriving from this multiple marginalization could be avoided if each operator had a null price margin on long-run costs (Tirole, 1989), but this is not the case of either INTELSAT or of private or public Signatories or terrestrial TLC operators who all enjoy high returns. In fact, the cost of international satellite communications currently represents a small fraction of the tariff charged users which is made up mainly of taxes or TLC operators' profits.[8]
- The anti-competitive arrangement of not licensing separate satellite systems which could compete without restrictions with ISO services has been confirmed up until 1997. As a consequence, separate satellite systems, e.g. Telesat in Canada and AUSSAT in Australia, created on a national basis to extend the communications network, rather than utilize ISO's capacity for domestic needs, were permitted to provide capacity to neighboring countries but not to compete with ISOs for services between third countries.

---

[7] The actual rate of return on Signatories' investments is higher than the ISO's because each Signatory freely sets its own tariffs for service resale to end users (OECD, 1992).

[8] Also in the USA, where one of the most favourable tariff systems exists, the unit price paid to INTELSAT (unit cost+INTELSAT unit margin) is less than 10% of the end user charge; 15% is the COMSAT share and the remaining 75% is the terrestrial carrier(s) revenue (OECD, 1988).

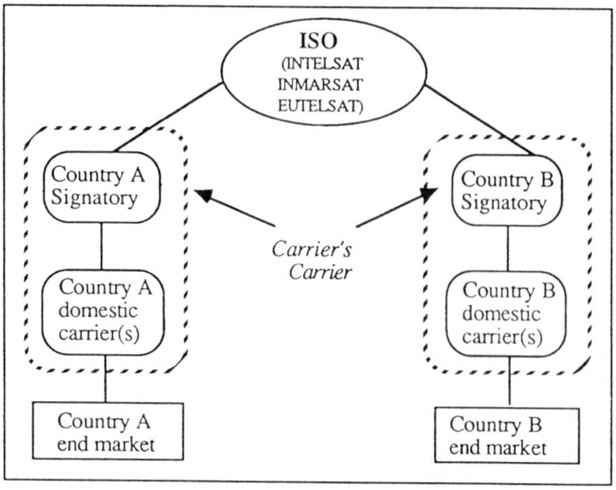

Figure 2.1. Cooperative organization of the international satellite service market.

Recently, a number of trends, briefly outlined in the following, have begun to exert competitive pressures on the cooperative structure of the international service markets.

2.3.1 Appearance of national separate satellite systems

The worldwide trend toward a gradual liberalization of communication markets has also had some effects on satellite services, with the appearance of nationally licensed systems able to compete to a limited extent with ISOs' services in regional markets.[9] In 1990, the FCC granted final authorization for the construction and launch of the Orion Satellite system and, in 1991, granted final operation licence for transatlantic communications, subject to the restriction of not carrying any public switched voice traffic, the primary revenue basis for the INTELSAT monopoly.

2.3.2 Competition in the national terrestrial segment and carrier privatization

There exist a number of conflicts of interest between an ISO and its Signatories, which in some cases are the self-same PTOs: first, Signatories retail the space capacity of different ISOs (e.g. European PTOs Signatories of a regional ISO: EUTELSAT, and an international ISO: INTELSAT); second, some of them operate terrestrial networks capable of supplying substitute services; third, some of them possess their own satellite systems. The effect of these conflicts are becoming more relevant, because some Countries have, at least partially, privatized their Signatories and PTOs that, consequently, have assumed a more market-oriented attitude.[10] Furthermore, Signatories with substitute terrestrial and satellite systems may use their

[9] Some examples of new national systems, independent from the ISOs, are the USA PanAmSAt system, which, after negotiation with INTELSAT, was allowed to provide a fixed number of circuits for international services in competition with the self-same INTELSAT, Luxembourg's Astra system, whose licence is confined to the provision of broadcasting services in Europe, Spain's Hispasat, for communications between Spain and the Dominican Republic and, lastly, the German DFS for services between Germany and Luxembourg.

[10] For example, Telefonica, the Spanish Signatory of all the three ISOs, is 65% privately owned and possesses the separate satellite system Hispasat.

position within ISOs in an anti-competitive manner against both non-Signatory satellite systems and ISOs themselves. A sort of Prisoners' Dilemma with many players arises in which it is unilaterally profitable to behave opportunistically if others cooperate (OECD, 1992).

2.3.3 Bypass of ground segment gateways and service providers

The satellite service market has experienced both the appearance and fast diffusion of independent satellite networks (mostly based on VSAT antennas), completely disconnected from the terrestrial network, less expensive and more efficient than the latter for point-to-multipoint communications (Stehman, 1992), thanks to their rapid deployment, quick and easy addition of new locations, international reliable high-speed data transfer (Gribbon, 1991). In response to new user requirements for global communications the distinction between national and international services is always more difficult to apply and more and more value-added service providers are entering the market, leasing the space capacity from the space segment providers.[11]

2.3.4 Proposal of new global mobile services: the challenge by LEO systems

The new personal communication services, with global roaming supplied through the proposed non-GEO systems (see Table 2.1), can be considered additional competitors in the mobile market currently dominated by terrestrial operators.

However, terrestrial mobile cellular technology, in so much as it offers better voice quality, cheaper equipment and lower tariffs, could in the long run relegate LEO MSSs to niche markets, where they will be able to exploit the special features of satellite communications: fast and complete geographic coverage, service provision (not only voice traffic) to wide areas which lack cellular service due to insufficient traffic density or areas with no telecommunications service at all. MSS technology is also expected to enter markets now served by fixed terminals, such as VSAT, because, in those cases concerning short and random (low rate) data transmission in large operating areas, mobile terminals and services are or will be cheaper than fixed ones (Commission of the European Communities, 1992) (Lorenz, 1992).

The diffusion and profitability of the innovation will depend on three key factors: (a) business globalization, (b) growth of global communications, and (c) unfulfilled demand of telephonic services. Consequently, target markets are: (i) international roaming (IRIDIUM), (ii) complementing terrestrial cellular systems (GLOBALSTAR), and (iii) developing countries and remote areas (CALLING System).[12] Also those services supplied as additions to personal communications (paging, position determination, data transmission) could play an important role in determining the profitability of a new system.

At present, the potential demand of the new satellite systems is generally assessed at 5–10% of the demand of terrestrial cellular services, but in the short run time and space windows opportunity, represented by lags in capacity provision and diffusion time and lack of roaming ability, are expected to play a major role.

---

[11] Consider, for example, international outsourcing services to business customers using VSAT technology such as the VISA International global data network operated by France Cable and Radio owned by France Telecom.

[12] Calling Communications Corporation (owned by AT&T-McCaw Cellular Communications Inc.) is in the process of submitting an application to the FCC for a constellation of 840 LEO satellites, with an assessed investment of $6.5 billion, designed to have as its primary market telephony for developing nations lacking modern communication services and for remote parts of industrialized nations (handheld terminals).

Table 2.1
New satellite systems for personal communication (FCC Applicants)

System	Sponsor	No. of Satellites	Key Services	Market size* (in million subscribers)	User charges#* ($ per minute)	Coverage	Start of service or launches*	Estimated investment (billion $ 1991 base)
Small LEO								
ORBCOMM	Orbital Comm. Corporation	26	data	n.a.	n.a.	global	1995 (s)	0.2
STARNET	Starsys	24	data	n.a.	n.a.	USA -> North Atlantic and Europe	1995 (l)	0.2
VITASAT	Vita	2	data	n.a.	n.a.	global	1994 (s)	n.a.
Big LEO								
IRIDIUM	Motorola	66	voice, data, paging	1.8 (after 5 years)	3	global	1998 (s)	3,4
GLOBALSTAR	Loral Qualcomm	24 -> 48	global roaming (extention of terrestrial cellular)	7	0.3	northern hemisphere -> global	1999 (s)	1.7
ARIES	Constellation Comm. Inc.	48	wireless services	0,9 (by year 2000)	n.a.	global (no global roaming)	1997 (s)	0.3
MEO								
ELLIPSO	Ellipsat Corporation	12 -> 18 -> 24	cellular	n.a.	0.4-0.5	US/northern hemisphere -> southern hemisphere	1996 (s)	0.6
ODYSSEY	TRW	6 -> 9 -> 12	voice, data, paging	2.3	0.6	northern hemisphere -> global	1997 (l)	1.3
GEO (?)								
PROJECT 21	INMARSAT	4 (12/15 MEO)	voice, paging, RDSS	2,5 (by year 2000)	1	global	2000 (s)	2

* As announced by proponents.
\# Excluding access user costs (terminal + subscription) and local distribution charges depending on the lenght of the tail connection, which greatly differ among systems according to design features.

In Europe, the cellular market, until now fragmented by national borders, is being opened to competition and to continental integration and roaming ability through the GSM standard. GSM diffusion, however, will depend on a set of heterogeneous factors, such as the saturation of analog cellular systems, cost of the user terminal, still greater than the analog one, and lack of coordination in the start-up of the service among the European Countries that have adopted the GSM standard. All these factors will hinder a fast, homogeneous European roaming, which could be a market opportunity for a possible very early MSS system.

In the USA, on the contrary, there is poor cellular coverage and lack of roaming facilities, because mobile services are provided by many proprietary, non-standardized digital cellular systems; therefore the USA is expected to be a primary market for LEO services.

In both cases, the competition represented by private satellite system could be very dangerous for the incumbent mobile firms and for PTOs due to service differentiation, potential by-pass capabilities, and technical incompatibility. This threat is also felt by the European terminal industry which would like to gain a world market for the GSM standard.

On the other hand, ISOs could be the competitive weapon that PTOs could use against their mobile private competitors, or could play on the mobile market arena by themselves.

The proposed new satellite systems have been classified according to their capability or lack of offering real time voice services. "Big LEO" systems, which include both the Low and the Medium Earth Orbit constellations (see Table 2.1), offer the full range of MSS characteristics (global coverage, handheld receivers, earth stations connected to each other and to the fixed network, voice communication, position fixing, paging, fax, low bit rate data communications), and are forecasted not to enter the market before the year 2000, in spite of more optimistic start-up dates promised by systems proposers. This is due to regulatory barriers, nation-by-nation approval process after spectrum and orbital position assignment, and technological barriers, which have an impact on cellular handsets and service prices.

However, "Small LEO" systems (see Table 2.1), involving less risky investments and providing a more restricted class of services (global coverage, handheld terminals only for very low data rates and paging, store and forward messaging with no possibility of realtime voice communication), are expected to be launched in the second half of this decade.

All the new projects, with the exception of PROJECT 21, are US initiatives and, since the USA market is potentially the most receptive with respect to LEO services, the first steps towards licensing are being taken in the USA. After WARC-92 (see section 4) the FCC set up two separate regulatory procedures for Big and Small LEOs, one for frequency allocation and the other licence granting. In the meanwhile, experimental licences regarding all Small and Big systems have already been granted, with the clear intention to grant operational licences at a later date.

The FCC will probably issue final rules governing the new communications services in the fall of 1993 and grant operating licences by 1994.

In this interim period, Big LEO competitors are looking for investors and strategic partners. The most ambitious of the new proposed systems: Motorola's $3.4 billion IRIDIUM system, is, at present, also the most credible, as it has already announced to have secured $800 million in initial financing from an international consortium (see Table 3.1) and plans to seek another $800 million within the next two years.

The previously described trends and events have had a set of relevant effects, although still limited, on the organization of the international satellite service markets and, in particular, on those strategies ISOs are actuating in order to defend their market power against the initial attempts of competitive entries. EUTELSAT continues to preclude end user direct access to its Signatories but is oriented towards allowing multilateral controlled agreements among Signatories for providing capacity to foreign markets. INTELSAT is assuming a more commercial orientation by: allowing its Signatories to have an investment (ownership) share greater than their capacity utilization level; entering for the first time the capital market; signing long-term contracts with PTOs; and raising the aimed rate of return. Lastly, INMARSAT has become directly involved in the race for entering new MSS markets by promoting PROJECT 21. For this reason, it has been accused of unfair competition with the private sector as it will have the advantage of an easier solution to those regulatory problems involved in global coverage and terrestrial network interfacing. This "unfair" advantage is due to the fact that often the PTOs are INMARSAT Signatories and at the same time able to influence national regulators.

3. COOPERATIVE MANUFACTURING AND SERVICE PROVISION IN TIME-BASED COMPETITION

As a consequence of its technology, the SCI is naturally segmented in three wide markets: the launch service market, the equipment market (that can be split into space and ground segments) and the service market. The analysis of the competitive framework of the service market must not fail to consider the vertical relationships among subjects operating in the different segments of the value system (Porter, 1980). In other words, according to Porter's model, the value chain of a service provider is only the last component of a larger value system (represented in Fig. 3.1) including input providers and output users.

The new MSS can be regarded as *composite goods*, or bundles of complementary goods, in which: the contribution of each component to both the service function and quality is distinct and recognizable; the production of the components of the bundle can require specific technologies and abilities (favouring vertical disintegration of production); and the choice of the components of a bundle fitting some end user needs is somewhat flexible (Castelli and Leporelli, 1992). Consequently, as far as the new MSSs are concerned: (1) the *time-to-market* (i.e. the time involved in project development and product marketing); (2) the *cost* of the project (mostly fixed and sunk, directly influencing the break-even tariffs, given the potential market); and (3) the *quality* and the *scope* of the service (which affect users' willingness to pay and their preferences for the MSS and its substitute services), are the final result of the contributions and supply strategies of different actors, public and private, performing or influencing different elementary activities of the product life-cycle.

In the following the nature of competition of the MSS industry is discussed, stressing in addition to the rivalry among actors engaged in the market, also the competitive forces constituted by the bargaining power of the input providers (that is, at the base of the vertical relation organization), and by the threat of new entries (which depends on the entry barriers and on the potential competitive reactions of the incumbent firms). As for the threat of substitute services, this paper limits itself to the previous considerations on terrestrial cellular services (section 2.3).

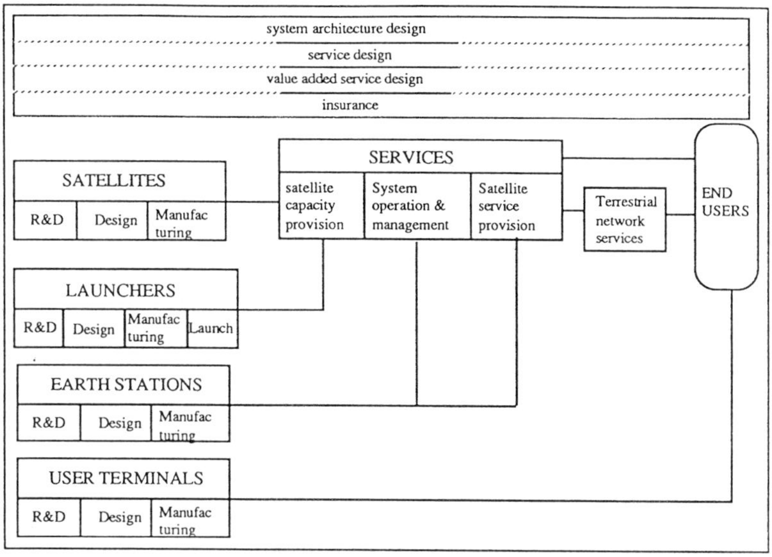

Figure 3.1. Value system of the new MSSs.

3.1 Time-based competition: the irreversible effects of lead time on competitors

The economic relevance of being the first entrant in a market presenting structural entry barriers for late comers makes the entry time a fundamental strategic decision for the positioning of the firm in the industry, and also for its future competitive options (Porter, 1980).

According to the technological description of the SCI given in section 2, a set of entry barriers can be singled out, regarding respectively the manufacturing segment and the service segment.

Production of satellites, earth stations, and user terminals requires high endogenous sunk costs (Sutton, 1992) affecting future product quality and supply capacity, such as R&D and design costs. In addition, this production presents significant learning economies which are the cause of the not easily closed gap between the European industry and that of the USA. A decisive advantage of the incumbent firms is constituted by the variety of uses of R&D and project results (public production input) in conjunction with the positive externalities of the public demand (civil and military) due to experience curves and the possibility of reuse on the commercial market projects financed by government agencies.[13]

The most important entry barriers in the MSS markets are, or could be: (a) natural resources scarcity; (b) configuration indivisibilities in small markets and therefore non-profitable supply for many competitors; (c) clientele base effects (when user learning by using or incompatibility of new service equipments give rise to costs of switching suppliers, as it may happen given that each LEO project has its own specific user terminal, often incompatible with the terrestrial cellular); (d) incumbent's advantages in finding large amounts of investment capital; and (e) operation licences and patents protecting new technologies from imitation in fragmented markets and absence of international standards. An entry barrier peculiar to

[13] For example, the Global Positioning System (GPS) provided by a US military satellite network and now available for civil use (Commision of the European Communities, 1992).

MSS markets is the length of time required for designing, financing, licensing and developing a new system. This can be considered only a temporary barrier, granting a temporary monopoly power to the first entrant, but when added to fill economies and user switching costs, it can become insuperable.

In conclusion, markets for new MSSs are not contestable (Baumol, Panzar and Willig, 1982), i.e. they are markets in which the threat of entry of potential competitors is not sufficient to prevent the incumbent firms from making positive profits, unless regulatory constraints on tariffs are imposed (Rochhia, 1990).

The first mover advantage in profiting from a new MSS market has given rise to a race among potential competitors for the control of specialized assets and for preempting the market through the commitment constituted by investments in indivisible space capacity.[14] Moreover, with the proposed access techniques (CDMA and TDMA) only a small number of MSS systems can coexist in the allocated frequency band, and this, added to the investment amount and to the relative scarcity of gateways to the terrestrial networks, could result in the creation of dominant positions if regulatory agencies allow the market to decide (Commission of the European Communities, 1992).

From another point of view, if the relevant market allows only one or a limited number of suppliers to break even, as seems to be the case of LEO satellite systems offering worldwide coverage (Commission of the European Communities, 1993), time-based competition, when it actually takes place, causes duplication of investments prior to the emergence of a "winner". This market failure requires specific regulatory interventions. In particular, in the USA, the FCC grants licences to single phases or development activities (such as experimentation licences) and limits the allowed investment in each phase. In this way several systems can be conceptually designed and experimented, while only the finally selected systems will be completely built and deployed, so that both the advantage of competitive innovation is obtained and wasteful duplicated investments are avoided.

The worldwide economic and political importance of the proposed global MSS systems necessitates a new global regulatory regime under which such systems would be licensed on the basis of equitable access possibilities rather than on the old principle of first come, first served.

3.2 Market segments and vertical relations: cooperation among actors and competition among systems

A particular feature of the SCI that decisively contributes to the definition of the firms' strategies is the inversion, with respect to all other TLC technologies, of the ratio between the total sales of the equipment market (commercial satellites+ground stations+user terminals) and the total sales of the service market.[15] The progressive evolution of the satellite communication market towards the direct supply of services directly to the end user, and the continuous terminal cost decrease, are furthering the growth of the ground equipment market (OECD, 1992)

The fact that profit opportunities of the space and ground satellite equipment markets depend on the diffusion of satellite services, as has already been verified in the past, has led

[14] The effectiveness of this strategy is confirmed by the reiterated charges against INTELSAT of creating excess capacity in order to preserve its market power from deregulation.

[15] In 1991, in the USA the sale of commercial satellites and earth stations was approx. twice that of the service sales, while for other TLC markets this ratio was approx. one to five.

manufacturers to vertically integrate the capacity or application provision segment. This strategy is aimed both at granting the demand and indirectly favouring the growth of their main activity sector through the diffusion of services, and at entering markets with high potentiality of growth and profitability.[16] A further incentive to the integration in the downstream market for the manufacturers could be constituted by the convenience to increase production beyond the external demand level in order to derive the optimal benefits from scale economies.[17]

Strategies shaping vertical relations are the key to analysing the evolution of the SCI structure in the near future. The proposals of innovative satellite systems for global personal communication services all originate from vertically integrated firms, of which the most well-known system is a typical example: the IRIDIUM system, considered by many a vehicle by which Motorola can obtain worldwide leadership in the terminal provision industry (Commission of the European Communities, 1992).

For the most part, competitors in the new MSS industry take the form of organizations centred around a leading figure: the *key investor* or the *technology provider*, i.e. the entrepreneural subject who: perceives a profit opportunity related to his main competencies and markets, conceives the system and, on the basis of the project, integrates vertically some activities and makes strategic alliances with subjects owning complementary competencies or controlling accesses to markets. The rationale is that an innovator who owns, rather than buys, complementary assets is also able to capture the benefits deriving from an increased demand of these assets determined by the innovation (Teece, 1986).

In this entrepreneurial role PTOs are increasingly substituted by space industry and radio equipment firms. However, PTOs, or, in their stead, ISOs, will still have a role to play in the future, both for the influence they can exert on the national regulatory and international coordination processes, and as the main channel to reach end users. In the words of Robert Kinzie, IRIDIUM's chairman: "*We want them (INMARSAT) to offer our product to their customers. They have the store that serves the market, we want them to give us the shelf space and sell our product.*" (Manuta, 1992). In fact, Motorola though still trying to obstruct Project 21, by requesting that the FCC impede Comsat participation, has repeatedly proposed the merging of the two proposed systems or, alternatively, has offered part of its IRIDIUM system to INMARSAT for use over international waters.

In Table 3.1, the corporate structure and the equity and non-equity investment shares, current and expected, of the two major Big LEO projects are given, in order to show the cooperative production process and the composite-good nature of the MSSs that will be developed and distributed through the coordinated efforts of a variety of subjects.

An anticipated entry strategy regarding new MSS markets may be better put into action by coalitions of specialized firms, rather than by a single firm that would have to create within itself the capabilities for performing many high-tech, time-consuming, investment-intensive specialized activities. Consequently, time-based competition gives a further incentive to cooperative agreements allowing better exploitation of time-cost trade-offs.

---

[16] In 1991, the USA leading satellite manufacturers, Hughes (32%) and G.E. (23%), owned over half of the domestic satellite capacity, while the remaining share was owned by GTE (35%) and AT&T (10%) who manufacture and market ground segment equipment. European and Japanese manufacturers have adopted the same integration strategies and US manufacturers have entered overseas service markets through participation in local service companies (OECD, 1992).

[17] The US company Hughes Communication, for instance, in 1991 directly supplied the space capacity of 9 satellites of a series in which they had already sold 39 (OECD, 1992).

Table 3.1
European participations in selected LEO project coalitions

System	Key investors	Technology providers and Partnerships		Other European opportunities for value added	
		European	Non European	Industrial	Services
IRIDIUM	IridiumInc. (system ownership and operation) SHAREHOLDERS Motorola (34%) Japanese group (15%) Saudi Arabian groups (15%) Raytheon (5%) Sprint corp. (5%) STET (5%) OTHER EXPECTED EQUITY INVESTORS - Gateway operators (up to 20 worldwide) - Indipendent primary service providers	- British Aerospace (spacecraft components) - Matra Marconi (power supply boxes) - Deutche Aerospace (GaAs solar panels)	- Motorola Satellite Communication Inc (spacecraft construction) - Raytheon (antennas, $300M contract) - Lockheed Missiles & Space (satellites, $700M contract) - GE Management and Data Systems (operation center software) - Com Dev Ltd (inter-satellite transmission links) - Hutchinson Telecomm (landbased)	- Launch services - GSM gateway switch manufacturing - Local production of handset by Motorola and licencees	- Operation of the gateways to the terrestrial networks - Resale and distribution
GLOBALSTAR	Loral Qualcomm Satellite Services Inc. SHAREHOLDERS - Loral Aerospace Corporation - Qualcomm Inc. - Alcatel NV - Alenia Spazio	- Alcatel Espace - Aerospatiale - Alenia Spazio - DASA - Service distributors - Passive investors	- Service distributors - Passive investors	- Satellites (contract), - Launchers (limited competition) (50-60%) - Gateways (locally negotiated contracts) (50%) - Terminals (open under Globalstar type approval) (25-35%)	- Gateway operation - Resale and distribution

In the emerging MSS market, in which technological and institutional changes modify the expectations on the evolution of the competitive framework and on the strategic firms' options, the cooperative interactions among supply subjects can assume a wide variety of configurations. They range from the less restrictive price transactions, to selling and using licences of assets and technologies, to joint ventures, to the more restrictive full vertical integration, according to the extent of specificity and appropriability of the innovative investments required of each partner. These vertical cooperative agreements can have a series of compatible strategic objectives (Contractor and Lorange, 1988), some of which are treated below.

- *Risk and financial exposure sharing*, in the development and future management of a new, commercially and technically untested, satellite system in a context of uncertainty over the long-term international regulatory order of the space infrastructure (Commission of the European Communities,1992). The client-supplier relationships in the production chain of new MSSs cannot be mediated by the market when large specific and irreversible investments are required in conditions of uncertainty of the output demand, as in the case of new conception equipment designing and manufacturing.
- *Production rationalization and scale economies exploitation*, through the allocation of the production activities to subjects able to perform them at a minimum cost, thereby also enjoying greater scale economies as a consequence of a greater production volume of specialized subjects. This strategic trend could lead, for example, in a hypothetical common Euro-American project, to the allocation of satellite and earth stations manufacturing activities to US actors and of launch activities to European actors, who control more than 50% of this market.
- *Exchanges of complementary skills and technologies and of patents or licences*, required by the scope of the technological challenge (involving the design and production of new types of satellites, ground stations, and user terminal equipments, the individuation of more efficient frequency band exploitations, etc.) which is beyond the technological skills and financial means of a single subject. This is the case of the R&D alliances or of joint project development and service management and marketing. Moreover, by joining single know-how and operation licences, it may be possible to enter the market more quickly or with improved service quality and variety, or to have access to new markets.
- *Access to a global market*, which requires national partnerships for supplying services in different countries and, in presence of regulatory entry barriers, makes joint ventures or strategic partnering between global service providers and national or regional gateways a suitable tool to avoid blockage power of the national regulation.
- *Industrial growth*, by firms that are not leaders in their own industry, as in the case of some European space industry firms which by participating in US technology based systems can benefit from learning by doing and spillovers.
- *Competition blockage and market preemption*, when a cooperative agreement is made with a (potential) competitor, as may probably happen for some joint ventures developing LEO projects, due to technological, frequency allocation, or market size matters.

In Figure 4.1 the macro activities of the development and supply process of services based on new (non-geostationary) satellite systems are shown in order to point out the key features of the arena in which the rival coalitions compete: government support to industry through public R&D funding, external effects and spillovers to industry, and blockage power of international organizations and local regulatory agencies.

4. CHALLENGES TO REGULATION PROCEDURES AND INDUSTRIAL POLICIES

Global satellite services have an international market, use spectrum resources that are allocated internationally, but are regulated at the national level and heavily affect national manufacturing industries and technology goals.

Markets for international satellite services involve sovereign nations, each free to choose its own domestic regulatory regime and industrial policy and to pursue its own aims in international trade policies and in the participation in the decision processes of the international organizations charged with defining and managing rules for the exploitation of common natural resources. Differences in goals between countries lead to conflicts of interests, but even when countries have similar goals they can fail to coordinate their policies.

At a domestic level, public agencies are involved in at least two important ways in satellite market development, in addition to the role they play, in some countries, as service providers. First, they act as market regulators, and in various ways restrict market transactions (access to the market, resource utilization, price and quality of supplied services, technical standards, etc.) in order to obtain socially efficient behaviour by profit-oriented subjects with market power. Second, they pursue system-wide economic goals through industrial and technology policy, promoting R&D and those innovative projects that are socially desirable but might not be commercially viable for private subjects.

At an international level, government agencies interact in the different bodies of the International Telecommunication Union (ITU), and in regional organizations, such as the Conference of European PTOs (CEPT), for frequency spectrum management and world-wide standards setting, and within the negotiation of the General Agreement on Tariffs and Trade (GATT), for international service trade.

Coordination problems arise at both levels. At a domestic level between regulation and public policy or between goals of national and international policy, and, at an international level, among national strategies because in an integrated world one country's economic policies affect other countries, giving rise to reactions (and retaliations) in the wide spectrum of the industrial sectors (Krugman and Obstfeld, 1991).

In this section, some of the specific problems and goals of public intervention (market regulation and industrial policy) in the satellite industry are briefly illustrated and the ways in which the decisions of the supply agents and of the national or international public bodies are affected by the conflicts among these goals (allocative efficiency, distributive equity, innovation support), and by the international coordination procedures that should take care of some of these conflicts, are analysed.

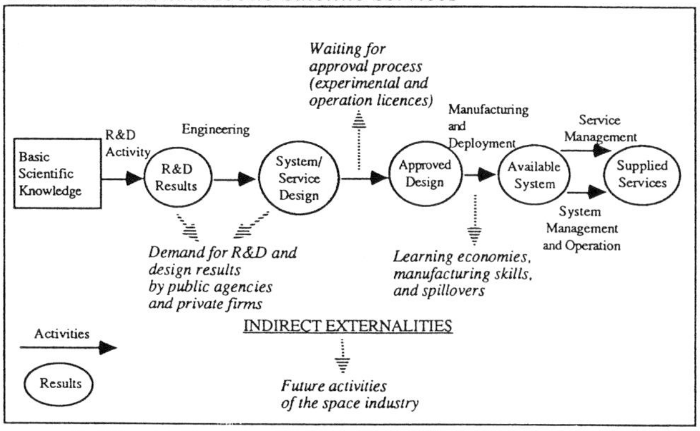

Figure 4.1. Macro activities in the life-cycle of the new MSSs.

4.1 National regulation and worldwide coordination in the frequency assignment process

The allocation of spectrum and orbital position is restricted by a complex international coordination process aimed at granting equitable access to the resources to all nations, guiding the pre-operational choices and the investments of the different administrations that cooperate to supply international communications services, and protecting authorized users from interference.

The International Telecommunication Convention and the Radio Regulations establish procedures for the:

(i) *allocation* of frequency bands to earth or space radio communication services;

(ii) *allotment* of radio frequencies or channels, within the framework of multilateral agreements, for usage by one or more administrations in one or more countries or geographical areas;

(iii) *assignment* of radio frequencies or channels for usage, under given conditions, granted by national administrations to operators of a radio station.

The International Frequency Registration Board (IFRB) is the permanent Body of the International Telecommunications Union (ITU) charged with information gathering and dissemination, technical analysis, and coordination process. World and Regional Administrative Radio Conferences (WARC and RARC) periodically approve, on the basis of a one nation, one vote rule, updates to the ITU Table of Frequency Allocations and Radio Regulations resulting from technical studies performed by ITU (through CCIR and CCITT)[18] and requests by national administrations (CCIR, 1988) (Maral and Bousquet, 1980).

Phases (i) and (ii) take place at an international level (global or regional), while phase (iii) is under the total control of national agencies. The main consequences of this multistage coordination and assignment process for the strategic opportunities of the SCI supply agents concern the powers given to national administrations, i.e. to the governmental department or service that represents a nation in the assignment process. Even when, as in the case of the US FCC, this agency has only a regulatory mission, and much more so when the government is

[18] Respectively, Consultative Committee on International Radiocommunication (CCIR) and Consultative Committee on International Telegraph and Telephone (CCITT).

represented by a PTO, national interests and points of view may prevail over abstract allocative efficiency issues at different levels of the allocative process.

With regard to the frequency allocation stage, consider, for example, that the last world conference, WARC-92, approved worldwide frequency allocations to LEO systems after strong pressure by the American delegation which was concerned with furthering the interests of private US companies engaged in LEO proposals. On the other hand, European countries successfully defended worldwide allocations to terrestrial mobile services, compatible with the European GSM standard, in the hope of a worldwide success of the European manufacturers (Sung, 1992) (Stiglitz and Blanchard, 1992).

In a subsequent stage, the providers of international services may be blocked by the need to obtain an operational licence in each of the countries in which they may operate. A further obstacle, even if only temporary, consists in the relevant and discretionary lead time needed to complete the advance publication, coordination, and notification process necessary to bring into use a frequency assignment. From this follows the competitive advantage of systems able to survive at a national or regional level relative to systems that need a wider market to break even.

Finally, note that countries that are members of ISOs are subject to parallel coordination processes within these organizations when separate satellite systems are proposed and that the possible economic harm to Signatories must be assessed. In other words, an incumbent has to agree to new market entries.

4.2 Technology and regulatory policy motivations and problems

Different types of public interventions in new MSS markets, during the various phases of R&D, project development, and supply, are to be hoped for in order to avoid inefficiencies in a variety of situations.

4.2.1 Allocation of scarce natural resources to efficient usages and users

The frequency band allocated by WARC-92 for commercial applications of mobile communications based on Small and Big LEO systems (Manuta, 1992) (Stiglitz and Blanchard, 1992) and the orbits of a particular class are scarce resources, due to the problem of interference between radio channels or orbital paths. Consequently, they need a rationing process that, in the short run, is compatible with criteria of static social efficiency, and, in the long run, does not hinder the development and application of new technologies which allow a more intensive resource exploitation. In other words, the choice should not only be limited to currently available alternative uses of the spectrum, but should also take into account how current allocations could restrict future possible uses (today it is practically impossible to revoke a licenced allocation).

In this context, regulation authorities must choose allocation mechanisms which ensure that resources are assigned to the most socially valuable use. The main difficulty derives from the lack of well defined property rights and therefore of a market for frequency spectrum and orbit positions (Spulber, 1989), which being universally common resources are often shared according to a first come, first served principle and, moreover, with no flexibility in the allocation (Benzoni, 1990) (Benzoni, 1991). Even if property rights were assigned to private subjects and a market was created for these resources, it is not sure that the market would be able to select the most efficient projects in terms of costs and spectrum usage. The market allocation could be inefficient because the privately most profitable usage of a resource, that determines its market price, may not be the socially most valuable usage. Furthermore, residual externalities (e.g. interference) and transaction costs (e.g. coordination of decentralized technological choices) (Hart, 1991) could critically affect the efficiency of the market allocation. The fact that natural resources are considered common property necessitates a centralized assessment of the social value of a proposed use for a given resource.

4.2.2 Time dimension effects

In the MSS market, the regulatory intervention could be justified in relation to another specific element: the time dimension, which adds complexities to the allocation process. It is not necessarily true that the first available project is the best one from a social point of view (it could be exceedingly costly or spectrum wasting); moreover, an incumbent system freezes orbital and spectrum resources for a long period of time and the sunk investment of the incumbent firm makes it difficult to recontract expired allocations. Consequently, a first come, first served policy for spectrum allocation has draw-backs that should be compared with the positive effects on innovation of time-based competition.

4.2.3 Sustainability and efficiency of possible market configurations

The high sunk costs and the relevant overcapacity typical of the SCI, deriving both from indivisibilities and from incumbents' market preemption strategies (Rochhia, 1990), make it necessary to control and limit the incumbents' market power, in the absence of competitive pressures from substitute technologies.

The market regulatory authorities must decide if the market allocation mechanism can solve the problem of selecting the most socially efficient industry structure or, on the other hand, judge if it is better to regulate access to the market (in other words, they need to evaluate whether a market is a natural monopoly and if so, whether it is a sustainable natural monopoly). Moreover, they should assess if the market is contestable, i.e. they have to decide if potential competition is sufficient to obtain socially efficient price and quality behaviour by the incumbent firm or if, alternatively, it is better to explicitly regulate also these variables. In the real world, answers to these questions are not clear-cut because not only market failures, but also market regulation failures have to be taken into account (Spulber, 1989). For example, theoretically wasteful cost duplications could be accepted if competition is considered preferable with respect to regulation as an instrument for market power control.

The currently proposed MSS systems widely differ in terms of architecture, scope of services, geographical coverage, integration with terrestrial networks, etc. Furthermore, they could operate in competitive environments which differ in terms of price and quality of substitute services (other MSS systems, terrestrial mobile systems), in terms of contractual relations with PTOs, and in terms of service access modalities for the end users. It is then necessary that government agencies be able to evaluate the economic viability of specific projects in the different possible scenarios. A methodology is then needed to estimate the costs of the system and of its potential ability to attract subscription and usage demand for given rate levels sufficient to break even. A joint evaluation of the proposals of new satellite systems by a global regulation authority would allow obstacles due to incomplete information on cost structure and market demand to be overcome, for the purpose of avoiding the additional social costs deriving from the supply of a given output by an inefficient number of firms or by inefficient firms. It would be even more important to be able to obtain rough-cut estimates of the cost, as a function of the main design parameters (geographical coverage, capacity, service scope and grade), and of the corresponding demand, as a function of the most important determinants of user behaviour, in order to evaluate not just specific designs but also general system concepts and industry configurations.

4.2.4 Technical and economic viability of time-phased introduction of systems for MSSs

A more detailed analysis of the system viability would require a treatment of the dynamics of the system deployment. If a system could be deployed in sequential stages and be able to supply, already from its early deployment stages, a subset (in terms of geographical coverage, scope and quality, capacity, etc.) of the designed services, this would modify the project cash flow, as a function of time. Most importantly, it would change the strategic options of the supplier in the game he plays with his competitors.[19]

This opportunity, however, must be carefully assessed from both a technical and an economic viewpoint. In particular, the project risk reduction is affected by the fraction of the cost that can be postponed and, possibly, avoided, and by the level of performance of the partially deployed system. The gradual introduction of new services requiring irreversible and specific investments could be socially efficient, in so far as this allows the experimentation of technologies and market answers, avoiding waste of resources. From a strategic point of view, the adoption of a system concept prone to time-phased deployment could modify the expectations of the competitors regarding the agent's commitment and thereby lower the barriers to entry and competition.

4.2.5 Interaction between regulation and industrial policy

Regulation, industrial and foreign trade policy interact in a number of ways within the SCI. Imposition of standards, at regional or international levels, favours firms that first create and diffuse the standards and that in their innovative activity could have benefited from public support. A barrier to foreign trade, on the one hand, is a subsidization to national industry, protected from foreign competition, but, on the other hand, prevents domestic firms from enjoying spillovers and positive externalities derived from operations in bigger and more advanced markets opened by trade liberalisation. Conversely, the liberalization of the service segment coupled with the protection of the equipment segment could be a policy aimed at indirectly promoting the growth of the national manufacturing industry (Analysys Report, 1993). R&D subsidies or temporary monopoly power granted to the innovator could favour technological change and increase the international competitiveness of industry (Stehman, 1992).

5. OPPORTUNITIES AND OPTIONS FOR A EUROPEAN INDUSTRIAL POLICY

Since all the LEO system proposals are of US origin (even if many European companies participate in some of these projects) it is easy to foresee a US world dominance in new MSS operations. European countries, whose satellite equipment firms currently cannot enter US markets (KPMG, 1993), will most likely try to slow down the diffusion of LEO systems. They

[19] From Table 1 it is evident how many of the Small and Big LEO projects have been designed to deploy the space capacity in two or more subsequent phases. The FCC has been called upon to assess technical and economic viability and to licence the various phases of the project.

will do so also with the aim of avoiding possible interference with their fixed services and protecting their investments in terrestrial cellular or regional MSS programs. So, at least in its initial phase, the MSS market is expected to be a regional market and to expand in the USA.

The FCC procedure of licensing not only the actual service but also the experimentation, building and launching of the system, is leading to the licence granting of the current proposals well before the global conditions for equitable access to frequency spectrum and worldwide standardization, necessary to global coverage and roaming, have been achieved. In addition, the WARC-92 allocation of frequency spectrum in the 2 GHz band in the USA only, has given to the United States the possibility of favouring the domestic start-up of the new LEO systems from 1996 to 2005 denying use of this spectrum to potential foreign competitors. All this will probably result in the imposition of US system and service concepts as standards, impeding the competitive development of various projects and leading to the dominance of the US industry and regulation to the detriment of others.

The European delay in the space segment and earth segment infrastructure and capacity supply as well as the strategic and economic importance of the space activities give rise to unsolved public policy dilemmas for the EEC Member States. On the one hand, they must decide either to develop a common global space programme (from R&D support to fixing competition rules to undertaking common projects) or to leave satellite communication to national policies; and, on the other hand, whether or not to open internal markets or to partially protect them, in an attempt to promote the growth of a competitive European industry.

In the past, European countries have supported their aerospace industries through public demand policies managed by national agencies and ESA, PTOs and EUTELSAT. As a consequence, they have achieved technological capabilities for the development of state-of-the-art and innovative products. However, the dimension of the market served by the US companies and their cumulated production (governmental and defence programmes gave rise, in the USA, to a demand ten times greater than the demand of the commercial market), i.e. their scale and learning by doing economies,[20] give them a competitive advantage in open market conditions (all the orders of private organizations and 70% of the world market for delivery of commercial satellites from 1992 to 1997 are directed to North American prime contractors (OECD, 1992)).

There are strong interconnections between public and commercial markets: the R&D programmes undertaken for the development of public funded space systems have a direct application in commercial projects; both in government and commercial markets launcher systems, space vehicles and equipment can be the same or very similar (Commission of the European Communities, 1992).

In this context, the large-scale investments required by the proposed high-performance MSS could be the occasion for further industrial policy initiatives aimed at bridging the competitive gap between the European and the North American industry. Therefore, the (relevant) costs of European projects could be allocated in part to the industrial policy goal of improving the productivity of the European high-tech industry.

However, similar goals could also be pursued with different strategies (for example, external growth or strategic alliances, such as the Space Alliance, among European and American companies). The effects of the different possible lines of conduct must be comparatively

[20] Due to 15 years of experience (and 150 operational satellites) before the start-up of European industry in the early 1980s.

evaluated, taking into account the complex and composite nature of the results obtained by each of the agents involved in the projects.

At the moment a European MSS project is not envisaged, but European public agencies and companies need, in any case, to decide how to be involved in current projects, both in the development and in the operational life-cycle stages and both from a regulatory and from an industrial policy point of view.

The following problems in the process of competitive growth of the European space industry are widely recognized:

- unregulated time-based competition among existent, US-led, projects would result in a market structure that is not necessarily efficient from a welfare point of view and that could harm the growth and competitive prospects of the European space and telecommunication industry;
- it is not clear how regulation procedures at the national level could be coordinated to take into account the multinational character of satellite service markets and the time-phased availability of the competing systems;
- European bodies involved in regulation and industrial policy have only recently begun to design common strategies in the satellite sector which take into account several, possibly conflicting, goals such as the deregulation and liberalization of competition in the service markets, the survival of the European space industry, and the improvement of its competitiveness;
- in principle, several strategies are possible for European public bodies and firms, from strict protectionism to "laissez faire" (Analysys Report, 1993), with the promotion of coordinated all-European projects, and the support to European members of transatlantic consortia as middle ground possibilities.

6. SUMMARY AND CONCLUSIONS

This paper identifies and analyzes the major economic, regulatory and public policy issues related to the development and market entry of new personal MSSs provided through LEO satellite systems.

The relevant entry barriers, due to sunk costs, natural resources scarcity, heterogeneity of national regulations, international coordination problems, greater for late comers than for innovators, and the scope and specificity of required technological skills have constituted a strong incentive to cooperative strategies within international coalitions engaged in a time-based competition.

National industrial policy and foreign trade goals often interfere with the allocative efficiency aims of regulatory authorities, and the powers and the level of dependence on the government of these authorities widely differ among countries.

The competitive game is deeply influenced by public intervention, not only because of regulatory procedures but also due to ties between government (civil and military) and commercial markets. First, a high public demand allows the exploitation of cost advantages determined by scale economies, which is also the case in an open market framework. Second, the results of space projects publicly financed have direct applications in commercial projects, because launch systems, satellite and other equipment can be the same or conceptually very similar (Commission of the European Communities, 1992). Therefore the large investments required by the new LEO systems could constitute an opportunity for industrial policies to improve the domestic industry's international competitiveness.

In summary, the authorization of a satellite service with international coverage is the result of a cooperative benefit sharing game: each country exercises its power in order to maximize the flow of benefits (in terms of consumer surplus, national industries profits, other possible political and social objectives) accruing to it.

The appearance and the envisaged diffusion of global satellite services underline the necessity for new approaches by national regulatory authorities and international coordination bodies whose decisions have global effects, in spite of the fragmentation of national sovereignties and interests.

The promotion of a coordinated development of the new services and markets, with objectives of global efficiency, requires a number of public policy decisions that involve the choice of appropriate trade-offs among a set of goals. The most important decision areas and related criteria are the following:

- *Spectrum sharing*
  guarantee equitable access to natural resources
  take into account possible future systems generation
  harmonize national assignment procedures

- *R&D*
  coordinate efforts in supranational bodies

- *Standardization and intellectual property rights*
  solve problems related to the intellectual property of the innovation
  codify global standards prior to the diffusion of new systems and services

- *Market access regulation*
  coordinate national regimes of operational licence granting
  grant experimentation licences in order to select the most efficient projects

- *Network interconnection*
  take into account, in the licence granting and market regulation processes, the competitive (local loop bypass) and cooperative (complementary service supply) interactions of new systems with existing terrestrial networks.

At the present time, three different scenarios can roughly speaking be envisaged:

(1) unresolvable difficulties in solving coordination problems would imply a protectionistic framework, reflecting present-day fragmentation in national regulatory and public policy goals, with separate licence-granting authorities in separate markets;

(2) INMARSAT Project 21's victory in the race would favour an internationally cooperative regulated framework, modeled on the ISOs' agreements, with a cooperative service supply by a world-wide MSS consortium and national distribution by regulated gateways;

(3) rapid liberalization of telecommunication service trade and market access would determine a multilateral deregulation framework, favouring the development of multinational projects, without borders across markets and with international regulatory agencies granting global licences for global markets.

REFERENCES

Analysys Report, 1993, Copyright Commission of the European Communities, The impact of non-European competitors on a deregulated European satellite telecommunications market, report for CEC DGXIII.

Baumol, W.J., J.C. Panzar and R.D. Willig, 1982, Contestable markets and the theory of industry structure (Harcourt, Brace, Jovanovich).

Benzoni, L., 1990, La rareté a-t-elle un prix?, Le Communicateur, Été, pp. 67–87.

Benzoni, L., 1990, Le spectre herzien: Bien public ou bien privé?, Colloque ENST/MIT.

Castelli, F. and C. Leporelli, Ottobre 1992, Qualita' e competitivita' nell'offerta di servizi: Il coordinamento delle politiche di offerta in risposta ai mutamenti tecnologici ed istituzionali, Atti del IV Convegno AiIG, Bari.

CCIR, 1988, Satellite communications handbook, ITU, Geneva.

Ciarniello, A., 1992, Global mobile satellite systems, COST 227TD(92)39.

Commission of the European Communities, 1990, Green book on satellite telecommunications, COM (90) 490 final.

Commission of the European Communities, 1992, Hearings on non-geostationary mobile satellite systems (Low earth orbiting satellite systems: LEOS).

Commission of the European Communities, 1992, Mobile satellite telecommunications services: Low earth orbiting satellite systems, ECC-Discussion Document No. 92–09.

Commission of the European Communities, 1992, The European Community and space: Challenges, opportunities and new actions, COM(92) 360 final.

Commission of the European Communities, 1993, Communication from the commission and proposal for a council resolution on satellite personal communications, XIII/F1-C LEO-COM2c/PV.

Contractor, F.J. and P. Lorange, editors, 1988, Cooperative strategies in international business (Lexington Books).

Convention Establishing the European telecommunications satellite organization "EUTELSAT", 1984, Supplemento ordinario n. 1 G.U. n. 335.

Gribbon, J., 1991, VSAT Networks fill global banking needs, Satellite Communications, pp. 33–38.

Hart R.F.G., 1991, Orbit spectrum policy: Evaluating proposals and regimes for outer space, Telecommunications Policy, pp. 63–74.

Howes, K.J.P., 1992, Fiber versus satellites: The competition heats up, Via Satellite.

KPMG, 1993, Copyright Commission of the European Communities, Barriers to European industry trading in the US satellite communications market.

Krugman, P.R. and M. Obstfeld., 1991, International Economics: Theory and policy, Harper-Collins Publishers Inc.

Lorenz, M., 14-17 June, 1992, The European satellite communication industry—market dynamics and deregulation, 9th ITS International Conference, Sophia Antipolis (Cannes).

Manuta, L., 1992, WARC World impact: New systems get spectrum assignments, Satellite Communications, pp. 26–30.

Mara, G., J.J. De Ridder, B.G. Evans and M. Richharia, 1991, Low earth orbit satellite systems for communications, International Journal of Satellite Communications, n. 9, pp. 209–225.

Maral, G. and M. Bousquet, 1986, Satellite communications systems (John Wiley & Sons, New York).

Noam, E., 1992, Telecommunications in Europe (Oxford University Press).

OECD, 1988, Satellites and fibre optics, Information Computer Communications Policy.

OECD, 1992, Directorate for science, technology and industry, Satellite communication: Structural change and competition.

Pelton, J.N. and M.S. Snow, editors, 1977, Economic and policy problems in satellite, (Praeger).

Pelton, J.N., March 1993, The race is on, Satellite communications, pp. 30–35.

Porter, M.E., 1980, Competitive Strategy: Technique for analysing industries and competitors (The Free Press, New York).

Poskett, P. and P. McDougal, 1990, WARC 92: The case for mobile satellites, Telecommunications Policy, pp. 355–363.

Rochhia, S., 1990, La structure des marchés des communications internationales par satellites, Revue D'Économie Industrielle, n. 51, pp. 52–74.

Roller, L.E. and L. Waverman 1991, The optimal structure of the European satellite operation industry, report submitted to ESA.

Seitz, P., March 8–14, 1993, Big LEO negotiations slow to progress, Space News.

Snow, M.S., 1986, Cost structure in international satellite communications: statistical estimation of natural monopoly in Intelsat with implications for business planning and public policy, 6th International Conference on Forecasting and analysis for business planning in the information age, Tokyo.

Stehmann, O., 1992, Facility based competition in Europe: The case for a separate satellite system, Telecommunications Policy, March, pp. 135–146.

Stehmann, O., March 1992, Facility based competition in Europe: The case for a separate satellite system, Telecommunications Policy, pp. 135–146.

Stiglitz, M.R. and C. Blanchard, 1992, Frequency allocations accommodate new commercial applications, Microwave Journal, pp. 28–29.

Sung, L., 1992, WARC-92: Setting the agenda for the future, Telecommunications Policy, pp. 624–634.

Sutton, J., 1992, Sunk costs and market structure (The MIT Press).

Teece, D.J., 1986, Profiting from technological innovation: Implications for integration, collaboration, licensing, and public policy, Ricerche Economiche, Vol. XL (4), pp. 607–643.

Tirole, J., 1989, The theory of industrial organization (MIT Press).

The Race to European Eminence
E. Bohlin and O. Granstrand (Editors)

Direct Foreign Investment in Telecommunications: Comments

Eli Noam

Professor, Columbia University Graduate School of Business
Director, Columbia Institute for Tele-Information

Why is direct investment in telecommunications an issue now? The historical context provides answers. On the supply side a general trend towards liberalisation has unshackled PTOs and led to their reorientation into non-territorially oriented companies. For example, Sweden's Televerket changed its name to Telia—there is no territorial reference anymore, meaning that it is not necessarily a Sweden-oriented enterprise but an internationally oriented firm. Such companies are entering global activities. At the same time, on the recipient end of direct investments, there is significant reorientation too. Developed countries began to liberalise and open up to outside investment and in the less developed world there is great pragmatism about who can come in and also about national ownership. As these two trends come together, they raise the question: is it a brief infatuation or a match made in heaven?

International direct investment has always existed, especially in the early years of telecommunications. It was only later that telecommunications came to be largely national, territorial and governmental. But in the early decades of telecommunications, there was much Bell-company investment in Canada, Sweden, France, Great Britain and several other countries. Later ITT ran the Spanish and Chilean systems. Most of these involvements disappeared and the question is why? Are the same cyclical dynamics going to reassert themselves again?

Among the reasons for abolition of foreign ownerships in the first era of internationalization are, firstly, economic sovereignty and, secondly, the performance of those companies abroad. These forces will be important at this current stage of internationalization as well.

To develop this argument further, some disaggregation of foreign direct investments and countries involved is necessary. Let us first address the developing world.

After gaining independence from colonial powers, the original trend was clearly to have one's own telephone system. Today, it is becoming more acceptable to think and talk about privatisation and direct foreign investment. But there is a serious divergence of expectations here: while the LDCs expect investment and expertise, the international telecommunications companies expect revenue generating operations. In a past conference a speaker from Argentina spoke enthusiastically of how Spain's Telefonica was going to bring in money and build up the Argentinian system. The Telefonica speaker, on the other hand, stressed how this company, under pressure to improve operations in Spain itself, would bring in expertise, while investment funds have to be generated locally.

Faced with such divergent expectations, a certain disappointment is therefore inevitable. Developing countries expect a first-rate telephone system within a few years. But the amounts of money necessary are prodigious. To get a sense of this magnitude: the German—formerly West German—telecom organisation must invest 60 billion marks directly in East Germany in one decade, plus, probably, 10 billion marks worth of expertise. This is about 40 billion dollars for a region of about 15 million people, or approximately 10,000 dollars per household.

Obviously that is not an economic calculation. Its purpose is multifaceted, yet a private firm is not going to put 40 billion dollars in 10 years into just one country, which wasn't so backward to begin with, merely to get the telephone service up to par. Especially if it can't repatriate its profits.

Furthermore, it will be difficult to self-generate profits locally. Telefonica plans to make money in Argentina and plough it back into the telephone network. This sounds attractive but consider the applications. Prices are either set by competitive markets, or they are set administratively. If they are set administratively, they are invariably politically sensitive. It is hard enough in the American regulatory system, involving home-grown American firms. Now imagine the same situation where nationalism is part of the discussion. Suppose an American company wanted to increase prices by 20% in Argentina. Political demonstrations and calls of "Yankees go home" would ensue. And while it can be argued that the profits will go back into the network, realistically, the interests of the people and companies who have to pay telephone bills will have a powerful voice in the matter.

Related to disappointed expectations are restrictions on the investing companies. These are typically regulated companies like US West or BT that have certain obligations at home. Therefore, if they do something abroad, the question invariably asked in their home country is "how come you've got money left over to invest abroad?". The example of Telefonica emerges again as a good one. Telefonica was starting to go into Mexico, Argentina and Chile, when people in Spain started to say "wait a moment, we've got lousy service in Spain, how come you have time and money for Latin America?". It is not easy to get away from it all.

The problems and barriers notwithstanding, there are several reasons for direct investment. In Latin America, foreign investment was driven by debt swaps. Countries were so heavily in debt that they traded in their debts and basically sold national assets. Eastern Europe is also in a historically unique period. These countries are relatively advanced, but not in their telecommunications infrastructure, and want to upgrade quickly. They also experienced a certain historical resentment against government solutions. That sentiment is fading—partly as expectations are disappointed.

An additional reason is the opportunities of new services. But in this area, the market segments that tend to be profitable and easy to do, such as mobile communications, are also coveted by local investors. If they do not have the expertise, presumably they can get it through turnkey solutions from foreign manufacturers. A related argument for local ownership is that those specialised segments such as cellular communications are needed to make money that can be put into basic service.

And what about direct investments among the industrialised countries? Here, the barriers are falling and the possibility of cross investments is good. The primary reason is that economies of scale are reasserting themselves. In the past, the incumbents used arguments of economies of scale and natural monopoly to defend national monopoly, but not to advocate supranational companies. Therefore, there were e.g. separate Belgium and Luxembourg telecom organisations. Once the barriers are lowered, expansion and joint ventures become possible. This way leads to the more powerful PTOs like France Telecom or Deutsche Telekom to expand while the smaller PTOs form alliances with one another.

However, there are several problems with this type of expansion. As long as the companies are near-monopolies, they will be regulated in some way, only now they would operate across multiple jurisdictions. The experience in the United States has been that once you have multi-jurisdictional entities you get into complicated issues of allocation of costs and revenues. These multi-jurisdictional ventures can be used to shift costs and revenues in order to show lower profits.

This encompasses not only carrying transmission, but also issues of content that are affected by national sensitivities about media and cultural policy. Other difficult regulatory issues arise in the United States; it is a politically sensitive question whether the Bell Companies should be in information services. Now, imagine if one of these Bell Companies was Japanese-owned. Similarly, Europe has a sensitivity about American media domination.

In most instances, direct investments among developed countries are not possible yet due to remaining barriers. Often, joint ventures are pursued largely because a company wants to get its foot in the door but cannot access the market directly. Therefore, if the ratio of joint ventures to companies going into the country directly is high, it would indicate that there are still significant barriers, and that one has to partner one's way into a market rather than invest one's way in. What the developed countries need, perhaps, are also not direct investment of the sort that says that BT is going to buy 20% of MCI, but perhaps more of the kind that permits direct competition, where France Telecom goes into Germany, Germany Telecom goes into Sweden, and so on. The result is a greater challenge to the existing organisations, with less co-operation and more competition.

In the equipment market, for a long time companies wanting to sell equipment in country X had to promise that they would also build a factory in that country or establish a joint venture. This requirement will decline in a competitive environment, if carriers have to keep costs of procurement down.

One final observation: developing countries need basic telecommunications more than they need advanced services. Yet they often praise the technologies and organisational model of the national telecom organisation. But this may be the wrong model. Focusing again on the history of telecommunications in the United States, large parts of the country were not originally served by the Bell system because they were rural areas. Small, independent, decentralised organisations, many of them co-operatives, put together fairly simple telephone systems to serve these areas. Finland is another example of this model. It has small local co-operative telephone companies working very well. What the national organisation does is that it links them together, sets some of the standards, co-ordinates the international communications and takes care of unserved areas.

Therefore, just as important as sending investment dollars overseas is the transfer of different organisational models, which would permit a decentralised evolution of telecommunications based on local initiative and control, assisted by foreign technology and investment. To think that a private profit-making Western company could effectively run the rural infrastructure in far-off places is not realistic in economic, organisational or political terms.

To conclude, the forces that reversed the previous era in internationalization – economic sovereignty and low economic performance – will reappear and curb the internationalization process in developing and rural areas. Furthermore, internationalization moves in developing countries will be scrutinized by regulators and increasingly meet political opposition. Therefore, the recent internationalization process may slow down quicker than current trends suggest, although not necessarily come to a halt.

The Race to European Eminence
E. Bohlin and O. Granstrand (Editors)

International Competitive Dynamics: Implications for Domestic Regulators

David McCarty

Department of Telecommunication, Michigan State University,
409 Communication Arts, East Lansing, MI 48824-1212, USA

The restructuring of European telecommunications and the resulting market dynamics have created a number of major regulatory challenges. The interplay of national and European Community policy, technological innovation, market conditions, business developments, and the promise of substantial further regulatory reform has resulted in an uncertain and rapidly changing telecommunications landscape. The central challenge for regulators across Europe is to establish an appropriate framework of incentives and safeguards which can help to guide the transition from monopoly to effective competition in a fair and efficient manner. In particular, domestic regulators will need to balance the interests of captive rate payers and efficient competitors against those of large, sophisticated users who can increasingly utilize private and competitive alternatives to the public telecommunications operator (PTO) and the PTO's need to respond to the increasingly competitive environment, both domestically and internationally. It is important that such regulation be proactive, that it function as an ongoing process which not only recognizes but actually incorporates evolving market dynamics.

The purpose of this chapter is to explore the political-economic dynamics of and to articulate key regulatory issues arising from the restructuring of the European telecommunications industries. The first section provides a context by describing some of the many factors influencing the formation of domestic telecommunications regulation. The second section explores how the interests of various players interact in terms of the different strategic games in which they may be involved. The third section discusses the economic aspects of the major issues facing domestic regulators.

1. FACTORS INFLUENCING DOMESTIC REGULATION

The gap between theories of perfect competition and the actual working of real markets can be great. However, both economic theory and "real-world" conditions influence regulatory developments. This section describes important characteristics of the European (and global) telecommunications markets and industries for the purpose of highlighting some of the political and business dynamics which may influence the development of domestic regulation in different countries.

1.1 Market conditions

Investment opportunities for telecommunications operators vary widely from country to country and from market to market. For example, only four European countries—Germany, France,

Italy, and the United Kingdom—have markets large enough to make them "interesting" to foreign suppliers. Even among these, one can easily prioritize investment opportunities based on the market, regulatory, and cultural conditions unique to each country. In this respect, Germany is perhaps the most attractive, because of its relatively liberal regime and the huge investment requirements needed to upgrade the former East German Länder. In France, on the other hand, competition has been strictly limited and the digitization of the public network has already been largely completed. The UK clearly has the most liberalized of the national European markets, but intensive competition has reduced profit margins to "barely sustainable levels". This leaves Italy as the second most attractive market during the next five years, because of a major modernization program upon which SIP has embarked (Crowther, 1992).

These conditions also allow one to guess which national PTOs will be in the best position to pursue foreign investment opportunities in the near future. For example, SIP and Deutsche Telekom are unlikely to make foreign expansion a priority, given their domestic investment needs. British Telecom, on the other hand, has a much greater incentive to seek market opportunities outside its saturated domestic market, and France Telecom is at least well positioned to take advantage of those foreign market opportunities which may present themselves.

Rather than continuing to focus on protecting the national PTO, some countries may see a strategic advantage in liberalizing their domestic markets more quickly. Such a policy is likely to be related to a nation's historical role in international trade. In this light it should be no surprise that the Netherlands and the UK are both among top choices for locating pan-European network hubs (Heywood, 1991). The ability of different countries to pursue such a strategy varies and would have differential impacts on domestic conditions. Such a strategy would be easier for smaller countries and those further along in rebalancing tariff structures.

Obviously, as conditions change in these national markets, incentives and opportunities for these players to participate in foreign markets will change, as well.

1.2 Restructuring the PTO

As the privatization of PTOs and competition in infrastructure and basic services come to pass, regulators will have to deal with the reality of dramatic restructuring of their national PTOs. The US experience is instructive in this respect: thousands of jobs were (and continue to be) lost as AT&T and the regional Bell companies have sought to cut costs, and consumer advocates have become increasingly strident in their claims that the American phone companies have squandered money on foreign ventures at the expense of domestic ratepayers.

Such developments help strengthen the resolve of certain players, such as labor unions, to fight the privatization of national carriers and the liberalization of domestic markets. A strike by French unions over the proposed privatization of Air France and similar concerns relating to efforts to change Germany's constitution to allow the privatization of Deutsche Bundespost Telekom are potent examples of the political difficulties of restructuring.

1.3 Counter evidence

While proponents can point to many examples of the benefits of privatization and competition, those opposed can also find counterexamples. The fact that France Telecom completed its dramatic modernization program as a state-owned monopoly might serve as one. A recent survey of users in the UK, suggesting that early dissatisfaction with a competitor may work to damage the credibility of competition in general (Pundyk, 1993), is another example of the ambiguous evidence that regulators must evaluate as opposing parties attempt to support their claims.

1.4 International alliances

The recent wave of joint ventures among major carriers has implications with respect to the dynamics of both international and domestic competition. For operators, cooperative ventures are a way to reduce the risk of expanding into foreign markets. Not only do they minimize capital investment, but they also lead to improved knowledge about international markets. Not even the largest carriers in the world have the resources to own and operate their own facilities on a truly global scale. Joint ventures "enable participants to give the impression of being truly global, despite owning no physical assets outside their own borders" (Campbell-Smith, 1991, p. 25). However, while such alliances may strengthen the ability of specific carriers to compete in specific markets, they may impair domestic competition in the long run by eliminating potential competitors.

1.5 Investment opportunities in Eastern Europe and the Third World

Telecommunications restructuring has not been limited to the developed nations of Western Europe. Countries in Eastern Europe, in Asia, and in Latin America have also privatized their national carriers and sought the benefits of competition. While the need for investment to upgrade and expand the telecommunications infrastructure in these countries is comparable to that in Western Europe, the demand for services and the potential return on investment is often much greater.

As a result, European PTOs are seeing opportunities to invest in the developing nations. At the same time they are coming under pressure to invest in their domestic infrastructures and pursue market opportunities in other developed nations. On top of this, domestic competition and privatization will create strong pressures to cut costs. To the extent that domestic regulators have jurisdiction over these carriers' investment and pricing decisions, they, too, will have to weigh carefully the potential risks and rewards of the carriers' global strategies.

1.6 Protecting the "flagship" carrier

One last political factor will be mentioned at this point. That is the possibility of national regulators taking strong protectionist positions with regard to their domestic PTOs. One study of the banking industry (Hilton, 1992) argued that "old fashioned nationalism in Western Europe is far from dead" and that some countries have made it an explicit policy objective "to defend the home market from invasion by other EC giants" (pp. 51–52). Government regulators might wish not only to protect their national PTOs domestically, but also to help further their interests abroad. It is easy to see that in such a case regulators might attempt to "drag their feet" in opening key domestic markets in order to give the national "flagship" carrier time to solidify its domestic position and/or establish itself abroad.

2. AN "ECOLOGY OF GAMES"

The factors described above can be seen as contributing to the outcomes as well as constituting the outcomes of the continuing interplay of technological innovation, economic incentives, business decisions, and legal and regulatory decisions. These multidimensional dynamics of the European telecommunications markets can be thought of as interrelated games, in which outcomes result from the complex interaction of different players employing various strategies in hopes of achieving specific objectives. The various players (e.g., operators, users, regulators, equipment vendors, competitive service providers) bring different resources and levels of skill to bear in playing these games, and their strategies and "moves" are constrained by the legal-regulatory rules in force in the various arenas, or jurisdictions, in which these games are played.

The notion of "interdependence" is important here in three respects. First, in making its own strategic decisions, each player takes into account all the other players' likely strategies and their likely responses to the perceived strategies and "moves" of others. Second, "coalitional dynamics"—that is, the ways in which players align themselves for and against specific issues in specific games—influence players' strategies and, in turn, influence outcomes. Third, players are typically involved in numerous games in various arenas (business, political, legal-regulatory) at the same time, meaning that an outcome of one game can easily influence players' strategies and moves in other games. This last point is what Long (1958) referred to as an "ecology of games".

This section examines some of the fundamental kinds of games being played in European telecommunications. A full elaboration of all aspects of all of these games would be beyond the scope of this chapter. The examples here are intended merely to illustrate some of the ways in which the interests of major players interact to influence domestic regulation.

2.1 Regulators vs regulators

In many cases the objectives of EC regulators have been in conflict with the objectives of national regulators. For example, EC policymakers have generally pushed for more rapid liberalization, while national governments and regulators have resisted (the UK being the main exception). This should not be surprising, given that the national officials are much closer to the political and economic consequences of restructuring. Moreover, EC and national regulators have somewhat different constituencies and concerns, with the EC being more attuned to pan-European issues and the concerns of multinational corporations. (Similar tensions have been apparent in the US between the Federal Communications Commission, which has tended to be more sensitive to national issues and the concerns of interstate users and operators, and the state-level regulators, who have been closer to local and intrastate users.)

2.2 Large users vs PTOs

The demands of large, telecommunications-intensive users—especially multinational corporations—for technologically sophisticated services and/or lower prices often place them at odds with national PTOs. As these users rapidly adopt technological innovations in hopes of gaining strategic advantages, their demands for reliable, high-speed, digital network services may outpace PTOs' abilities to supply them. In the multinational dimension, these users demand not only that these same services be available in all the countries in which they do business, but also that they be technologically compatible and comparably priced. In the economic dimension, these users may increasingly seek lower-cost private and competitive alternatives to the PTOs.

Thus, large users challenge PTOs in domestic regulatory games by pressing for greater PTO investment in new technologies, for rebalancing tariffs to their advantage, and for greater liberalization. These users challenge PTOs in EC regulatory games by supporting pan-European common technical standards (harmonization) and, again, by pressing for faster and more extensive liberalization.

2.3 Operators vs regulators

While regulators might be thought of as "referees" in games between users and PTOs or among different operators, they can also be seen as players in their own right. For example, the interests of PTO competitors and large users may conflict with the objectives of regulators who wish to protect national PTOs by restricting or delaying competition. Or the desire of PTOs to respond to competition may conflict with regulators' "universal service" objectives. (This last point is further elaborated below.) At the EC level PTOs have generally opposed EC initiatives to expedite liberalization and harmonization.

2.4 Operators vs operators

Games among operators may be cooperative or competitive. In strategic alliances such as those between MCI and British Telecom and between France Telecom, Deutsche Bundespost Telecom, and AT&T, operators seek the cooperation of other operators to share the costs of establishing global presence and/or to "neutralize" potential international competitors ("Three's a crowd," 1993).

Operators play competitive games when they attempt to enter each others' markets or when they fight for customers in markets where they are already present. Internationally, operators may also compete against other operators to establish footholds in new or expanding markets (for example, in developing countries).

2.5 Other games

Two other kinds of games deserve mention here. First, tariff rebalancing may put small/rural users in conflict with large users. Tariff rebalancing is generally thought of as raising rates for the former in order to reduce rates for the latter. Obviously, this puts these two customer classes in conflict.

Second, privatization games may put labor in conflict with management. The managers of private companies have an obligation (to shareholders) to cut costs and streamline operations—especially in the face of increasing competition. Unfortunately, this usually translates to lost jobs, as former monopolies find that they were overstaffed and as new technologies reduce the need for labor.

3. THE CHALLENGE FOR REGULATORS

The preceding sections have provided an overview of the kinds of factors and dynamics which influence domestic regulatory policy in Europe. This section argues that the major challenge facing domestic regulators in this context is to establish—quickly—a framework of incentives and safeguards that encourages efficient competition, protects domestic ratepayers, and allows the national PTO to respond to an increasingly competitive environment. It suggests that the cornerstone for such a regulatory framework is a "politically acceptable" cost-allocation methodology, and that regulation must be an ongoing process which responds continually to evolving market dynamics.

3.1 Cost-based pricing

Because the telecommunications industries in Europe evolved as protected state-owned monopolies—often more concerned with social policies than with economic efficiency—pervasive economic distortions and inefficiencies became embedded in their cost and rate structures. These distortions now influence the nature and consequences of competitive entry. Until these distortions and inefficiencies are phased out, it will remain uncertain exactly which telecommunications markets might actually be able to sustain effective competition and exactly when, if ever, regulation might no longer be needed.

Nearly everyone agrees that prices for telecommunications services in Europe must be "rebalanced" in order to bring them in line with the underlying costs of providing them. Cost-based prices minimize entry by inefficient competitors (domestic or foreign) and provide benchmarks against which to evaluate the PTO's pricing responses to competitive entry.

However, "cost-based pricing" is meaningless without some mechanism to allocate common costs to specific services. Most economists argue that there is no economically justifiable way to do this—that all cost-allocation methods are arbitrary and subject to political manipulation. Although this may be the case, it is important to establish *some* cost-allocation method which is at least politically defensible. (So-called "stand-alone cost" allocation methods (e.g., see Gabel, 1991), though not immune to criticism, may represent a good starting point.)

This is perhaps the key challenge for domestic regulators: to establish acceptable cost-allocation methods and accounting procedures which can ensure rational and relatively efficient "cost-based" pricing, while also ensuring that the dominant firm (the national PTO) is prevented from engaging in anticompetitive cross-subsidization of competitive services. It is crucial that European PTOs and their regulators begin immediately to address issues of cost cutting and rate rebalancing. Ideally, "rebalanced" tariffs based on streamlined PTO costs should be largely phased in by 1998, when competition for basic voice services is supposed to begin in earnest.

3.2 Regulatory safeguards

The national PTOs will remain "dominant firms" for the foreseeable future—especially if given the regulatory flexibility to respond to competition. It is unrealistic to expect that competition in all of these firms' markets will be sufficiently effective to constrain their market power in the absence of regulation. While the introduction of cost-based tariffs is an important first step, this will still leave unresolved important issues relating to subsequent competitive entry, universal service, and access and interconnection.

For example, if regulators continue to force PTOs to cross-subsidize service to high-cost rural areas internally, the high-profit markets generating the subsidy will continue to attract competitors—even though they may not be as efficient as the PTO in serving those markets. On the other hand, if the PTO cuts prices to cost, the source of the subsidy will disappear. "Rebalancing" tariffs also implies "deaveraging" them, meaning that the price of a 200-kilometer call between two (high-cost) rural communities should be higher than the price of a 200-kilometer call between two urban centers. This is the crux of the universal service dilemma: if the PTO is to be expected to subsidize high-cost customers internally, it will remain vulnerable to competition in those markets which are supposed to generate the subsidy. Perhaps the best way to deal with this problem is to externalize the subsidy in the form of a surcharge or tax on all telecommunications services (including private systems), through access charges on those who interconnect with the PTO, or through direct payment of the subsidy from general revenues.

As specialized competitive service providers (e.g., long distance carriers, cellular and "microcellular" operators, and competitive access providers) increasing by demand interconnection with the PTO, they should expect to pay access charges which reflect the cost of their use of the PTO's network. The calculation of these charges will not necessarily be straightforward, since it will once again involve the allocation of common costs. Nevertheless, the principle of cost-based pricing must apply here, as well.

In those cases where the PTO itself is providing a service (e.g., long distance, cellular and microcellular, local access) in competition with other operators, the PTO must be required to incorporate these same access charges in its own price. Otherwise, the PTO would have an unfair advantage which would thwart competition in the affected markets.

The kinds of measures discussed here may cause some competitors and customers to be negatively impacted in the short run. In the long run, however, they should go far toward safeguarding the interests of efficient competitors, the PTO, and all telecommunications customers.

3.3 Foreign investment opportunities

If the kind of rate structure and safeguards described above can be implemented, then the national PTO should be neither hindered nor unfairly advantaged in pursuing investment opportunities abroad. That is, if the PTO is neither solely responsible for ensuring universal service to high-cost customers nor able to generate monopoly profits from its captive ratepayers, then it should be free to pursue foreign market opportunities on the stand-alone merits of each business case. In other words, a regulatory framework of cost-based tariffs and appropriate safeguards should effectively insulate the PTO's domestic responsibilities from its ability to enter attractive foreign markets.

This approach also implies that the PTO should be free to enter other domestic markets (e.g., video programming), as well. So long as the regulatory framework can insulate captive ratepayers and PTO competitors from the financial effects of success or failure in these other markets, there is no reason not to let the PTO compete in these markets on its own. (Incidentally, in return for assuming the entire risk of such ventures, the PTO should be allowed to keep any profits or bear any losses itself, rather than transfer them back to domestic operations.)

3.4 Adaptive regulation

Just as markets are not static, neither should regulation be static. The regulatory framework described above should be considered only a starting point—an initial set of tariffs and safeguards which should help to "level the playing field" in anticipation of the introduction of competition in basic voice service markets in 1998. As new firms enter these markets and as technological innovations reduce costs, conditions will change. Markets in which the PTO may initially appear to retain substantial monopoly power may in time become effectively competitive. Certainly, changing cost structures will necessitate changes in tariffs.

The nature and extent of competition will vary widely across markets and will continue to change as entrepreneurial firms continue to seek out profit opportunities through the introduction of technological, managerial, and marketing innovations. As a result, regulators must be willing and able to monitor continually these changing market conditions and to adapt the regulatory framework as changing conditions warrant.

4. SUMMARY AND CONCLUSIONS

Domestic regulation of telecommunications in Europe is evolving in an environment of rapid technological change, economy-wide market restructuring, and contending political interests. The interplay of these influences can be seen as an ecology of games, in which PTOs and other operators, large and small users, national and EC regulators, and others employ various strategies in multiple arenas (e.g., the European Commission, the courts, the marketplace) to achieve their objectives.

This chapter has argued that, despite the complexity and political sensitivity of this environment, domestic regulators must move expeditiously to encourage their PTOs to streamline operations and implement "rebalanced," cost-based tariffs. In addition, they must implement specific regulatory safeguards to protect domestic ratepayers and competitors from potential abuses of residual monopoly power. A properly designed regulatory framework will not hinder the national PTO from pursuing attractive investment opportunities abroad, nor will it allow the PTO to finance such foreign ventures "on the backs of" domestic ratepayers. Domestic regulators must play an active role in monitoring market dynamics and revising specific regulations in response to new conditions.

The introduction of competition into former monopoly markets can bring many benefits. However, market dynamics alone may not necessarily be trusted to bring about optimal market structures without the visible hand of regulation. The challenge for regulators is to apply the right touch: neither so light that it fails to remedy market failures nor so heavy that it stifles entrepreneurial spirit.

The role of regulators during this transition from monopoly to competition is to define markets, monitor market conditions, deregulate markets when appropriate, and constrain monopoly power when necessary. Regulators must remain cognizant of political pressures, as they attempt to balance the contending interests of different players seeking to advance their own positions. Regulators must appreciate the complexity and dynamism of today's global telecommunications markets, anticipating the impact of new technologies, user demands, and international alliances. They must establish regulatory guidelines which help markets to work, without trying to micromanage the competitive process.

The transition to fully competitive telecommunications markets will be long and difficult. Enlightened regulation may help to make it a little less so.

REFERENCES

Campbell-Smith, D., 1991, Telecommunications, The Economist, October 5, pp. 5–36 .

Crowther, S., 1992, European market entry. Presentation at the 1992 national communications, Forum, Chicago, October 12–14.

Gabel, D., 1991, An application of stand-alone costs to the telecommunications industry, Telecommunications Policy, February, pp. 75–84.

Heywood, P., 1991, Where to hub Pan-European networks. International Telecommunications, McGraw-Hill, DataPro, February.

Hilton, A., 1992, Mythology, markets, and the emerging Europe. Harvard Business Review, November–December, pp. 50–54.

Long, N., 1958, The local community as an ecology of games. The American Journal of Sociology, Vol. 64, November, pp. 251–261.

Pundyk, J., 1993, Reality check, CommunicationsWeek International, August 16, pp. 16–17.

Three's a crowd, 1993, The Economist, November 13, pp. 72–77.

The Race to European Eminence
E. Bohlin and O. Granstrand (Editors)

Competing Visions for World Telecommunications: Europe and the Global Evolution of Industry Structure

David Allen

316 Heaths Bridge Road, Concord, MA 01742, USA; East-West Center, Honolulu, Hawaii

The globalizing telecommunications fraternity is rife with competing visions for its own future. How do we choose among three such visions, each associated with a major carrier? In this paper, we consider the full dictates of network evolution—innovation together with standardization. Instead of standard competition policy, networks that serve the evolution of their *communities* require a careful mix of structure and dynamics, according to this logic. The carriers' three structural visions sort by how well each supports such policy. The paper concludes by reviewing in this light the recent announcements/reports—by BT-MCI and about AT&T-DBP Telekom-France Télécom—then asks about next steps.

1. INTRODUCTION

At least three distinct visions of the global future for telecommunications are competing to be accepted within the industry. The past five years have seen telcos stir dramatically beyond their borders. In Europe that has meant steps to the pan-European regional network. Early this summer (1993) global steps also began to take concrete form, with the BT-MCI announcement. Now in the early winter, the press reports that another alliance between Europe and the US is under discussion. AT&T is huddled at its Brussels office with Deutsche Bundespost Telekom and France Télécom, the core of continental Europe.[1,2] Along the way, a contest for the hearts and minds of the telecommunications fraternity has broken out. Particularly three visions for global telecommunications are vying to shape the choices that we will make ahead, as the industry navigates a fundamental sea-change.

The three visions differ over the relative status of individual players as against collaborative groupings—such, in other words, is the turf for the contest among the visions. This question of divided but coordinated resources, of balance between parts and a whole, echoes in the new BT-MCI venture. Though coordinated as a joint effort, the two are dividing the world between them—MCI takes the American hemisphere and BT the rest. Indeed, the issue

[1] At the time of the BT-MCI announcement, AT&T also announced a global joint venture, WorldPartners. DBP Telekom and France Télécom also have a joint venture between themselves, named Eunetcom. The Swedish, Dutch and Swiss carriers have also formed Unisource.

[2] Among the largest carriers, only one does not have its global intentions accounted yet: NTT has so far been constrained from international carriage by domestic regulation. Nonetheless, it was one of the first to establish a full-fledged international presence, with offices around the world.

arises for anyone who grapples with constructing an organization—as encapsulated by one of the first modern economists, Adam Smith, in his model for organizing a pin factory with divided parts but coordinated whole. As we proceed, we will find ourselves asking how to think about what is at stake here.

2. THREE COMPETING VISIONS FOR GLOBAL TELECOMMUNICATIONS

Implicit in the three competing visions are different future structures for the global telecommunications industry. What are these different outcomes, and how do they compare with each other? Who espouses which position?

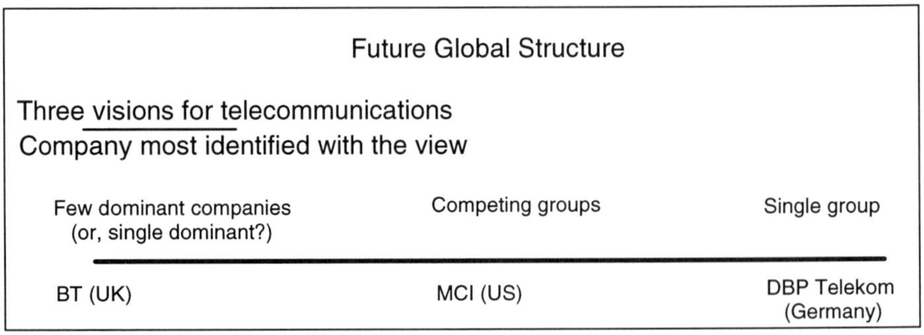

Figure 1.1. Future global structure.

The three visions are widely identified with three operators—BT, MCI and DBP Telekom. However the philosophical positions which lie behind these views are just as widely debated by other industry members, certainly by the other firms that are active globally. What is more, the purpose here is to establish contrasts among the *outcomes* implicit in these visions. We will not look at the detail or nuance which is necessary to propound each view. In fact, the alliance that opens this paper is between two of these operators with differing positions—that will certainly be grist for our work below, as will be the other, prospective alliance.

BT, in the UK, is identified with a view that sees a few companies emerging to dominate the global industry. It is even possible to infer the rise of a single dominant company, around the world. The mechanism to reach this result is a classic competition among combatants, and competition will define relations among any survivors. BT acknowledges, though, that there will be some coordination at the physical transport level, where one network interconnects physically with another.

In contrast MCI, in the US, has proposed a future in which consortia compete with each other. Individual companies gather into groups, which then compete with other groups to shape the future of global networks. Relations within the groups are marked, in MCI's terminology, by "cooperatition"—a combination of collaboration and competition internal to the group. Almost as if to put a stronger stamp on this vision, MCI has proposed a directly parallel approach for personal communication systems/PCS within the US. MCI has proposed that PCS in the US be developed by competing national consortia, including a mix of competition and cooperation internal to a group.

Finally DBP Telekom, in Germany, has brought forward a proposal for a new European structure in telecommunications. Though not yet so active as a template for global networks, this conception stands in the wings on the global stage. While a brief description fails to convey its several facets—as brevity fails each of these three views—a fair rendering of the proposed structure is, I believe, that all the principal actors join in a single group. Relations within the group are consciously a mix of competition and cooperation.[3]

So future industry structure in the BT vision is a few dominant companies, or perhaps a single dominant company; for MCI, it is competing groups of companies; and for DBP Telekom, a single group. Superficially, the structural outcome seems to differ across the visions by whether the units are individual companies or groups of companies. When we look a bit more closely however we see that the two extreme positions, BT and DBP Telekom, both (potentially) result in a single entity. The difference between them is whether individual companies are joined together by integrating into a parent organization or, instead, by forming into a collaborative group. In other words, the contrast between these structural extremes centers on the relationship between the individual company and the group.

Nominally there appear to be two entries from Europe and one from the US. Does the BT-MCI announcement suggest the divide is in fact between Anglo-American and European? We will come back later, to discuss new alliances.

3. CHOOSING A VISION: NETWORK EVOLUTION AND A POLICY CYCLE

If the present tussle, at some level, is between Anglo-American and European ideological positions (with, so far, no substantive entry from Asian ways of thinking) how does a thoughtful industry participant make a choice? Europe at this time is a major laboratory for the expansion of community, which globalization would further expand to world scale. Certainly, pan-European networks to support such a wider sense of community are necessarily subject to forces at the global level. Nonetheless, Europe's step toward a *regional* network may suggest some of the concerns which a choice among *global* visions must address.

- A more integrated telecommunications system for Europe faces (among others) two fundamental requirements, but in tension with each other. (1) The new community looks toward a universal infrastructure which will seamlessly interoperate across intra-community borders. Particularly for the more sophisticated intelligent services, integration to make the flow seamless will require collaboration. However, (2) tariff structures have become dependent on a tradition of higher rates for greater distance and particularly higher rates across the same intra-community national borders. Competition among the community's telecommunications firms—*not* collaboration, certainly—is the vehicle chosen to break this mold and bring lower rates that will encourage a greater flow of traffic across the new community.

---

[3] The views at each of the three carriers were discussed in early spring 1993, in a series of visits to each headquarters. In turn, these were a part of visits to a wider set of all carriers. For BT, numerous published reports of statements by BT Chairman, Iain Vallance, were confirmed by discussion at BT's strategy group, London. Discussions with MCI International were held in Rye Brook, New York. Discussions with DBP Telekom's international directorate were in Bonn; see also an address delivered at the Münchner Kreis Congress, 24–25 November 1992.

This only samples the concerns, but perhaps the apparent tension marks a degree of significance for these two.

If these are some of the concerns, on what basis do we make a choice among the competing views of future industry structure? I take as starting point a page from what we know about the evolution of networks. Evolution is, after all, the generic for community/network expansion, in Europe and throughout the world.

3.1 Innovation and standardization, competition and consensus

To enjoy continuing innovation in a network requires, we know, an iteration, a round—a repeated cycle between innovation and standardization.[4] An innovation, by its nature, breaks the network's connection;[5] to reap the benefits of new network technology, consensus about the network of choice must re-assemble. The policy implications are profound. The experimentation essential to the innovation phase is probably best supported by competition. But the new consensus for the other half of the cycle requires the opposite of competition.[6,7]

Two recent cases, both between Europe and the US, illustrate the pitfalls of neglecting either phase of the cycle.

- In digital cellular, Europe easily reached consensus on a standard while the US has instead produced hotly contested technical proposals. Europe demonstrated its capacity to reach consensus on a common approach, while competition in the US promoted experimentation and new ideas. But each lacked the other's strength, for the other phase of the cycle, and each is paying the price. Questions have been raised about the quality of Europe's GSM standard, while the wrangling in the US seriously slows or even prevents adoption of a common approach.
- High definition television/HDTV showed the same pattern. Again, Europe demonstrated its ability to agree on a standard. And again, US scrappiness brought forth winning technical proposals. In this case Europe's HD MAC standard has already been eclipsed by the US scheme for digital transmission.

However, the US recently brought together its competitive HDTV proponents into a "Grand Alliance," with agreement on a common approach. Is the US learning the other phase of the cycle? An attempt to answer that question highlights one crux of the new policy appropriate to these dynamics.

At this crux we find the switchover from competition to consensus. Prior to that point of inflection we can, and must, enjoy all the benefits of competition that most thinking about evolution holds dear. But if we are to enjoy the benefits of connecting our community of

---

4 For a directly parallel approach, see Leary and Sines (1993). Leary and Sines speak in terms of a "standards wheel (with a cycle that spins faster or is braked and slows)."

5 Like many others first encountering this discussion, you may ask how is it that the connection is really broken. E-mail is one case to consider. In the early stages of e-mail, those who became accustomed to it were frustrated, on occasion, by the isolation on their island of interconnection—indeed such frustrations continue today. Discussion of information have's and have-not's is another case, whether the have-not's are developing nations or just the house down the street which is still waiting for a network installation. User push for open systems is another case.

6 The annealing of metal has been suggested as a metaphor. The metal is alternately heated, when it is more malleable, and then cooled, when it becomes stronger.

7 For a full and detailed treatment of this subject (Allen, 1991).

interest by means of new technology, the mode must shift. Rather than rely on competition, we must take direct responsibility for our collective welfare—as a group, take responsibility—and sort out, from among all the new ideas bruited during the experimentation phase, those which will meld together to serve the group best.

This was exactly the task set for what became the HDTV "Grand Alliance" in the US. After a strong competition which produced several opposed proposals, the backers of those separate ideas were asked to combine them into a single "best" standard. Did they succeed? The resulting omnibus proposal is actually a collection of guidelines. The competing HDTV proposals in the US were particularly divided between the use of interlace or progressive scan.[8] Television set manufacturers argued that television viewing is best served by interlace; computer manufacturers said, on the other hand, that progressive scan is essential for closer-to-the-user computer viewing. The "Grand Alliance" made provision for both. Does this dual provision represent legitimate needs for variety from different constituents in the community? Or is this just political horse-trading that only protects some entrenched old-technology interests, in this case manufacturers of the less expensive interlace method?

3.2 A social choice *versus* protectionism

So, the social choice that must be executed at this crossroads between competition and consensus also raises the specter of protectionism. Decisions by a group open the door to decisions that favor *just* a group. Fortunately, however, now we are in a position to see—so far only conceptually, not yet with a numerical metric;[9] but at least conceptually—what separates undesirable protectionism from a necessary social choice, a choice on which vital function of the relevant community depends. Protectionism (in the network case[10]) sustains interests with passé capabilities and blocks access for new ideas. Legitimate social choice, by contrast, melds old and new ideas into a single common approach that will best serve—until, of course, the time is ripe for the next iteration of innovation and standardization.[11]

The crucial distinction between protection versus essential choice for the community is rarely perceived, if at all. The resulting confusion has regularly led into a morass of disagreement and opacity about useful policy direction, not infrequently on a global scale. Since the distinction between the two must be drawn in an exchange among community members, there will always be opportunity for those inclined toward protectionism to try and obfuscate. But simply to understand the framework for discussion—that best choice and protection are opposed to each other—is the first step. And success at instilling this approach holds the prospect for a community to enjoy benefits both from innovation and from connection.

---

[8] Progressive scan produces a complete picture in a single sweep down the screen. Interlace takes two sweeps to complete the picture, the first sweep drawing every other line in the picture, the second sweep filling in—"interlacing"—the remaining lines.

[9] Mathematically-based tools for network analysis (one formulation of the subject here) are beginning to emerge, however. For one enjoyable recent review see Maggioni (1993).

[10] For a step toward broader application of the ideas—intermediate cases, that is, which do not require the strict interconnection of the network—see the reference in note 7 (also note 18).

[11] Though not emphasized here, the turn to another spate of innovation is also an inflection, one in which the prevailing consensus must be set aside to allow further experimentation (or in terms from the text, protectionist inclinations could try to use "inertia" from the current standard). Schumpeter dubbed this inflection, of course, to be creative destruction.

3.3 Variety within the standard

The policy indicated is not as simple, however, as a temporal sequencing between competition and consensus for, say, fixed periods. We see already with the HDTV story in the US how agreement on a standard may serve as a sort of window that enables on-going diversity. There the debate between television viewers and computer users illustrates variety across a population, within the underlying standard.[12] In addition there may also be variety across time. Generations of technology, such as the fax G-series, typically represent incremental improvements which follow some basic technical breakthrough.[13] One label for these two types of diversity-within-the-standard is "variety in space and in time." Eventually the (old) breakthrough technology, which underlies a standard, is itself stirred to renewal; ideally a whole new iteration between innovation and standardization then takes place.[14,15]

The result is cycles (for the nested, "lesser" innovations) within cycles. This is one characterization of the interdependence that is found widespread among technologies. A policy with temporal sequencing was already not a simple story. Now we see that to enjoy the most savory technical fruits in our networked lives requires policy agility indeed. With numerous technologies, almost too many to name, each at a different stage and interdependent with other technologies … we must be sharply competing on some fronts, sagely agreeing on others, all at the same time, and inevitably at some times, doing both with the *same* party on the other side of the table.

3.4 Competition, in the evolution of networks?

Clearly, the policy formula is more complicated than just "competition." But, as we noted above, competition is integral to Europe's policy plans for the future, and indeed competition is the heart of a new creed for liberalization that has caught hold around the world. Focusing just on competition itself for a moment, what can we say about competition when it is applied to this network industry?

Fuller treatment will wait for a forthcoming book,[16] but three summary points are worth noting:

- In fact after about a decade of trials, competition—as conceived neoclassically—has not worked very well in telecommunications. In the US for instance, recent work argues that long distance is actually an oligopoly and that price reductions are the work of regulators' shifting between carrier and user access charges (Taylor and Taylor, 1993) (Huber, Kellog and Thorne, 1992). In other major cases, NTT in Japan has lost quite a significant share, but decreasing prices have been a function of the regulator's maintaining a price differential between NTT and the competing New Common Carriers. BT in the UK still holds greater than a 90 percent domestic share. Australia and New Zealand are newer cases without as much experience, but substantial doubts have already

---

[12] Scalability is a form of this diversity.

[13] Extensibility is used to describe the more predictable features of such a path.

[14] "Breakthrough" and "incremental" emphasize the differences—"more basic" and "less basic" would be equally accurate.

[15] For another discussion of variety within the standard, (see Allen and Gilbert, 1993).

[16] A version of this paper appears in the Harvard Journal of World Affairs, forthcoming. In an appendix, that article provides a slightly fuller treatment of the competition question.

been raised about the sustainability of competition in New Zealand. In Finland on the other hand, a case like almost no other, relative equals face each other and competition appears to operate closer to theory.

- The competition that we *have* seen to be vital, in one half of the evolutionary scenario, is decidedly not the commodity competition of neoclassical thinking. Rather, competition that serves a period of experimentation, and encourages innovation, is a competition of ideas à la Thomas Kuhn and shifts in paradigms. To complicate the picture further, surely successful technical innovation also requires collaboration—for spillovers among related technologies, for cross-fertilization between large R&D laboratories and small research boutiques, or simply for the social atmosphere that imbues many creative workplaces.
- Finally, as we see in Sweden (or indeed in telecommunications across Scandinavia) there appear clearly to be other ways besides competition to deal with what usually is deemed the main problem in the first place—the monolithic organization that so often dominates telecommunications, regardless of geography or polity. Sweden for example, though operating with a quite dominant service provider, has overall perhaps the most highly rated telecommunications system in the world – and with entirely "competitive" prices!

Telecommunications policy indeed cannot be summarized simply as "competition."

3.5 Making the choice of vision

Our purpose is to choose among the three visions for global telecommunications. We have described a policy scenario for network evolution. How does the expansion of community and network, which is projected by each of the three visions, grow from our evolutionary scenario?

Widening the sense of community surely depends on an underlying set of social and political forces. But just as surely that occurs in a joint unfolding with the communications tools which would help to knit such a larger community together. In both cases, the essential event is a joining together by smaller sub-communities. That linkage is tautological for the social expansion of community; for the network, linkage was implicit in the process of forming consensus around a new standard. Linkage across communities is the mechanism.

In the evolutionary scenario, linkages form for a consensus, then to enable the next experimentation the links are broken and the group fragments; the group re-forms, then re-fragments; re-forms, then... *Expansion* of a community/network—compared with network evolution for an already established group—is the special case of linkages that have not been forged before, in earlier iterations of the evolutionary cycle.

Before we move to answer our main question, we should note a fundamental feature of the architecture in this process. There is nothing magical about our notion of community. The entity that we have been calling community is not necessarily the community of nations, or the community of companies in an industry, or of residences in a neighborhood, or of members of a family, or of... It is as fluid as its members choose, and it could be as small as two individuals. The three-decade history of UNIX illustrates well, with the step toward a world agreement in the mid-80's through X/Open, then the split again in the late 80's into UNIX International and the Open Software Foundation, and now the recent turn yet again back to

X/Open.[17] Perhaps the basic metrics for these dynamics in the innovation-standardization sequence are the *size* that bounds the largest consensus and the *speed* of transit through a cycle. Together, these parameters determine who enjoys benefits and when—they measure what is at stake, in this approach.

Now, how do we choose among the visions for global telecommunications? Well, how do we implement the policy cycle that encourages fruitful evolution of the network? The final description, on the page before—where we are competing and agreeing, on multiple fronts, even at times with the same party for both competitor and partner—is demanding indeed. Does the DBP Telekom vision, of a single group, offer a structure with the potential to realize this demanding regimen? That depends, pretty clearly, upon relations within the group.

3.6 DBP Telekom's single group—Structured for success

The structure itself is what the contradictory policy demands require. When the occasion calls for consensus the group can *act as a group*, and when more independent action is indicated the pieces can *fragment into separate units*. The "nesting," of units within a group,[18] makes possible the re-forming and re-fragmenting that successful evolution—and expansion—require.

Relations within the group determine whether the group is actually so capable. Is the leadership core vacant, with leadership "handed around the group," so that there is a peer basis for participation—even though some may be more equal than others? That seems essential, both for successful operation as independents and also for the willingness to re-form into a group. It also seems the character of a workable democracy.

The essence of the arrangement is to be capable of operating *both* as units *and* as a group. That puts a premium on strong identity, and boundary definition, for individual units; equally it puts a premium on the ability to link together, across those strong boundaries. Perhaps we can see now the one common thread that does run through all use of competition in dynamic evolution. That is the style of competition. If companies (and people) are to organize in this both-contest-and-agree world, then they can only accept plus-sum conflict. Some competitive environments emphasize the detriment of other competitors; resources are expended just to demolish the opponent. That would be intolerable for those who had also to work with each other. Plus-sum competition focuses instead on doing better than the other side, even hoping the other side does well too, since that can make for greater wealth all around.

Thoughtful policy is only as good as our ability to implement. This "nested" organizational structure seems to be the right start for what is a complex and demanding—but prospectively quite rewarding—policy. We can also see that this approach is indeed the alternative method we need to deal with the large, potentially dominant organization. The nested structure, together with the new policy, aims to be both monolith and fragments, at the same time.

---

[17] Because computing does not strictly require interconnection across machines, in the way that networks do, but interconnection can nonetheless be powerful, computers make an important intermediate case.

[18] This is not, strictly speaking, the decentralization of organization theory (nor descended from Ronald Coase and Oliver Williamson, either). The parts and the whole are expected, instead, to cycle repeatedly through fragmentation and formation.

4. REPRISE: THE NEW ALLIANCES

4.1 BT-MCI

Do the announced details of the BT-MCI alliance suggest any insight into BT's and MCI's positions on these matters? We could look for evidence of BT acting on a view consonant with its vision of competitive winner(s). BT will control 75 percent of the new company. The investment is from BT into MCI. In board cross-memberships BT takes three and MCI one. Or we could look for MCI following its vision of "cooperatition." MCI sells only 20 percent and intends to remain independent. Indeed this is not the first significant minority investment in MCI, IBM once holding 16 percent.

These are straws in the wind. Though another straw suggests a different possibility. While forming the joint venture BT and MCI also carefully divided the world between them (as this paper recounted at the outset). MCI is even acquiring BT North America, seemingly to complete the separation. This is just the arrangement we have framed above—strong identity for component units, at the same time linkage between them. Is there reason to imagine migration toward such a view, over time, among *all* actors globally?

4.2 Migration, across the spectrum of global visions?

BT's vision (the extreme interpretation, that is, with just a sole survivor) surely encapsulates one stage in an historical process. In the US, around the Kingsbury Agreement after the turn of the century, AT&T vanquished its rivals, often by acquisition, and dominated American telecommunications for 50 years. In Finland a single long distance carrier emerged in the 30's, unchallenged until now.

MCI's vision clearly is intermediate between that view and the single-group structure we have explored here.[19] Further, for each of its global groups, the MCI approach also includes regional, feeder groups which flow traffic into the global ensemble.[20] (Again, this structural detail is reinforced in MCI's proposal for PCS in the US, where each national group is to include distinct local licensees.) The internal structure for any MCI group parallels, in other words, the DBP Telekom single group.

In effect, the DBP Telekom architecture specifies the internal structure for any one MCI group. The difference between them is competition among groups, in the MCI world. And BT has seen fit to pair with MCI. Will future history record a migration from the historically-founded view of competing companies, a migration facilitated by the intermediate position, and onto an evolved view that aims for the single-group structure, where competition and consensus consciously mix?

Another question must be asked beforehand. The single group looks, at least superficially, to be similar to the collegial correspondent relations that so far have structured international telecommunications. Is there any important difference? Does the vision of a single group entail any change, really, from the correspondent relations of the past?

---

[19] Interestingly, some of the MCI public discussion about its blueprint seems to echo the AT&T *before* divestiture.

[20] BT's and MCI's careful separation of their new venture into geographic regions seems to follow nicely from MCI's prescription for the local component of organization.

4.3 AT&T-DBP Telekom—France Télécom

The prospective alliance under discussion among AT&T, DBP Telekom and France Télécom offers some evidence. The company identified with the single-group vision, DBP Telekom, along with its partner France Télécom, reportedly[21] have rejected the loose sort of affiliation they see in AT&T's earlier global proposal, WorldPartners. Instead, apparently, they want a stronger union.

The loose affiliations that define the correspondent relations of the past assume no social rules-of-the-road, beyond the narrow focus necessary to deal with the business in hand. However, a functioning social group does develop broader, often implicit, guides that will shape behavior across the group. Culture, or social contract, are commonly-used descriptions. Conventions about internal governance are particularly central.[22] The single-group vision entails, it appears, a transformation of correspondent relations into those of a new group—what now is seen as *outside* converts into an expanded definition of the *inside* to create, or expand, a sense of community for the companies.

4.4 "Nested" structure, not dominance

Does this second alliance seem to spur, or deter, a migration of views toward the single group?

The historical AT&T, as said, dominated its territory when the chance arose (a description that would also fit its many monopoly brethren around the world). But the main news report of the proposed alliance[23] cites fears that AT&T will continue its dominating ways. Here, we have not looked in detail at AT&T's vision, or model, for future telecommunications—the reported fears may be groundless in fact. But we can see that tendencies to domination, whether from AT&T or from another player, are anathema to successful operation of the new scheme. Success at shifting the focus back and forth between individuals and the group depends fundamentally upon individual integrity for each player (what was described above as strong identity and boundary definition).

In fact half the news report is devoted to the prospect that Europe may draw back from an uncomfortable alliance with AT&T, apparently to build instead a strong European regional network around DBP Telekom's and France Télécom's Eunetcom. How would our new policy guide us here? If we interpret DBP Telekom's vision of a single group in terms of MCI's structural detail at the regional level, we see that the "nesting" actually involves at least *three* levels. The highest level group becomes a group of *groups*, which is then comprised of individual companies.

Successful policy depends, I believe in other words, on the formation of a strong intermediate *regional* identity, which is then able to contribute in a *global* forum. Nothing else seems likely to account adequately for the diversity around the world. The dynamics in such a multi-level world are challenging—perhaps we are not surprised at the complexity, as the arena expands to global scale. Mastering those dynamics does hold out to us the fruits of joint technical and social evolution.

If Europe does draw back and build its distinct regional network,[24] AT&T is reportedly not

---

[21] Communications Week International, 8 November 1993.

[22] Parallels with the network firm are intentional, of course. Again though, the policy discussed here also includes a continuing iteration between the whole and the parts, the group and the companies.

[23] See note 22.

[24] By the time you read this, you will know how this turned out, at least in part.

willing to accept a minority position in the enterprise. But rather than consider a minority position, the possibility for migration toward a single-group view suggests a different course.

Asia, and possible steps toward an Asian community, are also much in the news now; perhaps NTT, among others, might take a role on the network side. If that began to flower, there could be three regional building blocks in place—Europe, Asia and the Americas—to begin in earnest the assembly of a global system. AT&T, along with each of the others, would have strong ties within regions as well as links across regions. Such a relatively balanced tripartite division would diminish threats of dominance and elevate the prospect for a community of communities.[25]

5. SUMMARY AND CONCLUSIONS

Among the visions for world telecommunications that try to win our hearts and minds, which do we choose—a few/one dominant company or competing groups of companies or a single group? Network evolution—innovation with standardization—leads us to a policy that combines contradictory regimes—competition with consensus. The structure offered by the single-group vision can support the sophisticated dynamics necessary for this policy.

Competition by itself is not a satisfactory policy, and trials of classic competition have in fact not worked very well in this industry. Instead a carefully calibrated mix of structure and dynamics is required—"nested" structure to facilitate a dynamic iteration, where competition (particularly for ideas) is vital but where consensus must also alternate with competition and is equally vital.

Recent alliances offer some evidence for a migration of views, across the industry, toward the single-group vision. However the path forward is not well demarcated, and there are plenty of contrary impulses. If we master the challenges of this demanding new arrangement—complexities that seem unavoidable as a global world emerges—communities with strong boundaries will be able, as the most fundamental hallmark of our success, to open to new ideas, even if originating from the outside. The threat of dominance, which competition was hoped to eradicate, may actually be turned back only by this new policy. Nowhere is the imperative—as dictated by the pitfalls to be sidestepped *and* the fruits to be relished—more palpably stark than in telecommunications.

REFERENCES

Allen, D. and J. Gilbert, 1993, Standards and convergence: New realities, Pacific Telecommunications Conference, January 18.

Allen, D., 1991, Between innovation and standardization: The evolving network, first presented in Turin, Italy, April 18.

[25] Societies which do not substantively benefit from the new telecommunications have been left out of this discussion—and they constitute a significant portion of world population (Lamberton, 1993). But that omission does not betray contempt, only that adequate treatment will demand a larger text. Any community of communities is not complete until those many, many voices are also heard.

CommunicationsWeek International, 8 November 1993.

Huber, P., M. Kellog and J. Thorne, 1992, The Geodesic network II: 1993 Report on competition in the telephone industry, The Geodesic Company, Washington, D.C.

Lamberton, D., 1993, Globaloney: The impact of regions on the future of emerging markets in information technology and trade, Pacific Telecommunications Review, vol. 14, no. 4; also in Telecommunications: Bridge to the 21st Century, ed. Meheroo Jussawalla, AMIC, Singapore.

Leary, M. and W. Sines, March 29, 1993, Open systems: Impact and import, white paper, International Data Corporation.

Maggioni, M., 1993, Network analysis of regional industrial dynamics and local economic policies, Society for the Advancement of Socio-Economics Conference, March 28.

Taylor, W. and L. Taylor, 1993, Postdivestiture long-distance competition in the United States, American Economic Review, May, Vol. 83, No. 2, p. 185.

The Race to European Eminence
E. Bohlin and O. Granstrand (Editors)

The Internationalization of Telephone Operators: Survival in an Integrating World[1]

Dimitrios Ioannidis

Stockholm School of Economics, P.O. Box 6501, S-113 83 Stockholm, Sweden

The changes in the markets for telecommunications services are expressions of the renegotiation of two different contracts, the contracts betweeen national environments and telephone operators and the international contract between telephone operators. Present and future changes in telecommunications force telephone operators to cooperate more closely. International expansion of telephone operators is achieved primarily through international strategic alliances. International alliances account for a substantial share of the alliance portfolios of the firms studied. We also show that international strategic alliances focus on specific regions. The interface between the U.S.A. and Europe is a critical strategic area. Integrations between national environments and between firms are parallel processes that sometimes reinforce each other, but often also come into conflict with each other. Organizations of all kinds, whether nations or firms, have an ultimate goal of survival. We present three different strategic options regarding the integrative developments in Europe: integration by domination, federation and amalgamation. Different firms have different visions; so do different national environments. Who will be the survivors, the dominant, or the more genuinely cooperative?

1. INTRODUCTION

International integration, demonopolization, deregulation or reregulation and technological development make national borders transparent, and make national telephone firms interact with each other. They have to acknowledge the existence of other firms and enterprises, as competitors on their national markets and on new markets, but also as potential partners. Telephone operators are, in most countries, national firms. Their links with their respective governments are a source of support. However, it is imperative for telephone operators to examine, question and redefine their positions in relation to their national environments[2], other national environments, emerging transnational environments and foreign national firms.

[1] Financial support for this work was kindly provided by Telia, "Sixten Gemzéus Fond", The American Scandinavian Foundation and "Jan Wallanders och Tom Hedelius' Stiftelse". The valuable comments and support of Raymond Vernon, David Allen and Francine Bavay are gratefully acknowledged.

[2] The national structures that surround those firms, consisting of the owning, controlling, coordinating and policy-making institutions, but also of cultural and value systems.

The story of early telecommunications in some countries is one of private enterprise and competition. In most countries, though, telecommunications were operated by public monopolies because of welfare considerations as well as because of national security interests. However, even in the cases where telecommunications services were private and open to competition, they subsequently became nationalized and monopolized. The operation of telecommunication networks became rather quickly regulated so rigorously that the markets for telecommunication services were characterized by private monopoly or oligopoly or by public monopoly. There was a need for international coordination, regarding e.g. transnational infrastructure and exchange, but the operation of telecommunication networks remained, until quite recently, a strictly national affair.[3] There was an unwritten contract between the national environment and the telephone operator(s), stipulating the rights and obligations between them. There was also an international contract, reflecting an international logic regarding the coordination of the interfaces between national networks. This was a cooperative logic, which ultimately became a cartel, as it became clear that international traffic was a very substantial revenue source for the telephone operators and their owners, i.e. their governments.[4]

The conditions on which the contracts between national environments and telephone operators and the international contract were based are now changing. National demonopolizations, deregulations and the evolution of transnational structures, such as the European Union, are expressions of these changes. There is an ideological shift in the developed world, from the view that governments should intervene and control for the well-being of society which provides for individuals to the liberal view that markets should be set free for the well-being of individuals. The shift from national to international, and even global, perspectives is further underlined by technological changes.[5] The internationalization of telephone operators is an expression of the renegotiation of these different contracts. Internationalization is a way for the telephone operators to distance themselves from national environments that drastically change the conditions on their home markets, forcing them into competition. Internation-

[3] In 1865 Napoleon III of France hosted a meeting which became the starting point for the International Telegraph Union. In 1932 the name was changed to the International Telecommunications Union, as radio transmission was added to its tasks (Krasner, 1991). However it is important to note that international traffic always accounted for a very small share of call volume. Surprisingly enough, international traffic still accounts for little more than one percent of the overall call volume. Since telecommunications services were organized along national lines, this meant that the right to be able to communicate was intended to be used within the national borders. International calls still subsidize national calls, in a system in which geographical distance has little or nothing to do with transmission costs.

[4] Cowhey (1990) comments that this cartel lasted long: "For over a century, it defied the laws of economics that caused other cartels to crash..., the regime helped create what is now estimated to be a $550 billion market in voice, visual, and data communications services and equipment. The regime was in fact a political invention so successful that it eventually disappeared from sight."

[5] The infrastructure used for transmission of communication is becoming less physical. Wireless communications provide infrastructural solutions that do not necessarily require the same amount of resources to build and operate them (which was one of the reasons for the strong connections between national environments and telephone operators), nor is the space in which wireless communication is transmitted as limited as the surface (which was one of the reasons why national environments could exercise control over telephone operators). Regarding the latter point, it should be pointed out, that we now see the beginning of a revolution regarding land-based transmission as well. Fiber optics provides inexpensive ways of creating infrastructures with enormous capacities. Firms that already possess land-based infrastructures, like cable TV companies, water companies and electricity companies, have begun to use their networks for telecommunications too.

alization is also a way of establishing a new international logic between themselves. Strategic alliances with other firms are important parts of their strategies.

We will present some empirical evidence of the internationalization of telephone operators as it is reflected in strategic alliances. The emerging internationalization presents deep challenges to executives, policy-makers, regulators and the general public, causing tensions in national and international environments. We will therefore raise issues relating to the interaction between national interests and internationalization in the telecommunications industry. In light of this we will suggest the main strategic options that telephone operators face.

2. INTERNATIONALIZATION AND STRATEGIC ALLIANCES

International expansion of telephone operators is achieved primarily through international strategic alliances. We will give some evidence of this and show that such alliances focus on specific regions.

We have analyzed the strategic alliances of 21 telephone companies around the world, working under very different institutional and structural conditions. In Japan we chose NTT, the major operator, and two long-distance operators, IDC and KDD. In Europe we chose the national carriers, Telekom of Germany, France Telecom of France, Telefonica of Spain, the STET group of Italy, Telia of Sweden, BT of Great Britain, as well as the competitors of BT regarding long-distance telephony, C&W, and cellular telephony, Racal. In the United States we chose the three international operators, AT&T, MCI and Sprint (in the case of Sprint we included GTE), the Regional Bell Operating Companies as well as the mobile operator McCaw. These firms have different network positions in their national environments. Some of them, like AT&T and British Telecom, are the old established monopolists that have to acknowledge new competitors. C&W and MCI are examples of those competitors. McCaw and Vodafone represent mobile telephone operators using new technology that challenges the old operators, focused mainly on land-based transmission.

The data come from the SATI database. SATI contains 4,808 alliances between 1,926 firms, initialized in the 1980s. The alliances are characterized by form, purpose, and related technology or application area. SATI contains national as well as international alliances. The following table shows that international alliances account for, on average, 53 percent of the alliance portfolios of a selected group of 21 telephone operators.

Table 2.1
National and international alliances

Company	Percentage of International Alliances	Percentage of National Alliances
IDC (Japan)	83	17
KDD (Japan)	51	49
NTT (Japan)	48	52
Telekom (Germany)	82	18
France Telecom (France)	97	3
Telefonica (Spain)	94	6
Telia (Sweden)	94	6
STET (Italy)	89	11
C&W (UK)	79	21
BT (UK)	85	15
Vodafone (UK)	58	32
MCI (USA)	20	76
US Sprint (USA)	22	77
AT&T (USA)	27	71
Bell South (USA)	32	68
US West (USA)	26	71
Pacific Telesis (USA)	34	66
Ameritech (USA)	24	76
Nynex (USA)	29	71
Bell Atlantic (USA)	22	78
McCaw (USA)	19	81
Average	53	46
Average Japanese Firms	61	39
Average European Firms	85	14
Average US Firms	25	74

We see that international alliances form a substantial share of the alliance portfolios of the firms studied, although with some regional variations. In general the telecommunications industry is quite alliance-intensive. Wells & Cooke (1991) found that "An ever larger proportion of corporate assets are being placed into alliance relationships. For the individual firms, the crucial question is whether alliances are used successfully as competitive weapons, because collaboration in this sense is *not* cartellisation but an extension of competition." The old international logic of telecommunications was quite cartel-like, though. It is not altogether clear that the final outcome will be so different from the old logic.

Internationalization and globalization are almost synonymous concepts, and their most important element seems to be a link with the U.S.A. This is not surprising as a substantial part of the world's telecommunications traffic is directed either from or to the U.S.A.. We analyzed our sample of 21 telephone operators and found that alliances between American and non-American firms constituted 47 percent of the alliances, while intra-European and intra-American alliances constituted respectively 14 and 32.

Table 2.2
Alliances within and between Japan, Europe and the United States

	Number of Alliances	Percentages
Alliances Between Japanese Firms	3	2
Alliances between European Firms	27	14
Alliances Between American Firms	62	32
Alliances between Japanese and European Firms	10	5
Alliances Between European and American Firms	71	37
Alliances Between Japanese and American Firms	19	10

Wells & Cooke (1991) claim that "Markets are spatially defined as are the barriers to entry and exit; frequently...partners in strategic alliances provide geographical market access to specific, geographically bounded markets. An existing customer base (installed base) is an important corporate asset, particularly when that base is technologically locked in to the firm's product. In a more general sense it is often a political necessity to have a domestic partner when seeking to penetrate new national markets." The distribution above is an expression of this. Europe and the United States are two important geographical areas to cover for telephone operators, and 37 percent of the alliances in the sample are between the United States and Europe. The European market is important because of the integrative attempts. The renegotiation of the contracts between national environments and national firms in Europe opens up opportunities for telephone operators from other parts of the world. The United States is a key market, as much of the international telephone traffic is either directed to or from the United States. (A link with the U.S.A. market is important in all industries that are being globalized, e.g. the airline industry. One of the main reasons behind the failure of the Alcazar alliance was that KLM had a strong alliance with United, which the other airlines did not trust and did not want as an American partner to Alcazar.)

International strategic alliances are expressions of the renegotiation of the old international contract between telephone operators, as well as of the contract between national environments and national telephone operators. Such alliances are also expressions of integrative forces. These forces are driven by technological as well as by institutional changes. In this process, national survival and firm survival are crucial issues.

3. NATIONAL AND FIRM SURVIVAL

In pursuing a strategy of international expansion the firm is affected by its relationship with its national environment. Administrations and governments can be subject to quite different conditions and restrictions, determining their inclination and ability to support a national firm. They may be promotive, supportive or obstructive. Promotion and support can be expressed on the domestic as well as on the international level, while obstruction is more a domestic phenomenon. Administrations and governments still have an impact on the sources of competition within a market. They can limit entry into a national market, directly or indirectly. The ways that are available are through pricing, cost structures, and stimulation or restriction of output. Other important ways are through affecting standards and through distributing rights to infrastructure or space on and in which infrastructure can be built.[6]

The ability of national environments to promote and support national firms does not depend only on their nationalistic inclinations. The internal character and structure of a national environment can affect its power in potential conflicts on the international scene and determine its ability to promote a national firm. A highly fragmented and uncoordinated national environment is more obstructive than promotive. Institutional conditions are a filter which moderates or also possibly veils nationalistic ambitions. They determine the attitudes and actions of administrations and government.

There are internal and external institutional conditions. The internal conditions include the character of the relationships between firm, administrations and government on the one hand, and the character of the relationship between administrations and government on the other hand.[7] The existence of several national firms providing similar services is another important internal institutional condition. It can be easier for a government to be positive as well as promotive with regard to a national firm's international expansion if the firm is unique. The internal institutional situation can be used by administrations and government to veil nationalistic ambitions.[8] Among the external conditions, the most important is the degree to which primarily the administration has committed itself internationally to allow competition on the national level. Another external condition of importance is the position of the country as such internationally. If e.g. the country is a member of a relevant international organization, it might be in a position to promote and support its national firm.[9]

How sincere is the move towards international integration? Will national survival and national competitiveness cease to be goals for national environments, and be replaced by concerns for wider communities of nations? Will the survival and competitiveness of strategic national firms attract less of the attention and concern of national environments?

---

[6] Promotion of the national firm need not be particularly overt. Some argue that the introduction of competition through a duopoly functioned as a way of strengthening and preparing BT for the future.

[7] In Europe, there are three basic solutions to this. There are quoted companies (in the UK, Spain and Italy), state-owned corporations (Sweden, Denmark, Greece, Ireland, Holland, Finland and Portugal) and autonomous entities within a state structure (Germany, Belgium, France, Luxembourg and Norway). It might seem that these different solutions reflect differing levels of government and state involvement. This is of course true; there are big differences between quoted companies on the one hand and autonomous entities within a state structure on the other hand. However, the quoted companies have not been entirely cut off from the government and the state. The Italian government holds a controlling stake in the national operator and the British government has a golden share in BT, basically to limit foreign ownership.

[8] Telia of Sweden applied for a license to operate overseas traffic to and from the U.K., following the breakup of the duopoly. However, it took the British authorities one year to issue a draft license. In the meantime other firms that applied for similar licenses received their draft licenses earlier; these firms are mostly American. Telia complains about the time it has taken to acquire even a draft license. It should also be pointed out that C&W of the U.K. is already involved in Sweden with overseas traffic for corporate as well as for private users.

[9] By forcing other countries to adopt different things that are harmless or even beneficial for the national firm. Cowhey (1990) notes that "If the states interested in regulatory change have significant international market power..., they may successfully challenge the existing international regime. In the telecommunications industry for example, the United States, Japan and the United Kingdom had enough market power to stimulate global reform when they unilaterally changed their national telecommunications policies. This view is shared by Krasner who claims, regarding international communications regimes and disagreements regarding basic principles and norms, that "Stronger states have simply done what they pleased."

National(istic) ambitions may be difficult to distinguish, as many industries affected by international integration are also currently going through demonopolization processes. The actions available for national(istic) behavior are many. Technical and institutional conditions can be used in this respect. Clearly visible actions, like standard setting and allocation of radio frequencies, are not hard to distinguish, but there are also ideological and sentimental actions of importance. National environments differ in their ambitions, as well as in ability to pursue their ambitions.

Some anecdotal indications can be advanced in this context. Mr. Lars Ramqvist, chief executive of Ericsson, said according to the Financial Times, October 29, 1993, that Ericsson "was losing orders around the world because it lacked the political connections of its American, Japanese and European rivals. He even exclaimed that 'We could easily double our revenues today if we had the same political and financial muscle as our competitors'." On the other hand, according to the Times, Ericsson is absorbing more than half of Sweden's export credit support. This was apparently not enough; according to the Times he claimed that the Swedish credit support was "dwarfed by the financial and political support its rivals had access to". Ericsson recently emerged as the lowest bidder for a telephone system in Saudi Arabia. However, according to Mr. Ramqvist, "other factors such as political influence could swing the contract away from Ericsson". Interestingly enough, in Sweden the Foreign Office decided that its data communications should be handled by BT. The decision was based on price considerations, but also on quality dimensions like security and safety. The Swedish State Delegation for Telecommunications Issues has signed agreements with France Telecom regarding data communications. As voice communication will be liberalized, similar agreements will follow. Now, these developments should not necessarily be interpreted as signs that the Swedish national environment is turning its back on its national firms. The contracts awarded to foreign firms have a short-term character. It is thus possible for Telia to be competitive, and to provide what they were not able to provide previously, when the contracts are renegotiated.

Another indication of the role of national environments is the failed merger between Volvo and Renault. Objections from the big institutional owners, and from prominent members of the Swedish business community, made the merger impossible at least for the moment. The objections were centered around the fact that Renault is a state-owned company, deeply embedded in the French national environment. Even if Renault were to be privatized, it was evident that the French government would keep a golden share.

Firms might request and need the support of their national environments on one level, while on another level they might be in a process of liberating themselves from the attention of the very same national environments. Usually, there are some obligations that follow with the support given by national environments to these firms. The obligations may come into conflict with the overall strategies of these firms, which most often are in the process of becoming "ordinary" firms, primarily concerned with their individual survival and well-being rather than with the well-being of citizens and society.[10] Another reason why firms might want to liberate themselves from their national environments is connected with the ability of national environments to provide support and protection. Vernon (1977:1) asserts that "This is

---

[10] Moreover, there could be adverse effects of such a special relationship from the viewpoint of the firm. The frequent reorganizations of BT are attempts to "clean" the organization of influences from the time when postal and telecommunications services were provided together in one organization.

a century in which technological forces have pushed nation-states together and sharply reduced their autonomy." Firms, even in public service and network industries, are increasingly distancing themselves from national ownership, control and coordination.[11] This development complicates the relationship between these firms and their national environment. In light of this, what strategic options are available for national firms?

4. STRATEGIC OPTIONS

The contract between national environments and their telephone operator(s) is being questioned by both. It is far from clear that the concerned parties are prepared to let go of each other completely. They are highly ambivalent regarding this.[12] This development is a product of changes that are not unique for the telecommunications industry. These changes include increasing competition on the national as well as on the international level, demonopolization and internationalization and the integration of markets and nations. Competitiveness and survival of nations and firms are the underlying issues that both inspire these changes and make them difficult. The integration of national environments and national firms is an ongoing process. The old logic is changing, or more correctly it is is being questioned. There is a declared objective, involving the integration of national environments and firms: that national environments within Europe are to move closer, and ultimately merge, in one way or another. There are three strategic options for national telephone operators in this integrative process: integration by domination, integration by federation, and integration by amalgamation.

Integration by Domination

One of the present national firms will become the transnational and trans-European firm. This vision is compatible with the strategic vision of e.g. BT. Since the early 1990s BT has believed that in the future there will be three or four truly global carriers. Apparently, the ambition is that BT will be one of them. Domination for one firm will mean death for other firms. How likely is it that organizations like national telephone operators will be allowed to die, either from competition or from being swallowed by a foreign firm? This probably depends on the time horizon. In the long run it will probably be feasible, but now and in the foreseeable future it will probably be difficult, at least for telephone operators from Western Europe and North America. BT has tried repeatedly to form alliances with a series of European national telephone operators, but did not succeed. BT's potential partners objected to giving up too much of their autonomy. Finally, BT formed an alliance with MCI, the American long-distance operator. Some would claim that the 20 percent stake BT bought in MCI

---

[11] This is reflected in name changes. British Telecom, e.g., changed its name to BT in order to relax associations with a specific national environment. Another example, although with a slightly different touch, is the name change of the Swedish national carrier Televerket to Telia. This followed the change from Televerket, which was an autonomous entity within the state structure (in Swedish "affärsdrivande verk"), to Telia, which is a state-owned corporation. In the long run Telia will be privatized. The management of Telia was not altogether happy with this name change since Televerket was a name that was recognized and highly respected in Sweden as well as abroad. However, keeping the suffix "-verk" would have been an unfair advantage for this organization once it was not only "corporafied" but privatized as well.

[12] Mr. Iain Vallance, chairman of BT, calls his organization a "complex hybrid—part national utility, part international competitor", according to the Financial Times, December 6, 1993.

was expensive. On the other hand, it gave BT considerable influence over MCI.[13] Domination has its cost, and the agreement between BT and MCI was costly. Some commentators did indeed observe that only a virtual monopolist like BT could afford such a deal.

Integration by Federation

An alternative to domination is federation. National firms will survive as parts of a transnational or trans-European structure, representing a division of labor which can be based on functional qualities as well as on the simple fact that they supposedly are experts on their local markets. This strategy is based on cooperative strategies that leave space for some type of autonomy. This could perhaps explain why it has proved to be so popular.

The Unisource joint venture between Telia of Sweden and the Dutch and Swiss PTTs is an example of this. The three firms have formed a joint venture that is dominated by none of them. From the viewpoint of those that choose the dominant strategy there are some difficulties connected with choosing a federative strategy. Who will own the different customers, and how will eventual profits be divided? On the other hand, one clear advantage of this strategy is that it is more of a continuously developing process. BT's problems with finding a European partner could be explained by the difficulties with settling, once and for all, the balance within a partnership. It is interesting to contrast Unisource with the Alcazar alliance, between KLM of the Netherlands, Swissair of Switzerland, SAS of Scandinavia and Austrian Airlines, which failed during the autumn of 1993. The national background of the firms is similar, but these firms tried instead to decide extensively on many intricate issues, in order to prepare a full-scale merger. This proved quite difficult, and the alliance failed. Interestingly enough, another merger attempt also failed during the same time period, that between Volvo and Renault.

Unisource is expanding and according to the Financial Times, December 3, 1993, Telefonica of Spain is discussing entry into the alliance. France Telecom and Telekom of Germany also seem to be developing a partnership that resembles Unisource in its structure. The relative success of alliances like these can also be measured by the interest that AT&T has shown in both Unisource and the alliance between France Telecom and Telekom. It is essential to remember that BT tried to form an alliance with AT&T and failed.

Integration by Amalgamation

Integration by domination and integration by federation are strategies that we can see developing. However, in the long run a third integration strategy is possible as well, integration by amalgamation. The transnational or trans-European firm will be a new structure representing the mixture and the disintegration of the old national firms. Integration by federation could be the first step towards amalgamation.

---

[13] According to the Financial Times, August 25, 1993, BT has "a veto on any purchase by MCI of a business worth more than 20 percent of its market capitalisation, and on any purchase of a non-core business worth more than 5 percent." The Financial Times comments: "Although such restrictions are not uncommon in the US, they reflect BT's concern at the vulnerability of its investment should MCI wish to change its strategy, which hitherto has centred on building itself up as a powerful long-distance and overseas competitor to AT&T in the telecommunications business."

5. CONCLUSIONS AND SUMMARY

The Financial Times, December 2, 1993, published a photograph showing how the French recycled paper industry dumped truckloads of waste paper in Paris in protest at imports, on the eve of recycling talks by European Union environment industries; the trucks were marked "Oui au recyclage français". Now if recycled paper can arouse such sentiments, imagine the nationalistic power and potential of telecommunications services. In this paper we have presented some empirical findings on the cooperative strategic activities of telephone operators. We have also discussed the strategic options available regarding the increased integration between telephone operators as well as between national environments; integration by domination, federation and amalgamation. It is not within our ambitions to predict which of them will prevail. However, we can see that national survival and firm survival are crucial issues that will determine the outcome.

REFERENCES

Cowhey, P.F., 1990, The international telecommunications regime: the political roots of regimes for high technology, International Organization, 44, 2, Spring.

Financial Times, August 25, October 29, December 2, 3 and 6, 1993.

Krasner, S.D., 1991, Global communications and national power, Life on the pareto frontier. World Politics, 43: 336-366.

Vernon, R., 1977, Storm over the multinationals, The real issues (Harvard University Press, Cambridge Massachusetts).

Wells, P.E. and P.N. Cooke, 1991, The geography of international strategic alliances in the telecommunications industry: the cases of cable and wireless, Ericsson, and Fujitsu Environment and planning, Vol. 23, 87-106.

The Race to European Eminence
E. Bohlin and O. Granstrand (Editors)

Asymmetries, Competition and Internationalization: Telecommunications Operators and the Future Single European Market

Philippe Barbet and Laurent Benzoni

Economics Department, École Nationale Supérieure des Télécommunications, Paris, France

This paper examines competition and internationalization of former natural monopolies in the future single European market process. We try to address the following questions:

- What are the determinants of competition between national natural monopolies and how important are size and efficiency asymmetries?
- What is the role of deregulation and privatization in the competition between national natural monopolies?
- Do asymmetric deregulation and privatization affect the process of internationalization of European Telecommunications Operators?
- Are strategic alliances an alternative to Direct Foreign Investment?

1. INTRODUCTION

The European telecommunications market is deeply affected by the completion of the single European market and worldwide deregulation. For each firm or operator, the size of the market will increase as national markets are replaced by a single one. Most economic studies indicate that this trend will lead to an improvement in general welfare (Dixit & Norman 1980, EEC 1988). The welfare impact of the single market results from the combination of two effects: a competition effect and the exploitation of economies of scale (an efficiency effect). Moreover, the legal status of national operators is probably going to converge together with the liberalization of this market. This contribution examines the conditions for the realization of a European oligopoly in the telecommunications sector. It stresses the fact that operators are not symmetric in size, efficiency and their position on the road to deregulation and privatization. We will emphasize the role of deregulation in determining future international competition between former national natural monopolies. Finally, we will try to show that the development of strategic alliances can be considered as a substitute to Foreign Direct Investment (Bhagwati, 1982) when asymmetries of status between telecommunications operators prohibit FDI operations.

2. ASYMMETRIES AND COMPETITION BETWEEN NATURAL MONOPOLIES: A STATIC AND COMPARATIVE ANALYSIS

The first theoretical analysis comes from Dixit & Norman's (1980) contribution to the analysis of the consequences of liberalization with increasing returns to scale. The model suggests a relationship between the increase in the market size and the total number of efficient firms operating in the enlarged market. The basic assumption is that the economies of scale are not fully exploited in national markets and, after enlargement, the surviving firms will be larger and exploit these economies. The ensuing decrease in prices is due to two different effects, an *efficiency effect* because of the increase in the size of the market, and a *competition effect* because we move from a collection of natural national monopolies to a situation in which the new natural oligopolistic competition consists of more firms, each of which can better exploit economies of scale.

The results are consistent with the general methodology of the reports produced for the European Commission concerning the economic impact of the realization of the single market (EEC 1988). The model supposes that firms are identical and the "small operators" are added to form an operator with the standard size. This concentration is part of the strategy of "small country" operators.

The Dixit-Norman approach is useful in understanding changes in the oligopoly equilibrium in an internationally open market. Unfortunately, it is not fully appropriate to explain the changes in telecommunications after the completion of the single European market. Finally, if the model gives the final optimal number of firms surviving in the market after the opening of competition (a natural oligopoly equilibrium), it doesn't say anything about the process of selection itself (Barbet & Benzoni, 1991).

The second approach is focused on the second point: competition between natural monopolies. It comes from the work of W.W. Sharkey (1982, p. 157) in a chapter entitled "*Some complications in a dynamic natural monopoly market*". The author shows that the sustainability of a natural monopoly is not certain for all current and all future outputs as demand is growing over time. The condition for sustainability is that the production function of the supplier is so flexible that the latter can price along its average cost curve at every moment of time. Sharkey supposes that two changes must be studied.

A — The dynamics of demand

A firm which anticipates growth in future demand will build a larger plant than needed for the present period. In the short run, the average cost is high due to overcapacity and a smaller firm can compete in the market. Furthermore, as noted by Sharkey, it is difficult to take full advantage of economies of scale by expanding a plant by small increments because of the existence of sunk costs. The old plant cannot be exchanged at zero cost for a new one of the optimal size in every period.

B — Changing technology

Technology can change over time and a potential outsider may have access to new technology with lower average costs. In this case, insiders will try to secure a stable future demand by contracting with their customers. The stability of these contracts is difficult to maintain due to pressure from the customers and moral hazard problems for the suppliers because the technological changes depend on the research activities of firms.

Following Sharkey's approach, it is possible to show the different situations that can occur when both demand and technology are changing (Barbet & Benzoni, 1991). We suppose two countries (1 & 2) with a natural monopoly in each market before competition (t). The characteristics of each market are the size of demand (D1 and D2) and the efficiency of each supplier given by the long-run marginal cost (MC1 and MC2) , In the case presented below (Figure 2.1) the "large country" (D1>D2) compensates for its technological disadvantage (MC1>MC2) by its size advantage. Before the opening of the market (t+n), it is cost-efficient and has the "first mover advantage". If this supplier is selected by the market to become the common natural monopoly, it can be shown that the market fails to produce a good selection process.

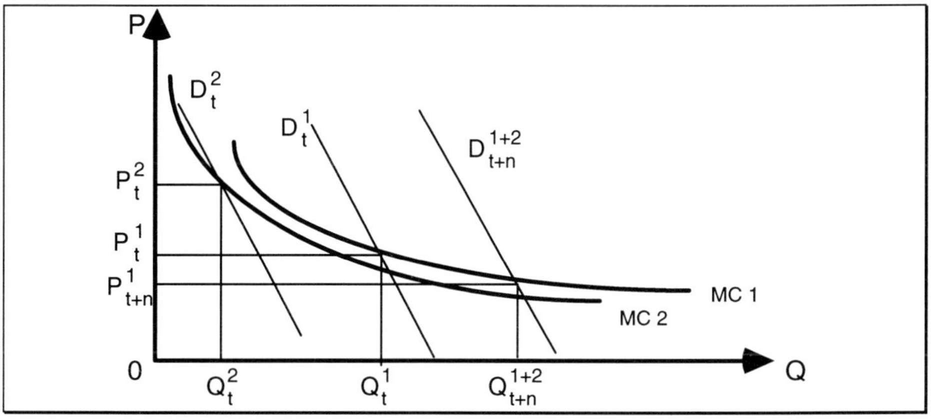

Figure 2.1. Size and technology asymmetries: competition is not efficient.

The ability of a firm to invest very rapidly in new markets after the implementation of a single market is the key to competition between natural monopolies. This ability to move seems ultimately more important than the real efficiency (in terms of technology) of the firms. If we consider that the flexibility of Public Telecommunications Operators in Europe is more or less linked to the degree of deregulation of the sector, this deregulation policy, initiated by public intervention, is not neutral in competition in the telecommunications sector in the case of the creation of a single market.

3. THE DETERMINANTS OF POTENTIAL COMPETITION BETWEEN NATURAL MONOPOLIES

European operators differ not only in the size of their national markets but also in technology and regulation. The combination of these dimensions gives a good appreciation of their competitive strength and their ability to become part of a future European oligopoly.

3.1 The size of the demand and rate of penetration

These dimensions give an appreciation of the initial size of the markets and the possibility of extensive growth (Figure 3.1). We can consider three kinds of national markets: first, the market of large EC countries, generally with a medium rate of penetration (about 50%) with a number of main lines above 20 million. The second type is characterized by a medium-range rate of penetration with "small" demand (below 13 million potential customers), and concerns most other EC countries. The third type consists mostly of EFTA countries with a high rate of penetration and small demand (below 10 million).

The size of the operator is considered important in an internationally open market where competition between operators becomes a reality. The search for greater size leads some countries to concentrate the structure of their national supply. In countries with functional and/or geographical monopolies (Denmark, Italy) there is a strong incentive to concentrate supply and the next step will probably be to set up international alliances. For their part, countries with limited national demand are already engaged in international alliances to become major players in the European oligopoly. Unisource is an example of a tripartite alliance between Televerket (Sweden), PTT Netherlands and Swiss PTT.

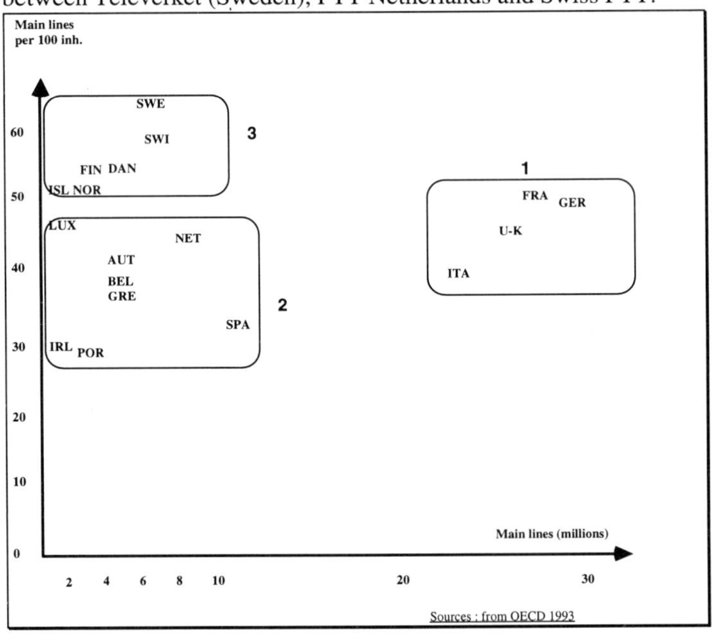

Figure 3.1. Demand size and rate of penetration.

Due to the completion of the single European market, there is a strong incentive for small countries to concentrate their national markets and make alliances to become part of a future oligopoly.

3.2 Technological and financial efficiency

Size is not the only dimension to be taken into account in understanding the establishment of the European oligopoly. Efficiency is an important dimension in the competition process. It is obviously very difficult to have a perfect analysis of the efficiency, both technological and financial, of European operators. We will assume that the penetration of digital technology in the network gives an appreciation of the modernization of the network (Figure 3.2).

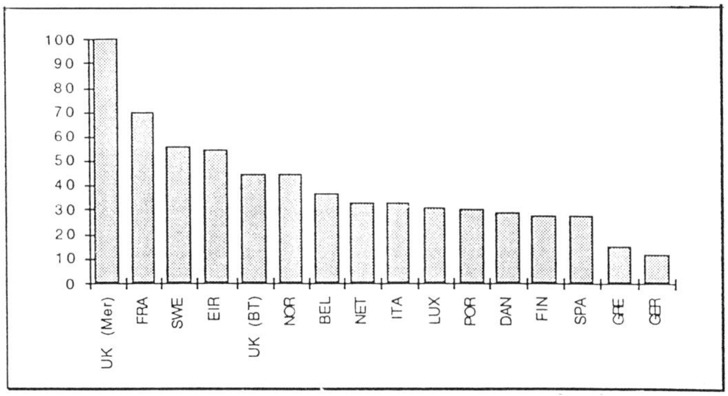

Figure 3.2. Rate of digitalization (%) for 1990.
Source: OECD 1993

Financial performances (turnover and profitability) show great advantages for large countries, especially the UK which enjoys very high profitability (Table 3.1).

Table 3.1
Turnover and profitability of main European operators in 1992

	Turnover (M US$)	Net profit (M US$)
British Telecom	23 134	3 934
DBP Telekom	28 438	-74.8
France Télécom	20 519	303
SIP	15 680	392
Telefonica	11 278	819
PTT Nederland	8 884	947
PTT Suisse	8 636	-31
Televerket	5 691	102

Source: IDATE 1993

This capital flow will be needed partly for the modernization of national networks, but will probably also be used for offensive operations (such as entry into European and US networks). The position of Germany is peculiar, as this country has to mobilize its financial capacities for both the digitalization of its network and the equipment of the eastern part of the country. This weak position can explain a policy of defensive alliance (Eunetcom with France Télécom). The size and financial performances of the other European operators (except France Télécom) do not allow them to have major offensives against other European operators. The only case of a very offensive strategy concerns Telefonica (the Spanish operator) in Latin America.

3.3 The degree and rate of deregulation and privatization

The degree of deregulation of telecommunications infrastructure in Europe is very different and is proceeding at contrasting speeds (Figure 3.3). Coming from a national administration in most European countries, the telecommunications sector is on the road to liberalization and privatization.

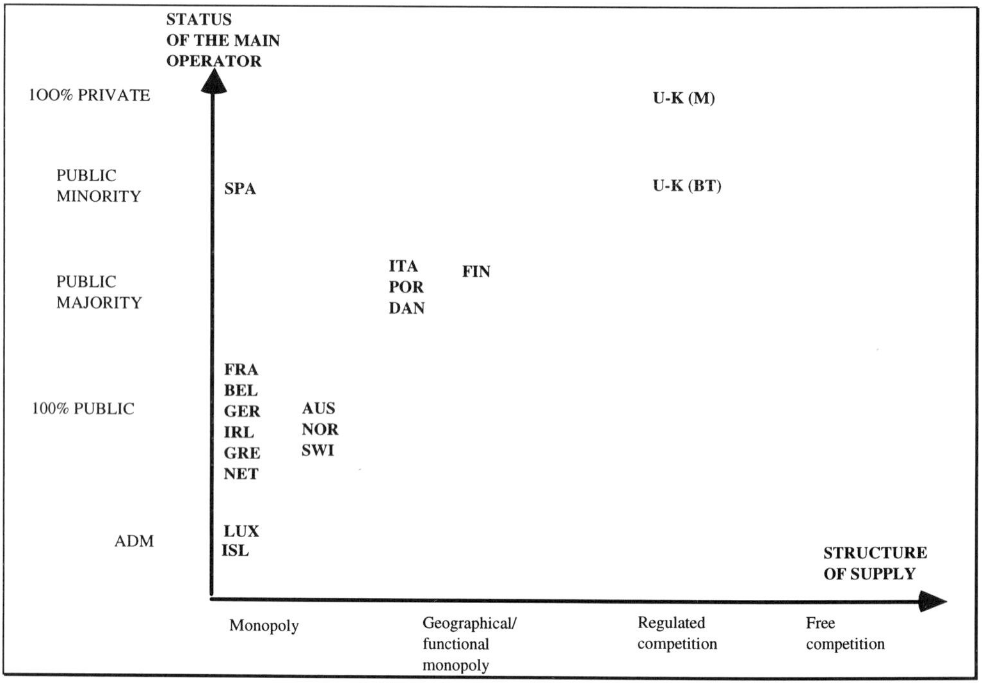

Figure 3.3. Deregulation and privatization of telecom infrastructures (mid-1993).

Liberalization of telecommunications is imposed by the European Community and each country must adapt. Privatization is a political decision and the government can use it for different goals: the first one is to reduce public deficits, as is mainly the case in Latin America but less so in Europe; the second one is to raise money for construction and modernization, as is the case of some countries such as Greece, Portugal and, to a lesser extent, Germany; the last one is to improve the opportunity of the operator to act as a firm in a free market. Deregulation and privatization in the United Kingdom give a good example of what can be called "strategic deregulation and privatization". British Telecom can take full advantage of its status to become very active in the field of international telecommunications services. The completion of the single European market in telecommunications and the realization of an oligopoly raises the question of the stability of the future market structure. The question is to know to what extent asymmetries in deregulation and privatization can destabilize this market structure. One can assume that concentration and alliances in Europe will lead to a limited number of operators in competition. If some asymmetries persist, a kind of Stackelberg structure with major operators and small ones can exist, and raises the question of the stability of this kind of oligopoly.

4. INTERNATIONALIZATION OF EUROPEAN OPERATORS: FDI VERSUS ALLIANCES?

Internationalization of telecommunication networks by FDI is linked to privatization. When a country sells part of its network, it seeks the entry of efficient actors. FDI is actually mainly limited to the privatization of developing countries, and most of the investments made by European TOs are in Latin America (Argentina and Mexico). These operations are asymmetric in the sense that there is no demand for reciprocity from countries which open their markets to foreign investment. Even a publicly owned operator like France Télécom can enter a new private market. These countries generally seek an efficient operator to rebuild their network, whatever its juridical status.

In Europe, it is possible to forecast that privatization will lead to FDI in countries such as Greece, Portugal and Eastern Europe. Even if these countries want to improve their networks, the question of the status of the foreign operator entering with FDI obviously matters in the decision of whether or not to privatize (e.g. in Greece). These countries do not seek reciprocity but keep a majority in the capital of the new operator. They are seeking a mix between FDI and alliance.

Concerning large developed countries (Japan and USA), major FDI of European operators without reciprocity is probably impossible. British Telecom has made a significant step toward internationalization by buying 20% (which is the upper limit) of the capital of MCI and by the creation of a joint venture with its American partner. Due to demand for reciprocity by American trade authorities, this operation has been very difficult to undertake. The private status of British Telecom obviously made this operation possible. France Télécom and Deutsche Telekom, still publicly owned, would probably have failed to enter the capital of MCI without reciprocity for an American operator to enter their capital. Asymmetries of status are a major problem for FDI operations in the telecommunications sector.

An alternative to FDI is strategic alliance between telecommunications operators. With strategic alliances, there is no need for symmetry in the legal status of the partners, and more important are symmetries in size and efficiency. If we look at the two major alliances in the European market, we see that the statuses are quite different, even if they are on the road to a future privatization.

Table 4.1
The UNISOURCE alliance

Operator	Legal status of the partner	Privatization	No. of main lines (X1000)	Rate of digit.(%)
PTT Neth.	100% public	1994	7395	33
Televerket	100% public	?	5900	56
Swiss PTT	Administ.	?	4184	28.2

These operators are quite close with respect to their size and efficiency of the network. Unisource reaches the size of a "large operator" on the European market without major asymmetries between the partners.

DBP Telekom and France Télécom have a comparable size and status but the privatization of DBP is likely to come sooner than France Télécom's. The major asymmetry existing between the two operators lies in technological and financial efficiency, and we suggest that this alliance is probably more defensive than offensive and part of the realization of a strong European oligopoly. Meanwhile, this operation is submitted to European law on concentration (Art. 86 of EEC treaty).

Alliances are an alternative to FDI when asymmetries still exist between national operators. The question to raise is to know whether these alliances are the first step toward mergers and more traditional FDI and globalization in the future.

5. CONCLUSION

This article presents a general theoretical framework for analyzing competition between natural monopolies. It underlines the fact that multiple asymmetries affect the economic results of internationalization of European operators. In particular, asymmetries in the status of operators and the speed of privatization and deregulation play a major role in the internationalization of the sector. We suggest that FDI requires a relative homogeneity of status to deal with the question of reciprocity in international relations. We conclude that alliances can be seen as an alternative to FDI when those asymmetries still exist. We do not conclude whether these alliances are the first step toward mergers or a stable market organization.

REFERENCES

Amendola, M. and J-L. Gaffard, 1988, La dynamique économique de l'innovation, Economica.

Barbet, P. and L. Benzoni, 1991, Rendements croissants et concurrence internationale: vers la déréglementation stratégique, Revue d'Economie Industrielle, N 55, 1 trimestre.

Baumol, W.J., J.C. Panzar and R.D. Willig, 1988, Contestable markets and the theory of industry structure, 2nd ed. (Jovanovich Pub., Harcourt Brace).

Bhagwati, J., 1982, Import competition and responses (Chicago University Press).

Braeutigam, R.R., 1989, Optimal policies for natural monopolies, Handbook of industrial organisation, 2 vol. (North Holland, Amsterdam).

Curien, N. and M. Gensollen, 1992, Economie des télécommunications: Ouverture et réglementation, ENSPTT et Economica.

De Drouas, F., D. Henriet and E. Turpin, 1991, La logique des achats d'équipements de réseaux, CNET-MEE, 1990, Communications & Strategies.

Demsetz, H., 1968, Why regulate utilities, Journal of Law and Economics, pp. 158-176.

Dixit, A.K. and V. Norman, 1980, Theory of international trade, Cambridge Economic Handbooks, 1st ed.

EEC, 1988, Research on the cost of non-Europe.

IDATE, 1993, Les services de télécommunication dans le monde, Analyses industrielles.

Khan, A.E., 1970 and 1971, The economics of regulation: Principles and institutions, 2 vol. (John Wiley, New York).

Müller, J., 1988, The benefits of completing the internal market for telecommunications in the Community, research on the cost of non-Europe, Vol. 10, Commission of the European Communities.

OECD, 1993, Perspectives des télécommunications, Paris.

OMSYC, 1992, Key figures for the world telecom market.

Panzar, J.C., 1989, Technological determinants of firms and industry structure, Handbook of industrial organisation, 2 vol. (North Holland, Amsterdam).

Reid, G., 1979, An analysis of the firm, market structure and progress, Scottish Journal of Political Economy, Vol. 26, 15-32.

Reid, G., 1987, Theories of industrial organization (Basil Blackwell).

Scherer, F.M., 1980, Industrial market structure and economic performance (Rand MacNally, Chicago).

Schmalensee, R. and R. Villig, 1989, Handbook of industrial organisation, 2 vol. (North Holland, Amsterdam).

Sharkey, W.W., 1982, The theory of natural monopoly (Cambridge University Press).

Sharkey, W.W., 1987, Natural monopoly, New Palgrave Dictionary of Economics (MacMillan Press).

Challenges for Management

The Race to European Eminence
E. Bohlin and O. Granstrand (Editors)
1994 Elsevier Science B.V.

Reorganization of European Telecoms: The Cases of British Telecom, France Télécom and Deutsche Telekom*

Rudolf Pospischil

Deutsche Telekom Headquarters, P.O. Box 2000, 53105 Bonn, Germany

1. INTRODUCTION

Most telecommunications operators have changed their organizational design during the past ten years more than once. Since January 1, 1993, France Télécom and Deutsche Telekom have run their businesses within a new organizational framework. British Telecom (BT) changed its organization on April 1, 1991. On the one hand, the new organizational designs of the telecommunications operators reflect the changes in their environment, regulation in particular, as well as in technology. On the other hand, their organizational design is a result of the strategic approach adopted by management. Due to the fact that networks are the prerequesite for the provision of services and that they account for far more than half of the costs of a telecommunications operator, their organization is crucial for the success of the whole organization.

Often the definition of the organizational design of a company is seen as a more or less mechanistic activity. Sometimes it is focused on the question "Who is who in the company?" However the purpose of this paper is to analyze different organizational designs with regard to Alfred Chandler's famous expression "*structure follows strategy*" (Chandler, 1962, p.14). This paper gives a comparison of the old and the new organizational designs of the three operators.

How does management identify the best organizational design for a given company? There are many different options. With regard to the context of a company some solutions seem to be reasonable, others not. The context can be reduced to four important categories or *contextual dimensions*: technology, size, environment and strategy (Aldag and Stearns, 1991). So besides strategy, other factors are determining the organizational design of a company.

Technology does not only cover the machinery to transform an input and to produce an output but also includes know-how of the workforce. In our case, all the telecommunications operators use almost the same technology to produce quite similar products or services. The technical system of a telecommunications operator is both complex and determines the work or production process. Obviously that is true for all three carriers.

All three carriers are either monopolistic or dominant in their markets which are mainly their domestic countries. Therefore the *size* of the three organizations and their markets and not only the size of the organizations are compared. In terms of their territory, economic structure, population and density of population, France, Germany and the UK are not very differ-

* A modified and longer version of this paper appears in Telecommunications Policy, Nov. 1993.

ent. As a whole, BT, France Télécom and Deutsche Telekom are very similar in size or belong to the same league. Therefore there is no reason why they should have a different organizational design with respect to their size.

One generation ago the *environment* of all three carriers was almost the same. They operated as public administrations. Their task was to supply telephony service, some other services and products of minor importance such as telex or customer premises equipment. The demand for their products grew slowly but constantly. Due to their monopoly there was no competitor and a serious debate about their monopolies was not even an issue at the time. The technology did not change very fast either. All this can be summarized as a stable environment. During the last decade the environment changed a great deal. But the winds of change did not blow with the same force all over Europe. The changes brought to the telecommunications market in the last decade made it less stable and more dynamic. It is expected that the market will become much more dynamic due to the abolition of the remaining monopolies and increasing competition as a result therefor. Today the British market is the most dynamic market followed by the German and French.

Strategy seems to be the crucial contextual dimension for the analysis of the organizational changes in the three telecommunications operators. Theory explains management as a systematic process covered by five subsequent activities: planning, organizing, staffing, directing and controlling, (cf. Mackenzie, 1969). Thus the management first defines a strategy or plan, and then implements the strategy by designing the organization. Due to the fact that the size of their organizations and the technology used are very much alike, and their environments also tend to be increasingly similar, the managements of British Telecom, France Télécom and Deutsche Telekom differ mainly in their strategies when designing their future organization. However for a transition period their different environments could be another reason for different solutions.

2. THE FORMER AND THE NEW ORGANIZATIONAL DESIGN

In this section the actual and the former organizational structure of the three carriers will be discussed. When describing these organizations the focal point is their core business. Therefore no specific information on support functions like logistics is given or discussed. Despite the fact that mobile communications are becoming more and more important, their organization will not be discussed. All three carriers have divisionalized their mobile business either within the organization or as a separate company.

2.1 British Telecom[1]

When BT introduced its new organization on April 1, 1991, one slogan was used to accompany and explain the organizational change: "We put our customers first". Indeed this was not only a slogan but also one of BT's values.[2] With regard to different customer groups BT's old

[1] Cf. Eurodata Foundation, Databook 1991; Leadbeater, 1991; McClelland, '1990

[2] BT has defined five key values for all employees:
- We put our customers first.
- We are professional.
- We respect each other.
- We work as one team.

organization was indifferent, or almost indifferent (see Figure 1). BT's three operating divisions "BT International", "BT UK" and "Communications Systems Division" had the following tasks:

- "BT International" was responsible for BT's international networks and services.
- "BT UK" was responsible for all inland networks and services and supplied customers with equipment.
- The "Communications Systems Division" had three main tasks. First, it was responsible for a portfolio of products, systems and applications for worldwide sales. Second, its responsibility covered mobile communications and third, it managed BT's customer premises equipment business.

With its reorganization BT wants to focus on the customer's needs (see Figure 2). That is the main purpose of the reorganization called "Project Sovereign" communicated internally and externally. Simultaneously this reorganization shall give BT a leaner organization through reducing its staff by about 40,000 including some 5,000 middle management positions. Altogether "Project Sovereign" was set up to create a structure for BT's vision "To become the most successful worldwide telecommunications group". A comparison of the old and new approaches vis-à-vis the customers reveals the differences.

- BT used to focus on two markets, the domestic and the international. Now there are two main customer groups: business and personal (or private). In addition to the two major markets or customer groups both the new and old organizations have a third category comprising a set of products which either do not fit into the two major units for business reasons or which are handled separately for legal or regulatory reasons. These are mobile communications, operator services, yellow pages, and visual and broadcast services. Within the new organization "Special Businesses" takes care of these businesses.
- While the former solution was to integrate sales, marketing and production, the new approach introduces a division between sales and marketing on the one hand and production, which provides the network platform, on the other hand. The provision of networks is not divided into domestic and international networks as it was before. Now the networks are provided worldwide by one department called "Worldwide Networks". This department has no access to the market, which is served by "Personal Communications" and "Business Communications". "Worldwide Networks" supplies "Personal Communications" and "Business Communications" with the network platform necessary for their tasks.
- Another major difference between the old and new organizations of BT is the creation of a department for the management of BT's product portfolio called "Products and Services Management". In the old organization there was no such high-ranking and powerful strategic apex. This department is an instrument for more than the coordination and controlling of the product portfolio of BT. Its responsibility covers for example the fixing of product prices.

Despite the fact that BT uses the expression "division" for "Personal Communications" and "Worldwide Networks" and so on, BT does not have a divisional design. BT's organizational

- We are committed to continuous improvement.

design is to be characterized as a functional design. The principle for a functional design is to separate the following basic functions: strategy, research and development, procurement, sales and marketing, production, personnel, finance. The following comparison shows how BT's organizational design matches the functional approach.

Basic functions	*BT department*
strategy	"Products and Services Management"
research and development	"Development and Procurement"
procurement	"Development and Procurement"
production	"Worldwide Networks"
sales and marketing	"Business Communications"
	"Personal Communications"
personnel	"Group Personnel"
finance	"Group Finance"
not attributable	"Special Business"

But BT does not have a purely functional design. The comparison given above shows that "Special Business" is an exception as it is not attributable to one function. There are more deviations from the purely functional design. These become particularly evident when analyzing the organizational design of the departments. Most of them incorporate personnel and financial functions. That means that different organizational designs are applied to different levels of hierarchy within the organization. Thus the incorporation of the personnel and financial function in the network department leads to the use of the word division because BT has given its departments some divisional attributes, but these remain subordinated to the central or group functions "Group Personnel" and "Group Finance". However, BT's principal structure is a functional one.

2.2 France Télécom[3]

In the past years the organizational design of France Télécom was changed frequently. Years before Deutsche Telekom, the public administration France Télécom created specific departments for sales and marketing as well as for international business (see Figure 3). Now, with the actual reorganization (see Figure 4), France Télécom is pursuing three main goals:

- The headquarters shall be focused on strategic tasks. Operational tasks executed until the end of 1992 within the headquarters are being separated and transferred to new internal service centers. Thus the staff of the headquarters will be reduced from about 5,000 to less than half this number.
- The levels of management will be reduced by one. According to the political and administrative structure of France with its 22 regions France Télécom had 22 regional headquarters reporting directly to the CEO. Within the new organizational structure the management of the regions is incorporated into the headquarters, i.e. the regional headquarters have been abolished. Now five regional managers are responsible for 48 operational directorates. This integration of the regional management into the headquarters changes the spans of control significantly. While the units directly reporting to the CEO have been

[3] (Cf. Billard, 1991; France Télécom, 1993)

reduced by 17 to about half the former number, the span of control for the regional managers has increased from sometimes two or three to an average of ten.

- A clear definition of the line authority and the functional authority shall lead to a less bureaucratic organization. The major functional departments ("National Networks", "Sales and Marketing", "Human Resources", "Finance and Budget", "Information Systems") now have a functional authority vis-à-vis the "Operational Directorates". The line authority vis-à-vis the "Operational Directorates" is now exercised by the regional line managers placed on the level of executive officers. The local units report directly to the "Operational Directorates".

The organizational design of France Télécom is less functional and more divisional than that of BT. While the functional departments of BT have both a line and functional authority to their regional or operational units, the functional departments of France Télécom hold only a functional authority vis-à-vis these units. France Télécom can be classified as an organization with a hybrid design that includes divisional units (regions) as well as functional units (e.g. "National Networks").

2.3 Deutsche Telekom

Compared with the two other organizations, there is no such powerful person like a CEO or chairman of the board at Deutsche Telekom. Deutsche Telekom's chairman of the management board is "primus inter pares", but he has no line authority vis-à-vis the other members of the management board. The old organizational structure of Deutsche Telekom (see Figure 5) was very similar to that of France Télécom. On the one hand there were functional departments and on the other there were the division-like regional headquarters. The regional headquarters did not report to one person but to almost all functional departments. This means that several functional departments exercised a line authority vis-à-vis the regional headquarters. But at the same time the functional departments bypassed the regional headquarters by approaching the local units directly. The main characteristic of Deutsche Telekom's organization was a high degree of standardization of work processes. Mintzberg describes such an organizational design as a machine organization, (cf. Mintzberg). With its new organizational design (see Figure 6) Deutsche Telekom pursues three main purposes:

- Production will no longer be separated into the two functions of switching on the one hand and transmission on the other. The new "Networks" department will integrate both functions. Almost all other network operators, including BT and France Télécom, have such network departments which comprise both transmission and switching. A look back to the old organizational structure in which switching was handled separately from transmission may explain the low degree of digitalization of Deutsche Telekom's local switches. The economies of scope resulting from organizing switching and transmission in one department could not be reached due to organizational disparities.
- Sales and Marketing has been split into three distinct departments serving different customer groups. These departments are "Private Customers", "Business Customers" and "Key Accounts". Furthermore, each of three departments is responsible for the management of its own portfolio of products and services. The development of new services and transmission and switching systems is the task of the new "Development" department.

- The regional headquarters will be given a new organizational structure and new tasks. Germany is divided into six regions. Each region has three regional headquarters responsible for "Networks", "Business Customers" and "Private Customers" respectively.

These specialized regional headquarters report to the three related departments at the central headquarters. This reduces the number of regional headquarters from 23 to 18 and creates a transparent and clearly defined management pattern. With regard to the operational core in the local units reporting to the regional headquarters, there will be less emphasis on the standardization of work processes and more emphasis on the standardization of outputs, i.e. management by objectives.

3. THE ACTUAL DESIGN OF THE NETWORK ORGANIZATION

The design of the network organization was not changed in all three cases to the same extent. Major changes occurred in the case of Deutsche Telekom. BT's organization was changed to a greater extent than that of France Télécom but to a lesser extent than that of Deutsche Telekom. With regard to the given alternative solutions for an organizational design of the networks, this chapter will treat the current organizational designs rather than the former ones.

3.1 British Telecom

BT's networks seem to be concentrated in one department, namely "Worldwide Networks". A more detailed analysis will show that not only "Worldwide Networks" but also "Personal Communications" is responsible for networks. An analysis of BT's networks with regard to their organizational design (see Figure 7) comprises the following steps or levels:

- The basic split within "Worldwide Networks" is between international and domestic networks. The domestic networks are then divided into a line function for planning, construction, operations and maintenance on the one hand and specialist staffs to set the rules and standards on the other.
- The domestic networks are separated into two distinctive parts: the access and the core network. The access network is not handled by "Worldwide Networks" itself. Planning, construction, operations and maintenance for the access network are carried out by the "Personal Communications" department. However, the general rules for planning the access network and its equipment are determined by specialist staffs in "Worldwide Networks". The idea behind "Personal Communications" being responsible for the access network is that the quality of the access network is crucial for customer satisfaction. Thus a high level of quality should be a competitive advantage over other network suppliers. The sales and marketing staff should therefore not only provide the customers with the access network, but also take charge of its planning, construction, operation and maintenance, because they know the customers' problems better, (cf. Allen and Lisle, 1991).
- Finally there is a regional split. The UK is divided into the London area and the rest of the country, which is again subdivided into several zones.

A more detailed analysis of the different tasks allocated to the above-mentioned organizational units would reveal a much more complex picture. There are, for example, specific rules for the use of new technologies such as fiber cables in the access network. It is therefore important to emphasize that this paper only outlines the general rules rather than the exceptions.

3.2 France Télécom

Whereas mainly one department is responsible for BT's networks, three organizational units are in charge of the networks of France Télécom. The division into three units is based on the network architecture (see Figure 8).

- The international networks are planned, constructed, operated and maintained by the "International" department. This department does not only deal with networks but includes sales and marketing abroad as well.
- The domestic networks are divided into trunk and local networks. The "National Networks" department has its own central and regional organization for planning, construction, operations and maintenance of the trunk network. For the local network the "National Networks" department has a functional authority.
- The local networks are covered by the "Operational Directorates" which report to the five regional executive officers. The rules and standards for the work on local networks, done by regional and local units, are defined by the "National Networks" department.

3.3 Deutsche Telekom

Deutsche Telekom has learnt a lot from both BT and France Télécom as well as from other network operators. The main features of the design of its network organization (see Figure 9) are:

- There is a division between domestic and international networks. The international networks are planned, constructed, operated and maintained by the international part of the network department. Contrary to France Télécom's solution, sales and marketing for international networks are not done by the same department.
- For domestic networks a clear distinction is made between functional and line authority. The headquarters for domestic networks can be described as specialist staffs that set the rules and standards while the regional units can be described as middle managers and the branch offices as the operational core.
- Within the headquarters as well as within the local units, planning is divided into long-term planning (network strategy) and short-term planning (operational planning). Network strategy is composed of strategies for the local network and the trunk network. Operational planning, construction, operations and maintenance are grouped according to the access network on the one hand and the core network on the other hand.

4. COMPARISON AND EVALUATION OF THE THREE SOLUTIONS

A comparison of the former and actual organizational charts reveals the following differences: (a) the classification of the finance department, (b) the importance given to international business or networks, (c) the different solutions for separating the domestic network, (d) the division of line and functional authority and (e) the segmentation of the market and its translation into an organizational structure.

A comparison of the *grade of hierarchy given to the financial departments* shows a significant difference between British Telecom on one hand and France Télécom and Deutsche Telekom on the other. In the case of BT the department "Group Finance" is not subordinated to the "Group Managing Director" that directly reports to the "Chairman of the Board". In the case of France Télécom and Deutsche Telekom the finance department is placed at the same level as the other executive officers. With regard to the ownership of the three organizations this difference becomes evident. BT's shares are held by many private and institutional investors as well as by the British Government. The non-governmental investors which hold more than half of the shares have bought the shares for they believe the shares have a good value compared to others. For an organization, which was a public administration in the past, it is crucial to learn the rules of the capital market. Therefore it is not surprising that BT has given the financial department such a high-ranking position and it would not be extraordinary if France Télécom or Deutsche Telekom changed their structure in that sense once they become private companies quoted on the stock market.

Only in the case of France Télécom is there an executive officer responsible for *international networks or business*. British Telecom used to have such a responsibility in the former organization, but it has now vanished. Deutsche Telekom did not and still does not have it. How can these differences be explained by the underlying strategy? In the past, BT divided its business into national and international markets. Now, the focus has shifted from these markets to the customers and services. Therefore a differentiation between the national and international sector cannot be found at the level of the executive officers but at the level below. At a first glance this shift seems to devalue the international business, while the opposite is true. BT acts no longer in separated domestic and international markets but in a worldwide market as a global corporation. France Télécom still divides its business into two almost separate markets on the one hand and promotes its international activities on the other. Therefore France Télécom has an international department at the level of the executive officers. Deutsche Telekom, which is a latecomer in the international business compared to BT and France Télécom, should by now—with its reorganization in early 1993—have an international department at the level of executive officers. Obviously this would give international business a strong support. However, Deutsche Telekom has decided to overcome the headstart of the others by a leap to reach the state-of-the-art organization, i.e. to follow the approach that telecommunications is a global business.

The *design of the domestic network organization* differs in two ways. The first observation is that only in the case of Deutsche Telekom is the management of the domestic networks the sole responsibility of one executive officer. In the case of British Telecom there are two executive officers, one responsible for "Worldwide Networks" and one for "Personal Communications". In the case of France Télécom the responsibility lies in the hands of six executive officers, one for "National Networks" and five for the regions. These divisions of responsibility are particularly interesting with regard to conflicts. It is hard to predict conflicts in almost new organizational structures. But BT's as well as France Télécom's executive officers responsible for "Worldwide Networks" and "National Networks" have great power because they at least have a functional authority in all domestic networks. That was not the case in Deutsche Telekom's old organization, when two executive officers were responsible for networks and thus had almost equal powers. The second observation is that there are two models to divide up the domestic networks. British Telecom's dividing line is at the main distribution frame which separates the access network and the core network. France Télécom splits between the local switch and the trunk switch, i.e. there are local networks on the one hand and trunk networks on the other. In the case of Deutsche Telekom the organizational chart does not divide the domestic network. The reason therefore is not that a division does not exist, but that this division is not a guideline for the organizational design at the three levels taken into consideration. First, Deutsche Telekom distinguishes between the trunk and the local network for the planning of the network structure. Second, for the short-term tasks of planning, constructing, operating and maintaining the network, a division between the access network and the core network was introduced. On the one hand this is to make sure that the structures are changed by almost neutral planners not blunted by day-to-day operations. On the other hand this will divide the crucial operational work into two very different tasks of the access and core network.

For large organizations it is almost impossible to live without rules and standards for the operational core. These are defined by specialist staffs that have the functional authority to implement their rules and standards. For all three organizations there is a clear pattern of *line authority vis-à-vis functional authority*. This means that a pattern for line-staff conflict has been installed. Such a clear pattern brings advantages in comparison to the old structures of Deutsche Telekom and France Télécom. It used to be almost impossible to define a clear structure of line authority and functional authority with regard to their regional headquarters.

The *segmentation of the markets* in residential customers and business customers becomes evident in the case of British Telecom and Deutsche Telekom. Both have divided sales and marketing into several parts aimed at different customer groups. Deutsche Telekom goes a step further in this direction when assigning to these sales and marketing departments the tasks of services and product management for their customers. In the case of BT, services and product management are not separated but combined for business and private customers in a central department headed by an executive officer. Such a high ranking and central function are also found in the case of France Télécom which contrary to British Telecom and Deutsche Telekom has no customer-specific departments for sales and marketing. Hence, Deutsche Telekom gives its sales and marketing departments more leeway to be able to achieve success. This makes the organizational design of Deutsche Telekom less functional than that of BT.

5. SUMMARY AND CONCLUSION

Over the past ten years most telecommunications operators have changed their organization more than once. British Telecom, France Télécom and Deutsche Telekom have implemented major organizational changes during the past three years. On the one hand the new organizational designs of the telecommunications operators reflect the changes in their environment, regulation in particular, as well as in technology. On the other hand their organizational design is a result of the strategic approach taken by management. Due to the fact that networks are the prerequisite for the provision of services and that they account for far more than half of the costs of a telecommunications operator, their organization is crucial for the success of the whole organization. A comparison of the old and the new organizational designs of the three operators reveals (1) a trend towards functional organizations in the case of British Telecom and Deutsche Telekom, (2) major changes in the management of the regional units, (3) three different approaches to designing a network organization, (4) a centralized strategic apex in the case of British Telecom and France Télécom, and (5) a powerful financial department in the case of British Telecom.

In view of the internationalization of the telecommunications business, more attention must be given to a new organizational solution, namely the division of a global network managed by one network operator into three parts: *domestic, international and intranational.* The domestic network is the national network in the country where the carrier has its home market. The international network connects the domestic network with other national networks. The intranational networks are those national networks operated in foreign countries. Only domestic and international networks could be identified for the three organizations under consideration. *Intranational* networks did not show up. The reason for this is not that intranational networks do not exist. Today intranational networks are handled often separately from the network departments for legal and management reasons. Intranational networks constitute different legal entities that are often joint ventures. Therefore they are not completely under control. Thus it is not possible to incorporate them in a line function. Either managers of the network department or the financial department are in charge to act as the owner of the participation in the companies running the intranational network. There is some evidence that the cooperation of former domestic carriers in the field of global networks and services will create for the organization of *international* networks similar solutions as for intranational networks.

REFERENCES

Aldag, R.J. and M.T. Stearns, 1991, Management, 2nd ed. (South-Western publishing, Cincinnati, Ohio), pp. 283–292.

Allen, P.A. and P.H. Lisle, 1991, Planning the access network in a changing environment, British Telecommunications Engineering, Vol. 10, pp. 8–12.

Billard, J., 1991, Organization—An ABC guide to France Télécom's new status, in Telecom France, international edition of La Revue Française des Télécommunications, Oct., No. 13, pp. 6–11;

Chandler, A., 1962, Strategy and structure: Chapters in the history of the industrial enterprise (MIT Press, Cambridge, MA).

Eurodata Foundation, ed., Databook 1991, London, Chapter UK BT

Leadbeater, C., 1991, Corporate reorganization: BT plugs into a cultural revolution, Financial Times, 27th March.

France Télécom, ed., 1993, Les télécommunications, Paris, pp. 125–143.

Mackenzie, R.A., 1969, The management process in 3-D, Harvard Business Review, Vol. 47, November/December, p. 87.

McClelland, S., 1990, A sovereign state: British Telecom restructures, Telecommunications, Aug., pp. 65–68.

Mintzberg, H., The machine organization, in H. Mintzberg and J.B. Quinn, eds., The strategy process: Concepts and contexts (Prentice-Hall, Englewood Cliffs, NJ), pp. 264–280.

Telecommunications Policy, Vol. 17, November 1993, pp. 603–621.

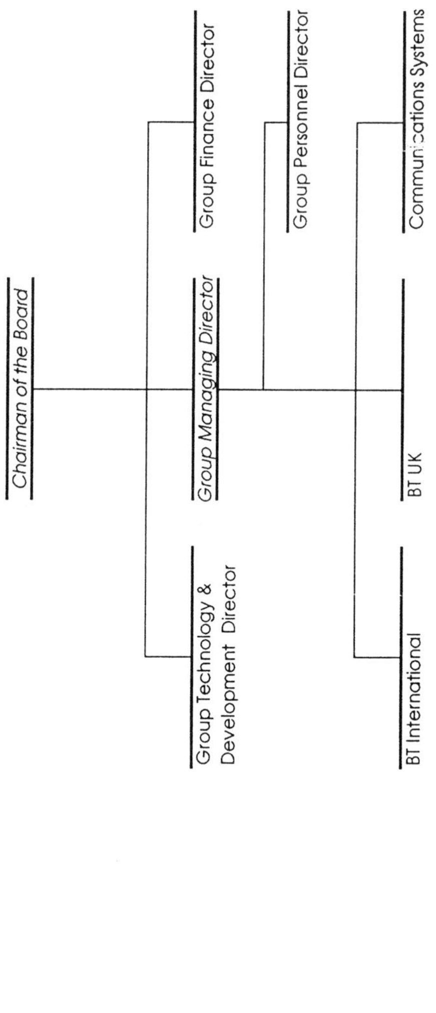

Figure 1. BT's old organizational structure

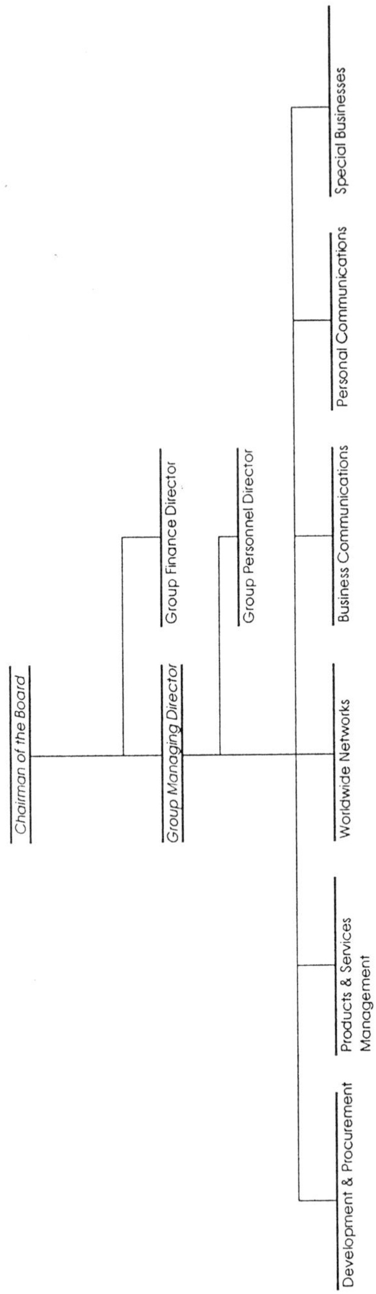

Figure 2. BT's new organizational structure

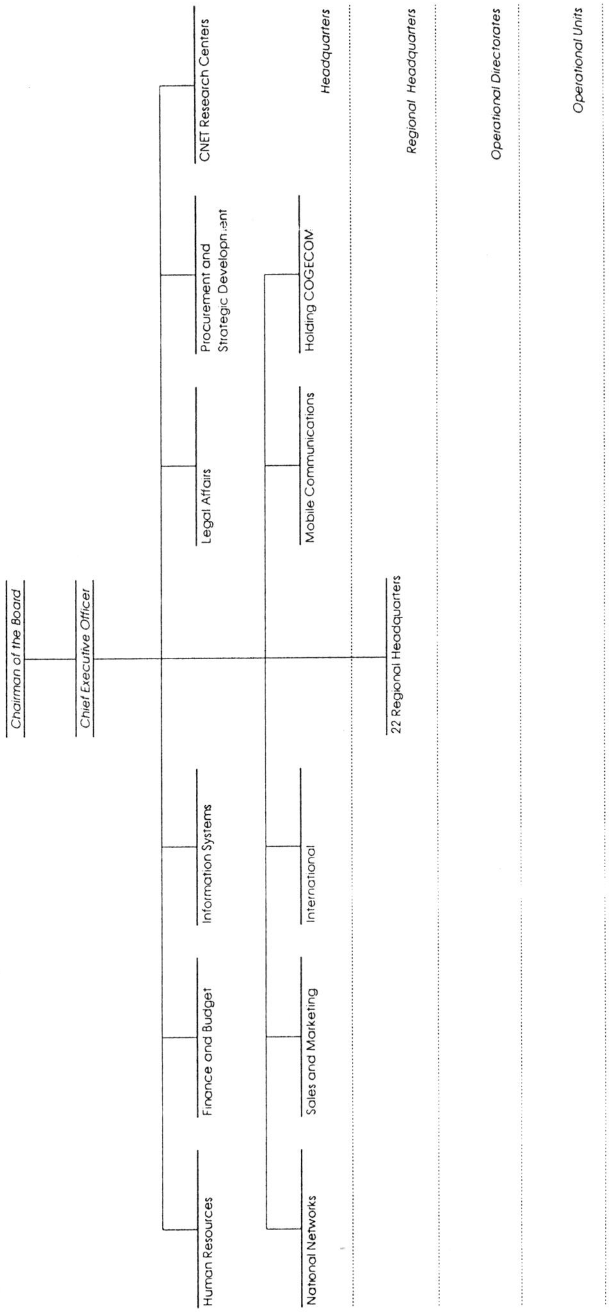

Figure 3. France Télécom's old organizational structure

Figure 4. France Télécom's new organizational structure

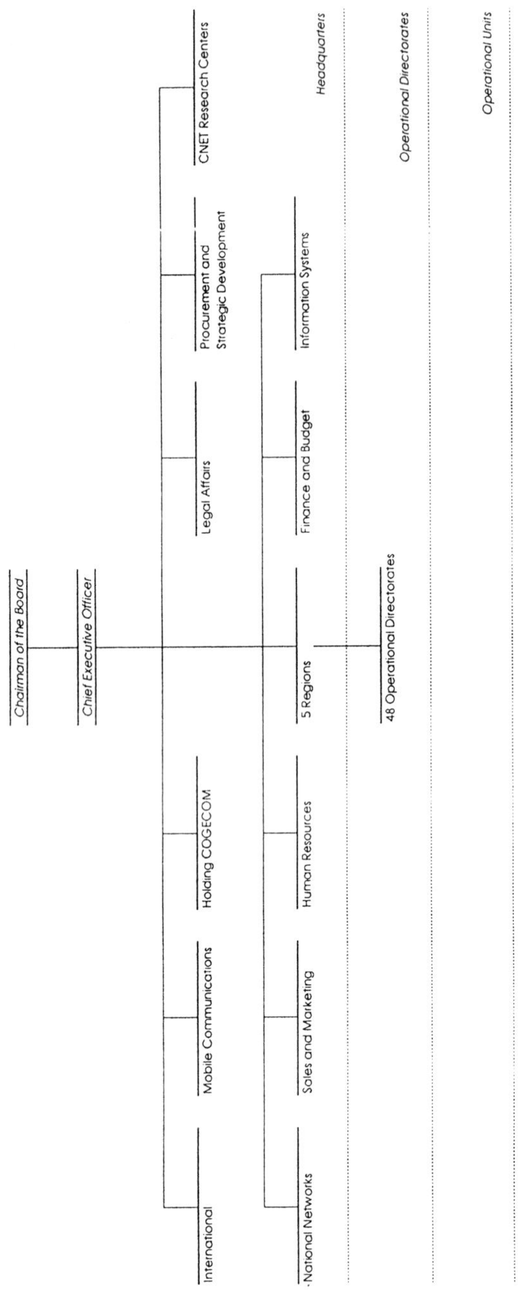

Figure 5. Deutsche Telekom's old organizational structure

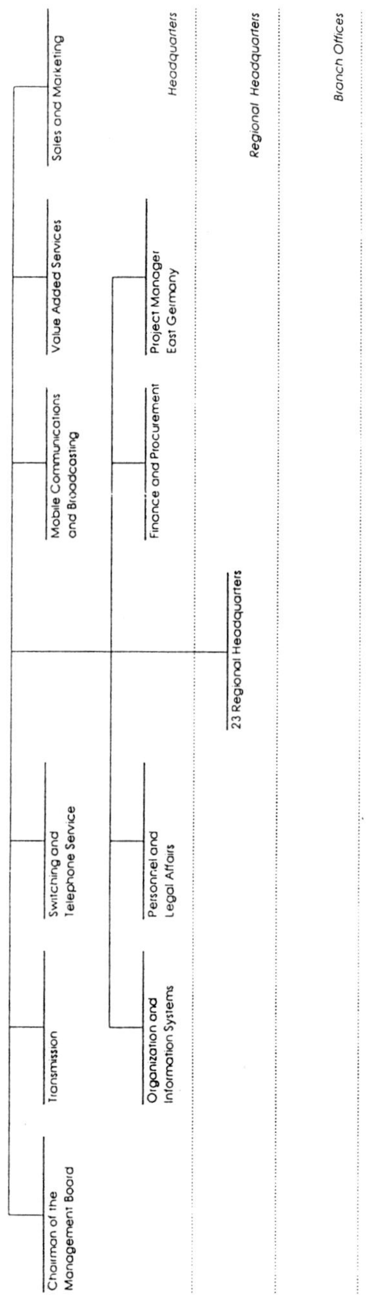

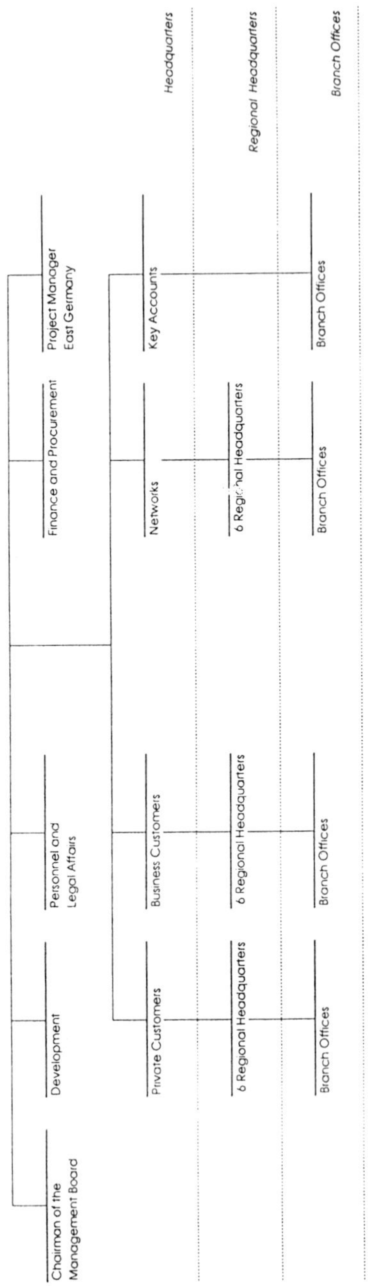

Figure 6. *Deutsche Telekom's new organizational structure*

Figure 7. The organization of BT's networks

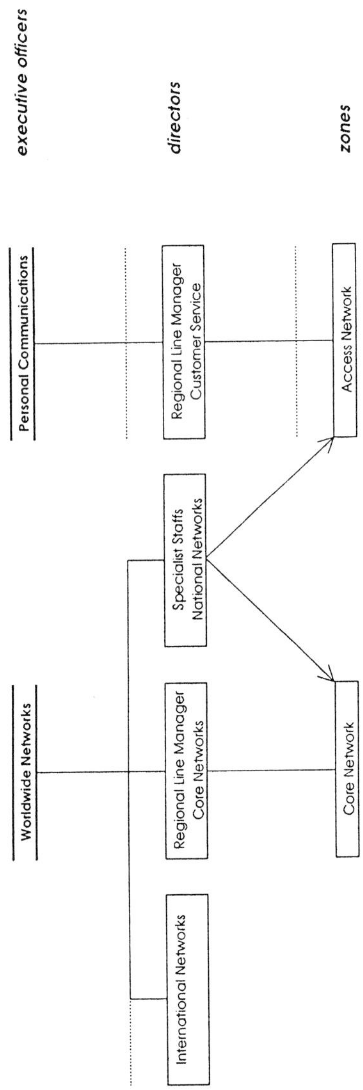

Figure 8. The organization of France Télécom's networks

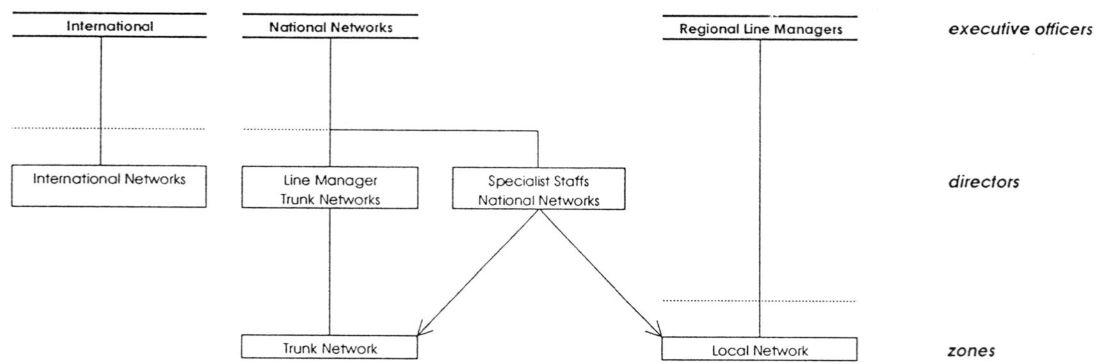

Figure 9. The organization of Deutsche Telekom's networks

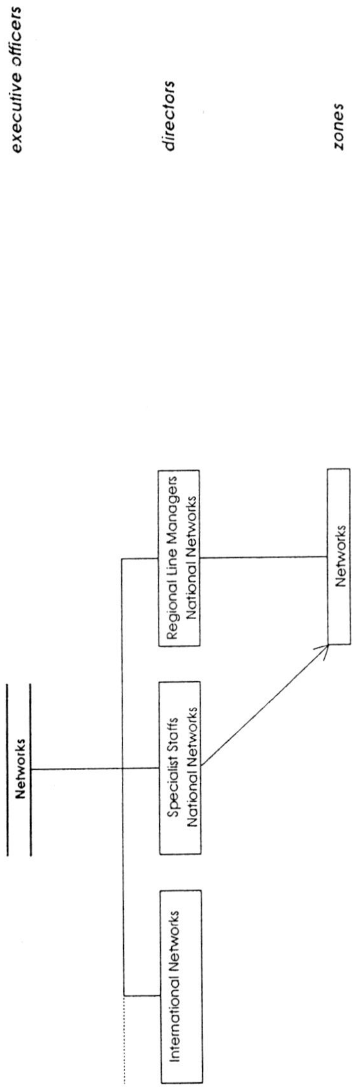

The Race to European Eminence
E. Bohlin and O. Granstrand (Editors)

Managing Network Investments: Illustrations from Telcos in Europe, Japan and the United States

Erik Bohlin

Chalmers University of Technology, Department of Industrial Management and Economics, S-412 96 Göteborg, Sweden*

This chapter addresses the question of how the investment process is structured and managed in several telcos in Europe, Japan and the United States, with special reference to how the investment process should be structured in an increasingly internationalized environment. A simple typology of each player's overriding investment model is constructed. For the future development of investment management models, two main options are identified: the product dimension vs the production dimension.

1. INTRODUCTION

The telecommunications industry is in a state of flux. Rapid and radical changes have occurred, matched by few other industries—from a once placid and stable situation with regulated monopolies to a state with foreign direct investments by other carriers, privatizations, changing industry boundaries, entrepreneurial new entrants and increased importance of proprietary technologies.

In this new environment, a re-examination of business and management practices is natural. Given the importance of investments in the telecommunications network, in terms of both spending and effects on business carried out, particular attention is required for the management and organization of network investments. Many questions could be posed in this context. How should investments be carried out in this new environment? How should investment planning be structured and managed in the future? In what directions should it change? Furthermore, given the increasing degree of internationalization and foreign direct investments of telcos, new challenges arise. Should foreign network planning be integrated with domestic planning? If so, how much and in what way? Cases in point are BT's and AT&T's reorganizations in 1991 of their international and domestic operations, both of which created a platform for stronger integration between the two, although the particular organizational contexts differ widely.

* Many thanks to the interviewed individuals who gave their time and shared their experience. Many thanks also to my academic colleagues who have provided comments and support in the preparation of this paper, especially Alain de Fontenay, Ove Granstrand and Ola Johansson. Lars O. Gustavsson, Telia and Dr. Rudolf Pospischil, DBT, also provided insightful comments. Support from Telia (formerly Televerket) is gratefully acknowledged.

This brief paper will bring perspectives on how several telcos in Europe, Japan and the United States have designed their network investment processes from their respective points of departure, based on an interview series conducted in 1992. Although there are notable differences in emphasis in the way investments are planned and executed, there is a common thread of search for new frameworks. Two elements are focused on. First, the planning context, in which each company conducts its investment management, is briefly outlined. For each player, a simple typology of the overriding investment model is constructed. Model is here used in a rather loose sense, meaning the overriding principle, around which network investments are planned and organized within each company. An important finding here is that organizational design is a possible and often important variable for the way investments are structured and planned. Second, the main dimensions that come into play with each player's investment model are noted. Based on the above, two options for the future development of investment planning are discussed—product orientation and production orientation (in part, this is a contrast of the AT&T vs the BT way). Although the empirical cases are mainly based on the organization of domestic network planning, the options are also useful for a discussion on strategic organizational design along the international dimension.

2. CHARACTERIZATION OF INVESTMENT MODELS

2.1 AT&T: Granularization model

AT&T has over the past decade undergone several organizational changes that have affected network planning and investment management. The main objective of the changes has been to create an organization increasingly focused on services and customers, rather than being exclusively centered on the technologies and the network as in the monopoly era. At the beginning of the 1980s, network investments were mainly classified and analyzed according to the respective network facilities (switching, transmission, signalling etc.) but at about the time of the divestiture, network investments were increasingly managed in a service dimension. Since the large organizational change in 1987, where strategic business units (SBUs) were formed according to services offered and customers served, *granularization* has become a key word. Network investments are, to the extent possible, analyzed and broken down according to services (internally denoted 'products') and justified as complete business cases, involving both revenue streams and costs. Obviously, still many joint costs and network investments of an infrastructural type remain. These are planned by a separate entity but the power to decide over these joint investments and the obligation to finance them reside with the SBUs. Since the late fall of 1991, SBUs encompass both the domestic and international dimensions, since the International Operations Division (IOD) at that time merged with the domestic communications unit, forming the Communications Services Group (CSG). The principal motivation for this change can be found in AT&T's often stated mission—to become a global leader in communications. As the mission is accomplished, the separation of the domestic from the international becomes increasingly artificial, it is argued.

In sum, AT&T offers an interesting example of investment management evolution and organizational change, in which the service ('product') dimension has become paramount. What was earlier considered a joint network is as much as possible broken down into separate products with full business responsibility.

2.2 BellSouth: Network program model

BellSouth has worked in an ambitious way to analyze all long-term and medium-term domestic network investments in different programs, i.e. aggregations of independent projects into appropriate clusters. The programs are mainly classified according to the technology dimension—digitization of switches, scrapping of analogue equipment, fiber, remote units, etc. The programs are then used as input and guidance for the current planning. The programs serve to ensure that there are links between the general policies on technology deployment and the immediate investment planning. BellSouth intends to develop the program analysis further, achieving even closer connections between programs and current planning. The international operations are conducted in a separate unit, BellSouth International. In sum, BellSouth's network planning is centred around technology programs, hence the 'Network Program Model'.

2.3 BT: Network cost center model

BT's major reorganization in the spring of 1991 has had a significant impact on their network planning and investment management. In the new organization, international and domestic network planning has been centralized into one unit, Worldwide Networks, which prepares network plans based on the requirements supplied by the customer interfacing units, Personal Communications and Business Communications. The idea behind this organizational change is that the customer-serving organizations identify the customer requirements, which are then delivered to the network planners. The network planners then translate the customer requirements into construction plans, which the customer-serving organizations have to pay for.

Before the reorganization, domestic investments were planned by 24 districts that had profit and loss accountability for their complete set of operations. Now investments are planned and divided into three zones in Worldwide Networks. Furthermore, international investments were previously planned and decided by BTI but with the new organization, the activities of BTI have been integrated with the rest of BT. The principal motivation for this latter change was that it corresponded better with BT's mission to become a truly global carrier, similarly to the case of AT&T. (In fact, BT made this change about half a year before AT&T.) The organizational context of BT differs, however. In the 'new' BT, profit and loss responsibilities have been recentralized as the decentralized profit and loss was found to create adverse incentives and lack of coordination. As a consequence, the Worldwide Networks Unit is primarily managed on the basis of costs. In terms of network planning the impetus has been to keep an integrated unit that is focused on the technological dimension, hence the 'Network Cost Center Model'. The customer dimension is clearly not forgotten but primarily managed by the customer-serving entities.

2.4 France Telecom: Decentralization phase

France Telecom has undergone two recent reorganizations, in 1991 and 1993. The basic aim of the organizational changes has been to transform France Telecom from a government administration to a business entity with decentralized profit and loss responsibilities. In particular, the 1993 reorganization aimed to transform the activities of France Telecom's domestic central network planning units from being involved with operational planning to focusing on strategic planning. In the old structure, the regional units reported to 22 intermediate regional units, which in turn reported to the headquarters. This organization created a distance between the headquarters and the operational level, at the same time as the operational units were influenced by several headquarter departments, sometimes providing quite detailed operational instructions. Since there were so many intermediate units interfacing with the headquarters, coordination was hampered.

With the new organization, there are clearer intermediate levels of planning (strategic, tactical, current planning) and a much stronger emphasis on decentralization and regional profit & loss responsibilities. Furthermore, there has been increased work with financial goal setting and financial ratios to allocate investment budgets and make prioritizations. The main dimensions previously used for investment planning focused on the network technologies and were coordinated centrally. However, with the stronger emphasis on decentralized profit & loss responsibilities, complete sets of regional plans are increasing in importance. The international investments have been planned and executed in several separate units (the Directorate for International Affairs and subsidiaries such as France Cables et Radio, Transpac, etc). New challenges face France Telecom's international operations as it deepens its alliance with DBT. In sum, France Telecom offers an interesting example of an ongoing transformation of the old PTT-structure toward decentralized decision-making and investment planning.

2.5 GTE: Capital optimization model

The outstanding feature of GTE's investment management of its domestic network is the centrality of a mainframe optimization tool developed in the mid-1980s. The advanced algorithm optimizes investment capital with a large number of restrictions and handles GTE's complete network investment portfolio. One of the drivers of the development and management of this system has been the desire to ensure conformity for all of GTE's operations—GTE serves 40 states and has expanded by mergers and acquisitions. As part of establishing conformity of operations, GTE has reduced the number of independent regional units in several stages during the 1980s. GTE is now divided into 5 main regions, under which regional units are grouped.

The investment program analysis is classified into two parts: core programs (i.e. mainly investments in the local loop with the current state of technology) and corporate programs (new technologies and services). Within the programs, customer segmentation profiles are used for prioritizations. The programs are treated separately in the optimization algorithm, and the budget allocation between the two is determined by top management. The international operations are separated from the domestic ones. In sum, the most striking element of GTE's model for investment management is its emphasis on the optimization of capital spent.

2.6 New Common Carriers (NCCs): Network expansion model

Among the three Japanese NCCs—DDI, Japan Telecom, Teleway Japan—the overriding principle of investment management activity has been to expand the network as quickly as possible in order to achieve national coverage (Teleway Japan initially covered the Osaka-Tokyo-Nagoya region but has had to change strategy). Network planning is based on growth expectations and traffic volume estimations, and financial constraints are found through prognosticized profit & loss accounts, balance sheets and cash flow analysis. Investment projects are rarely analyzed by advanced algorithms. In sum, the most striking feature of the investment model of the NCCs is the focus on growth opportunities. Their investment planning behavior is probably shared by new entrants in telecommunications provision in other countries who focus their strategy on growth and network construction from scratch.

2.7 New England Telephone and New York Telephone: Integrative phase

New England Telephone and New York Telephone, both subsidiaries of NYNEX, have during 1991-92 reorganized their network planning so as to achieve greater integration between different technological options in the local loop. Investment planning for the local loop, the switching and interoffice network was previously made in separate units, but as new remote technologies and fiber optics have come into play, new trade-offs between transmission and switching have emerged. Hence, both New England Telephone and New York Telephone have found that specialization misses the opportunity to reap the new synergies and are currently in an 'integrative phase'. (Both BellSouth and GTE have also placed increased importance on integration in the local loop but have not changed their internal organizations equally forcefully.)

A special circumstance of New York Telephone is the regulatory context, specifically the 'unbundling' and 'collocation' mandated by the New York Public Service Commission.[1] Collocation requires that New York Telephone open its physical facilities—ducts, buildings, manholes, etc.—to such an extent that competing carriers can locate their facilities there. Unbundling requires that the local exchange network be broken down into separate functional building blocks that other telecom service providers can interconnect with, at regulated tariffs. These pressures have raised new questions about the fundamental business idea of a local carrier—e.g. does the future business reside in offering a range of different services ('service differentiation') or in providing transmission capacity and a network platform to other actors which in turn provide differentiated services (a retail vs wholesale function)?

2.8 NTT: Technology phasing model

NTT has a very strong technological base and heritage, as it had assumed leadership in R&D in cooperation with its equipment suppliers already before it introduced a new scheme of procurement in the early 1980s (the so-called 'NTT family'). Even under the new system, the specification of network facilities is determined by NTT. The centrality of new technologies is also apparent in the way investments in digital switching are managed—the headquarter departments have a very significant influence on the pace of investments. Actually, the whole program of investments in digital switching is managed by the headquarter departments. A similar approach is also apparent in NTT's plans for a B-ISDN network—also here the headquarter departments determine the overall pace and direction of investments in new technologies. The local entities are obviously advised to comply with the recommendations. However, both the cases of digital switching and B-ISDN have a regulatory context which in part explains the high involvement of top management. The rapid pace of digital switching deployment has been strongly influenced by the necessity to provide detailed traffic and number information to new entrants. Likewise, the B-ISDN investments are part of the 1995 government review of the NTT privatization being prepared.

NTT has also gone through organizational changes in 1990 and 1991. The main objective of these has been to reduce the hierarchy of the regional units responsible for the whole set of core telecommunications operations. There has also been an ambition to decentralize—the

---

[1] For an account of this facet of the New York regulatory situation, see Schwartz, G.G., Hoagg, J. (1992), "Telecommunications competition and diversification by local telephone service providers", paper presented at the 9th International Conference of the International Telecommunications Society, Sophia Antipolis; and State of New York Public Service Commission (1992), "Opinion and order on pooling, collocation and access rate design".

regional units are autonomous within the restrictions and guidelines set by the headquarters. Furthermore, headquarter units have been reorganized to create a stronger locus for strategic planning. In sum, despite the fact that NTT is indeed decentralizing its operations and network planning, the typology that most aptly seems to fit is 'Technology Phasing Model'. The headquarter involvement and control are, comparatively speaking, rather strong with respect to timing of entry into new technologies and the pace of investing in new technologies.

2.9 Telia: From object management towards management by objectives

Telia (formerly Televerket) has over the past decade gone from a focus on individual investment objects and individual investment appraisals to an increased focus on general goals and objectives for the network plans. Goal setting and goal formulation on different levels in the organization have been an increasingly important management style, where general goals are broken down in a top-down manner. Specifically for network planning, the central department develops both technical and business goals for a 3-5 year period, which the local regions transform into regional investment plans. The regional plans are then presented to the headquarters. However, it is emphasized that the development of regional investment programs is decentralized to the regions, except for the initial implementation of a new technology. The main dimensions of the programs are geographical turf and technological options, even though the use of customer segmentation profiles and prioritizations has been introduced.

There have also been some recent organizational changes in 1991-1992. First, the regional units with profit & loss responsibilities for the full output of telecommunications services were reduced in number from 24 to 8, creating new possibilities for integration and coordination of investments. Furthemore, the number of headquarter staff concerned with network planning was significantly reduced. The detailed planning was decentralized to the regions, and strategic planning was emphasized in the headquarters. Lastly, new approaches to investment management have been implemented, in which the aforementioned goal formulation is central. In sum, the investment management of Telia shares several features with other telcos such as reduction of headquarter involvement in detailed planning. However, rather unique is the so far consistent emphasis on centralized goal formulation and decentralized goal realization in place of aggregate investment programs and plans.

3. SUMMARY AND CONCLUSIONS

This chapter has brought perspectives on how several telcos in Europe, Japan and the United States have changed their management of network investments. It is notable that most have undergone quite recent changes in their investment process and organizational context. In several companies, the organizational design and context is an important variable for the way investments are structured and planned. There are several common trends, such as:

1. Decentralization of business decisions. This is particularly apparent in AT&T, France Telecom, NTT, and Telia, although the different companies have pursued this to varying degrees of depth and duration (France Telecom and NTT starting later). Notable is that BT for some time pursued extensive decentralization of profit and loss responsibilities among its former 24 districts, but recentralized these in 1991.

2. Organizational integration. There is integration both along the dimension of technology and along the international/national dimension. For the former type, the cases of BellSouth, GTE, New England Telephone and New York Telephone are most apparent, being driven by the need to reap the benefits of using new remote switching technologies with fiber optics in the local loop. For the international/domestic type of integration, the cases of AT&T and BT stand out.
3. Concentration of regional units and investment decision-making. Several of the firms studied have reduced the number of regional units (BT, GTE, France Telecom, Telia), which has allowed a closer link between the strategic investment planning and the detailed, operational investment planning, by strengthening the intermediate level. In terms of centralized investment planning, BT has gone furthest by reducing the previous 24 districts into 3 investment planning zones.
4. A service and customer orientation. Increasing emphasis on the services produced and customers served already in the network planning phase trend is most visible in AT&T even though most other companies certainly include such impetus.

Trends 1 and 3 (decentralization vs concentration) may seem to be at odds with one another but there is an interpretation that unifies the two. Previously, much operational and tactical planning was centralized to network planning units that provided detailed plans primarily oriented around technology, not customer needs or financial aspects. The operational units then carried out the plans. However, with an increasingly saturated market for basic telephony services (i.e. high penetration levels), increasingly demanding large users and competitive pressures, a stronger business orientation among telcos is needed together with adaptation to customer requirements and flexibility. To remedy this situation, decentralization of business responsibilities takes place. However, the decentralization cannot be implemented successfully based on the former type of organization. An intermediate layer is needed to fill the void between the centralized planning and the multitude of operational units. Hence, regional units are concentrated and merged so as to ensure sufficient coordination, sufficient size for meaningful business decision making, and sufficiently strong tactical (medium-term) planning in the new, decentralized organization.

Furthermore, trends 1 and 2 (decentralization vs integration) may seem to conflict but also here there may be a unifying interpretation. In the cases presented, the need for integration has been motivated both by technological changes and by market changes. In order to cope with the new situation, integration of previously separate activities becomes necessary to achieve overview and reap new synergies, creating centralization. Hence, change may call for centralization at the same time as decentralization is necessary for motivational considerations and a business orientation. Depending on the type and comprehensiveness of the change in question, there may be a complex interplay of integration and decentralization both over time and concurrently within an organization. In the cases presented, integration and decentralization have mostly been a parallel process (cf. international/domestic, local loop). The increasing importance of cellular and personal communications services, and the continual merging of computers and communications, and of media and communications, will certainly require new forms of organizational integration.

The structuring of aggregated investment programs also has several commonalities. It is evident that the technological dimension plays a key role in network planning in most of the telcos investigated. For those telcos with a strong regional/local orientation, the geographic dimension is also very important (France Telecom, NCCs, Telia). AT&T is different in that the service dimension is emphasized, but this may have been influenced by the type of network served—the long-distance—which permits service differentiation to a greater extent than the local loop, at least until now (cf. the New York Telephone situation touched upon above).

Given that internationalization increases and that international and domestic activities are increasingly merged and blurred, around which dimensions should the integrated investment management be structured? Rephrasing the question—how should the network be 'divided' to best serve investment decisions in an internationalized telco? Allowing for some hypothetical reasoning, the following two options seem most attractive for a future, heavily internationalized telco—in part contrasting the case of AT&T with BT:

Option 1: The product dimension. Following the path pursued by AT&T, services should be identified and treated separately, as much as possible. Investment output is the focus, and investment decisions should hence be tied to separate services to enable both revenue and cost analyses. The strength of this framework obviously lies in the close connection between the investment decision and the total economic analysis. The main weakness is that the efforts to divide into parts may reduce the sense of the whole.

Option 2: The production dimension. In the spirit of the OSI-model and BT's organizational context, another option is to orient the planning and investment management of the network as a production process. The network is here divided into layers which contribute to output. Investment decisions should then be tied to the appropriate layer to which they contribute. Implementing this option in an organization, it is possible to envisage the intermediary roles of delivery between layers and the transfer prices between these. This would serve to create an orientation toward the final output—the services—even though the intermediary functions are not directed to outside customers. Network planning would in this context have the role of infrastructure provider, obtaining input with respect to planning from customer-facing entities (or alternatively, entities organized around the various forms of final services delivered).

Although an overall assessment of the most fitting arrangement is a quite complex matter, it seems that the product dimension has some advantages for increasingly internationalized telcos. First, the product dimension places more emphasis on the final output rather than on the 'factory', the network. As an internationalized telco, the product orientation promises to better suit the terms and conditions of international operations—alliances, joint ventures, leased lines and collaboration. The production orientation is reminiscent of a monopoly era that was primarily organized around in-house production. Second, the product dimension is probably easier to manage, given the increasing importance of keeping a clear marketing profile. In that respect, effective marketing and fast delivery of innovative new services require close integration between network planning and customer interfacing units—a product focus provides this bridge. Third, a product orientation seems congruent with the experience of industrial multinational enterprises (MNEs) which increasingly have organized their global activities around business areas ('products' broadly speaking) rather than primarily dividing the activities into geographic turfs.

The Race to European Eminence
E. Bohlin and O. Granstrand (Editors)

Foreign Direct Investments of Telecommunications Operators in Less Developed Countries: Application to the Telefónica Experience

Juan Duran[a] and Fernando Gallardo[b]

[a] Centro Internacional Carlos V, Universidad Autónoma de Madrid
[b] Telefónica de España, Subdirección General de Planificación Estratégica

1. INTRODUCTION

In recent years, one of the novel aspects of the activities of telecommunications service operators has been internationalization. The world's leading firms are developing into multinational companies or MTOs (Multinational Telecommunications Operators) by adopting a multi-domestic or global strategy according to the course of action they take.

Globalization is achieved through policies of cooperation with other agents and by making Foreign Direct Investments (FDI), both in Developed Countries (DC) and in Less Developed Countries (LDC), although the focus on each group is different. On the one hand, the purpose of FDIs in DCs is usually to compete in new service or advanced corporate communication segments that are totally or partially open to competition. On the other hand, LDCs receiving FDIs are seeking to promote and modernize their telecommunications in order to establish a solid base that encourages economic activity.

This paper will focus on FDIs made by MTOs in LDCs. The subject will be discussed in general terms and will specifically refer to the evaluation and management of the country risk, followed by a detailed analysis of Telefónica's experience. Figure 1.1 outlines the elements of the FDI process, to be discussed more in detail in the following sections.

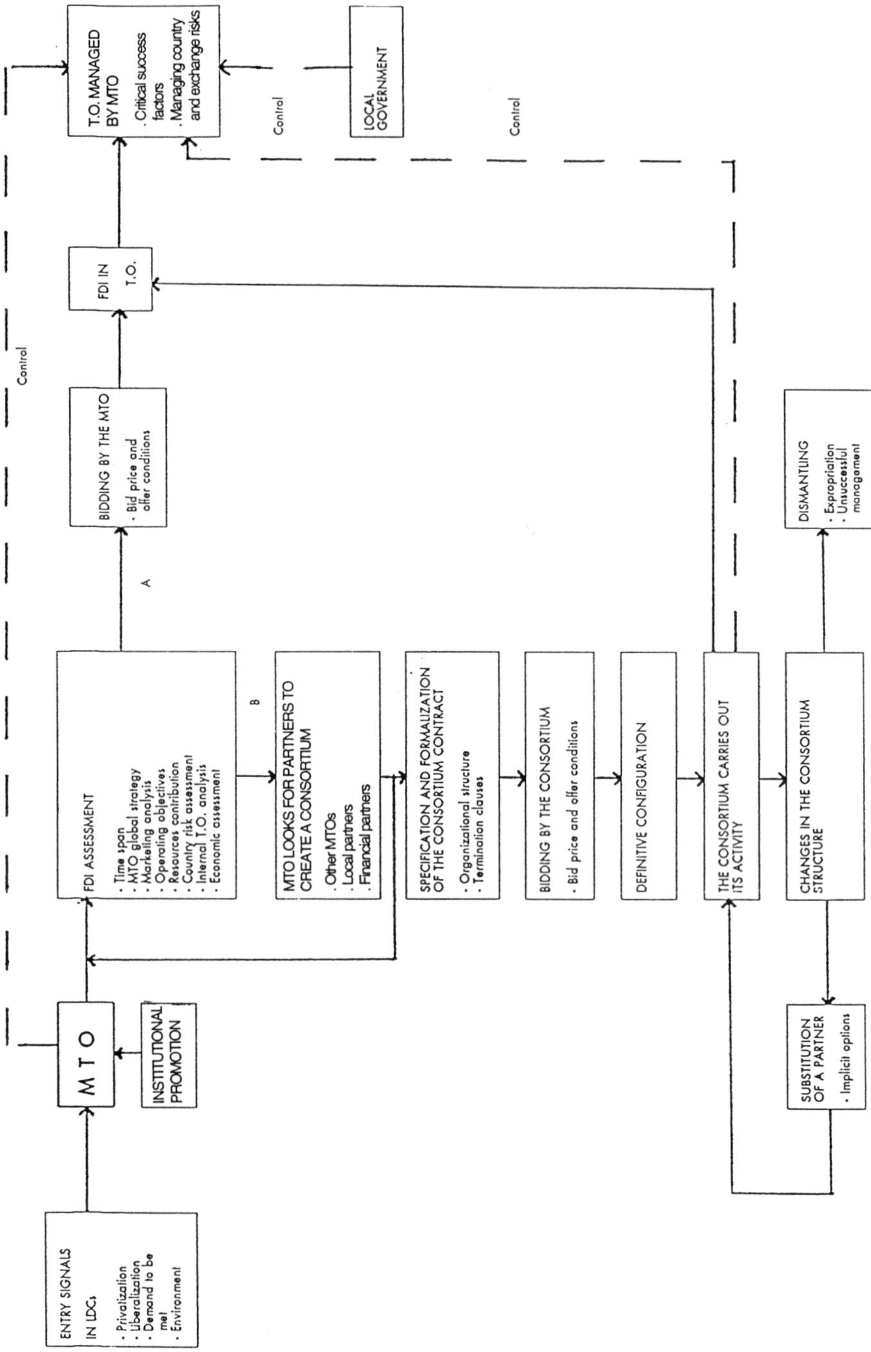

Figure 1.1. FDI in LDCs using a consortium.

2. DRIVING FACTORS OF FOREIGN DIRECT INVESTMENTS IN LESS DEVELOPED COUNTRIES

There is a very important driving factor behind FDIs both in DCs and in LDCs that is related to the increased rivalry perceived between operators. Losing a monopoly on a portion of the telecommunications services (which are those with the best future prospects) forces operators to recover abroad the market share that they have already lost or anticipate losing in the future. The legal impossibility of implementing a growth strategy in the country of origin through activity diversification also has the same effect. On the other hand, the disappearance of national monopolies will lead not to a situation of competition between many agents, but rather to a worldwide oligopoly in which there will be barely more than a dozen leading global service operators due to the existence of considerable economies of scale and scope. Therefore, those aspiring to be leading operators must start staking out their positions.

The decisive factors behind an MTO's FDI can be established within the context of Dunning's eclectic focus (1988). A company can begin an internationalization process if it enjoys specific proprietary advantages in the form of intangible assets (expertise) that are embodied in a certain cost leadership and in product differentiation. Internationalization must make it possible to maintain, and if possible improve, the company's competitive advantage. In this respect, an internationalization process is possible if there are corporate capabilities and proprietary advantages to carry it through, as well as the will to exploit them on a transnational scale. The internationalization method that is chosen (export, FDI, alliances, intercompany cooperation) will depend on the associated transaction costs. On the other hand, the economies of location (market size and structure, factor productivity, investment climate) will determine the geographical setting of the chosen internationalization process.

Corporate internationalization is closely related to a country's competence factors (Porter, 1990; Dunning, 1992) and to existing market rivalry or reactions (Knickerbocker, 1973; Graham, 1992). However, without institutional support either at the country or multilateral organization level, it would not be possible to explain the internationalization of certain companies and to a certain extent the strong growth of FDIs in recent years (Durán, 1992). In fact, institutional support facilitates an FDI by lowering barriers to international economic relationships, improving the investment climate, and reducing transaction costs by providing technical assistance and promoting investments and noncommercial risk coverage.

The MTO involved in operations of this kind gains a series of advantages, one of which results from economies of scale. The parent company that operates several networks succeeds in improving its negotiating capacity with suppliers, as regards both prices and technical standardization possibilities. It would also gain a better negotiating capability with other operators when participating in collaboration agreements or in international network projects such as underwater cables or satellite transmissions. Other positive effects stem from diversified sources of the group's income and increased Returns on Assets and on Shareholders' Equity. In order to increase turnover and profitability ratios, the managers must have full manoeuvering capability in corporate management. In addition, as we will see below, it is vital that specific incertitudes be previously clarified.

Likewise, the investing firm will obtain advantages stemming from an easier implementation of global services for its business customers that internationalize their operations in the countries where the operator has targeted its investment. Furthermore, in many cases, candidate countries for receiving FDIs made by different companies from the same country are usually the same ones that the operator has targeted. In addition, the larger customer base that is achieved helps to introduce new services.

3. ENTRY SIGNALS AND INSTITUTIONAL PROMOTION

The MTO will initiate a consortium establishment process if there are signals for entering the market. An early signal is a Government's intention to privatize shortly an operator of basic telecommunications services. If the country in question is a Less Developed Country, it is very possible that the Government wants foreign capital to improve the country's telecommunications. If the operator is from a Developed Country, what the Government may want is a limited exchange of shares with another operator from an advanced country that had already been acting as a strategic partner of the country's firm.

Liberalization is another important signal for entering mobile communications service and advanced corporate communications markets. If the Government decides to grant any additional licenses to operate these services, prestigious, well-known foreign operators will be the obvious candidates for obtaining them as they would improve service quality. In the event of total liberalization, such as data transmission in the Developed Countries, the MTO's position should be one of ongoing evaluation of opportunities for acquiring existing networks and for attracting customers.

A study of demand can also provide clues about which countries should be targeted for investment. If the basic telephone network is not well developed, a judgement which could be based on the ratio of lines in service per 100 inhabitants, it may very well be that the local Government is considering the possibility of finding an experienced partner in the business. In countries where data transmission is liberalized, a good entrance signal would be the existence of customers that have not found operators that can supply high-quality, fast, reliable services to them.

As a complement to the above, the country in which investments are to be made must have a suitable "climate" in which to conduct business. Not all countries lacking in telecommunications services are suitable for making FDIs. On the one hand, it is recommended that the country have a calm labor climate and a qualified job market. On the other hand, it is very useful to target investments in countries with which the source country has historical, cultural and linguistic bonds. Finally, it is essential that the country have a solid foundation for stimulating economic development and a stable political climate that makes expropriations unlikely.

In the face of signals for entering a market, the MTO must evaluate the profitability of an FDI made alone or as a member of a consortium. The essential reason why all operators decide on the latter option in almost all operations is risk attenuation. One of the fundamental principles of evaluation is that it must be made in terms of the operation's contribution to the company's global strategy.

Institutional promotion has recently emerged as a driving factor behind FDIs (Durán, 1993). It is an accepted fact that, generally speaking, the FDI contributes to the host country's development and is also a very important growth method for companies from the country originating the investment. Therefore, both source and receiving countries, as well as multilateral cooperation and development organizations, institutionally promote FDIs.

Institutional measures put into practice by potential FDI receiving and source countries include tax exemptions, capital subsidies, internationalization support plans for the country's firms, promotion of public firms that cover foreign exchange and country risks, and Reciprocal Investment Guarantee Agreements (RIGA). RIGAs are signed by two countries, one of which is usually a developed country and the other a developing country. These agreements are intended to encourage FDIs in LDCs in which the country risk has been eliminated or at least limited. At a supranational level and within the framework of the World Bank system, the MIGA (Multilateral Investment Guarantee Agreement) is noteworthy. This agreement provides for consulting and promotion tasks and also provides coverage for typical FDI risks.

The institutional promotion described above consists of a set of facilities that the MTOs could use more extensively in their FDIs, as they are a decisive factor when taking up positions in countries with an initially higher-than-average country risk.

4. CONSORTIUM CREATION AND OPERATION

4.1 Consortium foundation, bidding process and control outline

Once the suitability of making an FDI and of doing so within the framework of a consortium is accepted, the MTO must begin looking for partners. A fundamental issue is how to select partners. In the first place, it is essential that the partners have the same or complementary goals. In this respect, special care must be taken in choosing local partners. Likewise, it is advisable to appoint a dominant partner to carry the weight of management, lead negotiations with local political authorities, and settle possible conflicts that arise between the alliance's partners. On the other hand, the leading operator of the alliance will block the participation of another operator whose goal is to gain experience in that specific business. In order for the alliance to succeed, the nondominant operators must provide access to their own customers (banks, freight companies, etc.) or the know-how that complements the principal operator's weaknesses. Collaboration with a partner in one market while competing with it in another can tend to undermine the mutual trust that the partners must inevitably have in order for the alliance to succeed. The right choice of partners is one of the factors in the success of an operation.

Consortiums whose aim is to make an FDI in a basic telephone network or mobile communications operator in an LDC may include all or some of the following agents:

1. A telecommunications operator that leads the consortium.
2. Other operator(s).
3. Financial partners.
4. Local partners.
5. The local government.

The first four are members of the consortium, whereas the Government, though not a member, usually plays a decisive role.

The leading operator will carry the weight of creating the consortium. The goal it pursues through the internationalization operation is to seek new markets as part of a defensive strategy—because it faces competition in its domestic market—or of an offensive growth strategy to exploit its own competitive advantages in technological and management capability. This operator will be responsible for managing the business of the local operator and therefore, in addition to financial resources, it will provide technology, human resources and managerial expertise. As the principal partner of the consortium, it will assume responsibility for financial control and management.

The role that other operators, if any, will play in the consortium is similar to that of the leading operator, although there are some important differences. The goals they pursue on joining the consortium are to search for new markets and to learn about and gain experience in operating certain services. Their function is to collaborate with the leading operator in running the business. Just as the leading operator does, they must also contribute resources and exercise some control, although to a much lesser extent.

If there are local or nonlocal financial partners in the consortium, their priority will be to obtain an acceptable financial return as compared to their capital costs. A financial partner does not participate in running the local operator's business, although it may collaborate in providing access to financial markets. The resources contributed by financial partners to the consortium are exclusively financial, and the control they exercise stems from the established agent relationship.

Local, nonfinancial partners can participate in the consortium and can also have a presence in the operator's business because they hold a significant direct share. In most cases, the goal they pursue, apart from that indicated for financial partners, is to profit from their own business, either because they have interests in the supply of goods or services to the operator or because they are users in great need of telecommunications services. In addition to financial resources, they can help the operator run the business by providing internal contacts and knowledge of local markets.

Even though the Government does not participate in the consortium, it plays a fundamental role in its operation. Usually, when an international consortium takes out a share in a basic network operator, this involves at least a partial privatization of the local operator, which means that the principal agent negotiating with the consortium will be the local Government. As telecommunications are a strategic sector, the Government's negotiating position will be to defend national interests related to the country's economic development. As for consortiums whose aim is to participate in mobile communications networks, the Government also plays a relevant role since it must grant the license for operating mobile services, either by a monopoly or with limited competition. As it is not a member of the consortium, the Government will not make a direct contribution of resources. However, if the consortium leader effectively concludes the negotiations, the Government may commit itself to subsidizing unprofitable activities (such as extending the telephone service to rural areas) or provide certain tax incentives. In addition, effective negotiations with the Government and a subsequent framework for good relations may help to diminish the political risks of the operation. As regards control, the Government may exercise political control and in any event will act as the regulatory agency.

The consortium thus established must submit a specific tender to the local Government's bid request on the basis of a share package that will entitle it to manage the operator. Whether or not the license is obtained will depend partly on the monetary bid presented, but the operator's prestige and the combination of partners in the consortium will also have an influence.

The definitive consortium often takes shape after the license is obtained, in which case the consortium leader may renounce the license if it cannot achieve a suitable group of partners.

Once the partners are assembled, the consortium's organizational structure must be clearly defined and it must be provided with a personality independent of its component parts. Another very important task is to establish the clauses that provide for disbanding the consortium or for any of the partners to leave it.

The MTO will perform an ongoing evaluation of the operation's profitability through traditional NPV (Net Present Value) and IRR (Internal Rate of Return) methods. In order to apply these methods, general analysis bases must be established, i.e. the investment's time span (t), the relevant annual cash flows generated during the investment's lifetime (CF), the discount rate to be used (k), and the risks that should be considered (R). Thus, it is established that:

$$NPV\ (p) = F\ (t, CF, k, R)$$

where p is the investment project to be evaluated.

The operators or operator consortiums that make this kind of direct investment do not usually have 100% control over the subsidiary's capital. In almost all cases, the State or private groups in the country retain a significant percentage of the capital, although the operator making the investment will maintain control over the targeted company's management. Otherwise, the parent company would probably not be willing to make the investment.

In addition to expected dividends, the relevant flows should also include returns that may be generated by way of technical service, royalties, commissions, and net flows from current operations between the two operators, and the corresponding transfer prices applied. We have called these flows transfer flows (T). These flows are relevant to the FDI evaluation because they would not be generated if the investment had not been made.

On the other hand, the parent company must appraise the marginal effects on its own flows that will be generated during each period as a result of the foreign investment (MG). These could come from:

1. Reductions in the prices that it must pay for equipment as a result of an increased negotiating capability with suppliers, especially as regards public switching equipment.
2. Increased traffic with the zone where the investment has been targeted as a result of said investment. As both companies would be under the same management unit, it would be possible to more easily promote extension of international service facilities between the zones, especially corporate communications.

Thus:

$$CF = DV + T + MG$$

Definition of the risks affecting the investment is fundamental when evaluating the project. Three kinds of risk should be considered: the risk inherent in forecasting flows (ER), the exchange risk (FXR), and the country risk as regards both the expropriation risk (XPR) and the fund blocking risk (BR).

Thus, the vector that represents the associated risks for each time span period would be:

$$R = (ER, FXR, BR, XPR)$$

The discount rate that should be used to calculate the project NPV will be the weighted average cost of the parent company's financial resources.

The MTO's management of the local operator is subject to various controls. The MTO's own management will thoroughly control the management procedures implemented in the local operator, as regards both obtaining financial returns and the management ethics and fulfillment of other strategic objectives. Meanwhile, the other consortium partners will make sure that they obtain financial returns on their investment. The local government will be more interested in political type control and compliance with market regulations in order to constantly improve the country's economic and social conditions. (This, of course, involves assumptions about the government's rationality and honesty, which are questionable in many cases due, for example, to bribery and nepotism.)

4.2 Consortium structural changes

The option appraisal theory becomes especially important when a partner wants to withdraw from the consortium. The contract regulating a consortium can include an implicit sales option when the possibility of a partner withdrawing from the alliance is considered, and usually it is stipulated that the other partners (in our case the operator) be the first in line to make the purchase. It would be advisable that the contract explicitly state the method for appraising the share, which could be the book value, the stock-exchange capitalization value, or a combination of both. The value thus determined could serve as the option strike price, so that the selling partner would exercise the option if the strike price is higher than his own appraisal of the share, accounting for the opportunity's costs. On the other hand, the buyer would be interested in the operation if the strike price is lower than the appraisal, also made in terms of the opportunity.

The possibility of a negotiated departure from the alliance by any one of the partners, especially the financial ones, would mean an added attraction for these partners, thus making it easier and faster to form the alliance. In this sense, it may be of interest to include purchase or sales option contracts of the shares pertaining to the alliance partners, provided that the result of exercising the option involves acquisition of shares by the operator acting as leader (the MTO).

In any event, even in the absence of an option contract, a sales-purchase operation between partners may occur at any given time. This may be due to liquidity requirements of one of the partners, or simply to settle appreciations obtained by the financial partner. One of the reasons for the leading partner's willingness to acquire the share would be the differing worth of the share to be transferred at that time as appraised by the financial partner and the leader.

In many cases, the financial partner would be willing to pay an amount to ensure minimum returns and the leading partner would be interested in increasing its future share. In view of the different valuation of the underlying assets—the capital portion of the invested company—it may be of interest to purchase a sales option from the leader. On the other hand, the leading partner could be the one to buy a purchase option from the financial partner if the priority is to increase the firm's control in the future by ensuring a purchase price.

Almost all the alliances analyzed here are created for an indefinite period of time. However, subsequent problems can arise in connection with operating inflexibility, managerial conflicts, politicization, reluctance of one or more partners to reinvest profits, or changes in the political atmosphere that make it advisable for one or more partners to withdraw or to even completely disband the alliance.

5. CRITICAL SUCCESS FACTORS

In addition to running the business proper, the success of the alliance will depend on a series of issues that the partners must approach during the alliance establishment and negotiation stage.

The business conducted by the alliance must permit the exploitation of the dominant partner's proprietary competitive advantages, which in essence are technological or managerial skills and know-how. To do so, most alliances in which operators participate are of the horizontal type. The rest are vertical, either the upstream type to obtain advantages in equipment supply, or the downstream type to more easily penetrate liberalized markets. In both cases, the operator uses its power of monopoly, in the former as a key customer of the equipment supplier and in the latter as owner of the basic infrastructure used to provide advanced services. In no case should the alliance undermine the strategy of any of its partners.

If an alliance is to control a company being privatized, it is essential to negotiate well with the Government. This will require commitments in terms of service extension, quality and speed in addressing demand. In exchange for this, the leading operator will be able to "wrest" commitments from the Government regarding subsidies, tax exemptions, a stable billing framework, and limits on the country risk. The latter issue will basically depend on subsequent relations between the Government and operator, which should materialize into a risk evaluation and management system resulting in the adoption of coverage techniques.

A smooth flow of communication and trust between partners is also a critical factor for success. For this the partners must clearly define the extent of committed resources, the way to recover them, and profit sharing.

Likewise, it is indispensable that services be provided with acceptable quality levels and, in the case of advanced corporate communications in a liberalized setting, the quality levels must be in line with the references of the best operators. Otherwise, expulsion from the market would be definitive, to the benefit of another competitor or through the construction of private networks.

Whether or not the business is profitable will, aside from income and contingent factors (the most important would be to have a stable billing framework), depend on the level of operating costs. The investing company must make sure that it will be free to take over company management. In this respect, business profitability may be contingent on elimination of managerial inefficiencies. In the telecommunications arena, there are several very important areas that can be the source of potential inefficiencies: purchasing and service contracting policies, personnel policy, and investment policy.

As regards purchasing policy, the prices that the company is paying to acquire equipment and to contract outside supplies and services must be analyzed. At least initially, the company will not be in a strong negotiating position insofar as large switching and transmission equipment is concerned, as only a few multinational companies supply this kind of equipment. However, it is possible to detect concealed subsidies in prices paid to companies for other equipment and materials and for rendering of contracted services. In this case, the new management team must obtain the necessary guarantees to be able to eliminate, or at least cut back on, these practices.

As for personnel policy, it must be possible to freely allocate human resources. It may not be advisable to propose staff reductions in countries where unemployment affects a large part of the population, but it may be necessary to put functional and geographic mobility measures into effect. To this end, relationships with trade unions should start off on the right foot. In addition to the measures intended to improve allocation of existing resources, it is advisable to implement productivity plans, including incentives, in order to increase the possibly low productivity rates of these operators.

Many operators do not have a rigorous material investment evaluation and selection system. Returns on the financial investment made by the foreign operator will depend on the return on material investments. In this respect, it may be advisable to apply models successfully used by the foreign operator. However, it must be emphasized that the Government will almost surely require that some return-free investments be made, for example those related to extending services to rural areas. On many occasions, it may be possible to negotiate subsidies with the Government in order to fulfill the objectives of service extension as soon as possible and thus make this activity less costly. It would be very useful for the foreign company to have previously established a Telecommunications Plan that covers the longest possible time span and thus eliminates incertitudes.

6. ADVANTAGES OBTAINED BY FDI HOST COUNTRIES

In general, basic networks in the LDCs are operated by these countries' public firms or by autonomous organizations with no legal status that depend on the Public Sector (PTTs). Therefore, a shareholding by a foreign company entails at least a partial privatization of the firm or public organization, which introduces an important aspect to be analyzed: that the principal negotiator with the foreign company is the receiving country's Government, and in most cases the legislative branch's consent is required. This means that in the negotiation, the Government will defend national interests as opposed to private interests, once again revealing the strategic nature of the sector. If the operation is finally concluded, the receiving country obtains an important series of benefits.

In the first place, there are positive effects stemming from revenues obtained by the State as a result of privatization. Most of these countries have high public budget deficits, which can cause strains on some domestic financial markets that suffer from a lack of depth, range and flexibility. The entry of such additional revenues can partially ease these tensions. In turn, the State may lose control over the company but will do so in favor of a firm with much more experience in the telecommunications business and whose know-how is guaranteed by successful operations in the source country. In spite of the above, the State does not lose control over its country's telecommunications services because this business is usually highly regulated, which is achieved in several ways: by establishing billing frameworks, restricting competition, implementing medium- and long-term plans by defining commitments in matters of material investments and extension of the service to the country's less prosperous areas, etc. In some cases the State retains ownership of a large share of the company, although this may hinder management of the company by the directors of the foreign company that made the investment, who undoubtedly know the business better.

There is another important positive effect that can result from the technical materialization of the investment. This is the possibility of setting up the operation through a debt-equity swap (see Figure 6.1). With this transaction, the investing company will make the payment in two installments: one in cash (usually dollars) and another with bad debt acquired at a discount in secondary markets. The country receiving the investment will have decreased its foreign debt by cancelling part of it at a discount and with no outlay of foreign currency. It must be remembered that a shortage of foreign currency is usually a problem for these countries.

As regards the host country's Balance of Payments, there are also some obvious advantages. The current account balances of developing countries usually show considerable deficits, fundamentally due to the volatile nature of their exports. The foreign currency provided by an FDI helps to compensate the negative current account balance. Of course, these operations alone cannot solve this problem, although they do provide a certain degree of relief.

From the standpoint of company management, the participation of a prestigious, experienced foreign operator can result in benefits in the form of technology transfer and managerial ability, which leads to notable improvements in management of a sector that weighs heavily on global economic activity.

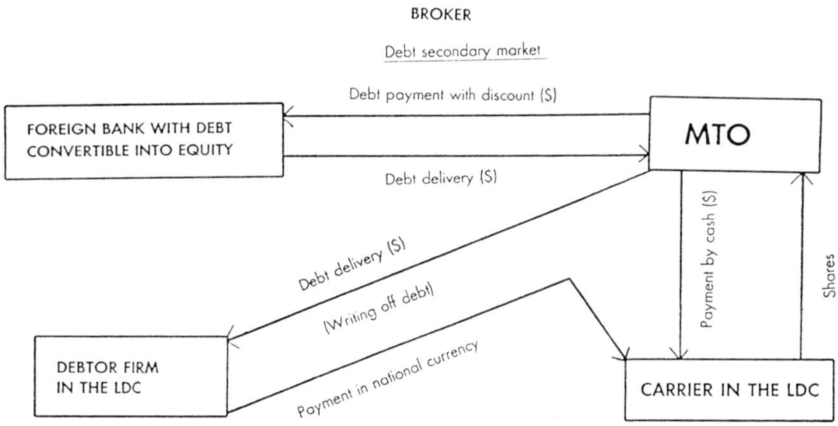

Figure 6.1. Debt-equity swap.

Furthermore, telecommunications in these countries are usually quite deficient, and this should be remedied as soon as possible because improved telecommunications are the key to development of the country's other economic sectors. It is a fact that without good telecommunications services, domestic companies cannot compete in international markets. On the contrary, a country with a good telecommunications infrastructure will attract other foreign companies that wish to make FDIs and will also encourage the creation of new domestic companies, with the resulting beneficial effects on the LDC's economy.

Finally, it is also worth mentioning the improvements in society's well-being. Aside from the indirect effects that would result from all the above, there would be almost immediate improvements in the ratios between telecommunications service equipment and the country's inhabitants, for example the ratio of telephone lines per 100 inhabitants, number of homes with telephones, and percentage of population with access to some telephone service modality.

Developed countries usually welcome FDIs in advanced corporate communications or mobile communications businesses. The advantages obtained stem from the higher level of competence gained, for which it is essential to have a high-quality service.

7. SPECIAL CONSIDERATION OF COUNTRY RISK

FDI decisions by a telecommunications operator must not only be based on economic evaluation criteria in the strict sense of the word, but must also weigh another series of risks associated with these decisions. A suitable environment must be defined within the corresponding legal, political, social and economic framework.

The country risk refers to the probability of expropriation or nationalization of the private sector, with or without compensation. Literature on the subject identifies this risk with foreign investment, although it should be extended to any private company regardless of its owner's nationality. This risk is determined by the historical background of each country and its cultural, sociopolitical and economic characteristics. The telecommunications sector is obviously exposed to this risk due to its strategic nature.

Nowadays there is an administrative component of the risk that concerns the impact of governmental actions and changes in a country's economic policy on the economic freedom of the private sector in general and on foreign subsidiaries in particular. These repercussions include restraints that may be imposed on mobility of productive factors, such as foreign exchange controls and foreign trade barriers. The direct, immediate impact on a company that has made a direct investment in that country would be the impossibility of repatriating dividends.

The political and administrative risks should be globally analyzed, although some agents in particular may be more susceptible to some types of risks.

The country risk faced by telecommunications companies that make FDIs is two-sided. On the one hand is the macrorisk, which is the probability of events occurring that may create or modify restrictions on all companies with foreign capital. The macrorisk can be measured by the existing country risk indices. These are useful for performing a preliminary analysis of the country targeted for an investment, as well as for determining the business "climate". There is a series of drawbacks (Kobrin, 1982): there is no generally accepted theory, subjective issues are taken into account, there are usually deficiencies in data collection, and forecasts are not possible. The available indices include the BERI (Business Environment Risk Index), the CPR (Country Potential Rating), and those published by specialized journals such as Euromoney and Institutional Investor. The other is the microrisk, which particularly affects the specific company making the investment. The country microrisk to which the company is exposed should be evaluated by the company itself or by outside consultants.

Country risk must be dynamically managed since it can vary throughout the operation's lifetime. Initially the extent of country risk is a factor that influences the decision to make an FDI, but once the MTO begins to manage the local operator it must enter the country risk in its information systems and design coverage policies to eliminate it or, as the case may be, reduce it. There are several general techniques available to the MTO to cover the country risk:

1. Insurance policies covering the country risk can be taken out. On the one hand, these policies can be taken out with source country firms, which are usually public firms although in recent years private companies have been proliferating. In Spain, this service is provided by the public firm CESCE (Compañía Española de Seguro y Crédito a la Exportación). On the other hand, the MTO can take out a policy with a supranational organization. Recently, the World Bank has created instruments of this sort through the MIGA to promote FDIs.

2. The very configuration of the consortium can also be a way of covering this risk. On the one hand, the consortium makes it possible to diversify the risk. On the other hand, the inclusion of local partners, aside from reducing the risk of expropriation, can help provide connections with the local administration and bureaucracy that can be useful in the future.
3. The risk can be decreased by integrating the operator's activities to the greatest possible extent into the local economy. In this sense, it is important to place as many orders as possible with local industry and service companies. As regards personnel policy, parent company directors should be limited exclusively to key posts and local employees should be allowed to participate in managerial tasks.
4. Likewise, it is very useful to make public statements referring to "the contribution to the national economy" in matters of material investments, supply of technology, and creation of local employment, both direct and indirect. The company's image can also be improved by financing social and cultural activities.
5. Local indebtedness mitigates the negative effects of expropriations and blocked funds. This measure also serves to reduce the foreign exchange risk.
6. Finally, more sophisticated financial techniques such as credit swaps, parallel loans and creation of instrumental companies could also be implemented.

Table 7.1
Country risk evolution

	September 1993	March 1993	March 1992	March 1991	March 1988	March 1987	March 1986
LATIN AMERICA							
Colombia	40.4	38.8	38.4	35.3	39.1	39.8	38.4
Venezuela	37.6	38.6	39.9	36.1	35.8	36.9	39.7
Brazil	27.8	27.7	27.0	27.4	29.4	35.5	31.9
Panama	20.9	20.4	18.5	17.9	28.6	30.7	30.9
Uruguay	34.2	33.7	32.0	29.8	28.4	27.5	27.5
Mexico	45.6	45.2	41.0	37.9	28.0	28.7	36.4
Paraguay	27.2	27.8	27.0	26.3	27.7	31.2	31.7
Chile	51.5	48.9	44.1	39.3	27.2	26.0	24.6
Argentina	32.6	30.5	23.6	19.4	24.8	24.8	22.7
Ecuador	21.3	20.8	19.8	19.0	23.2	26.2	26.4
Costa Rica	26.8	24.8	23.4	21.5	17.6	16.9	15.7
Domin. Republic	19.2	18.5	17.2	16.9	14.2	14.9	13.9
Peru	15.0	13.9	14.3	12.3	14.0	15.0	15.9
Guatemala	18.1	18.8	17.5	16.2	13.9	13.2	12.6
Honduras	15.6	15.7	14.1	14.6	12.9	12.7	11.2
Cuba	7.7	8.3	8.3	9.7	12.8	14.2	13.2
El Salvador	15.3	15.1	11.8	10.7	8.4	8.0	6.5
Haiti	8.0	7.2	7.2	8.0	8.0	9.9	9.2
Nicaragua	8.7	8.3	7.2	7.0	5.5	5.1	5.0
G-7							
United States	89.2	88.6	87.0	87.3	91.0	94.1	96.3
Japan	91.7	91.0	91.4	92.5	94.6	96.0	95.5
Germany	89.8	90.3	90.4	90.6	93.1	94.2	
United Kingdom	85.4	84.7	83.7	84.6	86.7	86.7	88.7
France	88.2	87.6	87.2	87.3	84.9	84.1	82.7
Italy	73.5	75.0	78.5	79.5	77.6	77.0	76.2
Canada	82.0	82.1	81.8	83.7	85.9	86.5	88.1
SPAIN	75.2	75.9	76.2	76.0	73.5	71.6	68.9

Source: Institutional Investor

8. TELEFONICA FDIs IN LDCs

In recent years, Telefónica has made important investments in companies that operate telecommunications services in other countries. They are all made through TISA (Telefónica Internacional Sociedad Anónima), a Telefónica subsidiary specializing in internationalization issues.

The goal of Telefónica's internationalization strategy is twofold: to develop a multi-domestic model through its investments in the Latin American area, and establish a presence in those international projects aimed at providing global services to multinationalized customers.

This section will analyze the investments in Latin America, which fit the pattern that we have been describing throughout this paper: they are FDIs in LDCs, the share is acquired by creating a consortium, and the country risk must be managed and covered.

8.1 Investments in Chile

The telecommunications privatization process in Chile began in 1987 and was consummated the following year with the sale of CTC (a domestic operator) to the Australian investment group Bond Corporation.

Telefónica's incursion into the Chilean market dates back to 1989 when it acquired 10% of Entel Chile, which provides international and long-distance services to the Andean country. Unlike Telefónica's investments in Argentina and Venezuela, this investment was made, not through a consortium, but rather through a holding (Inversiones Hispano Chilenas) owned 100% by Telefónica. In 1990, Telefónica increased its share in the Chilean operator to 20%.

Entel's growth is evident from the 1992 profit figures: 92 million dollars, or 14% more than the preceding year. Furthermore, Entel's shares currently sell for 28% more than the quoted price at the end of 1992.

One year after the Entel operation, Telefónica acquired 44.49% of CTC from Bond Corporation and holds the shares in the holding Telefónica Internacional de Chile. CTC is important because it operates 94% of all the country's telephone lines. For this reason, in 1990 Telefónica assumed major commitments as regards future development of Chilean telecommunications. In the following 6 years, it would have to invest at least 2,000 million dollars and thus succeed in doubling the 800,000 lines existing in late 1989, achieve integral automation of the country, modernize the network, install some 15,000 public telephones, and extend the service to 400 new towns.

CTC has known how to exploit the financing possibilities of international organizations. Specifically, in late 1990 it obtained a 10-year, standby credit line from IFC (International Finance Corporation). These funds were earmarked for extending the service to non-urban zones. On the other hand, the conditions set by international organizations on credits help to limit the country risk for the MTO.

Telefónica's business in CTC has been highly satisfactory to date. During the 1990-1991 biennium, a total of 190,000 new lines were installed, or the same number of lines installed in all during the 15 previous years. Profits have also been quite acceptable, with 184.4 million dollars in 1992, or more than 25% higher than the previous year. In addition, the current value of CTC shares is 51% higher than at the end of 1992.

Telefónica's dual share in two telecommunications operators in this country has involved a higher country risk, since there were some complaints about monopolistic practices. As a result, in mid-1991 Chile's National Treasury Office pronounced sentence against an increase of Telefónica's share in Entel to 20%.

Table 8.1
Telephone lines in Latin America

RANK	COUNTRY	LINES IN SERVICE (thousands)	PENETRATION (Lines in serv./100 inhab.)
1	Brazil	10,020	6.56
2	Mexico	6,025	7.81
3	Argentina	3,880	11.17
4	Colombia	2,750	8.74
5	Venezuela	1,599	7.95
6	Chile	997	7.45
7	Puerto Rico	1,033	27.34
8	Uruguay	451	14.5
9	Peru	594	2.6
10	Ecuador	520	4.54
11	Dominican Republic	385	4.75
12	Costa Rica	305	9.87
13	Cuba	366	3.41
14	Panama	229	9.85
15	Guatemala	223	2.08
16	Bolivia	221	2.36
17	Trinidad & Tobago	182	14.15
18	Bahamas	143	57.01
19	El Salvador	125	2.38
20	Martinique	122	35.88
21	Guadeloupe	118	34.70
22	Paraguay	112	2.61
23	Jamaica	97	4.00
24	Honduras	94	1.92
25	Barbados	83	31.92
26	Netherlands Antilles	63	25.11
27	US Virgin Islands	62	55.00
28	Nicaragua	52	1.34
29	Haiti	45	0.69
30	Bermuda Islands	41	65.66
	Others	195	
	TOTAL LATIN AMERICA	31,133	7.03

Source: Siemens, UIT, Telefónica de Espãna. Basis: January, 1992

8.2 Investments in Argentina

In June 1990, a consortium called Cointel and led by Telefónica was awarded a seven-year (extendable for another three years) concession to exploit as a monopoly the telecommunications operator in southern Argentina. The consortium, initially formed by Telefónica, Citibank and Technint (an Argentinian investment group) would hold 60% of the operator's capital and 40% would be retained by the State. The consortium was completed once the license was obtained and new financial partners were admitted. Each partner's share in the consortium was established as follows:

Table 8.2

Telefónica:	10.0%	Banco Hispanoamericano:	5.0%
Citicorp:	20.0%	Bank of Tokyo:	5.0%
Banco Río:	20.0%	Banco Comercial de la Plata:	5.0%
Technint:	9.9%	Aresbank:	4.1%
Banco Central:	8.4%	Banco Macro:	4.0%
Banco República:	7.5%	Banco Atlántico:	0.9%

The privatization and award to foreign companies of Argentinian telecommunications have not been free of difficulties. The left-wing parties and unions opposed the operation head-on. After overcoming the problems, the Argentinian President achieved his goal, although the company Entel was divided in two—the northern zone and the southern zone—and assigned to two foreign consortiums headed by STET and France Telecom (through its subsidiary France Cable) and by Telefónica, respectively. This solution helped to cover the country risk for the investor MTOs, since if the Argentinian government decided to globally expropriate or block funds, it would antagonize two investment groups including local partners. On the other hand, if it decided to act against only one of the parties, the other would input the increased country risk into its information systems which would affect the MTO's plans in the country, with resulting repercussions on the country's telecommunications development.

From a technical standpoint, the operation was implemented through a debt-equity swap, in addition to a cash payment.

The result of the operation to date is highly satisfactory. Telefónica's management has successfully fulfilled the goals established in the Argentinian Government's privatization specifications as regards service failures and quality. Furthermore, more than 350,000 new lines were installed in 1991 and 1992. Likewise, material investments during this period amounted to 800 million dollars. If we add to this the creation of some 1500 jobs, we can conclude that Telefónica's conduct throughout these years has made it possible to reduce the country risk.

Another aspect that limits the country risk is the weight—almost 20%—that the Telefónica de Argentina shares carry on the Buenos Aires stock exchange. An expropriation or fund blocking operation against the MTO would be detrimental to the Buenos Aires stock market.

In December 1991, Telefónica acquired an additional 1.95% of the capital of Telefónica de Argentina, the company's new name, on the occasion of the sale of part of the shares retained by the State. Proof of the smooth running of the company's business is the share price fixed for this new sale: 300% higher than the 1990 price. As of this date, Telefónica's capital share increased to 8%.

The operation has been worthwhile from the standpoint of financial profitability. In 1992, Telefónica de Argentina showed a profit of 220 million dollars, or 80.7% more than the preceding year. The company's shares are quoted at 50% above their value in late 1992.

The Argentinian Government's satisfaction with the two foreign consortiums became evident in 1993, when Movistar, a company 50% owned by Telefónica de Argentina and Telecom, was awarded the second cellular telephone license for the Buenos Aires area.

8.3 Investments in Venezuela

In November 1991, the Venezuelan Government partially privatized the country's telecommunications operator, CANTV. To do so, the Government invited tenders and several foreign groups responded. The license was awarded to the consortium Venworld; 40% of CANTV capital was allotted to it, while the State retained 49% and shared out the remaining 11% among the workers. Share percentages in the consortium were distributed as follows:

Table 8.3

GTE	51%	Electricidad de Caracas	16%
Telefónica	16%	Banco Mercantil	12%
AT&T	5%		

There are three MTOs in the consortium; GTE, and Telefónica to a lesser extent, are in charge of running the Venezuelan operator.

By privatizing and opening up the market to foreign MTOs, the Government intended to leverage the country's telecommunications and provide a short-term solution to service quality problems and the high level of unsatisfied demand. Therefore, the Government obtained the consortium's commitment to install 300,000 lines per year and to invest at least 800 million dollars a year. In turn, the consortium obtained approval to make automatic billing adjustments in accordance with the inflation rate and to progressively readjust structural rate imbalances during a period of nine years.

CANTV earnings in 1992 amounted to more than 240 million dollars, whereas the preceding year they barely passed the eight million mark, which clearly demonstrates how profitable the operation is.

8.4 Investments in Puerto Rico

In December 1992, Telefónica acquired 79% of TLD (Telefónica Larga Distancia) of Puerto Rico and took over operation of the company.

TLD provides international service to more than 40% of the country's lines. The rest of the international traffic is shared by the North American operators ATT, US Sprint, and MCI. Thus, Telefónica has access to the advanced market of North American telecommunications, although the country risk is very high as regulatory issues imposed by the FCC must be resolved.

9. CONCLUSIONS

As shown throughout this paper, internationalization of telecommunications operators is a matter of utmost importance for the future of these companies. By way of conclusion, the following issues are underlined:

- Even though they have been internationally active for quite some time now, telecommunications operators still do not have much experience in internationalizing their activities. Thus, internationalization is a recent phenomenon.

- Internationalization is determined by changes made in the sector, both economic (integrating economies) and regulatory (liberalization and deregulation being the most important).
- The operators' internationalization strategies must be guided by a suitable combination of cooperation and competition criteria.
- The world's leading operators are considering internationalization as one of their key strategies. Although there are differences in how they focus these strategies, it is important that they try to put a global strategy into practice in the future.

In connection with the FDIs made by telecommunication carriers for the purpose of operating basic public networks in LDCs, we emphasize the following:

- The telecommunications sector is one of the most important strategic sectors in order to develop the LDC economy and usually the State is the owner of the operators in these countries. Therefore, Governments play an active role in operator internationalization procedures.
- Alliances in general and consortiums in particular form part of most internationalization operations of this kind.
- Several positive signals for entering a market have been identified: privatization intent, sector liberalization, existence of a demand to be satisfied, and a suitable "climate" in the country for doing business.
- The following factors are critical to the success of an alliance: suitable selection of partners, selection of an activity that exploits the competitive advantages of the leading operator, effective negotiations with the Government, fluid flow of communication and trust between partners, and provision of high-quality services.
- Contracts that regulate consortiums may contain implicit options for selling shares.
- The MTO has to bear in mind the country risk. Given the importance of this subject, the MTO must include the country risk in its information systems and must design coverage policies to eliminate or decrease it.

The experience of Telefónica in LDCs is concentrated in the Latin American area. In this regard, we underline:

- The FDIs represent a small percentage of Telefónica's gross capital expenditure in fixed assets.
- Telefónica is getting a very important financial profitability from its investments.
- The country risk associated with Telefónica's FDIs is lower than the average in Latin America.
- Telefónica is fulfilling its commitments that it acquired with the Governments concerned.

REFERENCES

Buckley, J.V., 1992, Competition and modernization issues, Telecommunications policy, December, .

Dang-N'guyen, G. and T. De Swarte, 1992, Strategie et management: Le cas des opérateurs de télécommunications BT et France Télécom. Ninth ITS International Conference. Sophia Antipolis.

Dang-N'guyen, G. and D. Pouillot, 1993, Foreign direct investment in telecommunications in and out of France, ITS European Conference. Göteborg, Sweden.

Dunning, J.J., 1988, The eclectic paradigm of international production: A restatement and some possible extension, Journal of International Business Studies, Spring, pp. 1–32.

Dunning, J.J., 1992, The competitive advantage of countries and the activities of transnational corporations, Transnational Corporation, February, pp. 135–168.

Duran, J.J., 1992, Estrategia y evaluación de inversiones directas en el exterior. ICEX. Madrid.

Duran, J.J., 1993, La empresa multinacional en la configuración de la economía internacional. Centro Internacional Carlos V. Madrid.

Duran, J.J. and F. Gallardo, 1993, Assessing foreign direct investments for telecommunications carriers in less developed countries, ITS European Conference, Göteborg, Sweden.

Graham, J. 1990, Exchange of threat between multinational firms, The International Trade Journal, No. 3.

Knickerbocker, F.T., 1973, Oligopolistic reaction in multinational enterprise, Harvard University, Graduate School of Business Administration. Boston.

Kobrin, S.J., 1982, Managing political risk assessment (University of California Press).

Kok, B., 1992, Privatization in telecommunications—empty slogan or strategic tool? Telecommunications Policy, December.

Llerena, P. and S. Wolff, 1992, Inter-firm cooperative agreement in telecommunications: elements of an analytical framework, Ninth ITS International Conference, Sophia Antipolis.

Porter, M., 1991, The competitive advantage of nations (The Free Press, New York).

Thompson, S., 1992, Telecommunications privatizations and international capital markets. Telecommunications Policy, December.

The Race to European Eminence
E. Bohlin and O. Granstrand (Editors)

International Business Enterprises and Telecommunications: A User Perspective

Edward Roche*

Department of Computing and Decision Sciences, W. Paul Stillman School of Business, Seton Hall University, South Orange, New Jersey 07079, USA
Department of Information Systems, Stern Graduate School of Business, New York University, New York, New York, USA

1. NETWORKING IN THE MULTINATIONAL CORPORATION

Multinational corporations account for more than 80 percent of international trade, and taken together are very substantial users of telecommunications products and services. Using the estimations from Datamation, the data processing budget is usually 2% to 5% or more of company revenues, with expenditures on *data* communications being 10% to 15% of that amount. Adding together the annual revenues of the 2,000 or so largest multinational corporations, and adjusting for this ratio, yields a rather large figure. Even taking a sample of 45 manufacturing firms with annual sales greater than $1 billion and with at least 40% of their sales coming from outside of their home countries yields an average of $193 million average per firm for annual *data* communications expenditures alone. Multinational corporations have complex business installations in dozens of countries, and usually operate major data processing centers in all regions of the world—with some organizations having more than 100 major data centers. Linking together these data centers means that multinationals are also perhaps the largest users of telecommunications services in the world. They are indeed important "customers".

Although in the past, multinationals have introduced a great deal of national autonomy into their operations, today, they are being forced to adopt global competitive business strategies requiring considerably more coordination between geographically separated parts of the company, including overseas subsidiaries. In addition, the creation of new economic zones through European integration and the North American Free Trade Agreement is making it possible for multinationals to take advantage of economies of scope and scale in information processing by re-organizing many of their operations on a regional (rather than national) basis. If a new era of liberalized trade opens up as promised with successful conclusion of the GATT negotiations, including trade in services, then both manufacturing as well as the information and telecommunications-intensive service multinationals will face even stronger pressures to respond to competition on a global scale, leading to even more intensive use of international telecommunications circuits.

* Internet email address: E. Roche@rnd.stern.nyu.edu

A structural revolution is underway in the use of data processing and telecommunications networks in the multinational corporations in Europe. Many companies are consolidating their data centers. In one case, more than 300 data centers have been reduced to 8 "super-centers", with plans to reduce those 8 to only a single center by the end of the 1990s. Clearly this has both positive and negative effects on different nation states—some lose, some win big. Multinationals also face increasing pressures to build inter-organizational telecommunications-based strategic linkages in several directions: inter-subsidiary, subsidiary-supplier, and subsidiary-customer. These *internal* systems are being further supplemented by *external* linkages with business partners, e.g. joint ventures or other contractual relationships operating internationally.

Attaining the best levels of management in today's multinational means reaping large gains in productivity, simplifying systems development efforts through continued cost-cutting and data center consolidation, introducing new types of information services products on an accelerated product life-cycle basis, and developing the ability to quickly implement any inter-organizational systems required by the constantly shifting strategic alliances of the firm.

These challenges are placing new pressures on the telecommunications function. The operation of large-scale multiple-location international enterprises from only a handful of data centers will place a much heavier burden on telecommunications because more data and processing will be centralized in order to achieve economies of scale, whereas more access and end-user computing will continue to be decentralized. With processing and storage centralized but end-user access decentralized, telecommunications must bridge the gap. To paraphrase one successful vendor: "The telecommunications network *is* the computer".

1.1 Linking together data centers around the world

Today's multinational corporation places separate data centers in each country in which business is being conducted; in addition, several data centers are often found in some countries, i.e. if the firm has multiple manufacturing sites, or divisional operations. As a result, studies of geographical distribution of data processing centers generally involve more than 100 countries, and between 100 to 500 or more separate data centers.

The global structures of computer networking map loosely against the underlying control system of the enterprise. The simple subdivided single-unit enterprise evolved as more complex technology was introduced into manufacturing. Transportation companies tend to use the multi-unit configuration. Multi-unit multi-functional enterprise with a centralized functionally departmentalized structure was used by early integrated manufacturing companies. In the 20th century, the very large corporations have been using the multi-unit, multi-functional, multi-industrial structure with a decentralized multi-division structure at the heart of operations. This was extended to the multi-national enterprise in which an international group handles different geographical areas or products.

Due to the nature of different national regimes, historical development patterns, availability of equipment from vendors, nature of local business operations, and many other factors, different data centers are generally highly divergent in their hardware, software, particularly releases of different packages including operating systems, and applications. It is quite common, even within the same enterprise, for basic data definitions to be different from country to country, making the task of "integrating" the information system into a coherent global operation impossible or at a minimum very difficult, time consuming and expensive.

The underlying structure of telecommunications in the multinational corporation is complex. We may say there are two classes of telecommunications networks: internal and external. Internal networks can be further sub-divided into a) linkages between headquarters and subsidiaries, b) linkages between headquarters and regional centers, c) linkages between regional centers and subsidiaries. External networks can be distinguished into three classes: a) linkages with business partners (e.g. joint ventures, partially-owned subsidiaries), b) linkages with customers, c) linkages with suppliers. These different types of linkages display different characteristics. Internal networks are easier to control, and therefore tend to be more protocol-homogeneous and predictable; external networks frequently must use "least common denominator" connections (ASCII, EDI) in order to communicate effectively in a multi-vendor, multi-protocol environment. The hardware being linked together also varies in several dimensions: a) computer to computer, b) terminal to computer, or c) computer to intermediary computer (VAN provider) with different types being found in each of these linkages.

It is important not to underestimate the importance of external inter-organizational networks. Virtually all international business enterprises are tied together with other companies in a complex web of relationships involving licensing, servicing, joint ventures, consortia (*keiretsu, chaebol*), and other types of formal and informal contractual arrangements. These are extremely important relationships. One example is the European Airbus operation which is composed of different national partners from European states. We can see clearly that these linkages between different firms are critical to operations and long-term business survival of the company, and they are increasingly dependent upon sophisticated telecommunications and data processing linkages.

The geographic scope and variety of international computer communication networks is staggering. Agence France-Presse has a network connecting 129 countries, operating in 6 languages, and carrying text and images from 150 scanners over an X.25 system of Codex 6500 Series packet assembler-disassembler switches. General Electric Co. links all of its businesses worldwide with a digital network carrying voice, data and teleconferencing signals with on-demand bandwidth reallocation capabilities. Grand Metropolitan, which operates in a very decentralized fashion, is a heavy user of electronic mail systems for its coordination activities. P.T. Freeport Indonesia has set up a telecommunications network linking New Orleans with Jakarta, Singapore and a mine in Tembagapura, Irian Jaya, with voice, fax, data, and future teleconferencing capabilities. Holiday Inn has linked together its 330,000 rooms in different countries of the world with a private VSAT network with 200 downlinks supplied by MCI and Hughes Network Systems, thus cutting its costs by more than 7 percent annually. Upjohn is working with 20 other pharmaceutical companies to install videoconferencing facilities to link with the US Food and Drug Administration. GE uses 86 or more private videoconferencing facilities in more than 12 countries. These are only a very few of the many examples showing how international business is dependent upon international telecommunications for daily operations.

1.2 Changes to expect in the late 1990s

Multinational corporations will move towards stronger centralization of data processing, probably clustered around regional centers, such as "Europe", "North America", "Asia", etc. Also, telecommunications will make it easier for multinationals to rapidly project the business capabilities of their data processing systems—special applications—into remote geographical areas *without* having to build large supporting data centers. In general, as the velocity of information within the firm increases rapidly due to continued advances in telecommunications, the *information boundary* of the firm will extend further, particularly through the media of inter-organizational sytems. The ability of multinationals to provide increasingly complex levels of headquarters to subsidiary, subsidiary to subsidiary and inter-organizational coordination will continue to improve, hopefully leading to more economic efficiency all around. Multinationals will be able to make strategic structural changes in their data processing only as long as telecommunications costs continue to drop. If telecommunications tariffs remain high in certain parts of the world, e.g. Europe, then pressures to consolidate data processing centers will be mitigated under conditions where high telecommunications costs out-weigh any possible gains from consolidation. By the year 2000, examination of the physical distribution of computing resources may cease to be as useful as analysis of how information is distributed within the multinational corporation, since inexpensive telecommunications will make it easy to move information anywhere in the world it is needed without any noticeable delay.

2. OUTSOURCING TELECOMMUNICATIONS TO THE MNC

Telecommunications providers have responded to the demands of the multinational corporation. The recently-announced Financial Network Association, which is a joint venture of 12 national public telecommunications providers, half of which are European, is only one of many steps in the direction of providing multinational corporations with turnkey telecommunications services. There appears to be little fear on the part of telecommunications professionals that widespread use of outsourcing will endanger job prospects. Data processing departments have changed their philosophy in the last few years, moving away from being responsible for all types of services, towards adopting a strategy of working vigorously only on the most critical core business functions. In addition, strong efforts are being made to reduce costs, as outsourcing is generally cheaper, at least in the short term. In addition, the movement towards "right-sizing"—the transition from reliance on centralized mainframe computing towards use of client-server architecture based on LANs—has increased the need to link together the LANs themselves, thus further stimulating the search for new solutions. Most value added network (VAN) suppliers are able to provide both voice and data network services including peer-to-peer connections (which eliminates the need to convert application-linked data), electronic document interchange services (which converts everything to a standard format for transmission), and overall network management, including problem determination procedures (PDPs) and maintenance. Advantis company, set up by IBM and Sears Roebuck and Co., uses a TCP/IP backbone to link together Ethernet and token-ring LANs in a multi-protocol (SNA or X.25) environment to serve 9,000 customers and generate more than $1,100 million in revenues shortly after start-up. Inter-LAN connection capability can be critical to strategic business missions as evidenced by the General Motors C4 project to collapse product development time by, in part, linking together LAN-based engineering workstations between GM and supplier sites using identical systems and software.

Many global services are being provided by PTTs in newly emerging business arrangements. Most major PTTs offer their own version of global outsourcing capabilities, but new companies are being created as consortia between major PTTs, leading to a "cooperate and compete" scenario. Transpac, for example, represents one of the first times PTTs have themselves become entrepreneurs. The BT-owned Syncordia Corporation is targeted at providing turnkey telecommunications services in a variety of countries. It succeeded quickly in getting a US$12 million per year contract with British Petroleum, and another contract with Australian Air BP Company, a supplier of aircraft fuel around the world. PTT Telecom Netherlands and Swedish Telecom International have launched Unisource which will compete directly with Syncordia. Swiss Telecom PTT's monopoly was recently taken away to allow these global outsourcing ventures to also provide services in Switzerland.

AT&T has joined forces with Kokusai Denshin Denwa and Singapore Telecom to provide WorldSource, a service targeted at providing turnkey networks to multinational corporations. WorldSource allows the user to consolidate all of their telecommunications into one bill in a single currency. J.P. Morgan signed up with ATT global LAN networking service, a move indicative of the general data processing trend away from mainframes and toward client server architectures. Stattel, the information technology arm of the Swedish government, chose Transpac to be its data network services provider.

Consortia of PTTs are not the only organizations providing telecommunications outsourcing. Zale jewelry stores with 1,800 locations handed over its telecommunications contracting operations to the small firm CrossCom. In a celebrated case, Eastman Kodak Co. handed over its entire worldwide SNA network to IBM, and gave its voice and non-IBM data networks to DEC to manage. General Electric Co.'s GE Computer Services division offers network design and consulting, network maintenance and support service as well as network management. Large consulting companies such as Computer Sciences Corporation also provide complete telecommunications and data processing outsourcing, having recently signed one of the world's largest contracts with British Aerospace. Telecommunications equipment providers such as Timeplex which offers frame relay multiplexers and LAN bridges and routers also provides network management outsourcing services and is doing so for International Computers Ltd. in its North American network.

Some contracts are quite small. For example, Computer Services Inc. has been hired to link together 2 data centers with 170 regional banks in Kentucky using a T-1 backbone network fed by 19.2 KB multidrop analog lines for a doubling of network capacity with a slight reduction in overall costs for the banks. Larger contracts are exemplified by the SHL Systemhouse Inc. 10-year outsourcing deal for Canada Post Corporation. This deal supplemented the Systemhouse deal with Mexico's Ministry of Finance for $550 million in outsourcing. Motorola is outsourcing its network operations, domestic voice network and its peer-to-peer internetwork system. Gillette company decided to dump its use of foreign public data network (PDN) providers in favor of using British Telecom North America Inc. for global X.25 network services, with estimations that more than 35% will be reduced from telecommunications costs and total throughput can be increased by more than 300% by avoiding routing across X.75 network nodes.

Eastern Research Corporation predicted that in the U.S. alone, outsourcing fees will rise from US$4.8 billion in 1989, to US$20.7 billion in 1995. At that growth rate, outsourcing will be a very big business in the year 2000. The development of the NTT Plans service, as well as a variety of offerings from other providers, including AT&T, will continue this trend. The Yankee Group estimated that all of the Fortune 500 companies will evaluate outsourcing and 20% will sign outsourcing contracts during the 1990s. They predict a $29 billion market in 1990 expanding to a $49.5 billion market in 1994. The Yankee Group estimated that although many vendors promise up to 50% savings in costs, a saving of 15% to 30% is more common, based on their surveys.

These services for providing turnkey telecommunications to multinational corporations have been developed in recognition of the global networking and data processing needs peculiar to international business. They make it possible for the multinational to contemplate many new arrangements in its data processing configuration. Centralization is made less complicated. Better integration between subsidiary locations and headquarters is achieved, and new technologies such as groupware and teleconferencing can be added to internal networks to change the face of decision-taking.

Outsourcing is many times divided up into parts, with only a portion of the total operation given to outsiders. For example, Imperial Oil Resources outsources only its desktop computing support services. In other cases, entire networks are created to handle specific functional requirements. For example, the F/X Trader network supplied by Quotron to span 8 major financial centers and aid 4 of the world's largest banks is competing against Reuter Monitor Dealing Service, both integrated hardware-software-telecommunications networks.

In other cases, outsourcing is the result of the structural nature of the multinational corporation. When Unilever PLC, which is composed of more than 500 independent companies, decided to contract out its telecommunications to US Sprint Communications Co., a great part of the reason was that with such a diverse array of companies, it was not practical to build its own data network. Another example of structural constraint was when The Kendall Co. under pressure from a leveraged buyout was forced to outsource its recently completed AS/400 network to the IBM Corporation.

There may be a set of subtle disadvantages to outsourcing. Some worry about a creeping loss of control over the network. It is also difficult to anticipate the real cost savings over a long period of time. Presumably if a vendor became too costly, another could be hired in many instances, but cases of "switching costs" in this context have never been published. GTE consolidated its outsourced data processing for cellular telephone billing into a single third company, Systematics Telecommunications Services, Inc., so it appears possible to play off one outsourcer against another. There may be unanticipated problems with security and confidentiality of data. It is also possible that service levels might not be adequate, leaving the company with little mechanism for improvement, except paying out more cash than budgeted. Service and support for existing networks can be up to 20% of initial acquisition costs, and in that area, outsourcing may be a more expensive alternative, less able to respond to customer requests, and fostering an unhealthy fragmentation of responsibility. After the outsourcing decision is made, the company may experience a rapid decline in morale, followed by unexpected and unwanted resignations, requiring an injection of financial incentives to retain needed staff. There is, after all, much evidence that in many cases outsourcing can lead directly to unemployment for telecommunications professionals, particularly those who are older, or are without very specific technical skills. These concerns in general, however, may be more academic than real, judging from the rush towards outsourcing.

It is not altogether easy to understand how these changes will affect host countries as well as the internal control systems of the multinational corporation. The decreasing costs of telecommunications probably will change the *information access structure* of the multinational corporation. It will make it appear that remote data processing locations are closer than before and will move data storage away from subsidiaries and towards either headquarters or regional centers. The change in the geographical distribution of data centers will eventually have a profound impact on the operations of the multinational corporation of the late 1990s, but it will be years later before the economics of such transformation are understood completely.

3. LESSONS FOR SERVICE PROVIDERS

There is skepticism on the part of many users that international carriers are too bureaucratic and insensitive to be able to adequately provide the type of customer-centered services needed for outsourcing of global networking. The answer to this criticism can only be worked out over time. Although it is in many cases easy to make a convincing case for short-term savings, it is a continuing challenge to paint a long-term picture of stability, economy and flexibility for users, all of which are facing severe competitive pressures, rapidly shifting market conditions, and compelling requirements for rapid-fire adaptation to change in strategy. At the core of the problem is that a marginally less expensive solution for a cost which itself is only 10%–15% of data processing costs is of little advantage if it inhibits the overall ability of the corporation to respond to business competition.

In a world of competing super-providers, the struggle to lock-up the outsourcing contracts of the giant multinational corporations will be intense and difficult. The clear winners and losers will take time to appear, but it is likely the winners will have the following characteristics: (1) ability to offer services in most countries of the world, including emerging markets in developing countries to which an increasing amount of manufacturing is being transferred; (2) sophisticated billing and payment services which can handle complex currency problems and customized reporting consistent throughout the world; (3) ability to cost-justify outsourcing projects in terms additional to simple cost reduction, including identification with strategic business objectives of the enterprise; (4) ability to provide users with rapid configuration and capability changes upon demand; (5) ability to work closely with the data processing establishment, including any third-party vendors supplying outsourced services and consulting for information processing.

It must always be kept in mind that the purpose of building integrated telecommunications networks is not only to reduce costs: that function in actuality is secondary; but rather efforts must be made to support the strategic business mission of the international enterprise. If suppliers of turnkey telecommunications services get in the way of that broader objective, they will be slapped aside quickly.

PART D: COMPARISONS AND CONCLUSIONS

The Race to European Eminence
E. Bohlin and O. Granstrand (Editors)

Internationalization and Diversification of Technology-Based Services: A Comparative Analysis of 25 Telecommunication Service Corporations

Ola Johansson*

Chalmers University of Technology, Department of Industrial Management and Economics, S-412 96 Göteborg, Sweden

CHAPTER SUMMARY

The issues related to the phenomena of internationalization and service diversification in technology-based services are discussed, with reference to the case of the increasingly internationalized telecommunication services industry. The international activities of 25 large telecommunication service corporations (telcos) are classified and conceptualized, in order to identify and tentatively explain early patterns of internationalization and service diversification in the telco industry.

The technological pace of change is the major driver of, for example, new service concepts, liberalization and a more differentiated customer demand. Special features in the telco industry arises from telecom services being network-embodied, created at various levels, at different times and by various suppliers.

The paper summarizes patterns of international activities of telcos. Main findings are that early international activities are usually characterized by uncertainty, reflecting a cultural and geographical distance. The early stages of internationalization in the telco industry in the late 1980s were characterized by a major involvement in one of the main countries within a particular region, i.e. following the large multinational customers abroad independently of the liberalization process in the targeted country. Another pattern of internationalization is to invest in any country in a region wherever a window of opportunity (because of liberalization) arises. The latter is also common as a second stage for those telcos initially investing in one of the main countries.

Telcos tend to specialize in a few services, exploiting these in a number of new countries, and/or to pursue a broad set of activities—"trial-and-error". Several of the activities in newly liberalized countries could be seen as such learning activities (trial-and-error), sometimes outside the technology base. On the other hand, the empirical findings suggest that it seems reasonable to hypothesize that the management and technology base of telcos and the situation on the home market are essential factors explaining the direction and level of foreign

* The research support received from Telia is gratefully acknowledged. The research is part of the research program "Economics and Management of Technology" at the Department of Industrial Management and Economics, Chalmers University of Technology and has been carried out under the auspices of the Institute for Management of Innovation and Technology (IMIT). We also thank Erik Bohlin, Thomas Falk, Ove Granstrand and Staffan Jacobsson for valuable comments and support, and Sven Lindmark for the extensive work he has contributed in the gathering and compilation of the background data.

investment in the telco industry, especially for those telcos who tend to specialize in a few services. The RBOCs (mainly investing in mobile and telephony services), the US long-distance and the large European telcos (both mainly investing in international value-added services) provide illustrative examples of this behavior of internationalization.

1. INTRODUCTION

The chapter aims at contributing to the knowledge of internationalization and diversification of technology-based services through an analysis of the case of telecommunication. Technology-based services are here defined as services which require embodied and/or disembodied technology for their provision (Bohlin and Granstrand, 1994). The international activities[1] of 25[2] large telecommunication service[3] corporations in the world are surveyed. These companies will hereafter be abbreviated as *telcos.*

Internationalization of telecom services markets in the telco industry has increased rapidly over the last decade, even though substantial volumes of international voice services have been seen for a long time through contractual arrangements between the national monopolies. In addition, there is a huge expectation of revenue growth in international and (foreign-) domestic telecom services. More specifically, internationalization in the telco industry is affected by a number of factors (Granstrand and Johansson, 1993). Here it is suggested that the internationalization process has been predominantly influenced by three sets of factors: technological development and innovation in networks and services, change in customer demand, and liberalization[4] of telecom services markets (see Figure 1.1).

First, international telecom services have been facilitated by technological development and innovation in networks and telecom services over the last few decades. Technological development has decreased the cost/performance ratios substantially, which in turn has put pressure on price levels. The rapid development of network equipment also means that performance in telecom networks has been improved substantially. The development of network equipment is still ahead of standardization, which would mean that telcos should invest in R&D and keep close cooperation with few[5] suppliers. This has opened up possibilities for new telecom services and new solutions. In addition, the technological development has forced competition in telecom services to include other parameters than price, such as cost, reliability, predictability and controllability.

---

[1] In this chapter international activities are those included in the establishment and provision of international or foreign domestic telecom services, e.g. setting up an office, buying a stake in a cellular joint venture, acquiring a foreign firm or offering specialized solutions to customers in foreign countries.

[2] These telcos are Ameritech, AT&T, Bell Atlantic, BCE, Belgacom, Bell South, BT, Cable & Wireless, DBP Telekom, France Télécom, GTE, KDD, MCI, NTT, NYNEX, Pacific Telesis, PTT Telecom, Singapore Telecom, STET, South Western Bell, Telia, Sprint, Telstra, Telefónica and US West.

[3] The telco industry is in this paper considered as a service industry simply because it has characteristics similar to other service industries (see also section 2).

[4] The terms liberalization and deregulation are often used synonymously. However, here liberalization means a process of adjusting the regulatory framework to allow for competition (which might mean reregulation), while deregulation means a process of reducing regulation.

[5] In the telco industry, telcos normally have close cooperation with a single supplier and telcos use on average three suppliers of switches.

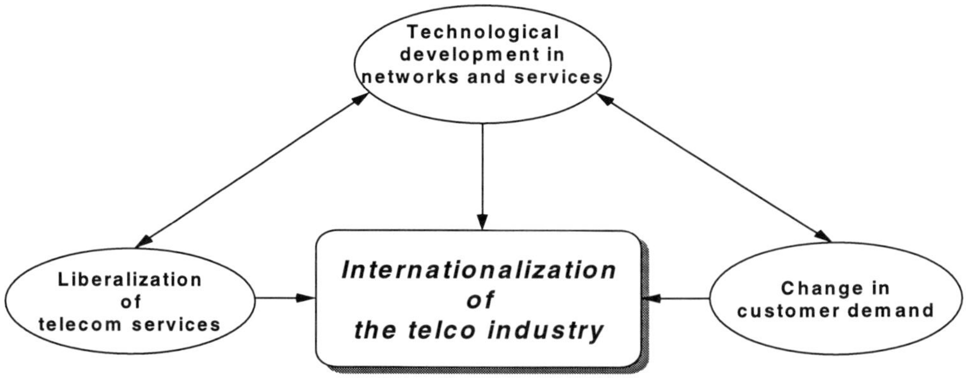

Figure 1.1.
Major determinants of internationalization and their interrelations in the telco industry.

Moreover, liberalization of telecom services (EC, 1993) has made it possible[6] to make foreign investments, while at the same time forcing telcos to make investments abroad because of increased international competition in the home market. New regulations determine the structure of the market (often through licenses and price levels), and define the role and market interfaces of the regulatory body. In addition, the former monopoly-based prices of telecom services represent incentives to new entrants.

Lastly, customers need more international access simply because they have generally become more internationally oriented. This argument is especially valid for large multinational companies but becomes more and more important also for medium-sized and small companies when new applications are not only used to improve efficiency but also to improve quality and create new services (Barras, 1986). As an example, the need for international access today includes not only voice services but also advanced datacom services. This applies to customers who have seen a rapid development in computer-based applications, and who are proactively demanding new and more advanced telecom services. For this customer segment (and other segments with datacom needs) other parameters than price are more important.

The purpose of the paper is to identify, in an explorative way, and tentatively explain patterns of strategies for internationalization of technology-based services in the telco industry. To fulfill this purpose, the structure of the paper is as follows. Section 2 of the paper outlines a conceptual framework for technology-based services and internationalization. Section 3 outlines the methodology used. Section 4 describes and classifies the behavior of telcos in terms of internationalization and diversification of foreign telecom services from the late 1980s until 1993. Section 5 analyzes the empirical findings of section 4 in a discussion. Finally, in section 6 conclusions are drawn.

---

[6] Actually, the telecom services market will be open for competition in the EU in 1998, but the race has already started because technological development has made it possible to circumvent the regulations to some extent.

2. CONCEPTUAL FRAMEWORK

The conceptual framework of internationalization in the telco industry needs an improved definition of the telecom service concept. The conceptual framework outlined in this section discusses the characteristics of telecom services and internationalization. The identified concepts will constitute a basis for analyzing the activities of internationalization in the telecom services industry further on in the paper.

2.1 Internationalization in the telecom services industry

Competition in each national telco industry has previously been independent of the state of competition (here monopoly) in other nations (*national competition;* Sölvell, 1987). Today, however, several telcos have investments and subsidiaries in other countries as well as services[7] to customers in foreign countries, which means that there has been a shift from national competition to *international competition* (Sölvell, 1987). *Multi-domestic competition* (Porter, 1986) is an intermediate or complementary structure characterized by foreign presence in countries where competition is still confined to the boundaries of the respective countries. This is often the case in the telco industry, since each country has its own laws for competition and even its own standardized telecommunication network. When major markets have become penetrated by foreign players, competition in each country is dependent on competition in all other countries. This means that the *competition* has become *global* (Sölvell, 1987) and that telcos may compete either on domestic markets or on global markets, or both.

In this paper the process of internationalization is defined as the process from national, through international and multi-domestic to global competition. In the process of internationalization of firms, it is especially the mobility of resources, rather than the number of firms, that creates competition (Cantwell, 1989). This view has been adopted here, which means that by *internationalization* is meant a process involving activities of making at least one investment[8] in non-domestic territories (in order to compete with foreign domestic or international/global telcos).

Actually, one could have been discussing market diversification instead of internationalization as Penrose (1959) does, when she proposes a market to be characterized by the kind of customers it serves. But, because countries in the telco industry have several unique features (such as regulation etc.), they are most preferably analyzed as markets. Thus, market diversification as an aggregated phenomenon in the telco industry is close to internationalization.

The relative degree, or rather the absolute level, of internationalization, may be determined in several ways. In this paper, the simplest approach has been used. The number of countries where each telco has activities[9] has been counted and used in order to determine the *level of internationalization.* At this stage the measurement seems appropriate, as the purpose is merely to separate the telcos into broad groups.

---

[7] These services are offered beside the traditional accounting procedures of traditional international (voice) services.

[8] This means that the contractual arrangements made for a long time between telcos in order to provide the possibility of making calls to foreign countries are not considered as part of internationalization.

[9] Activity means an activity on non-domestic territories, which includes providing a certain telecom service as defined in section 2.2, in cooperation with other telcos (e.g. for the purpose of joint development of services), setting up an office or a subsidiary, acquiring another telco, buying a stake in another telco, participating in fiber cable projects etc.

2.2 Telecom services

The offerings by telcos include several services[10] created in a network or infrastructure basically consisting of a set of transmission media, switches and databases. This means that offerings by telecom services are embodied in goods (Dunning, 1989) or rather in facilities as we would suggest. Moreover, telecom services are provided at various levels in the hierarchy, at different times and by various suppliers, which makes the characteristics of a telecom service rather complex. In addition, several offerings, apart from the embodied ones, are considered as services in the industry (such as special price packages or service packages for outsourcing of a firm´s internal telecom operations).

Goods-embodied services are often *intermediate services*, i.e. they are not directly consumed, but embodied in final products or other services. Dunning (1989:34) gives examples of services with this characteristic, such as transportation, telecommunication, public utilities, tourist facilities, consultancy services, trade and distribution, construction, insurance and banking, and some wholesaling and retailing. In particular, telecom services have this characteristic of, for instance, always being available to the customer. Telecom services have another distinct characteristic of being embodied in a network. In this paper we shall characterize a telecom service as specifically embodied in an infrastructure, i.e. *network-embodied.* Other examples of network-embodied services are transportation (such as railways and airlines etc.), gas, electricity and water supply. However, gas, electricity and water supply services only provide a service with "one variant", while telecom services display an infinite number of variants.

Another important characteristic of telecom services is that they are *technology-based services*, i.e. they are offered primarily by the efficient use of *technological knowledge* but also by the use of *management knowledge* in order to create services out of a set of *production and management facilities*. The technology base of a company is the intangible asset of the technological competence a firm possesses (see, for instance, Granstrand and Sjölander, 1990). Similarly to the technology base, the management base of a company will, in this paper, be characterized by the intangible asset of the managerial competence a firm possesses, e.g. the competence of applying the technology base to markets. Production facilities in telecommunications basically control and route calls on a network, consisting of transmission media (cables, satellite connections, micro-wave links etc.) and switches in a hierarchy by using information from databases. Management facilities are here supposed to mean all facilities but production facilities, such as personnel, personal computers, internal telecom systems, customer databases etc., which help management e.g. in marketing and supporting activities.

These facilities could be used either to generate a set of basic services[11] or a set of value-added services[12], or a combination thereof.

---

[10] Services are generally characterized by the intangibility, heterogeneity, perishability, customer participation, and simultaneity of production and consumption; see, for example, Grönros (1982), Normann (1984) and Carman and Langeard (1980). However, another view is increasingly adopted in the literature: Segel-Horn and McGee (1991), Levitan (1985), Levitt (1986) and Dunning (1989). This view differentiates services as being either location-specific (simultaneous production and consumption) or tradeable (which means a choice of location).

[11] See Chase and Hayes (1991) for a similar classification of services. However, they use the concepts *core* and *peripheral* services instead of basic services, and *amenity* services instead of value-added services. In addition, their classification is made without any connection to the underlying knowledge or facilities required to provide services. Furthermore, basic services are also conceptually blurring and need to be clarified.

Carman and Langeard (1980:9) suggested core services to have the characteristics of being the necessary output of a service organization in order to be present on a market. However, it should be emphasized that in this study *basic services* simply are those which are provided at a low level in the hierarchy of a generic network of production facilities, i.e. there are basic services in each kind of telecommunication network[13].

Value-added services, on the other hand, are those that are offered either in order to differentiate the basic services or those services that are offered by enhancing the basic services to become more application-specific to a customer segment[14].

For the purpose of this study[15], a *telecom service*[16] will be characterized by

- basic services (i.e. cable-television services[17], telephony services[18], mobile[19] services and satellite services)
- value-added services (i.e. systems maintenance related activities, systems integration software services, clearing house/billing services, directory/publishing services, integrated services digital network (ISDN services) / frame relay / ATM, electronic and voice mail, videotex / telemarketing, electronic data interchange (EDI) / video conferencing, managed data network services (MDNS), virtual private networks (VPN), outsourcing and consultancy services).

By adding new basic and new value-added services to an existing set of services (see for instance Penrose, 1959) the base of services offered is increased, i.e. *service diversification*. The degree of service diversity in general may be increased in several ways. This includes,

---

[12] The concept of value-added services is used in the telco industry in a similar way, even though it actually lacks a distinct definition there and certainly needs to be clarified.

[13] This definition of basic services is, thus, supplier-oriented. However, it is not really equal to the traditional definitions of basic services (see, for instance, Langley, 1986), which usually define basic services as the service provided to all subscribers in return for a basic rental and/or call fees. Anyway, the concept of basic services is still used here, until a more coherent definition is found, because it is fairly broadly used in the industry.

[14] This definition of value-added services is, thus, customer-oriented. Strictly speaking, however, basic services may also add value to a customer. Similarly to basic services, the concept of value-added services is used until a more useful definition is found, because it is such a broad concept which is commonly used in the industry.

[15] This definition of the concept is the conception of the author which has evolved through discussions and interviews (approx. 30) at Telia.

[16] These definitions need to be improved and consequently more research effort ought to be assigned to them (see, for instance, Mansell, 1988).

[17] Cable television (CATV) is considered a telecom service since it is becoming a strategic part of the telecom services network (trials are already being made in the UK) even though it is usually still a one-way (broadcast) communication of television. CATV services are offered on a separate network and have properties which make the network a substitute for other networks and consequently a substitute for basic services, such as telephony services.

[18] Telephony services are those that are created in the traditional telecommunication network, i.e. traditionally a switched or packet-switched analog signal (usually voice) which is transmitted. Any datacom service is considered a value-added service as it usually requires some kind of compilation in addition to the basic transmission.

[19] Most mobile services are telecom services in terms of having the characteristics of two-way communication. Paging forms, however, a special case because it is based on one-way (broadcast) communication, just as the original CATV networks are. The CATV networks have, however, options to include two-way communication which paging networks do not have today. Paging is, however, a substitute for some other mobile services in certain applications and therefore it is included here.

according to Penrose (1959:109), "increasing the variety of final products produced, increases in vertical integration and increases in the number of basic areas of production".

In the telco industry, we have similarly identified three general ways of service diversification. First, the scope of basic services could be increased, e.g. adding cellular or satellite services to the basic service telephony. Second, the scope of value-added services may be increased, e.g. adding advanced MDNS capabilities or VPN. These two possibilities of service diversification could be realized either within the existing management and technological base or outside this base. In the latter case, a new generic service is established and new adjacent management or technological knowledge has to be acquired. Third, vertical integration from or to value-added services increases the base of offered services. These new services may naturally also be established within existing fields of competence or outside the knowledge base.

The *degree of foreign*[20] *service diversity* is, in this paper, simply defined as the number of foreign services (with respect to the services included in the definition above of a telecom service) which a telco offers. This definition ought not to affect the validity of the concept, even though the actual degree of diversity is dependent on the classification of services. In order to consider the level of investment in foreign services, the *level of activity* for each telco has been subjectively assessed on a three-level scale (high, medium or low). The assessment is based on the number of countries where telcos offer foreign services, but most importantly on the actual or estimated relative (within basic and value-added services, respectively) resource investments.

3. METHODOLOGY

In this study, data have been collected mainly from the semi-monthly industry journal Communications Week International, but also from the consultancy firm Data Pro (company, country and service reports), from annual reports, from the other chapters dealing with foreign direct investments (e.g. Cave and Sharma, Dang-N'guyen and Pouillot, Elixmann and Schnöring, Kramer and NiShuilleabhain, Williams and Taylor) and from approximately 50 interviews conducted with top telco executives in Telia (approx. 30), with competitors[21] of Telia in Sweden (10) and with PTT Telecom (8), Unisource (2) and France Télécom (1). It has not been possible to make deeper case studies beyond these interviews within the existing time frame of this research. The collection of data has been concentrated on the last five years of international activities. A classification of the collected events in terms of telco, activity, country, time range and reference (with additional background data) has been used in order to create a database. This database has been the major source for the analysis below.

Due to the fact that large volumes of international activities only have been in evidence during the last five years (from 1988), it is reasonable to assume that the secondary data have a relatively high degree of validity, even though the articles might be distorted by telco announcements. Before this period of time, few major international activities took place in the telco industry. This means that the chosen concentration on the last five-year period is an appropriate time range. The two main sources (the industry journal and the reports from the

[20] Strictly speaking, this suggests diversification of foreign services to be a dimension of internationalization because it excludes diversification on the domestic market.

[21] These competitors are AT&T, BT, MCI, Transpac, GEIS, Tele2, Comvik, Nordic Tel and Dotcom.

consultancy firm) have been considered to cover enough of the international activities in the telco industry, since the purpose of the study is to identify, in an explorative way, and tentatively explain patterns of strategies, i.e. the process of internationalization and service diversification. Thus, the purpose is not to perform an exhaustive and exact market data collection. In addition, these two sources have different focuses, which makes them complementary as the industry journal has a focus on news (spots), while the reports have more of a long-term viewpoint (process). The 25 telcos in this paper have been selected with reference to turnover, geographical location[22] and international activities, which suggests that they represent the bulk of the international activities in the telco industry. Thus, the selected telcos have a large turnover in telecom services[23], they are located in the industrialized part of the world and they have had a fairly large number of international activities supporting the provision of telecom services.

It is reasonable to assume that the reliability of the sources is relatively high, as both the industry journal and the consultancy firm are (by telcos) considered as dominant and "up-to-date" concerning international activities in the telco industry. The data have briefly been tested against annual reports (1981-1991) and that of another industry journal (Telephony)[24] and the data were found to be accurate. In fact, other sources[25] of data for this kind of information have not been found to be usable. However, exact determination of points in time prior to the chosen period might be less reliable and the database is probably also incomplete[26] and, thus, not all activities by each telco are covered. Finally, additional data from more industry journals, more interviews and also scheduled questions to all major telcos would of course have made the database more reliable.

4. EMPIRICAL FINDINGS OF INTERNATIONAL ACTIVITIES

This section characterizes the general patterns of international activities in the telco industry. It outlines the international activities in accordance with the operationalizations given in section 2, and tries to separate the activities incorporating internationalization and diversification of foreign services.

4.1 Internationalization

The telcos are analyzed below in seven areas, according to the region in the world where they have their main activities: Europe (i.e. Western and Eastern Europe as well as the countries in

[22] Telcos are selected from the OECD countries (Belgium, Canada, France, Germany, Italy, the Netherlands, Spain, Sweden, the UK and the US (except from Switzerland because the Swiss PTT has not made any major international investments apart from the Unisource venture). Telcos from other important industrialized countries (Japan, Singapore and Australia) have, however, also been selected.

[23] This criteria explains why IBM, for instance, is not included in the study.

[24] Additional information from these other sources has not been found to the same extent as in the reports from the consultancy firm. Telephony is more American-oriented and the information in the annual reports concerning international activities is often very short and distorted.

[25] Examples are statistics from OECD, Telia etc.

[26] The database is continuously updated with new and additional information and is a part of the ongoing research project.

the former Soviet Union area); Middle East and Africa[27]; South East Asia (here including Japan and China); Australia (including New Zealand); North America (here including Mexico); and finally Latin America (here including South America and the Caribbean countries).

Global activities are found among some telcos with activities in *all* regions of the world[28]. These telcos have reached a relatively high level of internationalization (approx. 40 countries on average). In this study AT&T (the US-based telco and telecom manufacturer), Sprint (one of the long-distance carriers in the US, which was formed out of United Telecom), France Télécom and Cable & Wireless (the Hong Kong based British colonial telco[29]) will be recognized.

Sprint has various interests in several markets, with their global value-added services as a base. AT&T, on the other hand, is one of the world's largest companies and has a tradition as a manufacturer of telecommunications equipment of making foreign investments. France Télécom and Cable & Wireless are telcos with a history in the colonial countries France and the UK. France Télécom took care of the French foreign communications (in particular to Africa), often via satellite and submarine cables, while Cable & Wireless was responsible for the foreign communications of Britain, with its homebase in Hong Kong. France Télécom has activities also in Eastern Europe and Latin America. They seem to be most successful in Argentina (together with STET (Italy) and local investors) and in Mexico (together with South Western Bell and local investors).

Telcos *without* global activities generally have a lower level of internationalization (approx. 15-20 countries on average) than those with global activities. These telcos have, on the other hand, usually shown a major interest in one or several regions in the world. Below the presence of main actors in selected countries in each of these regions is described.

A large number of ***European*** activities are found for several telcos in the industry. In Table 4.1, it is shown that it was the large European countries which attracted the bulk of the interest in the late 1980s, i.e. France, Germany and the UK, even though the Netherlands and Belgium were also chosen rather early. Over the past years it is especially in the Eastern European countries that investments have been made.

The European telcos started major internationalization activities in the last few years of the 1980s. In Eastern Europe Telia[30] (Sweden) has made investments in, for example, Latvia, Estonia, Russia and Hungary. PTT Telecom (the Netherlands) has made some investments in Eastern Europe (e.g. in the Ukraine) and jointly with Telia formed Unisource (headquartered in the Netherlands) in 1991. Since 1991 DBP Telekom (Germany) has added new markets in the Commonwealth of Independent States (CIS), e.g. Russia and Ukraine (Elixmann and Schnöring, 1993). In the Western countries in Europe BT, C&W and France Télécom have been especially active among the European telcos over the last few years.

---

[27] There are, however, very few of the selected telcos that have activities in this part of the world apart from those with a colonial history, such as France Télécom.

[28] This means activities in all other six regions as defined above, i.e. the level of competition for the specific telco in one country is affected by the level of competition in most other markets (Sölvell, 1987). Consequently, it is not necessarily a telco with a "global service" that has a global coverage.

[29] C&W is classified as having global coverage even though they do not have any activities in Africa and the Middle East. In addition, they are treated as having the UK (where they also own 80% of Mercury, the second national carrier) as their country of origin even though they actually also have a base in Hong Kong.

[30] For more information about Telia, see Johansson and Granstrand (1993).

Table 4.1
Main actors with major internationalization (i.e. new countries) into European countries

Telco	I	<1986	1986-87	1988-1989	1990-1991	1992-1993
EUROPEAN: **BT**	30			France	Austria, Belgium, Czech, Denmark, Germany, Ireland, Italy, Netherlands, Spain, Sweden, Switzerland	
Cable & Wireless	50			Germany	France, Ireland, Italy, Poland, Russia, Spain, Sweden	Belgium, Germany, Latvia, Netherlands, Switzerland
DBP Telekom	20			Netherlands, UK	Austris, Belgium, Czech, France, Hungary, Poland, Romania, Russia, Slovakia, Switzerland	Belarus, Denmark, Estonia, Finland, Italy, Kazakhstan, Latvia, Lithuania, (Turkey)
France Télécom	35			Germany, Belgium, Italy, Switzerland, UK	Czech, Denmark, Finland, Germany, Hungary, Ireland, Luxemburg, Netherlands, Norway, Poland, Portugal, Romania, Russia, Sweden	Canada, Greece, Spain
PTT Telecom	10			Germany	Belgium, Czech, France, Sweden, UK	Bulgaria, Ukraine
Telia	10				Denmark, Estonia, Finland, France, Latvia, Netherlands, Norway, UK	Hungary, Russia, Switzerland
RBOCs: **Ameritech**	20			UK	Austria, Denmark, France, Germany, Italy, Netherlands, Poland, Spain, Switzerland	Norway
Bell Atlantic	15			Austria, France, Germany, Italy, Netherl, Switzerl, UK	Belgium, Czech, Russia, Spain	Ireland, Norway, Sweden
Bell South	20		France, UK	Switzerland	Belgium, Denmark, Germany, Netherlands, Poland	
NYNEX	20		France, UK	Gibraltar, Netherlands	Belgium, Czech, Germany	
Pacific Telesis	10			Germany, UK	Portugal	Belgium, Denmark, Spain, Sweden
US West	20			Czech, France, Hungary, Netherlands, UK	Belgium, Bulgaria, Germany, Lithuania, Malta, Norway, Poland, Romania, Russia, Sweden, Ukraine	
US LONG-DIST: **AT&T (services)**	40	UK		Ireland, Russia	Austria, Belgium, France, Germany, Italy, Poland, Netherlands, Russia, Spain, Sweden, Switzerland, UK	Denmark, Ireland, Kazakhstan, Portugal, Ukraine
Sprint	35			Germany, Netherlands, Switzerland, UK	Austria, Belgium, Denmark, Finland, France, Ireland, Italy, Norway, Portugal, Russia, Spain, Sweden	Poland

Legend: I=Level of internationalization, approximately rounded to closest five.
Note: The telcos in the table are selected from the 25 telcos included in the analysis as being the ones that are present in more than five foreign countries in Europe.

Several of the regional Bell operating companies (RBOCs) have also made significant investments in Europe: Bell Atlantic has several ventures in the Czech republic since 1990, even though they seem to concentrate on Western Europe; NYNEX has made investments mainly in the UK and France since 1987; US West was rather early (1989) with activities in the CIS countries (Ukraine and Russia) and Eastern Europe (e.g. Hungary, Poland, Romania,

Lithuania, Bulgaria and the Czech republic), even though they started in the UK in 1987; Pacific Telesis has investments in Germany, the UK, and more recently (since 1991) in Belgium, Denmark, Portugal and Sweden; Bell South has activities in e.g. France, the UK and Denmark.

AT&T and Sprint have shown a major European interest in the UK even though they are present in several other West European countries as well. AT&T also has investments in Kazakhstan, Ukraine and Russia.

In the last half of the 1980s, it was particularly Japan who caught the interest of the main actors in the ***SE Asian region***. Apart from KDD, Singapore Telecom and Telstra who belong to the region, it is France Télécom, the large British (BT and C&W) and American telcos (primarily the long-distance carriers AT&T, MCI and Sprint) who have been active in the period considered in Table 4.2.

Table 4.2
Main actors with major internationalization (i.e. new countries) into SE Asian and Australian countries

Telco	I	<1986	1986-87	1988-1989	1990-1991	1992-1993
SE ASIAN: **KDD**	20			South Korea, Singapore	Australia, China, Hong Kong, India, Malaysia, N Korea	Cambodia
Singapore Telecom	15			Japan	Australia, Hong Kong, Malaysia, Mauritius, Sri Lanka, Thailand	China, Vietnam
Telstra	20		Vietnam		Cambodia, Hong Kong, Japan, Laos, Malaysia, New Zealand, Pakistan, the Philippines, Thailand	India, Indonesia, South Korea, Singapore, Taiwan
***EUROPEAN*:** **BT**	30		Japan		Australia, Hong Kong, Thailand	Indonesia, Malaysia, New Zealand, the Philippines, South Korea, Singapore, Taiwan
Cable & Wireless	50	Hong Kong	Japan		Australia, China, South Korea	
France Télécom	35		Japan	China, Indonesia, Singapore	Australia, Bangladesh, Cameroon, S Korea, Thailand	Hong Kong, India
***RBOCs*:** **Bell Atlantic**	15				Indonesia, New Zealand, S Korea, Taiwan	Japan
Bell South	20		Australia	China	India, New Zealand	India
NYNEX	20			China, S Korea	Australia, Hong Kong, Indonesia, Japan, the Philippines, Singapore, Taiwan	Thailand
***US LONG-DIST*:AT&T (services)**	40			Hong Kong, Japan, S Korea	Australia, China, Indonesia, New Zealand, the Philippines, Singapore, Taiwan	
MCI	60			New Zealand	China, Japan	Australia
Sprint	35		Japan	S Korea	Australia, Cambodia, Guam, Hong Kong, Malaysia, Singapore, Taiwan, Thailand	China, Indonesia, Sri Lanka

Legend: I=Level of internationalization, approximately rounded to closest five.
Note: The telcos in the table are selected from the 25 telcos included in the analysis as being the ones that are present in more than three foreign countries in South East Asia.

The telcos from the SE Asian region have invested in most countries in the region, also smaller ones, showing no obvious geographical starting point. Most of the other telcos seem, however, to have started to invest in the late 1980s in one of the main countries in the region, i.e. Hong Kong, Japan and/or Singapore. From any of these countries, they (except for MCI) seem then to have made further investments in smaller countries. There are no major differences among the European and the US long-distance telcos. In addition, these telcos have over the last years engaged themselves in more and smaller countries in the region. Among the RBOCs, however, Bell Atlantic and NYNEX seem not to have started their activities in one of the main countries, but in one of the smaller countries.

In the ***Australian region*** there are a number of RBOCs who have made substantial investments (see Table 4.2). Ameritech, Bell Atlantic (both in 1990) and MCI (from 1989) have invested in New Zealand, while Bell South, Cable & Wireless (both in 1991) and MCI (1992) have investments in Australia. Bell South was one of the first (1987) telcos to engage in Australia.

In ***Latin America*** (or rather the Caribbean), Cable & Wireless has operated the local telecommunication networks in the former British colonies (see Table 4.3). In the last few years of the 1980s and the first years of the 1990s, a large number of licenses were issued in South America, e.g. in Argentina, Chile and Venezuela. In these countries, a large number of activities are found for the European telcos Telefónica (Spain) and STET (Italy). Telefónica has activities in Argentina, Chile and Venezuela, while STET has investments in Argentina, Brazil and Chile. Among the RBOCs, it is especially Bell South who has obtained cellular licenses in Chile, Argentina and Uruguay, while the US long-distance telcos also have several investments in the region. Examples of these are AT&T in Colombia, Venezuela and Brazil, and Sprint in Argentina and Brazil.

Apart from the North American telcos, it is particularly BT, Cable & Wireless, France Télécom and NTT who have been active in the ***North American*** region since the mid-1980s. In particular, it is BT who has made investments in several telecom services in the US and Canada (Williams and Taylor, 1993). France Télécom has a heavy investment in Mexico together with SW Bell since 1990. NTT also seems to have a large number of activities in the US. They have, for example, a subsidiary (since 1984) with related activities. Moreover, most telcos[31] have some activities in the US and, finally, the US telcos did not make any major efforts to enter Mexico[32] or Canada[33] up until 1990.

[31] In addition to the ones mentioned above they are, for example, DBP Telekom, KDD and Singapore Telecom.

[32] These US telcos are especially SW Bell, Bell South, Bell Atlantic, AT&T, MCI and Sprint.

[33] For example, GTE, SW Bell, AT&T, MCI and Sprint.

Table 4.3
Main actors with major internationalization (i.e. new countries) into Latin American countries

Telco	I	<1986	1986-87	1988-1989	1990-1991	1992-1993
***EUROPEAN*: Cable & Wireless**	50	Caribbean (16!)				
France Télécom	35			Mexico, Venezuela	Argentina, Brazil	
Telefónica	10			Chile	Argentina, Venezuela	Puerto Rico
STET	15				Argentina, Brazil, Chile	
***RBOCs:* Bell South**	20			Argentina, Puerto Rico	Mexico, Chile, Uruguay, Venezuela	
GTE	15			Brazil	Mexico, Venezuela, Dom. Rep.	
***US LONG-DIST*: AT&T (services)**	40			Mexico, Columbia	Brazil, Venezuela	
MCI	60				Mexico	Brazil, Colombia, Venezuela
Sprint	35				Brazil, Mexico, Guatemala, Puerto Rico	Mexico, Argentina, Venezuela

Legend: I=Level of internationalization, approximately rounded to closest five.
Note: The telcos in the table are selected from the 25 telcos included in the analysis as being the ones that are present in more than two foreign countries in Latin America.

4.1.1 Diversification of foreign telecom services

In this subsection, the diversification of foreign telecom services and the level of activity within each telecom service are presented. The description is carried out in accordance with the definitions of the telecom service concept outlined in section two.

There are several telcos who have a heavy foreign involvement in ***basic services***. Generally these telcos are local service providers in their respective domestic market. Their average level[34] of internationalization is approximately 20, which indicates that they have a lower level than those telcos that have a relatively higher number of value-added services. ***Value-added services***, on the other hand, have attracted the interest of several telcos, and several alliances have been formed during the last few years in order to provide these foreign services. Below, in Table 4.4, the process of diversification of foreign services for some main telcos is presented.

[34] The level of internationalization is simply the number of countries in which the telcos have activities.

Table 4.4
Main telcos' diversification (i.e. new services) of *foreign* telecom services

Telco	<1986	1986-1987	1988-1989	1990-1991	1992-1993
***EUROPEAN*:** **BT**		B:Sat, Tel VA:EDI	B:Mob VA:Mail, MDNS	VA:Con,ISDN, VPN, Out	VA:Sysint
Cable & Wireless	B:Tel		B:Mob	B:Sat VA:Con, EDI, ISDN, Mail, MDNS, VPN	B:CATV
DBP Telekom			B: Mob VA:EDI, ISDN, Mail	B: Sat VA: Out, Vidtex	B: Tel VA: MDNS
France Télécom	B: Sat	VA: Vidtex	VA: EDI, ISDN, Mail, MDNS	B: Mob, Tel VA: Con, Out, Sysint, VPN	
PTT Telecom		B: Sat	VA:ISDN	VA: Con, EDI, Mail, VPN	B:Mob, Tel VA: MDNS
Telefónica			B: Tel	VA: EDI, Mail	B: Mob
Telia				B: Mob, Tel VA: EDI, Mail, VPN	VA: MDNS
***RBOCs*:** **Ameritech**		B: Mob	VA: Dir, Mail	B: Tel VA: Sysint, Sysmain	
Bell Atlantic			VA: EDI, Mail, Sysint, Sysmain	B: Mob, Tel VA: Con	VA: VPN
Bell South		B: Mob VA: Vidtex	VA: Con	B:Sat, Tel	
NYNEX		VA:Sysmain	B: Tel, Mob VA:Con, EDI, Mail	B: CATV VA:Dir, Sysint, Vidtex, VPN	
Pacific Telesis		B: Mob,Tel VA: Mail	B: CATV		
SW Bell	B: CATV VA: Dir		VA: Con	B: Mob, Tel	
US West			B: CATV,Tel	B: Mob VA: Con, EDI, Mail	B Mob:
***US LONG-DIST*:** **AT&T**		B:Sat VA:EDI,Mail	VA:EDI, Mail, ISDN, VPN	B: Tel VA:Con, Out, Sysint, Vidtex	B: Mob, Tel
MCI			B:Tel VA:Mail	B:Sat VA:VPN, Mail, EDI, Con	VA:FR
Sprint		VA:Mail, EDI	VA:MDNS	B: Sat VA: VPN	B:Tel VA:Con
***SE ASIAN*:** **KDD**			VA: EDI,ISDN, Mail	B:Sat VA: Con	B:Tel
NTT			VA: Con	VA: MDNS, Out	B: Tel
Singapore Telecom		B: Sat	VA:EDI, ISDN, Mail	B: Tel, Mob	B:CATV VA: VPN
Telstra		B: Sat	B:Tel	B:Mob VA: Dir, EDI, Con, Mail	VA: Out

Legend:
B=Basic: *CATV*=cable television, *Tel*=telephony services, *Mob*=mobile services, *Sat*=satellite services
VA=Value-added: *Sysmain*=systems maintenance related activities, *Sysint*=systems integration software services, *Clear*=clearing house/billing services, *Dir*=directory/publishing services, *ISDN* = integrated services digital network, *FR*=frame relay services, *ATM*=asynchronous transfer mode services, *Mail*=electronic and/or voice mail, *Vidtex*=videotex/telemarketing, *EDI*=electronic data interchange (or videoconferencing), *MDNS*=managed data network services, *VPN*=virtual private networks, *Out*=Outsourcing, *Con*=consultancy services.
Note: Diversification is here defined as the adding of new foreign services to an existing set of foreign services.

Considering **basic services**, there are some patterns of diversification of foreign services which are possible to identify with reference to Table 4.4. One group of telcos has *started with mobile services and then diversified into telephony services.* Ameritech, Bell South and Telia belong to this group. Bell South has most evidently concentrated its effort on mobile, but also Telia seems to have a similar interest[35] with their investments in, for example, the Baltic states (1991-1992) and Hungary (1993). After the initial cellular license, Telia did also receive a license (through a joint venture) to operate telephony services in Estonia (1991) and in 1993 Telia got a license for international voice services to and from the UK. Bell South has made several cellular investments in South America (1988-1991), Europe (1989-1993), India (1992) and in the Australian region (1987-1993) where they also have obtained a license as the second national operator. Ameritech operates the cellular network of the British Virgin Islands since 1987 and acquired the PTT in New Zealand (together with Bell Atlantic) in 1990.

Some other telcos, mainly Cable & Wireless, Pacific Telesis and Telefónica, have instead *started with telephony services and then diversified into mobile services.* Cable & Wireless has a large number of activities in (domestic) telephony services, due to having been the telco for the British colonial empire. This means that they have somewhat specialized to some degree in voice telephony services in almost all countries in the Caribbean and Hong Kong. These telephony licenses have often also been followed by cellular licenses. More recently, Cable & Wireless has also made investments in telephony services in Australia, Russia and Sweden (all three in 1991). Over the last years, however, they have also received several GSM licenses in, for instance, Australia, Hong Kong (both in 1992) and Latvia (1993). Pacific Telesis, on the other hand, started their foreign activities with an investment in the international carrier JDC (Japan) in 1986 but have subsequently conentrated their investments on cellular in Germany (1989), Chile, Portugal (1991), Sweden (1993), Denmark (1993) and Japan. They also have paging licenses in Thailand (1991) and Portugal (1992). Telefónica has concentrated on telephony services in Latin America (1990-1992) but recently obtained a paging license also in Portugal (with Marconi in 1992).

Five telcos have *started with satellite services and later diversified into telephony services.* These telcos are AT&T, Telstra and two European telcos (France Télécom, DBP Telekom and PTT Telecom). AT&T established satellite services to, for instance, the UK and Germany rather early and has then invested in telephony services in Venezuela (1991), Kazakhstan and Ukraine (both in 1992), and in Canada (1993). Telstra started with an investment in a satellite joint venture in Vietnam in 1986 and later on invested in telephony services in, for instance, Saudi Arabia (1989) along with international telephony services in Russia and Kazakhstan

[35] However, in 1991 they formed the Unisource venture with PTT Telecom in order to provide value-added services.

(both in 1992). France Télécom has offered satellite services between the French colonies since the 1960s, and more recently (from 1990) has turned its emphasis to new ventures in mobile and telephony services. DBP Telekom withdrew from some mobile ventures in the UK (1989 and 1990) and actually started with satellite services within Europe in 1990. Over the last few years, DBP Telekom has concentrated on the former East Germany, providing telephony services in Belarus (1992) and Ukraine (1992). PTT Telecom started satellite services after the mid-1980s, but has in recent years concentrated on telephony services in Bulgaria and Ukraine (both in 1992).

Within basic services there is a final group which has *started with CATV and then diversified into mobile or telephony services.* SW Bell has been operating CATV in Israel since 1985 and has subsequently started to invest in CATV in France (in the late 1980s) and the UK (1989). SW Bell has made its largest investment in Mexico (1991), where it operates telephony and mobile services with France Télécom. US West started with CATV in Hong Kong and Hungary in the late 1980s and has later also invested in several CATV franchises in the UK (since 1990), France (1991) and the Nordic countries (through United Telecom). In parallel to these investments, US West has made investments in mobile (Czech, Hungary (both in 1989), the UK, Russia (both in 1990) and Japan (1991)) and telephony services (Czech (1990), Lithuania, Russia and Ukraine (all three in 1991)).

Among the **value-added services**, there are also some main patterns of diversification of foreign services that may be identified in Table 4.4. The main behavior is to *start with mail and EDI services and then diversify into more advanced datacom services* such as VPN, ISDN and MDNS. There are several telcos having shown this behavior. Most obvious are the behaviors of AT&T, BT and Sprint. All three of them started early (mid-1980s) by establishing mail and EDI services, and they are now offering a broad set of value-added services. AT&T acquired the British firm Istel in 1989, providing datacom services on their Accunet. Sprint has expanded their US based Telenet into other parts of the world and has formed SprintNet, which could be the largest data network in the world. BT has been active in this arena since 1989 when they acquired Tymnet, a US VANS provider. In 1991 they led an outsourcing project called "Pathfinder" with DBP Telekom and NTT. The deal was never closed, however, and BT started Syncordia, an outsourcing company, on their own. The base for Syncordia is the Tymnet network (now called GNS). In addition, BT recently bought a 20% stake in MCI initiating a close cooperation. BT has also acquired a private datacom network in Spain.

Other telcos with roughly the same behavior are Cable & Wireless, DBP Telekom, KDD, MCI, PTT Telecom, Singapore Telecom and Telia. Cable & Wireless and Telia were late starters (1990/91), while the others have usually started by offering international ISDN or VPN connections (in the late 1980s) along with mail and EDI services. Cable & Wireless has tried to establish a global digital highway. They have also been the leading actor in fiber cable projects in several parts of the world (especially in the Caribbean and South East Asia), but seem to have realized that they are too small to maintain this position on their own. Telia formed the Unisource venture with PTT Telecom in 1991 in order to provide value-added services in Europe. KDD has traditionally had several international services and has also made several fiber cable investments through alliances in South East Asia. MCI bought a 25% stake in Infonet[36] (which however will be disposed of because of the deal with BT) and has also

[36] Infonet is a global value-added network services provider, originally an American company which was jointly bought by a number of PTTs.

participated in several cable projects in the Atlantic and the Pacific. In addition, MCI has started a close cooperation with the Canadian telco Stentor (1992) for the establishment of an international intelligent network. DBP Telekom has also acquired a stake (16%) in Infonet (1990), and has several international value-added ventures together with France Télécom, such as Eucom and Eunetcom, even though their commitment in these ventures is questioned (Elixmann and Schnöring, 1993). Finally, Singapore Telecom ought to be mentioned as a telco which has increased its international activities over the last few years, not only in value-added services but also in mobile communications.

A special case is France Télécom, who started early with Videotex services but now offers the same value-added services as the large group above. France Télécom has also, through its subsidiary Transpac, acquired a majority of the shares in domestic data network operators in, for example, the US (1990), the UK, Italy (both in 1991), Sweden, Luxembourg (both in 1992), and a minority of the shares in Germany (1992) and Switzerland (1993) for instance. Another special case is telcos (Pacific Telesis, Telefónica and US West) who only offer mail and EDI services at this stage.

There is another group of telcos who have *started either with systems integration software and maintenance software or with directory/publishing.* Ameritech started with directory/publishing in Japan in the late 1980s, but has later on moved into software in, for instance, Chile, Colombia and Venezuela. Bell Atlantic has especially concentrated on systems maintenance in several European countries after it acquired Sorbus in 1988, a French-based computer maintenance company which is now based in the UK. NYNEX, on the other hand, acquired the UK-based BIS group in 1987, an international information technology services company. BIS was at that time present in 13 countries, and is the largest provider of applications software for the US financial services industry. The takeover of BIS has meant that NYNEX has short-term consultancy contracts in several countries in Europe and South East Asia. NYNEX also offers mail and EDI services today.

The identified behavior above of the six large groups of telcos does not take into account the ***level of activity***, either for a certain object or for a whole telco. In order to give an idea of the level of activity achieved by telcos in various services, Table 4.5 shows the level of activity for each telco on a three-level scale (high, medium or low) according to the definition of level of activity in section 2.

Within *basic services,* some European telcos and the RBOCs, especially Bell South, C&W and Pacific Telesis, have a deep involvement. Bell South and Pacific Telesis have concentrated on mobile services. Pacific Telesis seems to have been successful in receiving GSM licenses, while Bell South has been successful in receiving licenses especially in South America. Moreover, Cable & Wireless has heavy investments as being the telco for the British colonies, especially in the Caribbean islands.

In particular, Cable & Wireless shows a high diversity of foreign basic services. Cable & Wireless is present in all basic services. Examples of other telcos who have diversified to a somewhat lower extent are Ameritech, BCE and Pacific Telesis. Bell South and Pacific Telesis, both focusing on cellular, also have interests in telephony services, while Bell South has satellite services and Pacific Telesis cable television services as their third basic service. Telefónica and Telia have both made investments in telephony services and cellular services, but with a different focus.

Most telcos who provide the whole range of services on their home market also have some *value-added services* beside their basic services. Value-added services on foreign markets have attracted a high level of activity for several telcos as seen in Table 4.5.

Table 4.5
Level of activity of telcos in basic and value-added services, respectively, in 1993

	Basic services	Mobile communication service	Voice telephony services	Cable TV service	Satellite service	**Value-added services**	Systems integration software	Systems maintenance	Clearing house/billing	Directory/publishing	Consulting	ISDN/Frame Relay/ATM	Voice-/Electronic mail	Videotex/Telemarketing	EDI/videoconference	Managed data network service (MDNS)	Virtual private network (VPN)	Outsourcing
RBOCs																		
Ameritech		Low	Medium				Low	Low		Low			Medium					
Bell Atlantic		Low	Medium	Low			Low	Medium			Low							
Bell South		High	Low		Low			Low		Low	Low							
NYNEX		Low	Low	Medium			Medium			Low	Medium			Low			Low	
Pacific Telesis		High	Low	Low														
US West		Medium		Medium							Low							
South Western Bell		Low	Medium	Medium			Low			Low								
(BCE)		Low	Medium	Low				Low										
(GTE)		Low	Low		Medium												Low	
US Long-distance																		
AT&T			Low		Low		Low	Low		Low	Low	High	High	Low	High	Low	Medium	Low
MCI			Low		Medium						Low		High		Medium		Medium	
Sprint			Low		Low						Low		Medium		Medium	High	High	
European																		
Belgacom					Low						Low	Low	Low		Low			
BT		Low	Low		Low		Low	Low			Low	Medium	Medium		Medium	High	Medium	High
Cable & Wireless		Medium	High	Low	Low						Low	High	Low			High	Medium	Low
DBP Telekom		Low	Medium		Low							Medium	Low		Low			Low
France Télécom		Low	Medium		Low		Low	Low			Medium	High	Medium	Medium	High	Low	Medium	Low
PTT Telekom		Low	Low		Low						Low	Low	Low		Low		Low	
STET		Low	Low				Low						Low			Low		
Telefónica		Low	Medium									Low	Low		Low	Low		
Telia		Medium	Low		Low						Low	Low	Low		Low		Low	
South East Asian																		
KDD			Low		Low							Medium	Medium		High	Medium		
NTT		Low	Low								Medium		Low			Low		
Singapore Telecom		Low	Low	Low	Low							Medium	Low		Low		Low	
Telstra		Low	Low		Low					Low	Low		Low		Low			

Legend: *High* level of activity within each class of services (basic and value-added).
Medium level of activity within each class of services (basic and value-added).
Low level of activity within each class of services (basic and value-added).

Note: This is an assessment (by the author) of the level of activity which is done primarily with respect to the resources invested but also with respect to the number of countries where the telcos are present.

It is especially the US long-distance telcos AT&T, MCI and Sprint, and the European telcos BT, Cable & Wireless and France Télécom, as well as KDD, who have concentrated on various value-added services on foreign markets. They all show a rather high diversity of these services, although they have concentrated on different value-added services.

Thus, France Télécom and Cable & Wireless seem to have the highest diversity of foreign services among the selected 25 telcos.

Finally, in this subsection, we will penetrate the conventional wisdom in the telco industry of size of telcos being linked to the strategy pursued. An example is that telcos are traditionally assumed to have to be large in order to have a global focus and provide international value-added services. This is however not a clear case. When the average turnover of telcos in 1991 was calculated for each class of major market or service involvement, no major differences were found that supported this hypothesis. Exceptions are, however, SE Asia (relatively lower average turnover) and North America (relatively higher average turnover). In Table 4.6 the calculations are presented.

Table 4.6
The average turnover of major services/markets

Major services Major markets	**Basic services**	**Value-added services**	**Av. turnover in each region 1991**
Global	FT(21425*), C&W(4977)	AT&T(38805) Sprint (8780), FT (21425*), C&W (4977)	16731
Europe	Telia(6225), BellAtlantic (12280), BellS (14446), DBP Telekom (27169) USW(10577), PacTel(9895)	Bell Atlantic (12280), PTT Telecom(6043), MCI(8433), NYNEX(13229)	12058
SE Asia	Telstra(8648)	KDD(1637*),NYNEX(13229) Singapore Telecom (1300)	6204
Australia	Ameritech(10818), Bell S(14446)	Bell Atlantic(12280)	12515
Latin America	Telefónica(10643), STET(n.a)		10643
North America	NTT(44337), SW Bell(9332)	BT(25248)	26306
Av. turnover in 91	14500	12897	

Legend: ()=turnover in 1991 in MUSD (Local currencies converted using the average exchange rate in 1991). *=turnover in 1990. FT=France Télécom, C&W=Cable & Wireless, BellS=Bell South, PacTel=Pacific Telesis, USW=US West, Singa.=Singapore. Av.=average

4.1.2 Modes of entry by telcos on foreign markets

Telecom services are network-embodied and may be delivered either through the telcos' own establishments (with a corresponding contractual arrangement) or through some contractual agreements with other establishments. In the latter case, the telco uses, for instance, another telco's cable, leasing capacity on that cable. Thus, it is important to note that the telecom service is location-bound (Dunning, 1989:36), i.e. its delivery is dependent on the underlying equipment[37].

[37] This characteristic might be questioned by saying that radio transmission makes mobile services independent of the underlying equipment and, thus, those services would not be location-bound. However, radio transmission is still limited to a certain area (e.g. a cell).

Roberts and Berry (1985) propose classification of entry modes on foreign markets as well as new products and/or markets. Although they do not explicitly consider technology-based or network-embodied products, their strategies or modes of entry into new businesses seem to be applicable for describing the modes of entry by telcos on foreign markets (often entry with old telecom services).

The entry strategies or modes may, according to Roberts and Berry, take the form of internal development, acquisitions, licensing, internal ventures, joint ventures or alliances, venture capital or nurturing and educational acquisitions. Several of these modes, except[38] for venture capital or nurturing, have been noticed in the telco industry over the last decade. In the following paragraph we will discuss these alternative entry modes in the telco industry.

Internal development is traditionally taken to mean exploitation of internal resources. In this paper we will, more specifically, mean exploitation of internal resources by establishing offices or subsidiaries abroad. These offices or subsidiaries may work with R&D, marketing and distribution with no obvious relation to a specific new business (internal ventures, see below). They are majority-owned and thus internal. Most telcos in the industry have established foreign offices and subsidiaries in most industrialized countries during the last decade. Larger telcos generally have more foreign offices and subsidiaries while smaller telcos have fewer. However, there are some smaller telcos, such as Cable & Wireless, who also have a large number of foreign offices because of their international business orientation.

Acquisitions or takeovers of foreign firms are always linked to a business opportunity in a certain telecom service. Examples of acquisitions in the telco industry are: AT&T acquiring[39] Istel (datacom); BT acquiring Tymnet (datacom) and Mitel (PBX manufacturer) which they later sold; France Télécom Transpac acquiring Info AG (datacom) for instance; South Western Bell acquiring cable television companies in the UK; Ameritech acquiring Tigon (datacom) and directory publishing firms in Germany and Austria; NYNEX acquiring the BIS group and Perthcrest in the UK (consulting and software) etc.

Licensing is generally taken in the telco industry to mean the regulation of the number of competitors on the market, even though licensing is also used in the traditional sense of acquiring new technologies. Another common approach in the telco industry is to *lease* capacity on a competitor's network. This contractual arrangement may be similar to licensing in (which gives a right to use proprietary knowledge in a special technology) since the capacity leased is an essential part of a telco's business.

Internal ventures are intended to give clear and flexible rules for a new product or market to be developed within a separate entity within the existing company. This mode should be distinguished from internal development (with offices and subsidiaries) which has a more general approach. In the telco industry there are few examples of internal ventures, since most ventures are carried out jointly with other telcos in order to share risks. If the telcos are doing business abroad alone, it is usually through an acquisition or through a "compulsory"[40] consortium who tries to get a license for a specific telecom service.

---

[38] Roberts & Berry (1985) studied small technology-based firms and found venture capital and nurturing significant for those firms. This is however not the case in the telco industry, which consists of large (technology-based) firms.

[39] They also recently aquired McCaw (cellular) in the US for some USD 12 billion.

[40] Usually it is in fact necessary to join forces with, for instance, a local company or another telco in order to look more stable or simply to have the right political connections. This seems to be the case especially in cellular.

Joint ventures or alliances are the most common entry strategy in the telco industry. Usually it is necessary to join forces for various reasons as mentioned above, even though it might be risky due to potential conflicts among the joining parties. Alliances sometimes include swapping of shares, while joint ventures rather include the creation of a third and jointly owned firm. Typically, there are joint ventures or jointly owned companies for country-specific licenses or foreign domestic services, while alliances are typically formed in order to enlarge the network coverage of telcos. Some examples of joint venture-like activities in the telco industry include the following: developing countries (in Latin America, Eastern Europe etc.) as well as New Zealand are selling off most of their national carrier services to foreign interests. GSM licenses are offered in an increasing number of countries, and satellite ventures are found in VSAT and mobile satellite. Joint ventures are also found in directory publishing services, like the yellow pages. In addition, joint ventures are most often formed for investments in new fiber cables. Over the last decade, some twenty to fifty fiber (submarine) cables have been planned and most often carried out. These cables are usually used in order to provide international value-added services. Examples of alliances are: BT and MCI who joined forces in 1993; Unisource (PTT Telecom, Telia and forthcoming also Swiss PTT) in 1991; Worldsource (AT&T, KDD, Telstra and Singapore Telecom) since mid-1993; and in late 1993 the proposed alliance between AT&T, France Télécom and DBP Telekom. Examples of other alliances include FNA, GEN, GNP, ABFK, Pacific Partners etc.

Educational acquisitions are an internal form of venture capital since they are used for more or less the same purpose. However, this mode requires a higher degree of internal skills and usually also a higher level of financial commitment. Still several telcos have been found to use this entry strategy, "shooting at everything that moves", since it might teach them something—"trial-and-error". This is especially true for larger telcos like France Télécom, BT and several of the RBOCs, who have a strong financial position through their (domestic) monopoly. Examples are the investments in national carriers in several developing countries like the former Soviet Union and the investments by the RBOCs in cable television in the UK.

5. DISCUSSION

In this section the empirical findings are discussed and analyzed. The section summarizes the perceived behavior of internationalization and diversification of foreign services in a set of patterns.

Basic drivers of ***internationalization*** which influence international activities are (as pointed out in section 2): changed customer demand, liberalization of telecom services and technological development in networks and services.

Generally, the telcos show a behavior of either making entries into a major country in a region after which they make investments in minor adjacent countries, or making direct investments in any of those minor countries. The importance of a country relates to economic activity such as trade etc., but more importantly to the fact that large customers —multinational companies (MNCs)—usually have a major portion of their activities in one of those major countries.

The main behavior of the selected 25 telcos is to initially (in the internationalization process) *enter one of these major countries,* such as the UK, France and Germany in Europe, Hong Kong, Japan, Singapore in SE Asia and the US in North America. A closer look at Europe reveals that telcos (also the European ones) have started their internationalization into

Europe by investing in one of these three main countries. An interesting and important point here is that neither France nor Germany are regarded as liberalized countries. In SE Asia, a similar behavior is found for most of the studied telcos. The behavior suggests, thus, that most telcos have started their internationalization by following their customers (large MNCs) abroad. However, after the initial regional investments in one of the major countries, telcos tend to internationalize, by spreading to several other countries within the same region. This behavior is valid in Europe and SE Asia.

As an example of the "follow-your-customer" behavior, AT&T and Sprint both seem to have followed their customers abroad from the US. Examples of this include their investments in countries like the UK, Germany, Japan, Hong Kong etc. Most of the RBOCs have also followed their customers, usually to Europe, even though the competitive situation of the RBOCs might be different from that of their customers. BT and NTT have also shown a similar behavior, towards the US.

Another case in point is that, within Europe, there are several examples of European telcos, such as Telia, PTT Telecom and DBP Telekom (Eastern Europe), who have focused their international activities on countries in the "home region". Similarly, KDD, Singapore Telecom and Telstra also focus on their home region. This could be a reflection of the focus of their customers, but it might also reflect some kind of uncertainty with respect to geographical and/or cultural distance (which of course might be reduced by following their customers abroad). This "psychic distance" has been discussed by, for instance, Johansson and Wiedersheim-Paul (1975) as an important factor in the internationalization of the manufacturing industry.

In the telco industry, Europe and the US are, for example, favorable for the American and European telcos, respectively, probably because of limited cultural barriers. Another distinct example of low cultural barriers is Telefónica who has made several investments in Spanish-speaking countries such as Argentina, Chile and Venezuela. Moreover, the behavior of the telcos with a huge global involvement is often accompanied by several submarine fiber cable projects. There seems to be a tendency, by these telcos, to take positions in projects which will gradually expand their geographical coverage. The focus of AT&T and Sprint and others like Cable & Wireless and France Télécom, also has strong historical links. Cable & Wireless as well as France Télécom were both colonial telcos for the UK and France. They have both provided telegraph and telex links during the whole century. Now they are using these sites as strategic hubs for a successive geographical expansion towards a global network.

The other pattern of *investing directly in one of these minor countries* is the case especially in Latin America, where telcos have invested in any of the countries irregularly. However, in Latin America there is no country which has the same importance as any of the countries mentioned above. In this case it is rather the opportunities arising from liberalization that have determined the investment decisions.

In another respect, the foreign expansion of several of the US telcos is also a result of the regulatory situation in the US. Their investments abroad are triggered by their being prohibited by law to enter new markets inside the US. However, most important for most telcos in the world is the liberalization on foreign markets, which has created a race to obtain foreign licenses for operation. This is especially valid for Europe where the new cellular standard GSM has been launched with the requirement for two competing operators, most often implying market entries by foreign telcos. Liberalization or the absence of regulation in value-added services has also been a platform for the establishment of international value-added services, such as electronic mail and virtual private networks. In addition, liberalization on foreign markets may form an opportunity for "trial-and-error" before starting similar operations on the home market. The most obvious examples are the cable television projects by US West and SW Bell in the UK aiming at telephony services.

However, apart from the opportunities opened up by liberalization, there are several windows of opportunity that are triggered by other factors than liberalization, even though liberalization of course results in new opportunities in the first place. Market opportunities caused by liberalization will, however, decrease in importance over time, which makes windows of opportunity a more general case. These windows of opportunity on liberalized markets are usually a part of an industrialized and growing region in the world, such as the EU region. Opportunities are most often found in datacom services, typically value-added data networks, and in mobile communication services. These entry areas have traditionally been the only opportunity for telcos to learn from competition. Furthermore, telcos seeking these opportunities would probably show a more aggressive behavior in comparison to those who simply focus on markets within their psychic distance.

The third important determinant is the technological development in networks and services. It seems especially important to emphasize the digitization of networks as a determining factor behind several international activities, but also as a triggering factor for an increased pace of liberalization (France and Germany for instance). Actually it could be questioned whether the internationalization that has taken place until now would have been possible without digitization of networks, which has made it possible to inter-connect and separate telcos and networks. A closer look at the point in time, when the bulk of international activities started to appear, reveals that it was in the late 1980s. As hypothesized in Johansson and Granstrand (1993), this point of time might reflect a critical degree of digitization (in Sweden 20-30% of the central switches) which made it economically interesting to enter new markets. Moreover, the technological development in services has increased the capabilities of especially advanced (and application-specific) datacom services such as VPN, MDNS etc.

Service diversification shows a more ambiguous pattern than internationalization in general. The patterns shown in Figures 5.1 and 5.2 suggest that when the degree[41] of foreign service diversity is low, the telcos have either their main interest in basic services or no major interest at all. This pattern suggests that telcos either start to provide services on foreign markets by providing these basic services or perform a great deal of "trial-and-error". The activities by the RBOCs, for instance, show such an involvement in basic services. Some of these telcos seem to specialize in certain services, such as Bell South in mobile communication services. This suggests that the management and technology base of the firm and the services provided on the "home market" would influence the decisions on what services to focus on, even though these decisions often also reflect the actual foreign opportunities that arise.

When the scope of foreign services is medium (or on average) the telcos show no major pattern of behavior—any main service involvement might be found. This could mean that they have found what is success and what is failure. On the other hand, it could reflect a random result of trial-and-error with a broad set of services.

A high level of diversification of foreign services by telcos seems to indicate that they have their major interest in value-added services or no major involvement at all. This suggests that telcos concentrating on providing value-added services do not have these services as their

[41] The level is here defined as the number of telecom services (breadth) a telco provides. The telecom services used are the same as those presented in section 2.2, i.e. a maximum of 16 telecom services.

single business, but as complementary[42] to their basic service business. This also suggests that when telcos have a high level of diversification of foreign services and no major involvement, they also do a great deal of "trial-and-error". A concentration on value-added services is seen mainly for the US long-distance telcos and for some large European ones, which suggests that the technology base of telcos and their financial position on the "home market" reflect their international concentration.

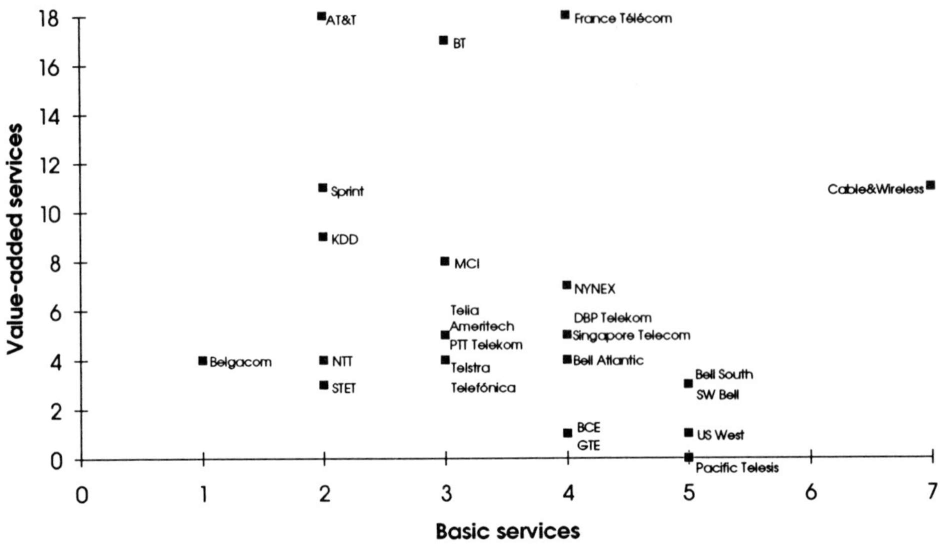

Figure 5.1
Level of activity in *foreign* basic and value-added services

Note: The figures of basic and value-added services are simply the sum of the three-level-scale assessments (made in Table 4.5) of the level of activity for each telco. Figure 5.1 above is intended to show the different concentrations by telcos as of 1993, not the exact number of basic or value-added services.

The identified patterns of internationalization and diversification (including both basic and value-added services) of foreign services are illustrated in Figure 5.2. Referring to the discussion above, it is reasonable to assume that a low level of internationalization is characterized by a low or medium level of diversification of foreign services. Telcos with mainly basic services generally show a lower level (number of countries) of internationalization than telcos with mainly value-added services. Higher (or more medium) levels of internationalization also seem to be characterized by a low or medium level of diversification of foreign services. Thus, as shown in Figure 5.2 below, telcos have a tendency to specialize[43] in few services, economizing on those in new countries.

---

[42] There are, however, also firms such as GEIS and IBM who offer (value-added) telecom services, only in the form of value-added services.

[43] All telcos are present in a number of countries equal to or greater than the number of services they offer.

On the other hand, telcos with mainly value-added services generally have a higher level of internationalization (sometimes a global involvement as in the case of AT&T) than those with mainly basic services. Thus, considering the discussion above, it is reasonable to suggest that a high level of internationalization will be characterized by a medium or high level of diversification of foreign services. However, those telcos still provide few services on several markets.

Most telcos offer approximately six foreign services in approximately 15 countries. MCI, Cable & Wireless, AT&T, Sprint, France Télécom and BT are most internationalized, i.e. they have activities in more than 30 countries. In terms of diversification of foreign services it is AT&T, France Télécom, BT and Cable & Wireless who have diversified most in the industry. Thus, AT&T, France Télécom, BT and Cable & Wireless have invested in several foreign services (mostly value-added) in a large number of countries. An example of the opposite is Pacific Telesis who concentrates on a few services (only basic), offering these in approximately 10 countries.

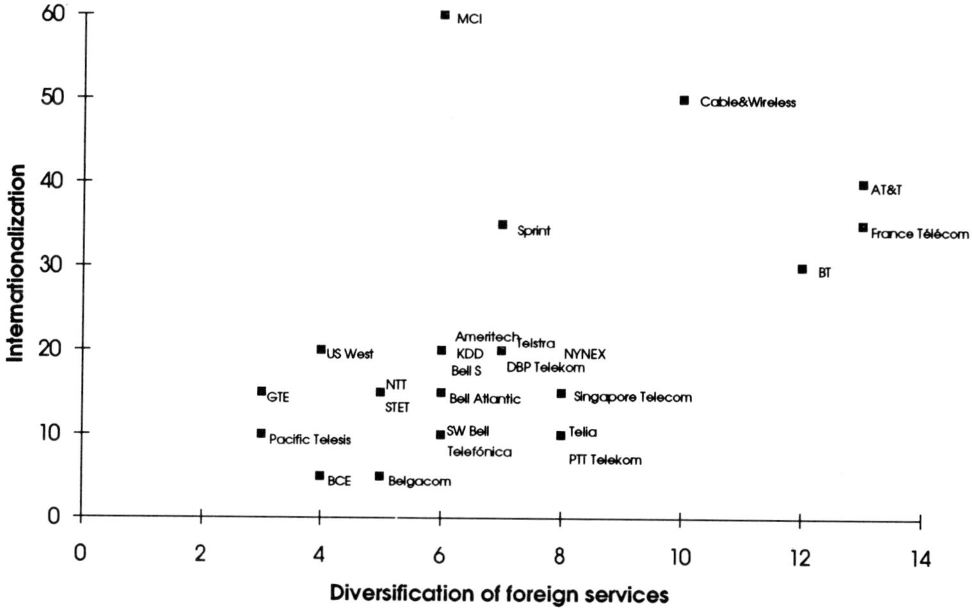

Figure 5.2
Internationalization and diversification of foreign telecom services

Note: Internationalization is the number of countries where each telco has been found to have activities. Diversification of foreign services is the number of services offered by telcos (according to the classification in section 2). The figure is intended to show different strategic behavior as of 1993, not the exact number of countries or services for each telco.

Finally in this section, the modes of entry and the question of size will be discussed. The characteristics of telecom services regulation have an impact on the choice of modes of entry. Basic services, for instance, are traditionally controlled through a number of licenses, particularly in mobile services. Usually these licenses are applied for by a joint venture in order to create a strong offer in the eyes of the regulatory body. In addition, it is not unusual to include a local company (not necessarily a telco) in the joint venture. Acquisitions seem, however, to be rare for telcos focusing on basic services. This is understandable because the basic services are protected by national laws. Acquisitions seem, on the other hand, to be more common among telcos, their main interest being value-added services. Several of these telcos have acquired their international network and their international offices.

Regarding the question of size, there is, at this stage, no clear evidence of value-added services being predominantly focused on by large telcos nor basic services favored by small telcos, or vice versa. In addition, there is no clear evidence supporting market focus to be linked with size. Thus, at this stage in the telco industry, there are small as well as large telcos investing and offering telecom services at certain markets independently of size. If major involvements in certain services or markets should correspond to strategy, and turnover correspond to size, small telcos would have equal possibilities of providing global services and have the same global coverage as larger ones. Of course smaller telcos, because of their relatively weaker financial strength, ought to have a greater interest in forming an alliance or a joint venture than larger telcos (see for example KDD and Singapore Telecom allying with AT&T in Worldsource and Telia, PTT Telecom and the Swiss PTT forming Unisource). Related hypotheses are that no telco is large enough to have a global focus without an alliance, and that no telco has the possibility of establishing these value-added services with a global focus within an appropriate time range. These issues of various forms of cooperation and modes of entry will be discussed in the next subsection.

Internal development seems to be equally used among all types of major involvements, even though many smaller telcos seem to focus on setting up offices and subsidiaries in countries which are of high priority for their customers' export, in order to provide them with value-added services.

6. CONCLUSIONS

This paper is part of a larger research effort focusing on technology-based telecom services. The data in the study are based on early events, which means that only some modest conclusions can be drawn.

Contractual arrangements are an essential and necessary part, especially if the foreign business is virtual private networks (usually established through alliances) in global markets or cellular services (usually established through joint ventures). Moreover, contractual arrangements, because of financial limitations and time constraints, form an incentive for cooperation in addition to the incentives arising from the interconnection of different telecommunication networks. Thus, the management of cooperations and competition becomes a key issue.

The main pattern of internationalization of the 25 large telcos in this study is characterized by an early entry into one of the major markets in a region in the late 1980s, i.e. to follow the large multinational customer abroad independently of the liberalization process in the targeted country. The main facilitator of those foreign investments seems to be the technological development in networks (digitization) and services allowing for circumventing the traditional solutions by the national PTT and initializing licenses for new services. A rather large group of telcos also seems to be bounded by uncertainty, reflecting a cultural and geographical distance—psychic distance. Another, not equally distinct, pattern of internationalization is to internationalize into any country in a region wherever liberalization may open up a window of opportunity. This behavior is especially found as a second stage of telcos who have followed their large multinational customers, but also among telcos investing directly in regions, such as Latin America.

Telcos tend to specialize in a few foreign services and/or pursue a broad set of activities—"trial-and-error". In early stages, the major involvement seems to be basic services, especially for telcos (such as the RBOCs and smaller European telcos) who offer such services with a local orientation in their home markets, even though trial-and-error with several services is also a common behavior here. Pacific Telesis provides a clear example of a concentration on basic services, while at the same time not being very internationalized yet. In later stages of internationalization, however, the major involvement seems generally to be value-added services on a more global market, offered by telcos (such as the US long-distance telcos, KDD and BT) who have a heavy international orientation and/or a strong financial situation in their home markets. Generally, these telcos (such as AT&T and BT) have acquired a large user or a datacom services firm in order to obtain their international network. Thus, it seems reasonable to hypothesize that the technology base of telcos in combination with the financial situation on the "home market" is essential as an explanatory variable for both historical and future strategic behavior.

Moreover, in spite of economies of scale, the data suggest no obvious link between size and major market or service involvement, i.e. both large (e.g. AT&T) and small telcos (e.g. Sprint) may show a global focus with international value-added services, when measured by the number of countries connected to the (global) network.

REFERENCES

Barras, R., 1986, Towards a theory of innovation in services, Research Policy, Elsevier Science Publishers B.V. (North-Holland), pp. 161-173.

Bohlin, E. and O. Granstrand, 1994, Introduction and overview, The race to European eminence in telecommunication, Elsevier Science Publishers B.V. (North-Holland).

Buckley, P.J. and M.C. Casson, 1981, The optimal timing of a foreign direct investment, Economic Journal, 91, March, pp. 75-87.

Cantwell, J., 1989, Technological innovation and multinational corporations (Basil Blackwell, Oxford).

Carman, J.M. and E. Langeard, 1980, Growth strategies for service firms, Strategic Management Journal, Vol. 1, pp. 7-22.

Casson, M.C., 1990, Enterprise and competitiveness: a systems view of international business (Clarendon Press, Oxford).

Cave, M. and Y. Sharma 1993, Foreign entry and competition for local telecommunications after the duopoly review, International Telecommunication Society conference, Stenungsbaden, Sweden, June 20-22, 1993.

Dang-N'guyen, G. and D. Pouillot, 1993, Foreign direct investment in telecommunications in and out of France, International Telecommunication Society conference, Stenungsbaden, Sweden, June 20-22, 1993.

Dunning, J.H., 1989, Transnational corporations and the growth of services: some conceptual and theoretical issues, UNCTC Current Studies, Ser. A, No. 9, United Nations, New York, 1989.

Dunning, J.H., 1991, The internationalization of the production of services: some general and specific explanations, GSM Working Paper, 91-11, Graduate School of Management, Rutgers University, New Jersey, 1991.

EC (European Commision of the European Communities), 1993, Communication to the Council and European Parliament in the consultation on the review of the situation in the Telecommunications Services Sector, COM (93) 159 Final, Brussels, 1993.

Elixmann, D. and T. Schnöring, 1993, Internationalization of the German telecommunications service market and strategic behavior of DBP Telekom, International Telecommunication Society conference, Stenungsbaden, Sweden, June 20-22, 1993.

Granstrand, O. and O. Johansson, 1993, Internationalization of Telcos—an early analysis of behavior and strategies, in M. Christoffersen and A. Henten, eds., The proceedings of the 7th Conference on European Communications Policy Research, EURO CPR'92.

Granstrand, O. and S. Sjölander, 1990, Managing innovation in multi-technology corporations, Research Policy, Vol. 19, No. 1, Feb, pp. 35-60.

Grönros, C., 1982, Strategic management and marketing in the service sector, Helsinki, Svenska Handelshögskolan.

Johanson, J. and F. Wiedersheim-Paul, 1975, The internationalization of the firm—four Swedish cases, The Journal of Management Studies, Oct., pp. 305-322.

Johansson, O. and O. Granstrand, 1993, Internationalization of the Swedish telecom services market, International Telecommunication Society conference, Stenungsbaden, Sweden, June 20-22, 1993.

Kramer, R. and A. NiShuilleabhain 1993, Cable & Wireless: services, investments and prospects, Internationalization of the Swedish telecom services market, International Telecommunication Society conference, Stenungsbaden, Sweden, June 20-22, 1993.

Langley, G., 1986, Dictionary, second edition (Telephony Publishing Corp., Chicago).

Levitan, S.A., 1985, Services and long-term structural change, Economic Impact.

Levitt, T., 1986, The marketing imagination (The Free Press, New York).

Mansell, R., 1988, Telecommunication network-based services: Regulation and market structure in transition, Telecommunications Policy, September, pp. 243-255.

Normann, R., 1984, Service management: strategy and leadership in service businesses (Wiley, Chichester).

Oscarsson, C. and N. Sjöberg, 1991, Technology analysis and competitive strategy—the case of mobile telephones, in C. Oscarsson, Technology diversification—the phenomenon, its causes and effects, Dissertation, Industrial Management and Economics, Chalmers University of Technology, Göteborg, Sweden, 1993.

Penrose, E.T., 1959, The theory of the growth of the firm (Basil Blackwell, Oxford).

Porter, M. J., 1986, Competition in global industries: a conceptual framework, in M.J. Porter, ed., Competition in global industries (Harvard Business School Press, Boston).

Prahalad, C.K. and G. Hamel, 1990, The core competence of the corporation, Harvard Business Review, May-June, pp. 79-91.

Roberts, E.B. and C.A. Berry, C.A., 1985, Entering new businesses: selecting strategies for success, Sloan Management Review, Spring, pp. 3-15.

Saviotti, P.P. and J.S. Metcalfe, 1984, A theoretical approach to the construction of technological output indicators, Research Policy, 13, pp. 141-151.

Segal-Horn, S. and J. Mc Gee, 1991, Services in turbulent times—internationalisation and globalisation for service companies, 17th EIBA conference, Copenhagen, Denmark.

Sölvell, Ö., 1987, Barriers to entry and foreign penetration, Dissertation, Stockholm School of Economics IIB, Stockholm, Sweden.

Williams, H. and J. Taylor, 1993, Competencies and diversification: the strategic management of BT since 1984, International Telecommunication Society conference, Stenungsbaden, Sweden, June 20-22, 1993.

The Race to European Eminence
E. Bohlin and O. Granstrand (Editors)

Summary

Erik Bohlin

Chalmers University of Technology, Department of Industrial Management and Economics, S-412 96 Göteborg, Sweden

1. INTRODUCTION

This volume represents an effort to deal with the emerging and increasingly important phenomenon of internationalization and foreign direct investments (FDIs) in the telecommunications service sector with particular attention to Europe. It offers perspectives on many of the critical issues involved, in-depth accounts and wide-ranging analyses of the internationalization moves and impacts on a selection of 14 European markets and telecommunications service providers.

The volume has been structured into two main sets of chapters, excepting the introductory overview and the final comparisons and conclusions. The first main set of chapters focuses on foreign entries and inward FDIs in a country market and on foreign activities and outward FDIs by telecom service providers with the country concerned as a home market, typically involving the traditional PTT. The second set of chapters raises various issues and challenges for regulation, policy and management across countries and actors, posed by the current industrial restructuring and internationalization.

Table 1.1 provides a summarizing overview of all the contributions in order of appearance, leaving the editorial ones out. In the following, the table headings and their relationship to the chapters are briefly discussed.

2. MAIN FOCUS

Regarding general focus, all chapters deal with internationalization of telecommunications although the chosen facet of the problem complex differs as well as the means to advance our understanding of the matter. In terms of general scope, the opening chapter by Kurisaki and the final one by Johansson merit attention. Kurisaki offers a broad overview of internationalization options and strategies complemented with a general discussion on the policy implications. Johansson provides an empirical analysis of the internationalization behavior of 25 telcos, in which he finds several patterns that aid our understanding of the internationalization process. In terms of regions analyzed, Johansson provides a truly global overview of FDI and internationalization.

Table 1.1
Chapter summary

Chapter Title Author(s)	Main focus	Main area(s) & actor(s) covered	Statistics related to foreign direct investments	Main findings & arguments	Outlooks, speculations & implications
2 Challenge to the Globalization of Public Telecommunications Operators—Toward a New Framework of Telecommunications Policy Y. Kurisaki	Provides an overview of the main internationalization activities of telcos and their policy implications.	OECD countries	No	Internationalization of telcos creates competitive pressures also in countries receiving FDI, forcing domestic policy and regulation to become more liberalized. Furthermore, policy makers need to develop a common, international framework to handle globalization impacts.	Internationalization of telcos will increase the speed of domestic liberalization.
3 Foreign Entry and Competition for Local Telecommunications in the UK after the Duopoly Review M. Cave Y. Sharma	Analysis of possible implications of developing local access competition, driven by changing technologies and costs, regulatory initiatives and foreign entry into CATV.	UK BT Mercury RBOCs	Inward: Ownership figures and market shares of CATV networks (cf. Table 5.1).	After the duopoly review, the scope of competition for local access has significantly widened, spurred by increasing investments in CATV and wireless. Different tariff systems for local access have been introduced that serve to differentiate the options to fit appropriate customer categories.	It is likely that BT will prefer price cutting to loss of domestic market share although the loss may still be significant over the next years. The UK will develop into a fruitful test-bed on the viability of local access competition.

4 Competition in the British Telephony Market G. Dang-N'guyen D. Phan	Delineation of BT's evolution of its international and domestic strategy, the competitive responses and the regulatory moves, particularly with reference to competition in the local loop.	UK BT Mercury (C&W)	Figures on BT's foreign acquisitions (section 1.2).	BT's international strategy has evolved from trial-and-error into a more focused multinationalization strategy. Regulatory and government pressures have initially been soft, giving BT leeway to learn (at great cost) from its international ventures and to dominate and manipulate the home market. However, regulatory pressures have increased, forcing BT to cut manpower and to face competition in the local loop.	The UK deregulation experience will continue to strongly affect policies in the European Commission.
5 Competencies and Diversification: The Strategic Management of BT since 1984 H. Williams J. Taylor	Critical analysis of BT's internationalization moves and evaluation of BT's competence base for market and service diversification.	UK BT	Various figures on BT's outward investments in manufacturing, PTO services, telematic services and network management (cf. section 4).	Although BT has remained true to its goal to become global, the strategies and alliances to accomplish this have been unstable and lacking apparent synergies in terms of technologies and markets.	The volatility of and search for strategic relationships among telcos resemble the past restructuring process of the telecom manufacturing industry, suggesting similarities in industrial evolution and restructuring.
6 Cable & Wireless: Services, Investments and Prospects R. Kramer A. NiShuilleabhain	A review of the range of investments, businesses and strategies, across the C&W group.	Europe Asia/Pacific Region, North America Carribbean	Various quite detailed figures of C&W's foreign activities throughout the paper.	C&W's business structure is an unusual and unique mix of local monopolies, new entrants into previous monopoly markets and global networks.	The C&W group is vulnerable to: a) takeover since its diversified portfolio lacks strong synergies and probably is undervalued b) currency fluctuations c) political shifts in the Asia/Pacific Region.

Table 1.1 (cont.)
Chapter summary

Chapter Title Author(s)	Main focus	Main area(s) & actor(s) covered	Statistics related to foreign direct investments	Main findings & arguments	Outlooks, speculations & implications
7 Internationalization of the German Telecommunications Service Market and Strategic Behavior of DBP Telekom D. Elixmann Th. Schnöring	In-depth account of the German telecommunications service market, foreign entry into the same and DBP Telekom's emerging foreign ventures.	Germany DBP Telekom	Inward: Extensive key data on foreign entrants (cf. section 2, Tables 2.2, 2.3). Outward: Detailed account of DBP Telekom (cf. Table 4.3).	The German market is in a transition period becoming increasingly open to foreign entry and competition. However, the regulatory reform of DBP Telekom is not sufficiently thorough for full-fledged international operations even though DBP Telekom has made several international moves, despite the huge managerial requirements posed by the German reunification.	Competition in the German market will be very intensive due to liberalization and the market potential. Achieving the goal to become a global player will be increasingly difficult for DBP Telekom within the current regulatory constraints and managerial caution.
8 Foreign Direct Investments in Telecommunications In and Out of France G. Dang-N'guyen D. Pouillot	Assessment of France Telecom's foreign activities and FDI inward to France.	France France Telecom	Outward: Various figures on investment and ownership of FT throughout the paper; Transpac (cf. Table 3.1).	France Telecom has pursued a broad internationalization strategy primarily involving multi-domestic operations but also "trial and error" and "learning by doing". France Telecom's strengths revolve around its traditional relationships with developing countries and the technical and management expertise accrued during its rapid domestic network expansion in 1975-80 and the introduction of new services in the 1980s.	France Telecom will focus its strategy, possibly establish vertical ties with Alcatel after privatization and deepen its alliance with DBT.

9 The International Economics of Tele-communications Services in Italy C. Antonelli A. Geuna	Account of the evolution of the complex structure of Italian tele-communications providers and manu-facturers together with a review of the fast, recent internationalization process of IRI-STET and foreign entry to Italy.	Italy IRI-STET Group	Outward ventures of STET (cf. Table 6.2, section 6.2). Key data on inward foreign entrants (cf. Table 6.3, section 6.3).	The complex telecommunications structure of Italy is being transformed due to technical change, liberalization movements and globalization. The IRI-STET Group has pursued a process of internationalization that aims to establish domestic operations in many countries. On the other hand, FDI inward to Italy seems mainly to be connected with foreign telcos pursuing a strategy that aims to establish global and proprietary services, particularly related to data communications using leased lines.	The Italian market will be divided into wired voice telecommunications, data and cellular communica-tions, the latter two dominated by foreign telcos. The former area will also be pursued by IRI-STET through out-ward FDI.
10 Internationalization of Telecommunica-tions Service Provi-sion and the Greek Privatization Debate Y. Caloghirou T. Darmaros	Assessment of the 1990-1993 developments in Greek telecommunica-tions, particularly the privatization and lib-eralization policies and ensuing debate.	Greece OTE	Inward: wireless (cf. section 4.3); privati-zation value (cf. Table 4.3).	Privatization of OTE, including take-over by foreign investors, has failed because of lack of political consensus, insufficient regulatory framework and a political short-term financial objective. Furthermore, the emerging internationalization process was not properly taken into account in the political debate.	A more evolutionary procedure to achieve privatization is needed, as well as forming a broad consensus with clear ob-jectives, not only for domestic ventures but also for the international ones.

Table 1.1 (cont.)
Chapter summary

Chapter Title Author(s)	Main focus	Main area(s) & actor(s) covered	Statistics related to foreign direct investments	Main findings & arguments	Outlooks, speculations & implications
11 Internationalization and Foreign Investment in Telecommunications Services in Portugal D. Lucena C. Seabra	Account of the prospects of internationalization and privatization of Portuguese telephone operations and of foreign entry into the Portugal market.	Portugal Marconi TLP Telecom de Portugal	Inward: wireless and VAS (cf. Table 3.1). Outward: ventures of TLP, Telecom and Marconi (cf. section 4).	Telecommunications in Portugal does not move rapidly with inward or outward FDI, excepting Marconi's investments in Latin America, former colonies and Asia. Instead, the state-owned telephone operators are focused on internal restructuring and preparing for privatization. However, current policy emphasizes national ownership which in turn will delay privatization.	The pace of change will continue to be slow although Marconi is expected to continue its pursuit of foreign ventures.
12 A Note on the Internationalization of the Netherlands O. Johansson	The recent internationalization process of PTT Telecom and the Netherlands is put in perspective by briefly analyzing structural properties of the Netherlands and PTT Telecom's strategies.	The Netherlands PTT Telecom	A few scattered data (sections 3 and 4).	Since the Netherlands is a small, dense and highly internationalized economy, PTT Telecom is vulnerable to foreign entry. Therefore, PTT Telecom has built international alliances and competencies to properly manage the new era.	The Netherlands will become increasingly attractive to foreign entry due to liberalization.

13 Internationalization of the Swedish Telecom Services Market O. Johansson O. Granstrand	Survey of the state of internationalization of the Swedish tele-services market, both in terms of inward and outward activities and in assessment of network-based barriers of entry.	Sweden Telia, the Kinnevik Group	Inward: key data on new entrants (cf. Table 3.1) Outward: Alliances and joint ventures of Telia and Kinnevik (cf. Table 4.1).	The recent inward internationalization process has evolved in two main stages, the first being characterized by isolated niche entries and the second by multiple and sequential trial entries, aiming for an increasingly wider scope and market coverage. Both Telia and the Kinnevik Group have many international ventures.	In analogy with the "establishment chain model", internationalization of telecom services will intensify, increasingly involving activities that were previously limited to the respective home bases.
14 Internationalisation of the Norwegian Telecommunication Market R. Brekke S. Jensen C. Riis	Account of the performance, structure and regulatory context in the light of the increasing internationalization of Norwegian telecommunications.	Norway Norwegian Telecom	Inward: CATV network ownership (cf. Table 4.1), key data on new entrants (cf. Table 5.1). Outward: Various ventures (cf. section 6).	Norway is still in a rather docile position in terms of outward and inward FDI. However, Norwegian Telecom is vulnerable to foreign entry and competition in outgoing telecommunications due to tariffs and high concentration of large users, both geographically and numerically. The strengths of Norwegian Telecom pertain mainly to mobile and satellite communications.	As liberalization and foreign entry proceed, Norwegian Telecom needs to form alliances with larger players to ensure long-term sustainability.
15 Inward and Outward FDI in Danish Telecommunications A. Henten K.E. Skouby	Account of the history, and recent structural, regulatory and strategic changes that facilitate especially outward FDI of Tele Denmark.	Denmark Tele Denmark	Ownership figures of inward and outward investments (cf. sections 3 and 4).	Tele Denmark is regionally focused on East Europe with technology/services in specialized areas (e.g. mobile, VANS). Copenhagen aims to become a telecommunications hub for the East-West transit traffic.	Internationalization will, especially for small operators, in general mainly involve focus on weaker regions with growth potentials and specialized services to particular market segments.

Table 1.1 (cont.)
Chapter summary

Chapter Title Author(s)	Main focus	Main area(s) & actor(s) covered	Statistics related to foreign direct investments	Main findings & arguments	Outlooks, speculations & implications
16 The Internationalization of Finnish Telecommunications: Motivations, Strategies and Activities J. Karpakka	An assessment of the Finnish prospects of inward and outward FDI in telecommunication services.	Finland TeleGroup Finland, Finland Telecom	Employees and number of offices of inward and outward FDI (cf. section 6). Ownership shares and alliances of outward FDI (cf. section 8).	So far, FDIs inward to Finland are scant despite an open market. However, both main actors, TeleGroup Finland and Finland Telecom, have pursued internationalization with increasing aggressiveness, at the same time as they are developing into domestic competitors.	Domestic competition will be increasingly fierce and outward FDI will expand. It is likely that inward FDI will mainly concern ownership and joint ventures rather than investments in the Finnish physical network.
17 Transformation of Poland's Telecommunications and Foreign Investment A.T. Jarmoszko	In-depth account of past and on-going structural and policy changes of telecommunications.	Poland TP SA	Inward: foreign funding (cf. Table 1.1); various network construction projects (cf. section 4); equipment production (cf. section 5).	The transition process towards liberalized telecommunications is complex and characterized by institutional pluralism and close political attention, particularly to broadcasting but also to telecommunications. Furthermore, an increasingly cautious approach toward market forces has been adopted and bottlenecks relating to international communications have been rapidly reduced.	The transformation from a politically administered system into a market place of telecommunications will take time, despite quite radical measures taken.
18 Foreign Investment in Hungarian Telecommunications F. Kiss	In-depth assessment of the prospects of foreign investment in Hungarian telecommunications, with particular attention to the legal-regulatory framework recently developed.	Hungary Hungarian Telecom (HTC)	Various figures on Inward: local telephone companies (cf. Table 6.1); wireless (cf. section 7); paging (cf. section 8); professional services (cf. section 9).	Despite positive prospects for investment in Hungarian telecommunications, actual foreign investment has been small in volume, mainly due to delays in establishing a clear legal-regulatory framework.	Given the recent emergence of the legal-regulatory framework, foreign investments will increase, spurred by the planned privatization of HTC and concessions in the wired and wireless network.

19 Options for the Reduction of Deficiencies in Telecommunications in the Czech Republic J. Cas	Analysis of the options available for rebuilding the Czech infrastructure as well as those of Central and East Europe (CEE) in general, with special emphasis on the mutual relationships between regulation and investments (including FDI).	Czech Republic SPT	Total foreign funding requirements (cf. section 3.2).	The needs of expanding the network and the deployment of digital technology necessitate the raising of capital, more than 4 BUSD of which 1 BUSD must be external (including foreign funding by international financing organizations, supplier credits but also financing from national sources such as banks and privatizations). The capital requirements pose several challenges for pricing and investment returns.	Liberalization and conformity with EC regulations can only come slowly since cross-subsidies and artificial rates must be kept to achieve sufficient internal funding and quickly ensure universal service.
20 The Race to New International Mobile Satellite Services P. Bonanzinga F. Castelli C. Leporelli	Characterization of the competition and pre-paradigmatic phase of new mobile satellite services (MSS). Regulators and industrial policy makers are faced with challenges posed by new satellite technologies.	United States Europe Mobile satellite service providers (e.g. INMARSAT, IRIDIUM, GLOBALSTAR)	Total investment in new mobile satellite systems (cf. Table 2.1) of selected players.	A competitive race to seize first-mover advantages in the emerging MSS market necessitates a new global regulatory regime. This regime must properly handle spectrum sharing, R&D, standardization and IPR, market access regulation and network interconnection.	Three scenarios are envisaged: 1) Protectionism and market fragmentation 2) An internationally cooperative framework 3) Multilateral liberalization, favoring the development of multinational projects.
21 Direct Foreign Investments in Telecommunications: Comments E. Noam	Assessment of the political and regulatory conditions, barriers, and problems for internationalizing telcos.	Developed & developing countries	No	The prospects of FDI in the developing world are conditioned by nationalistic interests (economic sovereignty) and low returns. In developed countries, home regulators question international ventures, and will do so increasingly.	The pace of internationalization may slow down due to unfulfilled profit expectations, multijurisdictional tensions and political reactions, both in the developed and developing world.

Table 1.1 (cont.)
Chapter summary

Chapter Title Author(s)	Main focus	Main area(s) & actor(s) covered	Statistics related to foreign direct investments	Main findings & arguments	Outlooks, speculations & implications
22 International Competitive Dynamics: Implications for Domestic Regulators D. McCarty	Analyzing tension between domestic regulation vs internationalization, and the evolution of complex games between a multitude of domestic and international players.	Europe	No	Domestic regulators must implement cost-based tariffs, protect domestic ratepayers and competitors from monopoly power, and allow the national PTT to respond to an increasingly competitive environment.	There will be increasingly similar conflicts between EC and domestic regulation in Europe as between FCC and state regulation in the US.
23 Competing Visions for World Telecommunications: Europe and the Global Evolution of Industry Structure D. Allen	Delineation of policy options to ensure innovative network evolution in an increasingly competitive telecommunications environment.	Europe North America Asia BT MCI DBT	No	Policy that emphasizes consensus and integration alternated with competition is more conducive to desirable network evolution. Recent alliances could configure an industry structure that will maintain the tension necessary between fragmentation and integration.	A policy solely geared toward competition might eventually lead to reappearance of dominant players, now on a global scale.
24 The Internationalization of Telephone Operators: Survival in an Integrating World D.E. Ioannidis	Delineating strategic options for internationalizing telcos and raising issues posed by national interests and concerns.	National states	No	The internationalization process is conditioned by the national stakes and interests which historically have been an integral part of telecommunications.	Internationalization of telcos is likely to be followed by nationalistic reactions from the general public and politicians.

25 Asymmetries, Competition and Internationalization: Telecommunications Operators and the Future Single European Market P. Barbet L. Benzoni	Analysis of the natural monopoly concept and empirical proxies in the context of internationalizing telcos and regulation.	West Europe	No	Small countries will engage in alliances & mergers to appropriate economies of scale. Emergence of international oligopoly is conditioned by the speed of privatization and liberalization.	Alliances may be a first step toward mergers and FDI.
26 Reorganization of European Telcos: The Cases of British Telecom, France Telecom, Deutsche Telekom R. Pospischil	Description and analysis of recent organizational changes of the three largest European telcos.	British Telecom France Telecom Deutsche Telekom	No	Trend towards functional organizations in the case of British Telecom and Deutsche Telekom. Centralized strategic apex in the case of British Telecom and France Télécom. Different approaches to design a network organization. Three parts of a global network are identified: domestic, international and intranational.	Stronger organizational integration between selected carriers, not only for multi-domestic operations but also for the purely international traffic, will develop as carriers become increasingly internationalized.
27 Managing Network Investments: Illustrations from Telcos in Europe, Japan and the United States E. Bohlin	Organization of network investment planning in an increasingly internationalized environment.	AT&T BT France Telecom GTE New Common Carriers NYNEX NTT Telia	No	Most carriers have gone through at least one organizational change in recent years. Common trends are identified.	For the heavily internationalized telco, an investment process centred on the product dimension seems most viable.

Table 1.1 (cont.)
Chapter summary

Chapter Title Author(s)	Main focus	Main area(s) & actor(s) covered	Statistics related to foreign direct investments	Main findings & arguments	Outlooks, speculations & implications
28 Foreign Direct Investments of Telecommunications Operators in Less Developed Countries: Application to the Telefónica Experience J.J. Duran F. Gallardo	The development of a framework for the process of FDI in LDCs and the case of Telefonica as an illustration.	Latin America Telefonica	Outward: Telefonica's ownership shares in Argentina, Chile, Puerto Rico, Venezuela (cf. section 8).	Appraising FDI in LDCs is multidimensional and complex. Special attention must be devoted to country risk. For FDI carried out in a consortia, five critical success factors are probed.	Management of FDI will increasingly become a key strategic capability.
29 International Business Enterprises and Telecommunications: A User Perspective E. Roche	How multinational corporations are creating new market opportunities for telecom service vs providers.	Large telecommunication users of OECD origin. Multinational corporations. International telecom service providers.	No	While computer access is being decentralized in MNEs, data storage is increasingly centralized to corporate nodes and data bases, forcing the MNEs to rely heavily on telecommunications. In response to the strategic importance of telecommunication, various forms of outsourcing are offered by telcos to handle technical and contractual complexities.	MNE will concentrate data processing around a few regional nodes. As telecommunications become strategic for MNEs, the performance rather than price of the outsourced network becomes paramount. Five critical abilities of telcos aiming to win the MNEs are identified.

30 Internationalization and Diversification of Technology-Based Services: A Comparative Analysis of 25 Telecommunication Service Corporations O. Johansson	Analysis of internationalization and diversification strategies of 25 major telcos. Tentative explanations of the early pattern of internationalization and service diversification are given.	Developed and developing countries. Main telcos in OECD countries.	Ventures by main regions and services starting from the mid-1980s (cf. section 4).	The early internationalization phase is characterized by entry activities firstly into the main markets within a region (e.g. Europe), regardless of level of liberalization and secondly to adjacent country markets—here interpreted as 'follow your customer' strategy. Furthermore, early stages of internationalization seem to be connected with 'trial and error' in terms of services and countries invested in, while later stages are characterized by global services, particularly value added ones.	The 'home' technology, market and home competence base are lead indicators of future international activities. There seems to be no obvious link between size and global coverage of value added services.

As explained already in the Introductory chapter, the FDI chapters were written on the basis of a common set of contents and questions. They all share the general aim of describing and analyzing the situation of inward and outward FDI in telecom service provision in terms of European markets and actors active in Europe. Moreover, when it comes to methodology, the FDI chapters are fairly even in approach. Secondary material as well as interviews have been utilized in the chapters but no sophisticated statistical analysis has been performed. However, there are variations in emphasis and content, reflecting the particular country/actor dealt with. Examples of such differences are, firstly, that some chapters give an in-depth account of both the country market and the incumbent actor (Elixmann/Schnöring in particular—see also e.g. Antonelli/Geuna, Lucena and Seabra, Brekke et al., Johansson/Granstrand). There are other contributions that mainly focus on areas (e.g. Cas) while others focus on actors (e.g. Dang-N'guyen/Pouillot). Secondly, some chapters are primarily descriptive (e.g. Henten/Skouby, Karpakka), while others seek to develop a coherent argument (e.g. Dang-N'guyen/Phan, Williams/Taylor). Thirdly, some contributions pay particular attention to the regulatory/political conditions and frameworks (e.g. Kiss, Cave/Sharma, Caloghirou/Darmaros) while others probe the historical context (e.g. Jarmoszko, Antonelli/Geuna).

Broadly speaking, the issue papers deal with policy and management options in an internationalized era, both in the developed and the developing world. Several papers focus particularly on the tensions, clashes and interdependencies between national and international interests (Noam, McCarty, Ioannides, Barbet/Benzoni, Bonanzinga et al.). In particular, Bonanzinga et al. amply demonstrate the role of new technologies (new satellite technologies) to bring forth new internationalization patterns and attendant regulatory tensions. The management challenges related to investments in domestic and internationalized telecommunications of telcos are highlighted in Pospischil, Bohlin, and Duran/Gallardo while Roche presents the strategic options and investment considerations facing the telco seen from the perspective of the large users, the multinationals.

3. ACTORS AND MARKETS COVERED

By design, the European market and players are mainly covered in the FDI chapters but the issue chapters are not primarily confined to the European scene although Europe is dominant. Among the FDI chapters, the actor vs market analysis is somewhat varied. Dang-N'guyen/Phan, Dang-N'guyen/Pouillot, Kramer/NiShuilleabhain, Williams/Taylor are particularly centred on actor analysis (mainly BT, France Telecom, Cable & Wireless and BT, respectively). The other chapters are predominantly focused on the country regions, although actor strategies are covered to a considerable extent in some (e.g Elixmann/Schnöring, Lucena/Seabra).

Regarding non-telco actors, there is considerable analysis of a particular new entry group in Cave/Sharma, namely the CATV industry. Moreover, long-standing, purely international actors, such as INMARSAT and EUTELSAT, are analyzed with reference to new satellite technologies (cf. Bonanzinga et al.). A group of central importance is the large and multinational user, analyzed by Roche. Regulatory and policy-making institutions surface in some issue papers (e.g. McCarty) although considerable attention is also given to these in some FDI chapters (e.g. Jarmoszko, Kiss).

Several papers contain explicit multi-actor accounts and comparisons that go beyond tabulations. Examples of such contrasts include Dang-N'guyen/Phan (BT vs Mercury), Johansson/Granstrand (Telia vs Kinnevik), Karpakka (TeleGroup Finland vs Finland

Telecom) among the FDI chapters and Allen (BT, MCI, DBT), Pospischil (BT, FT, DBT) and Bohlin (AT&T and other telcos) among the issue chapters.

Focusing on the dimension of country markets, large vs small industrialized countries is a common distinction. Large countries have been analyzed by Dang-N'guyen/Pouillot (France), Elixmann/Schnöring (Germany), Antonelli/Geuna (Italy), those by Cave/Sharma and Dang-N'guyen/Phan (the UK). Small countries have been analyzed by the Nordic authors (Brekke et al., Johansson/Granstrand, Henten/Skouby, Karpakka), by Lucena/Seabra (Portugal) and Caloghirou/Darmaros (Greece).

The Central and East European countries fall into a special category as they are far less advanced than most other European countries in terms of telecommunications penetration and technology. However, as the chapters by Cas (the Czech Republic), Jarmoszko (Poland) and Kiss (Hungary) amply show, there are many opportunities for significant future market growth. In the set of issue-chapters, the developing countries are also discussed (by Noam, Duran and Gallardo) which is to be welcomed as these raise particular challenges and problems.

Certainly, more European countries could have been involved. However, most main classes of countries and regions are covered: small, large, internationalized markets, saturated markets, growth markets, liberalized vs non-liberalized, the northern, eastern, continental and southern regions, etc.

4. STATISTICS RELATED TO FOREIGN DIRECT INVESTMENTS

As can be gathered from Table 1.1, many of the papers include statistics related to FDI and internationalization of telcos. Several variables are possible to use in this context: foreign investment spending, employees abroad, turnover abroad, offices abroad, number of switching nodes abroad, etc. Examples of papers that present such data rather comprehensively include Elixmann/Schnöring, Brekke et al., Antonelli/Geuna, and Johansson/Granstrand.

Ideally, the chapters would form a database for an aggregate assessment of FDI in the regions presented. With approximations developed from the database, one could then devise estimates for the whole European region. Unfortunately, the data are not yet sufficiently coherent or complete to achieve this. Instead, Table 4.1 aggregates and summarizes the incidence of foreign activities for the different countries by various measurements, as reported by the contributors.

The data limitations notwithstanding, the data are suggestive in several respects. First of all, there are indeed a large number of foreign entrants, even in countries that traditionally have been considered lagging in liberalization (e.g. Germany, Italy). The number of foreign entries could be interpreted as a compound of market attractiveness and (expected) liberalization. Second, compared to the annual volumes of revenue and investments in the whole European telecommunications sector, the figures are still small. However, the strategic emphasis, attention and prioritization as described in the various contributions support another view, giving the benefit of the doubt to some international ventures being motivated by PR and political considerations. Third, the most promising and accessible variable to include in a future research venture seems to be 'employees abroad'. This variable is easy to connect with internationalization theories, emphasizing cost advantages of multinationals in terms of competence, tacit knowledge and skill (cf. the internalization argument). With appropriate assumptions and extrapolations, such a figure can be transformed into an aggregate estimate of annual turnover and investment spending.

Table 4.1
Aggregation of chapter statistics related to foreign direct investments[a)]

Country	No. of foreign entrants[b)]	Foreign entrants[c)]	Main entry areas	Aggregate of suitable foreign entry indicator[d)]	Comments
Czech Republic	> 4	Bell Atlantic, PTT Telecom, Télédiffusion de France, US West	Cellular, data networks, paging, software services, VSAT	1 BUSD foreign funding needed by year 2000 for First Telecom Project	Privatization of Czech PTT is pending, with foreign participation encouraged.
Denmark	8	AT&T, BellSouth, BT, France Telecom, GEIS, Incentive, Nordic Tel, Sprint	Cellular, international services, VAS	n.a.	NordicTel operates GSM in Sweden and is owned by Pacific Telesis, Vodaphone and 3 Swedish investors.
Finland	4	BT, Infonet, Sprint, Unisource	International services (voice, data)	3 nodes	Infonet Finland is 90% owned by Telecom Finland.
France	9	AT&T, BellSouth, BT, DBP Telekom, EDS, GEIS, Pacific Telesis, US West, Unisource	CATV, data services, mobile data, trunk networking, VAS, VSAT	> 200 MFF yearly turnover	Only BT Tymnet turnover given.
Germany	> 30	Satellite service providers, telcos, VAS providers	Cellular, consulting, international services, VAS	> 3,000 employees excluding cellular and satellite. For these latter two, > 2 B DM invested by 1997.	The number of foreign satellite and VAS providers is not exhaustive, hence > 30 entrants.

a) Aggregation is subject to the reliability and completeness of the respective country market studies.
b) Foreign minority and majority owners are counted separately in case of domestic joint ventures and consortia. (Consortia, e.g. Unisource, are counted as one entrant.) In the case of several subsidiaries of foreign firm(s), counting is only based on the foreign owner(s).
c) The full set of company names is only given when the total is below 10.
d) Whenever the data in the chapters allow, the indicator has been adjusted to share of foreign ownership, i.e. to indicate approximate 'foreign equity'.

Table 4.1 (cont.)
Aggregation of chapter statistics related to foreign direct investments[a)]

Country	No. of foreign entrants[b)]	Foreign entrants[c)]	Main entry areas	Aggregate of suitable foreign entry indicator[d)]	Comments
Greece	3	France Telecom, STET, Vodaphone	Cellular	> 700 MUSD to be invested by 1997. Target No. of users 160,000.	OTE privatization is put on hold.
Hungary	14	Ameritech International, DBP Telekom	HTC privatization (US$875 million)	1.3 million main lines	Foreign entry indicators are mixed with projections due to HTC privatization in progress and pending local wired service concession awards at the time of chapter writing.
		US West	NMT 450 concession	50,000 subscribers	
		Denmark Telecom, Telecom Finland, PTT Telecom, Norway Telecom, Telia, US West	GSM concessions	(service starts April 1994)	
		Central Euro Telecom, Compagnie Generale des Eaux, Digitel 2001, Hungarian Telephone & Cable Corp., PTT Telecom, Swiss Telecom PTT, United Telephone Service	15 local wired service concessions	0.2 million main lines	
Italy	8	AT&T, BT, C&W, FT, GEIS, IBM, PTT Telecom, Telia	International services (e.g. data services calling card), VAS	> 200 employees	Foreign telcos mainly specialize in data services on leased lines.

a) Aggregation is subject to the reliability and completeness of the respective country market studies.
b) Foreign minority and majority owners are counted separately in case of domestic joint ventures and consortia. (Consortia, e.g. Unisource, are counted as one entrant.) In the case of several subsidiaries of foreign firm(s), counting is only based on the foreign owner(s).
c) Company names are given when the total is below 10.
d) Whenever the data in the chapters allow, the indicator has been adjusted to share of foreign ownership, i.e. to indicate approximate 'foreign equity'.

Table 4.1 (cont.)
Aggregation of chapter statistics related to foreign direct investments[a)]

Country	No. of foreign entrants[a)]	Foreign entrants[b)]	Main entry areas	Aggregate of suitable foreign entry indicator[d)]	Comments
Norway	16	AT&T, BT, MCI, Singapore Telec, Sprint, Unisource, CATV providers, VAS providers	CATV, cellular, international services (e.g. calling cards, leased lines, MNS), VAS	> 200 employees	AT&T, BT and MCI operate their Norwegian business through their Swedish offices in Stockholm.
Poland	> 20	Funding institutions, manufacturers, telcos	Network construction, manufacturing	> 500 MUSD foreign funding	Foreign entry is in collaboration with the incumbent, TP SA.
Portugal	> 15	Cofira, Iberfone, Pacific Telesis, LCC, Telefonica, Matrix Asia, Protocol Ventures, plus about 10 VAS providers	Cellular, paging, VAS	> 30 MUSD total investment	80% of FDI is in GSM by Pacific Telesis.
Sweden	9	AT&T, BT, C&W, France Telecom, GEIS, IBM, MCI, Pacific Telesis, Vodaphone	Cellular, data services, international services	> 200 employees	No restriction on foreign entry in telecommunications as of July 1, 1993.
UK	> 25	CAPs, cellular providers, international resellers, RBOCs, CATV providers	CATV with telephony, cellular, fixed link networks, international resale	> M£600 investments per year	Only aggregate investment figures of CATV.

a) Aggregation is subject to the reliability and completeness of the respective country market studies.
b) Foreign minority and majority owners are counted separately in case of domestic joint ventures and consortia. (Consortia, e.g. Unisource, are counted as one entrant.) In the case of several subsidiaries of foreign firm(s), counting is only based on the foreign owner(s).
c) Company names are given when the total is below 10.
d) Whenever the data in the chapters allow, the indicator has been adjusted to share of foreign ownership, i.e. to indicate approximate 'foreign equity'.

5. MAIN FINDINGS AND ARGUMENTS

The main findings and arguments allow a brief discussion around the drivers of the internationalization process, country markets, actor strategies, policy tensions and management challenges.

5.1 Drivers

Even though there is no paper that is explicitly geared toward the daunting task of assessing the drivers of the internationalization process in a comprehensive fashion, it is certainly possible to find suggestions of various causality influences in several chapters. For instance, Johansson emphasizes the role of technological change, liberalization and change in telecommunication user demand—a characterization general enough to fit the tenor of the whole book. Kurisaki focuses her account on the technical options and the artificial administrative rules that serve as a gateway for international entry. Noam points to the role of market saturation and excess funds available to the incumbent telcos in industrialized countries. The chapter by Bonanzinga et al. is suggestive as it strongly emphasizes the role of technical change in opening up a new market and new area of international competition in new satellite services, pointing to technological change as a prime driver of change. With such a view, liberalization, competition and changes in demand are the effects of technical change rather than the cause thereof, although the latter may be embedded in complex interactions and systemic relationships with the former (cf. the evolutionary interactions between technology push/demand pull).

5.2 Country markets

Leaving aside the causality dimensions for now, the chapters show a varied picture of the current and potential state of internationalization in the respective country markets. Among the markets with rather low current and immediate future activities are Portugal and Greece, belonging to the southern region of Europe. In these two, the political environment is currently not very favorably inclined to foreign ventures, excepting cellular. Among the Eastern and Central European countries covered, the number of foreign entries is increasing, particularly so for Hungary. However, the authors emphasize that a long-term view is needed by investors.

The Nordic markets are fairly open to foreign activities (particularly so for Sweden with its new legislation, passed in 1993) but the foreign entry activities in the region are generally not very intense (yet), perhaps owing to the size of the country markets. Again, as shown by Johanssson/Granstrand, Sweden has a number of foreign activities in telecommunications —by some claimed to be a foreign test market due to its liberalized environment and highly internationalized large users.

In the main markets of continental Europe, France and Germany, the foreign market interest is obviously high due to their size, despite the fact that liberalization has not proceeded very far. However, Germany displays a much larger foreign entry intensity and activity than France. A possible interpretation is that the marked difference between the two, as reported by the respective authors, indicates higher liberalization expectations for Germany, in terms of both pace and thoroughness, compounded by market attractiveness.

In the remaining two large European markets, Italy and the UK, the status of internationalization is different (the country market Spain has regrettably not been covered).

In the case of Italy, the foreign entry is particularly geared toward proprietary data services using leased lines. In the UK, however, competition and foreign entry are active and intense in every field—competition in the local loop has surfaced since the Duopoly review as an increasingly critical battlefield. Since the UK market can be interpreted as a test-bed and precursor for the rest of Europe, based on the early move into market liberalization, the UK experience promises that internationalization and foreign entry will both broaden and deepen in all areas of telecommunications.

5.3 Actors

In terms of actor strategies, the contributions are suggestive. Clearly, one can see trends of differences in strategies pursued and in the emergence of various strategic groups. The earliest case of a traditional European PTT going global is BT and much can be learnt from this case. Indeed, both Dang-N'guyen/Phan and Williams/Taylor find that BT has used its early moves to learn and experiment (not always successfully), developing its strategy from trial-and-error to a strategy mainly focused on global outsourcing. As Williams/Taylor note, BT's goal to become global has remained but the strategies and alliances to achieve this have been unstable and not primarily geared towards finding synergies in terms of technologies and markets. Hence, the competence base necessary for internationalization may not be easily apprehended. Dang-N'guyen/Pouillot also find that France Telecom is pursuing a broad trial-and-error strategy, moving into opportunities as they arise without a strong pattern of service/market discriminations or prioritizations. They argue that the competence heritage of France Telecom, conducive to its internationalization, mainly revolves around its relationships with its previous colonial allies, its technical and management expertise gained in its own rapid domestic expansion in the 1970s and from its introduction of new services in the 1980s. Similarly to France Telecom, Marconi also focuses on its former colonial allies (cf. Lucena and Seabra).

Another example of strategic differentiation is STET. Here Antonelli/Geuna show how the company has developed a multi-domestic internationalization strategy, primarily geared toward developing countries in the construction and expansion of basic telephony, in congruence with its primary activities in its home market (Duran/Gallardo offer a similar argument for Telefonica). For DBP Telekom, Elixmann/Schnöring argue that its experience in former Eastern Germany makes it particularly fit for internationalization in Central and Eastern Europe but question its prospects of becoming a truly global player due to the cautious approach of DBP Telekom's management and the limiting German regulatory framework. For small operators, Brekke et al. and Henten/Skouby suggest that the telco pursue specialized services (e.g. mobile and satellite) and niche market segments. Finally, as emphasized by Kramer/NiShuilleabhain, Cable & Wireless is somewhat of an exception to the search towards a strategic focus: C&W is rather a conglomerate and has almost a contradictory structure of business, involving local monopoly concessions, new entrants in former monopolies and an international digital undersea cable network.

Johansson's is the only paper that makes a comprehensive and comparative analysis of actor strategies. The main pattern of an internationalization strategy is an early entry into one of the major markets within a region (e.g. Europe), regardless of the state of liberalization, followed by investments in smaller, adjacent countries. Johansson interprets this pattern as a 'follow your customer' strategy. Moreover, Johansson suggests that a pattern of 'trial-and-error' is followed by an increased focus on market segments/services. In terms of success factors, Johansson argues that the competence base developed in the home market is critical.

5.4 Policy tensions

A common thread throughout several issue chapters is the policy tension between national and international interests. Although suggested courses of action vary, the chapters all point to one of the internationalization impacts that will undoubtedly play an increasingly significant role, displaying a multitude of complex interactions (and surprises). For instance, Bonanzinga et al. show how the emerging market for international mobile communications based on new, mobile satellite technologies brings conflicts of interest onto the international regulatory scene. In this respect, they offer insights into the relationship between technical changes, internationalization and regulatory tensions. Moreover, both McCarty and Noam raise the question of multi-jurisdictional conflicts between domestic and international regulators, similarly to the state vs federal tensions in the United States. Allen calls for a reconsideration of domestic competition policies and a search for an integrated policy framework, in the light of the visions of global dominance that some of the major telcos ascribe to. Ioannides suggests that national interests and stakes will increasingly surface, challenging the internationalization process and actor strategies. Similarly, Noam sees political heat and conflicts of interest mounting in developing countries that will preclude the international telcos from reaping large financial gains there. In fact, the chapter by Noam is the only one in the volume to seriously question the viability of the internationalization process, the argument mainly based on the ensuing political tensions. Finally, Kurisaki argues that the on-going internationalization process will cause a faster pace of domestic liberalization. However, Barbet/Benzoni reverse the causality relationship into the other direction, pointing to the fact that domestic liberalization and privatization measures in the countries involved condition the internationalization pace. They argue that without regulatory reciprocity between alliance partners it will be difficult to proceed with international mergers and far-reaching collaboration (cf. the limits on foreign ownership, e.g. the 20% limit in the United States).

5.5 Management challenges

The internationalization process also poses new management challenges, particularly in regard to network investments due to the high capital intensity and the role of technical change. A sign of increasing attention to management questions is the multitude of major reorganizations that have taken place over the past years among advanced telcos. Both Pospischil and Bohlin give a brief account of some of the organizational changes, primarily on the domestic scene, and provide some general trends with application to the internationalizing telco. One of the overriding trends seems to be a decentralization of business responsibilities combined with the creation of more centralized strategic planning. Previously, much of the operational and tactical planning was centralized but now a more hierarchical structure is pursued, with clearer division of labor and more operational flexibility in order to better respond to current customer requirements. Duran/Gallardo focus exclusively on the management of FDI in developing countries, citing some experience of Telefonica in Latin America. They emphasize that appraising FDI is a multi-dimensional and complex matter and that special attention must be devoted to country risk. Another capability of central importance is managing your customers, particularly the large multinational users. This capability can hardly be overemphasized, and Roche provides insight into this area, stressing, among other things, global coverage for outsourced services and excellence in networking capabilities instead of price leadership.

6. OUTLOOKS, SPECULATIONS AND IMPLICATIONS

The authors suggest several outlooks to the future, some of which are, by their very nature, hypothetical and speculative. Here a simple twofold grouping of the outlooks is provided, while a more in-depth outlook and discussion is given in the final and concluding chapter.

First, a common theme running through many of the outlooks is a sort of international division of labor, shake-out, bifurcation and segmentation of the actors involved in different strategic clusters, regions covered and customers served. The BT strategic learning experience is suggestive of similar strategic evolution processes among other telcos currently in such a trial-and-error phase. Hence, telco strategies can be expected to become more distinct in terms of overriding goals and means to carry this out. There are several dimensions around which such strategic positioning could develop: regions and customers served, telecommunications service offerings, R&D intensity, etc. Examples of emerging strategic positioning are offered by Dang-N'guyen/Pouillot who predict that France Telecom will focus its strategy, Elixmann/Schnöring who suggest that DBP Telekom will be most successful in the East European area, Antonelli/Geuna who believe that STET will pursue basic telephony in developing countries, and the advice by Henten/Skouby that small players should aim for specialized services and niche markets. (For Cable & Wireless, Kramer/NiShuilleabhain suggest that its conglomerate structure makes it vulnerable to takeover.) Hence, even among those intending to become global, strategic differentiation and search occur—suggesting that the telecommunications service market may become more heterogeneous with time. An important implication of this trend is that it is crucial to develop the following: a more finely tuned grasp of the telecommunications service concept, the attendant strategic differentiation concepts and an analysis of the technology and management base necessary to carry the strategy forward.

A second set of outlooks that is recurring pertains to the conflicts between national interests and the internationalization process. These conflicts will surface not only in regulatory and jurisdictional spheres but will certainly be felt in the pockets of some of the internationalizing telcos. The question is how much the nationalistic interests will surface and how deep they will be—will they, as suggested by Noam, delay the internationalization process? The questions are many and the stakes are indeed high. One thing is sure, though—the future will always hold surprises.

The Race to European Eminence
E. Bohlin and O. Granstrand (Editors)

Conclusions and Speculations*

Ove Granstrand

Chalmers University of Technology, Department of Industrial Management and Economics, S-412 96 Göteborg, Sweden

1. INTRODUCTION

In drawing to a close it is natural to step back and review the current and future relevance of the main theme and the title question of this book. Is internationalization of the telecom service industry and markets to be taken seriously? And if so, who stands the best chances to gain advantages by becoming MNCs? This final chapter will first summarize some of the findings from preceding chapters, then try to assess the conditions for continued internationalization. Given the likelihood that internationalization of telecom actors and markets will actually continue, some speculations about future developments are made, drawing also on received theory, the applicability of which to the telecom service industry is discussed. Finally, the chapter attempts to identify what characterizes a winning company in the hopefully positive-sum game of internationalization of telecom service markets.

2. SUMMARY

Let us first briefly recapitulate some of the findings in the preceding chapters, especially the previous two chapters by Johansson and Bohlin. Despite the fact that the post-war internationalization of telecom actors and markets, starting in the 1980s, is a recent phenomenon with a great deal of trial-and-error behavior, there are already a number of more general findings and indications that deserves to be pointed out, for example that:

a) (Pace) Apart from more or less singular earlier events, internationalization among typical telcos started in the mid-1980s, and grew particularly rapidly in the late 1980s in terms of number of foreign market entries. In summary, the pace of internationalization seems to have been influenced by the paces of liberalization in home and host countries respectively, the pace of internationalization of demand in various forms, the market opportunities offered (much based on new technologies), a certain urge for reciprocity and finally a bandwagon effect.

b) (Actors) Anglo-Saxon operators were dominating the group of first movers, with AT&T entering early in Europe, followed by the RBOCs, and C&W and the RBOCs entering early

* Comments on this chapter from Erik Bohlin, Ola Johansson and Sven Lindmark are gratefully acknowledged.

in Asia. There is currently a group of telcos emerging who explicitly try to globalize (e.g. AT&T, BT, NTT, DBT, FT and Unisource). As yet there is no clear indication that large size is positively related to the propensity to internationalize. Several telcos (e.g. C&W, FT) have adopted a strategy of internationalization and diversification simultaneously (which in other industries has proved risky). However, on average there seems to be a certain specialization, complemented by trial-and-error, among the actors in connection with internationalization. The home market and regulatory regime and the way technological and managerial competence have been accumulated have been important for the internationalization behavior of the various actors.

c) (Country markets) UK and France were early target markets in Europe, followed later by Germany. In general, the main countries in a region have been the early, primary targets for foreign entries, complemented by entries in smaller countries for special reasons, e.g. that they offer good learning opportunities for the foreign entrant. In summary the choice of country for FDIs seems to have been influenced primarily by three factors—the size of the host market, the learning opportunities it offers and its geographic and/or cultural closeness to the actor's other operations, which almost always means closeness to the home markets, at least initially.

d) (Services) Many FDIs have been made early on in CATV as well as in some basic telephony services, wired as well as wireless as a platform for further expansion. Thus, internationalization has often directly affected the traditional core business (telephony) and core technology (wire) of a PTT in its core market (home country). The common moves then of the foreign entrant in the foreign service expansion (i.e. diversification) process have been according to (→ indicate 'subsequent move to'):

$$\text{Mobile services} \rightleftarrows \text{Wired telephony services} \leftarrow \text{Satellite services}$$

That is, first entries have commonly been made in mobile, wired or satellite services while subsequent entries have been into wired services from satellite or mobile services, or into mobile services from wired services. This points to the importance of new technologies for foreign entry.

e) (Strategies) Cross-border joint ventures and alliances have been the most common foreign entry strategy. Acquisitions have been a rare strategy for foreign entry into core telephony services, but more common for foreign entry into data services. Foreign entry through greenfield operations has been a rare strategy in general. The cross-border cooperative mode of internationalization has been motivated by the need to save money and time and the need to overcome barriers to foreign entry, regulation-based or not.

f) (Investments) Although any rough approximation of the aggregate European FDI level does not indicate a large spending activity relative to total telecommunications investments as yet, actor strategies, commitments and market opportunities make an aggregate FDI size escalation credible and probable.

3. THE RACE TO EUROPEAN EMINENCE

3.1 Is there a race?

Without doubt the issue of internationalization has rapidly raised a significant amount of awareness and interest since the late 1980s. From the preceding chapters it is moreover clear that a process of internationalization in telecommunication services has rapidly gained momentum in terms of FDI decisions and actions by various actors in various countries. Thus, internationalization is not an ephemeral fad, although to some extent clouded with tactical PR announcements and moves. Nevertheless, internationalization is still by and large a marginal, although embryonic, element on world telecom service markets.

It is yet too early to evaluate the alleged pay-offs to the service users and providers from this internationalization. Moreover, it is not immediately clear that this internationalization process is sustainable. It is easy to become misleadingly impressed by current announcements and figures surrounding internationalization and conclude that the process will continue to evolve. Thus, one has to take a closer look into the prospects of continued internationalization.

3.2 Will the race go on or will internationalization be stopped again?

As mentioned in Chapter 1, internationalization and international competition in telecom services, typically telephony, are not a new phenomenon. However, for various reasons the early internationalization process, spurred by telephony developments in the US, was curbed and regulation transformed the evolving industrial organization into what essentially became an international cooperative network of local, public monopolies. (See further in particular Brock 1981, Bradley and Hausman 1989, Cowhey 1990 and Noam 1992.)

A natural question then is whether the current tide of internationalization—or in fact the current reinternationalization—will again be dammed up and diverted by regulation and/or other factors. As described in several of the preceding chapters, regulation conditions the pace and extent of internationalization, and the regulatory pendulum may swing.

However, the field of driving and hampering forces behind the current reinternationalization is different from the field of forces behind internationalization in the early days of telephony (cf. Figure 4.1 in Chapter 1). A thorough analysis of the complex nature and compound significance of these differences cannot be made here but a few of them may be pointed out.

First, there has been an apparently lasting shift in institutional powers surrounding telecommunication services. The nation state and their public institutions have become weaker relative to corporate powers, including the MNCs, in general. The cooperative institutions of nationalized telecom monopolies (ITU, CCITT etc.) have also been weakened relative to other international institutions with a wider scope (GATT etc.). The legislation and legal institutions supporting competition in general are moreover stronger today.

Second, internationalization in various sectors of national economies has increased during the whole 20th century and the international economic interdependence is much higher in the 1990s than in, say, the 1920s. In this connection there has specifically been a steadily growing demand for international and differentiated telecom services, e.g. within globalizing MNCs, and the significance of nationally confined telephony as a bulk, public service has declined in relative terms.

Third, technological innovations have by and large been fostering rather than hampering competition. Judging from history, this in itself is not necessarily a sufficiently strong factor for lastingly breaking up an industrial organization consisting of national monopolies. The century-old national postal monopolies in Europe mostly absorbed the radical innovations of both telegraphy and telephony without a major lasting disruption, only transforming themselves into PTTs and the like with essentially preserved status as national monopolies, in some cases after a period of internationalized competition as mentioned. National interests hampered an innovator like the Bell Company in internationalizing, and not only national interests in its host countries but also, and decisively, national interests in its home country where the public wanted the company to focus. A stable situation with reciprocal national monopolies remained.

Although such historic examples illustrate that even radical technological innovations have been accommodated in the wider communications service industry without "creative destruction", it could be argued that sooner or later technological changes and options will have accumulated in such a way that an old industrial organization structure has to yield. This argument is admittedly general and imprecise. Nevertheless, there are instances of technological innovations that are apparently undermining the analytical as well as empirical economic foundations of national telecom monopolies, and probably also their political foundations in the longer run. In general, technological innovations severely challenge the assumptions on which the argument for a given natural monopoly rests. More specifically, the convergence between computing and communication technologies, albeit slow and possibly misperceived, has prompted computer firms to successfully challenge telephony monopolies. Other technological innovations, independently or interdependently, similarly have led to severe challenges of the old structure of national monopolies. Digitization facilitates interconnectivity and thereby competition; computerized switching allows decentralized switching and network control by users; optical fibre transmission cost decreases and opens up possibilities for new entrants; radio and microwave communications allow bypassing; etc. In addition, the technical possibilities for a single telco to coordinate international operations are greater today as are the technical and managerial possibilities to run an MNC in general.

With a simple illustration one could say that if the PTTs were to absorb the major new technologies of the 20th century as well, they would have to extend their past evolutionary chain P (postal services) to PT (+ telegraphy) to PTT (+ telephony) by adding under their control at least R (radio), C (computing), and M (media), thereby becoming PTTRCMs or the like.[1] Simplicity apart, this suggests that total integration will become increasingly difficult and cease at some point. Reintegration, perhaps with different configurations, may then occur, of course. However, if disintegration has begun and there are elements of irreversibility, full reintegration will not occur. At some point this happens when technological options, irreversible by their very nature, become numerous and widely available to a large number of actors.

In sum, there is now an accumulation of technological options with vested interests that challenge the traditional organization of telecom service monopolies to an extent that apparently has no counterpart earlier in history. In this connection it is important to note that the different national interests vested in new technologies and their market opportunities have fragmented and thereby weakened the national interest attached to traditional telecom technologies and their protected markets. A major case in point is the divestiture of AT&T.

Fourth, there has been a swing in the recent decade or so in political economic ideology in many industrialized countries towards employing more market solutions and less administra-

[1] Early radio technology for telecom actually came to a considerable extent to be controlled by the PTTs (see Brock 1981).

tive approaches, notably state-oriented ones. This has led to deregulation (although with some reregulation) and liberalization of markets, subjecting them to competition, privatization of public companies, subjecting them to less public ownership and control. These new economic policies apply not only to telecom services but to many other service industries as well, many of which have been nationally regulated and focused by tradition (Dunning 1991). Although it is certainly possible that such a movement in economic policies could be reversed, the movement is significant for the internationalization of national telecom service markets for two reasons. First, the changes in the various national economic policies are much more similar and simultaneous than they used to be. Somewhat differently expressed, national economic polices have become more internationally interdependent and homogeneous. Thus, the contemporary simultaneity of the swing towards market solutions in various industrialized nations has allowed an emerging international reciprocity in competition in telecom services that could substitute for the traditional reciprocal monopolies with less national asymmetries hampering a transition. Second, once the new economic policies have triggered reinternationalization, the other factors (internationalized corporate powers, international economic integration, new internationalized demand, and technological options) come into play and mutually reinforce the process.

Fifth, there are many powerful corporate actors in the telecom and information industry at large, having wide-ranging expectations attached to internationalization. True, the volume of their FDI-based foreign business is usually still small in relative terms and there might still be attractive prospects for domestic growth (e.g. in building a national infrastructure for an information society—although it must be questioned whether this can also be confined nationally). However, domestic profit margins are generally shrinking and the long-term profit and growth prospects on world markets for the largest and most competitive companies are easily perceived to be far higher, given the large foreign markets, often lacking strong, local competitors. In addition, internationalization might be perceived to offer prospects of decreased dependence upon domestic, public authorities. Asymmetric expectations and competitive advantages among telcos will then destabilize any international cartelization and reinforce internationalization, once it has come underway. This also adds an element of irreversibility to the process.

Thus, as a conclusion, there are reasons to believe that the internationalization process of telecom services is sustainable and will not be reversed this time as it was once before. The race is on and is not likely to become stopped and cancelled. This does not mean that internationalization, once started, will proceed fast or in a similar fashion throughout the world. As evidenced in several of the preceding chapters, there are deep tensions between national interests, there are severe jurisdictional problems to be solved, and there are significant heterogeneities among countries, markets and actors. Besides, the interests of the old regime have been powerful and long-lasting, nurturing a culture that is not easily swept away. In various ways the continued internationalization process will be influenced hereby. Even if a PTT accepts, maybe even welcomes, the pressures for liberalization and internationalization, it will most likely try to influence its pace and direction. National interests will use infant industry arguments as well as related strategic trade arguments in order to ensure suitable protection. Although natural monopoly arguments are being challenged, they are not irrelevant and will be considered and reused. Schedules will be delayed, decisions will be postponed, entries will be restricted and so on. Limits to internationalization as well as the possibility of a major, local reregulation in favor of nationalization must still be kept in mind, especially when internationalization has gone far enough for denationalization to set in and/or has resulted in xenophobic perceptions to individual nations and actors. The forces at play are so strong, creative and complex that with certainty some actions and outcomes will occur that are presently inconceivable.

Nevertheless, the most probable future scenario is that of continued and reinforced internationalization of telecom service markets and actors. Some countries, markets and actors will be slow and lagging, others will be fast and lead the way, at least for some time until possibly overtaken by still others, and some will undeniably drop out. However, there might be a period of counterreactions and internationalization slow-down, at a point where national interests become severely threatened and/or the common over-expectations become unfulfilled and/or large, successful actors in e.g. the US, UK and Japan are pushing too strongly for a faster pace.

The impact of the internationalization process, fast or slow, can be manifold. It is likely to have strong and lasting effects upon market and industry structure, at least in the long run (see Section 3.4). The century-old traditional regulatory regime in telecommunications services has conserved its industry structure against the otherwise normally disruptive effects of technological changes as mentioned. The internationalization process is yet in its infancy, being dominated by a variety of trial-and-error moves among a variety of early and late movers on the international scene. The possibilities to economically evaluate these moves ex post are still slim while the expectations are high and often more optimistically biased than rational. Circumstances like these make the bed for overinvestments, disappointments, duplication, waste of resources and defensive reactions. The heavy hands of history and xenophobia will fight against far-reaching denationalization of the core network and core businesses in the major advanced countries.

However, even if major foreign inroads in basic services are fought off by local forces in general[2] at least four factors will still allow ample room for internationalization:

a) The existence of large, growing segments (e.g. mobile services, corporate communications).

b) The emerging markets in NICs and medium-developed countries (incl. Eastern Europe, Russia etc.) and LDCs.

c) The emergence of *service merging* and service integration (e.g. "edutainment", i.e. merging of educational and entertainment services) and inter-sectorial competition (e.g. with financial services, media services).

d) Cross-border alliances, joint ventures and acquisitions.

The interplay between internationalization in the telecom services and internationalization in the telecom equipment industries and possible future vertical integration or quasi-integration is yet unclear. However, it is far from unlikely that if a country has a sizeable domestic telecom service and equipment industry which is being severely threatened at both the service and equipment ends by foreign competition, then protectionism and procurement bias will recur, possibly aided by the new arguments in strategic trade theory (see Chapter 1).

---

[2] Note that several early foreign entries have in fact already been made in basic telephony services as mentioned.

Thus, if not one single race is on, a set of races is, although the not so invisible regulatory hand holding the starting pistol has been quite shaky, and these races are likely to go on. However, the race or races are not only towards European eminence. Although Europe is a major target market, some actors, mostly with Anglo-American origins, are more ambitious with more global aspirations. Still the eminence in Europe may be at stake, although as long as EC regulation is credibly pro-competitive, monopolistic dominance in Europe or a "Eurotel" constellation is not a likely outcome.

3.3 A paradigm shift in telecom services

A growing internationalization in telecom services could be viewed as part of a larger ongoing process of a paradigmatic shift in the whole telecommunication sector in various countries, essentially triggered by events in the US in the late 1970s and early 1980s. Internationalization of the telecom service markets and actors was perhaps not fully intended or anticipated in general at the outset of the paradigmatic shift, but nevertheless has become an integral part and reinforcing factor of it. (See also Chapter 1.) Table 3.1 summarizes some characteristics of this shift in paradigm and corresponding regime.

3.4 Some speculation about the future

As concluded above, internationalization of telecom service markets and actors is a recent but viable phenomenon which must be taken seriously as such by managers, policy-makers and analysts, as well as by customers and suppliers. It is then naturally important to try to speculate about the future and to the extent possible attempt to predict the developments that are likely, although still uncertain. A major contingency is then the regulatory and political development. As mentioned there are reasons to believe in continued, although delayed and distorted, liberalization favoring continued internationalization on an average. However, it is hazardous to attempt more detailed predictions about long-run regulatory changes. Moreover, there seems to be no real economic or political theory about regulatory dynamics in general that could be used, at least not to my knowledge.[3]

Given that the regulatory and political possibilities for continued internationalization are there, the next question is what courses it will take. This is obviously an important question, which will be elaborated upon to some extent here but mainly must be left for further research. The available empirical data, much of them presented in the preceding chapters, offer some indications, but by and large the empirical grounds are yet too weak for predictions.

---

[3] Of course, there are predictions, with or without foundations in fact or theory, about the evolution of politico-economic systems in general (e.g. by Marx and Schumpeter), but hardly specific enough for the purposes here. See, however, Cowhey 1990 on regime theory.

Table 3.1
Shift in paradigm in the telecommunication services

Old national monopoly paradigm elements (NMP)	**New international competition paradigm elements (ICP)**
National regulation	Supra-national regulation
Natural monopoly	Liberalization
Nationalization	Internationalization and "Glocalization"[1]
Public ownership	Privatization
Monopoly provision	Intra- and inter-sectorial competition
Strongly protected markets	Weakly protected markets
National strategic industry (civilian, military)	"Free trade"/strategic international trade
	Inward/outward FDI
Universal/public service	High margin market/segmentation (corporate customers, mobile services etc.)
Broad-based cooperation and international PTT cooperation (in ITU etc.) and harmonization regarding standards, revenue sharing etc.	Cross-border alliance cooperation Competition and cooperation mixing ("competeration"[2])
	Telecom arbitrage
	Fragmented standard-setting (regional multilateral)
Weakly protected technologies	Strongly protected technologies
Technology sharing	Proprietary technology

Notes: 1) Intertwined processes of globalization and local, regional agglomeration.
2) Intertwined processes of competition and cooperation.

What can theory and experience in other industries then tell us? First of all, received theories about internationalization and MNCs mentioned in Chapter 1 are in general more explanatory than predictive. Second, there is the objection that they are mainly based on conditions in manufacturing industries, conditions which by and large do not apply in service industries. Third, still another objection is that even if received theory is applicable in principle to service industries, the idiosyncrasies of telecom services will make its usefulness very limited.

Without going into detail here, it can be argued that received theory is applicable to telecom, far from fully of course, but to a considerable extent, not least because of the high technology and capital intensity of telecom services (see Chapter 1). The differences between the telecom service industries, other technology-based service industries, and other technology-based industries easily become exaggerated. True, there are major idiosyncrasies in telecom. Of major importance here are, first, the initial conditions in the internationalization process with a particular regulatory-based industrial organization, dominated by local and national monopolies, often with domestic preferred suppliers. Second, there is the traditional common operation of a global, physical network consisting of interconnected national networks, with an overlayer of inter-organizational cooperative networks. Third, the telecom sector has traditionally been perceived to be of strategic, national interest, civilian as well as military, and has as a consequence become politicized in the sense that economic and political decision-making has become strongly interwined.

At the same time the telecom service industry is technology and capital intensive and serves mass markets, factors which make the industry similar to several other industries. Besides, the idiosyncrasies described above are becoming weakened. Liberalization and internationalization are under way, the perception of strategic, national interest is changing,[4] and parallel international networks are being built.[5]

Thus, there are no strong reasons for rejecting received internationalization theory wholesale from the point of view of the telecom service industry. On more general grounds, Buckley et al. (1992) as well as Dunning (1992) reach the similar conclusion that there is no need for an entire new theory or paradigm for the internationalization of service industries compared to the internationalization of manufacturing industries. The actual mix of MNC advantages, pertaining to ownership, location and internationalization, in Dunning's OLI-paradigm[6], may differ between service and manufacturing industries but not the basic usefulness of the paradigm.[7]

Internationalization of telecom services (in terms of FDIs) is far less developed than in many other industries, including some service industries. From an economic theory standpoint, this would in principle explain why the economic advantages accruing to an MNC in telecom services, e.g. the mix of OLI-advantages, are not very large. But as demonstrated by Cowhey (1990), the reasons behind the absence of telco MNCs are to be found in the political arena rather than in economics. This points to a general limitation of received economic theory when applied to a politicized industry such as telecom services. At the same time the past political conditions have created some economic conditions that in principle would favor internationalization once started, e.g. the profitable asymmetric tariffs in international traffic (see Antonelli 1993).

Accepting a seasoned view of the applicability of received theory, what can it tell us? A thorough review cannot be given here, we can only offer a few comments. (See Chapter 1 for an overview of received theory and models.) First, one can note that, in the short run, the notion of psychic distance (which increases transaction costs) has actually to some extent influenced the choice of countries in the internationalization process of telcos. For obvious technical and market reasons, geographic closeness has also influenced any network expansion process, however decreasingly so. The theory then predicts that, in the longer run, psychic distance will become less important, relative to other factors, such as size of market. As a slight extension of the theory one may also hypothesize that psychic distance influences somewhat the choice of partners in joint ventures and strategic alliances. Especially cultural similarity is important for a stable and efficient functioning of a strategic alliance.

The establishment chain model also seems to be applicable to some extent. After having established since long ago what has in fact been a mutual sales agent relationship between telcos in international traffic, an internationalizing telco is likely to establish its own foreign

---

[4] For example, some countries are becoming less aspiring in their protectionism. Military considerations behind protectionism in telecom also become weakened. The relations between civilian and military technology are also changing, with a weakening of the connection through reversal of technological leads (e.g. in cellular telephony) from military to civilian applications, which also often diverge and limit dual use of new technologies.

[5] Similarities with other industries may also change, of course. For example, there will be less mass-marketing of telephony as a bulk service through increasing service differentiation; the capital and technology intensity may become particularly high in telecom and so on. But these are changes of degree rather than of kind.

[6] Dunning claims, for example, that transaction costs are generally higher for services than for goods. Other things being equal, this would favor the emergence of service MNCs.

[7] Still more authors emphasize the similarities between service and manufacturing industries in their internationalization process; see e.g. Guile (1988).

sales and market support offices, partly as a consequence of a "client following"-strategy.[8] A next (or parallel) step is to establish partially or fully owned or controlled foreign networks, with various arrangements (build, own, operate, transfer—BOOT etc., see e.g. the chapter by Kiss on Hungary). In this stage a stronger integration of the networks and businesses of several telcos is also a likely possibility, thereby deepening their traditional mutual export/import-relations in international traffic by means of mutual-equity participation (e.g. by "network swapping"), holding companies and the like. If such a group of "integrated network integrators" moreover engage in low tariff country hubbing, a special variant of Vernon´s international product life cycle model partially applies (see Chapter 1).

Still another step, yet not taken by telcos but predicted by an extension of the establishment chain model, is to proceed to internationalize R&D operations as well (apart from participating in international cooperative R&D programs), besides the marketing and production of telecom services. This step could be taken for both demand and supply side-oriented reasons, i.e. to support local marketing and production as well as to source technology and R&D inputs (see further Granstrand et al. 1992). Considering the massive investments in R&D made by leading telcos it is very likely that some of this R&D will be spread internationally as well for various reasons as the telcos internationalize. Alliances and joint ventures, some of them also aimed to share R&D costs and risks, will lead to internationalized R&D, as will as the so far less common acquisitions and mergers. Moreover, R&D-oriented FDIs tend to cluster to some extent to certain regions in the world (i.e. they "glocalize"). Several such regions are conceivably attractive to foreign telco R&D inside and outside Europe. As a result R&D will also globalize in the sense that it will be rationalized and managed on a global basis.

Many theories about internationalization stress the role of technology and innovations behind, at least in the longer run (see e.g. Cantwell 1989). Any company competence that combined with complementary assets brings benefits to many customers worldwide, benefits that are better appropriated by the company through internationalization than cross-border transactions, create a basis for becoming an MNC, especially since competences have significant economies of scale as well as scope. Technical competence, i.e. technology, offers particular possibilities for appropriation and economies of scale (through IPR, cumulation etc.) and thus becomes a key competence (together with management and marketing) for internationalization. This would mean that telcos with successful and appropriable R&D stand a better chance to be successful in internationalization, everything else being equal.

Disregarding differences in R&D productivity and innovativeness, which may be large indeed, this would mean that the largest R&D spenders among telcos would have the best chances in internationalization in the longer run, everything else being equal. Another prediction along these lines is that technology in telecom services will become privatized and protected, e.g. through more intensive commercial and strategic use of patents, as R&D becomes a major means for competition, domestically and abroad. The need for a telco to have complementary assets to be able to economize upon its technology assets will moreover make some degree of integration likely, not only in the form of owning or controlling critical parts of a network but also in the form of owning or controlling the supply of some critical equipment, which would mean at least a quasi-integration backwards. In the extension of this, also quasi-integrated R&D with equipment suppliers and other service providers, even outside

---

[8] A client following-strategy (a CF-strategy), that is "follow your customer abroad", is common in service industries in general, and has been common among telcos as well, at least initially. (The strategy has learning advantages but growth limitations.)

telecom, is likely to become an important consideration. However, this does not mean large-scale integration backwards into manufacturing. Neither does it mean full ownership of all parts of a network. The supply of know-how is generally more critical than the supply of bulk capacity.

Turning to more general theories, a Schumpeterian approach is of relevance for speculating about the future of telecom markets and actors.[9] In line with this approach, the most successful companies, in international competition in telecom services, will be the most innovative and entrepreneurial companies. The question is whether they will be old or new companies, telcos or originally non-telcos (e.g. computer firms, media firms). Without going into detail here, an argument could be made that organizational and managerial learning in conjunction with relative advantages in making technological transitions would favor incumbents in general (see Granstrand and Sjölander 1990). This would mean that old, experienced telcos with innovative R&D and entrepreneurial management stand the best chances to survive in the domestic and international Schumpeterian competition that is likely to follow upon liberalization.[10] Still this competitive process, if allowed to unfold, will give lots of room for new entrants, new as well as diversifying firms, in various service areas, room that is created by both new technologies and expanding markets. The latter is generally predicted for the telecom sector worldwide. However, the process of Schumpeterian competition will involve both innovation-based entries and exits, that is, with Schumpeter's well-acclaimed label "creative destruction". To the extent that there are cumulative incumbency advantages, this process will be dampened, at both the entry and exit sides. Regulatory protection can give similar effects. Certainly features in the telecom sector such as the high fixed and considerably sunk cost in the network, the systems nature, the traditional culture, the national interests, and so on will tend to counteract creative destruction, as will certain cases of market expansion. On the other hand new technologies (e.g. satellite and/or cellular), especially if in the hands of large, agile companies inside and especially outside traditional telecom, will stimulate creative destruction. Considering the huge R&D investments in telecom-related technologies, a great potential for creative destruction is there, even if there may be overinvestments or appropriability problems. However, more specific predictions on the type and degree of future creative destruction in telecom services are difficult to offer at present.[11]

Finally, one can ask what theories traditionally applied to telecom have to offer about internationalization. More specifically, could the theory of natural monopoly be applied on a global scale?[12] Perhaps so if its concepts and tools are basically changed, but hardly at present.

---

[9] Critics of Schumpeter claim he never presented a real theory. However, at the same time as one must admit that he was weak on theory, one has to acknowledge that he has spurred a number of scholars to develop his theoretical thoughts; see e.g. Phillips (1971).

[10] Schumpeterian competition in services would mean that the generation of new or enhanced services based on innovations and entrepreneurship, giving higher service performance-to-price ratios and customer value, is more important than plain price competition. That Schumpeterian competition will accrue in telecom services is a prediction that is especially important considering that there is a traditional focus in telecom on pricing matters, more or less assuming undifferentiated services ("plain old telephone services").

[11] Various positions on whether new, innovative firms or old, large firms will dominate in the longer run have also been taken in general by various scholars. Even Schumpeter shifted position from the former to the latter during his lifetime.

[12] Note that this theory is more normative than predictive, but since it has served and been served by powerful interest groups in the past it has appeared as predictive. For accounts of the theory, see Sharkey (1982), Berg and

One basic difficulty is that it is difficult to gather data on key concepts like cost sub-additivity (implying economies of scope), a difficulty which is aggravated on a global scale. An even more serious difficulty is how to accommodate technological changes enabling telecom service differentiation and service innovation, implying enhanced performance and new functionalities, e.g. customer mobility and/or new ways to economize on the frequency spectrum. In addition, technological substitution and Schumpeterian competition are always contesting a natural monopoly at some level, which is difficult to account for. A third difficulty is to design and implement supra-national regulatory bodies and measures for controlling a natural monopoly on an international scale regarding pricing, inefficiencies, principal agent problems etc. and stimulating its innovativeness.[13] Needless to say, the naturalness, if any, of past and current telecom monopolies has not extended beyond national borders so far. Nevertheless, it is conceivable that an argument of a natural monopoly type could be extended internationally in certain cases, supporting ideas e.g. about a mega-merger or a mega-consortium of telcos to create a PTT type of monopoly with a "network of networks" on a continental or a triad or a global basis (e.g. "Eurotel" in Europe or a "World-TT"); or supporting ideas e.g. about having a single consortium to supply satellite services linked to land mobile services in the local loop. However, apart from intellectual possibilities as well as difficulties to significantly extend the theory of natural monopoly to realistic international conditions, and apart from its prescriptions, the practical problems of implementing a more genuine global monopoly seem insurmountable for the foreseeable future, not least since such a notion would clash with the new paradigm in telecom services and deep-rooted political values in the industrialized world (e.g. freedom of speech). There is not currently a supra-national institution with sufficient powers to merge PTTs. The situation today is different from the historical situation when local telephony companies became nationally integrated, as described above. This does not mean that there will not be an ongoing concentration in the international telecom service industry, based on economies of scale, a concentration that very well could witness an emerging integration of e.g. European telcos.

What is less uncertain is that telecom service markets will be restructured and the telcos will become more service-specialized in order to reap economies of scale.[14] As a result international division of labor will increase, everything else equal, and there will be more differentiated competitive races, also as a result of a widening of total service diversity through newly developed services.

Tschirhart (1988), and Train (1991). The idea that natural monopolies should be nationalized goes back at least to J.S. Mill, before the advent of telephony.

[13] A natural monopoly could in principle be implemented without traditional regulation, but these possibilities seem even more remote to apply on a global scale.

[14] This does not preclude that telcos will provide a wider range of services in their own names, some of which are supplied by some kind of "original service producers", analogous to OEM firms. Service branding will also be of increasing relevance in general, and in that connection franchising arrangements may appear with proper control of service quality.

4. WHO ARE THE COMING TELE SERVICE MULTINATIONALS

4.1 Introduction

What will it take for an aspiring player to have a good chance in the race to European or international eminence? What strategies will be successful? These questions will be touched upon next. However, this is not the time and place for a detailed analysis with specific answers. The answers are not only worth multi-billion dollars if right; the possible answers are also multi-billion in number. The empirical basis for analysis is weak and the MNC literature is in fact scarce what constitutes successful strategies for internationalization. The notion of a single race with a single winner must also be questioned at the European as well as international level. In order to conclude this chapter, a brief outline showing strategic options and strategic requirements for competing telcos is given. These outlines will hopefully prove useful for continued analysis and trial and not-so-expensive errors.

4.2 Strategic options

The general strategic options for internationalizing a company in telecom services are numerous as shown in Table 4.1. The classic strategic dimensions, i.e. the product and market dimensions, need to be complemented, as well as the two additional dimensions, the competence and partner dimensions, which are of major importance to telecom service companies, see Figure 4.1. The general strategic options in Table 4.1 are based on MNC literature as well as on revealed behavior, much of it presented in this book. Detailed comments will not be given here. What is noteworthy in this context, however, is the need to consider a restructuring process of diversification in connection with internationalization. The principal strategic options for a telco of the traditional PTT type are then as in Figure 4.2. This is a simple illustration and one has further to distinguish between what is here called *outer diversification*, i.e. inter-sectorial diversification (e.g. into media and entertainment), and *inner diversification*, i.e. intra-sectorial diversification. Both the ongoing transition from the NMP-regime to the ICP-regime (cf. Table 3.1) and emerging new technologies fundamentally reshape the possibilities of reaping economies of scale and scope. This creates a need to redefine corporate identity, not only in terms of national identity but also in terms of scope of business activities with new boundaries replacing old ones within and between market and industry areas. Such conceptualizations and redefinitions play a fundamental role in shaping competition and competitive advantage. The multinational telcos of the future, and the industries they are operating in, will most likely be conceptually quite different from today.

Table 4.1
Overview of general strategic options for an internationalizing telco

Motives/ objectives	Geographic area	Country targets	Customer segment	Timing	Service area	Entry service area	Entry mode	FDI type
Growth Profits RoI Survival etc.	Globalization[2)] Triadization Continentalization Multidomestic expansion[2)] Regional expansion Scattered country expansion	Europe Precursory test markets Advanced ICs Rich NICs Near-NICs LDCs etc. Follow your customer Follow your supplier[1)] etc.	Corporate customers Professionals Households Teenagers Urban life-style Infra-structure etc.	Pioneer Early mover Early majority Late majority Laggard	Wired/wireless voice, data, text, image Mobile com Corporate com International infrastructure etc.	Mobile services Satellite services Cable telephony CATV Leased lines Telecom arbitrage/hubbing etc.	Alliances, joint ventures Acquisitions and mergers Greenfield Minority holdings etc.	Equity position (minor/major) Marketing/ /sales/liaison office Network (private/public/ old/new) service production Other infrastructural investments R&D etc.

1) This strategy is unusual, but is nevertheless a realistic option, especially since many telecom equipment suppliers are highly internationalized. To the extent that a telco can cumulate experience with supplier specific equipment, it may develop competitive advantage concomitant to the supplier.

2) These strategies refer both to geographic scope and mode of organization and coordination (see Chapter 1).

Table 4.1 (cont.)
Overview of general strategic options for an internationalizing telco

Alliance design	Competencies	Competitive means	Service/product focus	Outer diversification	Outer organization	Inner organization	Economies driving internationalization	Other strategic dimensions
Compatibility of objectives Geographically close Horizontal/vertical upstream/downstream Size similarity With/without dominant partner Triad coverage Service complementarities Overcome entry barriers/market access Risk/cost sharing Competence (resource) complementarities Network complementarities etc.	Managerial Entrepreneurial Technological Commercial Legal Political Financial etc.	Service performance Service price Service scope Ancillary offerings (e.g. technology transfer, credits, counter-trade)	Inner/outer diversification/specialization Service/product conglomerate/niche	Data services Financial services Education Entertainment Multi-media, Broadcasting etc.	Horizontal Vertical Forward/backward integration/disintegration, e.g. own network, own manufacturing, in-house R&D; Network organization[2] etc.	Extended domestic organization International federation of subsidiaries International division Global/continental service divisions Global service/region matrix organization Multi-domestic/global coordination	Economies of scale (static/dynamic) Economies of scope (cost/revenue based) Economies of speed [1] Transactional economies Locational economies	Ownership Management Finance Supply Technology Personnel etc.

1) This concept is introduced here to capture the costs and benefits of shifting the pace of a process as well as the timing of an event. Such costs and benefits could stem from early or late mover advantages in making FDIs, introducing a new service on the market or increasing the pace of internationalization in general. Economies of speed are related to economies of scale and of scope but are not necessarily a cause or consequence of such economies. (Economies of scope, scale and speed may be popularly seen to favor the survival of the fittest, fattest and fastest respectively.)

2) A quasi-integrated organization of closely cooperating companies with some kind of coordinating center.

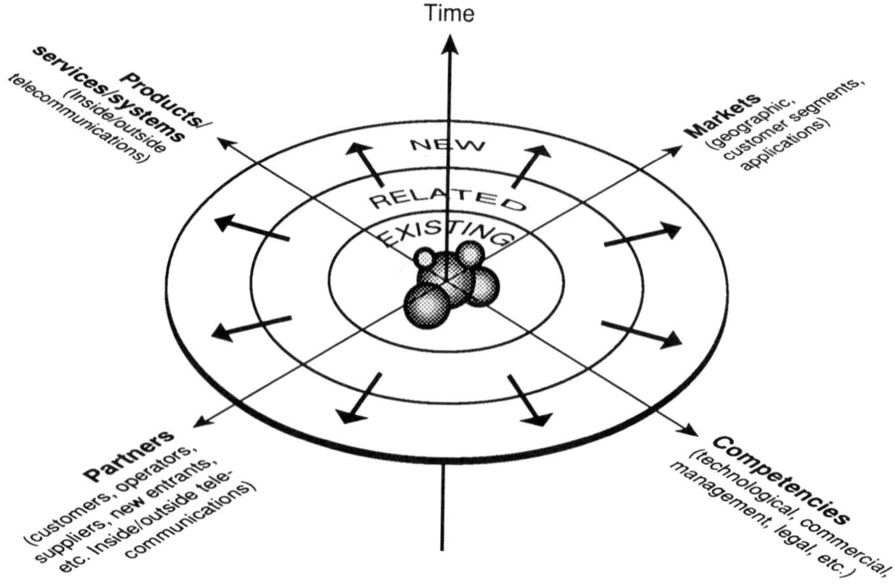

Figure 4.1
Major dimensions of corporate strategy

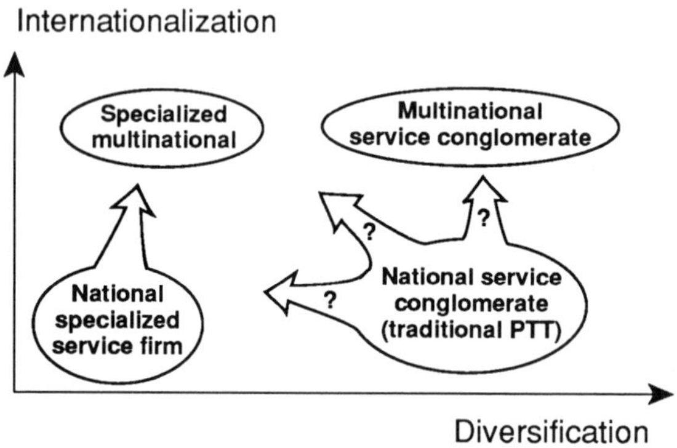

Figure 4.2
Principal strategic options for a telco or telco alliance

Note: 1) Common route for an internationalizing manufacturing firm (with diversification possibly following later).

4.3 Strategic requirements

A number of principal requirements for the strategic performance of an internationalizing telco could now be formulated as in Table 4.2, based on the preceding chapters, literature and analysis.

The requirements are fairly natural, but a few comments are in place. The managerial requirements are important and would involve a major restructuring of management and organization of a traditional PTT, which takes time and effort. A strategic commitment by owners and management to internationalization, in contrast to lip-service, is not easily achieved, especially if there is domestic public pressure for domestic focus, or if the domestic market is large and/or profitable, or if owners and managers are internationally inexperienced.[15] Managerial resources and a common corporate culture for internationalization are not easily built up, either alone or in alliances. In fact, management is likely to become more of a crucial bottleneck than financial resources.

Table 4.2
Strategic requirements for a candidate in the international telecommunications race[1)]

Management
Strong management base and learning ability in international business
Strategic commitment by corporate owners and management
Internationally business-oriented corporate culture
Entrepreneurial and commercial management skills
Skills in managing multi-cultural external relations (alliances etc.)
Technology
Innovativeness
Strong technology base with in-house R&D, technology acquisition and learning skills
Intellectual property rights
Marketing/purchasing
Good learning position in precursory markets
Key positions in expanding markets
Productive, long-term relations with internationalized and innovative customers
Productive, long-term relations with internationalized and innovative suppliers
Financing
Domestic financial strength
International financing ability
Production
Ownership and/or control of critical network parts
Politics
Culturally and politically well-positioned
Home market politically and regulatively conducive to and apt for learning and self-financing
Diplomatic skill in alliance formation
Lobbying effectiveness (domestic and foreign political skills)
Miscellaneous
Geographically well-positioned (e.g. close to NICs or medium and less developed countries)

Note: 1) These principal requirements apply to single telcos as well as to alliances.

[15] Future recruitments for the board and top management of PTTs will be an interesting indicator to watch.

Technological innovativeness is more or less necessary but not sufficient for successful and sustainable internationalization, as discussed earlier. This would require an overhaul of technology management in traditional PTTs, with increased attention paid to factors such as technology strategies, IPR matters, R&D productivity, technology acquisition, economies of scale, scope and speed in R&D, R&D coordination in alliances, coordination with innovative customers and suppliers etc. Early mover advantages put a premium on speed and positioning in key markets and segments, and on building long-term relations with customers.

The regulatory, nationalistic and oligopolistic nature of telecom puts requirements on political competences. Financial strength and/or financing skill is clearly needed. Fortunately, financial markets function in favor of telecom. Moreover, the rapid and turbulent development of the telecom sector gives a telco a premium on good learning ability and good learning positions in new technologies and new markets with trials and not-so-expensive-errors in various areas. It must be emphasized here that the requirements in Table 4.2 are not to be taken in a static, absolute sense but must be viewed as dynamically changing, relative to competitors. Strategic flexibility, avoiding being prematurely locked into a certain strategy mix but ensuring preparedness and speedy responses, together with ability to learn faster than competitors, then becomes important. To make rapid improvements from a low level is generally more important in the long run than starting from a high level with slow improvements.

Finally, applying (without further comment) what has been discussed to the headline question of this concluding section reveals that the typical multinational telecommunication carrier[16] in the future Europe is likely to be European, a telecommunication company or telco-dominated consortium, strong in R&D, innovative with comprehensive managerial abilities; etc.

Among the current players, who does best fulfill the requirements in Table 4.2? This question is left open here.

The answer may be—in the air.

[16] The term 'telecommunication carrier' is kept here as a reminiscence of the past. A better term for the future is 'telecommunication service company' (or 'tele service company') for which 'telco' has been used as a shorthand.

REFERENCES

Antonelli, C., 1993, Technological change and multinational growth in the new international telecommunications services economics, Dipartimento di Economia, Universita' di Torino.

Berg, S.V. and Tschirhart, J., 1988, Natural monopoly regulation. Principles and practice (Cambridge University Press, Cambridge).

Bradley, S.P. and J.A. Hausman, 1989, Future competition in telecommunications (Harvard Business School Press, Boston).

Brock, G., 1981, The telecommunications industry. The dynamics of market structure (Harvard University Press, Cambridge).

Buckley, P.J., C.L. Pass and K. Prescott, 1992, The internationalization of service firms: a comparison with the manufacturing sector, Scandinavian International Business Review, Vol. 1, No. 1, 39-56.

Cantwell, J., 1989, Technological innovation and multinational corporations (Basil Blackwell, Oxford).

Cowhey, P.F., 1990, The international telecommunications regime: the political roots of regimes for high technology, International Organization, 44, 2, Spring, 169-199.

Dunning, J.H., 1991, The internationalization of the production of services: some general and specific explanations, GSM Working Paper, 91-11 (Graduate School of Management, Rutgers University, New Jersey).

Dunning, J.H., 1992, Multinational enterprises and the globalization of innovatory capacity, in Granstrand, O. et al., eds., Technology management and international business. Internationalization of R&D and technology (John Wiley & Sons, Chichester).

Granstrand, O. and S. Sjölander, 1990, Managing innovation in multi-technology corporations, Research Policy, Vol. 19, No. 1, pp. 35-60.

Granstrand, O., L. Håkanson and S. Sjölander, eds., 1992, Technology management and international business. Internationalization of R&D and technology (John Wiley & Sons, Chichester).

Guile, B.R. and J.B. Quinn, eds., 1988, Managing innovation: cases from the services industries (National Academy Press, Washington D.C.).

Noam, E., 1992, Telecommunications in Europe (Oxford University Press, Oxford).

Phillips, A., 1971, Technology and market structure. A study of the aircraft industry (Heath Lexington Books, Lexington, Mass.) .

Sharkey, W.W., 1982, The theory of natural monopoly (Cambridge University Press, Cambridge).

Train, K.E., 1992, Optimal regulation. The economic theory of natural monopoly (The MIT Press, Cambridge, Mass.).

Index of Actors and Markets

Index of Subjects